T0228872

Finite Element Programs in Structural Engineering & Continuum Mechanics

It is a massy wheel,
Fixed on the summit of the highest mount,
To whose huge spokes ten thousand lesser things
are mortis'd and adjoin'd; which when it falls;
Each small annexement, petty consequence,
Attends the boist'rous ruin.

Shakespeare, *Hamlet* II, iii

ABOUT OUR AUTHOR

Carlisle Thomas Francis Ross was born in Kharagpur, India and educated in Bangalore at St Joseph's European High School (1944-47) during the closing years of the British Raj. After coming to England he attended the Chatham Technical School for Boys (1948-51) followed by part time education at the Royal Dockyard Technical College (1951-56) at Chatham, Kent where he served a five-year shipwright apprenticeship.

He proceeded to King's College, Newcastle-upon-Tyne (University of Durham), reading for a B.Sc.(Hons) degree in Naval Architecture (1956-59). During university vacations he worked as a part-time draughtsman at HM Dockyard, Chatham.

For the next two years he laid the foundations of his powerful industrial experience as a Designer in the Project Design Office at Vickers-Armstrongs (Shipbuilders), Barrow-in-Furness (1959-61). His outstanding work there was acknowledged by promotion to the position of Deputy Chief of the Project Design Office. He next worked as a research graduate in the Department of Engineering, University of Manchester (1961-62), where in 1963 he gained his Ph.D. for research in Stress Analysis of Pressure Vessels.

He brought his industrial experience from Vickers-Armstrongs into teaching, first as Lecturer in Civil and Structural Engineering at Constantine College of Technology, Middlesborough (now University of Teesside) (1964-66), and later to Portsmouth Polytechnic (now University of Portsmouth) (1966-71) as Senior Lecturer in Mechanical Engineering, where he still remains as Professor of Structural Mechanics.

His outstanding research in structural mechanics is based on computational methods, tested experimentally with colleagues. He has moreover made important discoveries on the buckling of ring-stiffened cylinders and cones under external pressure, and has also developed the application of microcomputers on finite element analysis. His outstanding contributions to engineering science were recognised in 1992 by the award of D.Sc. for research on Stress Analysis and Structural Dynamics by the CNAA, London.

FINITE ELEMENT PROGRAMS IN STRUCTURAL ENGINEERING AND CONTINUUM MECHANICS

Carl T.F. Ross, BSc, PhD, DSc, CEng, FRINA, MSNAME
Professor of Structural Dynamics
Department of Mechanical and Manufacturing Engineering
University of Portsmouth

WOODHEAD
PUBLISHING

Oxford   Cambridge   Philadelphia   New Delhi

Published by Woodhead Publishing Limited,
80 High Street, Sawston, Cambridge CB22 3HJ, UK
www.woodheadpublishing.com

Woodhead Publishing, 1518 Walnut Street, Suite 1100, Philadelphia,
PA 19102-3406, USA

Woodhead Publishing India Private Limited, G-2, Vardaan House, 7/28 Ansari Road,
Daryaganj, New Delhi – 110002, India
www.woodheadpublishingindia.com

First published by Albion Publishing Limited, 1996
Reprinted by Woodhead Publishing Limited, 2011

British Library Cataloguing in Publication Data
A catalogue record for this book is available from the British Library

ISBN 978-1-898563-28-0

DEDICATION

To my grandson

Daniel Steven Dryden

CONTENTS

Page

PREFACE

Introductory Chapter   1
    Loads   5
    Screen Dumping   6

Chapter 1   **Forces in Plane Pin-Jointed Trusses**   8
    1.1   Introduction   8
    1.2   Input Data   9
    1.3   Results   11
    1.4   Example 1.1   11
    1.5   Input Data File   13
    1.6   Output   13
    1.7   Screen Dump   14
    1.8   Example 1.2   15
    1.9   Results   16

Chapter 2   **Bending Moments in Beams**   18
    2.1   Introduction   18
    2.2   Data for BEAMSF   19
    2.3   Output   20
    2.4   Example 2.1   21
    2.5   Results   23
    2.6   Screen Dumps   23
    2.7   Example 2.2   24
    2.8   Results   24
    2.9   Screen Dumps   25
    2.10   Data for BEAMERF   26
    2.11   Output   27
    2.12   Examples 2.3   27
    2.13   Output   29

Chapter 3   **Bending Moments in Rigid-Jointed Plane Frames**   31
    3.1   Introduction   31
    3.2   Data for FRAME2DF   33
    3.3   Output   35
    3.4   Example 3.1   35
    3.5   Output   37
    3.6   FRAME2DF   38
    3.7   Example 3.2   39
    3.8   Output   40
    3.9   Screen Dumps   41
    3.10   Frames with a Combination of Rigid and Pin-Joints   43
    3.11   Derivation of the Stiffness Matrix for the Beam of Figure 3.8   43

3.12   Loading   46
3.13   Moment-Nodal Displacement Relationships   48
3.14   Deflection Distributions   50
3.15   Description of How To Use the Program FRAMERP   52
3.16   Computer Program   53
3.17   Input Data File   54
3.18   Example 3.3   55
3.19   The Input Data File FRAMERP.DAT   56
3.20   Results   57
3.21   Example 3.4   58
3.22   Example 3.5   58
3.23   Example 3.6   59
3.24   Screen Dumps   60

Chapter 4   **Forces in Pin-Jointed Space Trusses**   **62**
4.1   Introduction   62
4.2   Input Data   64
4.3   Output   65
4.4   Example 4.1   65
4.5   Output   68
4.6   Input Data File   68
4.7   Example 4.2   69
4.8   Results   71

Chapter 5   **Static Analysis of Three-Dimensional Rigid-Jointed Frames**   **72**
5.1   Introduction   72
5.2   Input Data   78
5.3   Output   81
5.4   Example 5.1   82
5.5   Input Data File   84
5.6   Results   85
5.7   Screen Dump   85
5.8   Example 5.2   86
5.9   Input Data File   87
5.10   Results   88

Chapter 6   **Vibration of Rigid-Jointed Space Frames**   **89**
6.1   Introduction   89
6.2   Continuous Reduction Technique   92
6.3   Zero nodes and other details   92
6.4   The Input Data File   92
6.5   Example 6.1   94
6.6   Input   96
6.7   Computer Output   97
6.8   Example 6.2   98
6.9   Data   98
6.10   Computer Output   99

	6.11	Example 6.3	100
	6.12	Input Data	100
	6.13	Computer Output	101
	6.14	Example 6.4	102
	6.15	Input Data File for Example 6.4	103
	6.16	Results	110

Chapter 7	**Bending Moments in Grillages**		**111**
	7.1	Introduction	111
	7.2	The Input Data	113
	7.3	Output	114
	7.4	Example 7.1	114
	7.5	Data for Example 7.1	115
	7.6	Results	117
	7.7	Example 7.2	118
	7.8	Input Data File for Example 7.2	119
	7.9	Results for Example 7.2	122
	7.10	Example 7.3	123
	7.11	Input Data File for Example 7.3	125
	7.12	Results for Example 7.3	126
	7.13	Screen Dumps	127

Chapter 8	**Vibration of Grillages**		**129**
	8.1	Introduction	129
	8.2	Input Data File	130
	8.3	Example 8.1	131
	8.4	Input Data File for Example 8.1	132
	8.5	Results for Example 8.1	133
	8.6	Example 8.2	134
	8.7	Results for Example 8.2	134
	8.8	Example 8.3	136
	8.9	Input Data File for Example 8.3	136
	8.10	Results for Example 8.3	137

Chapter 9	**Slab on Elastic Foundation**		**139**
	9.1	Introduction	139
	9.2	Input	140
	9.3	Output	142
	9.4	Example 9.1	143
	9.5	Input Data File for Example 9.1	144
	9.6	Results for Example 9.1	145
	9.7	Example 9.2	147
	9.8	Results	147
	9.9	Example 9.3	148
	9.10	Input Data File for Example 9.3	149
	9.11	Results	149
	9.12	Conclusions	151

Chapter 10   In-Plane Stresses in Plates   **152**
    10.1   Introduction   152
    10.2   The Computer Program STRESS3N   153
    10.3   Example 10.1   155
    10.4   Input Data File for Example 10.1   155
    10.5   Results   157
    10.6   Input Data File for STRESS3.DAT   157
    10.7   Stress Contours   158
    10.8   Example 10.2   160
    10.9   The Computer Program STRESS4N   162
    10.10 Data   167
    10.11 Example 10.3   169
    10.12 Input of Data File for STRESS4N   171
    10.13 Results   172
    10.14 Screen Dumps   173
    10.15 Example 10.4   174
    10.16 Results   177
    10.17 The Computer Program STRESS8N   179
    10.18 Data   183
    10.19 Example 10.5   185
    10.20 Input Data File for STRESS8N.EXE   187
    10.21 Results   189
    10.22 Screen Dumps   190
    10.23 Example 2.3   192
    10.24 Results   195
    10.25 Example 2.4   196
    10.26 Results   199

Chapter 11   Bending Stress in Flat Plates   **200**
    11.1   Introduction   200
    11.2   Input Data File   200
    11.3   Results   202
    11.4   Example 11.1   203
    11.5   Data   204
    11.6   Results   205
    11.7   Input Data File   205
    11.8   Screen Dumps   206
    11.9   Example 11.2   210
    11.10 Input Data File   210
    11.11 Results   213
    11.12 Screen Dumps   213

Chapter 12   Stresses in Doubly-Curved Shells   **217**
    12.1   Introduction   217
    12.2   Input Data File   219
    12.3   Example 12.1   220

12.4	Input Data File	221
12.5	Results	223
12.6	Screen Dumps	224

Chapter 13   In-Plane Vibrations of Plates   **227**
13.1	Introduction	227
13.2	Input Data File	227
13.3	Example 13.1	228
13.4	Input Data File	228
13.5	Input Data File for Example 13.1	231
13.6	Results	232
13.7	Screen Dumps	232

Chapter 14   Lateral Vibrations of Flat Plates   **237**
14.1	Introduction	237
14.2	Input Data File	238
14.3	Example 14.1	239
14.4	Input Data	240
14.5	Input Data File for Example 14.1	242
14.6	Results	242
14.7	Screen Dumps	243

Chapter 15   Vibration of Thin-Walled, Doubly-Curved Shells   **249**
15.1	Introduction	249
15.2	Input Data	250
15.3	Example 15.1	251
15.4	Input Data File for Example 15.1	251
15.5	Results	252
15.6	Example 15.2	252

Chapter 16   Stresses in Solids   **255**
16.1	Introduction	255
16.2	Data	257
16.3	Output	257
16.4	Example 16.1	258
16.5	Input Data File	258
16.6	Results	259
16.7	Approximate Checks	262

Chapter 17   Two Dimensional Field Problems   **264**
17.1	Introduction	264
17.2	Torsion	264
17.3	Heat Transfer	266
17.4	Steam Functions	267
17.5	Groundwater Flow	268
17.6	Electrical Field Problems	268
17.7	Magnetic Field Problems	269

17.8   The Computer Program FIELD3SF   269
17.9   Data   270
17.10   Output   272
17.11   Example 17.1   272
17.12   Results   273
17.13   Example 17.2   274
17.14   Results   277
17.15   FIELD4SF   278
17.16   Data   279
17.17   Results   281
17.18   Example 17.3   282
17.19   Results   284
17.20   Example 17.4   284
17.21   Input Data File   285
17.22   Results   287
17.23   Example 17.5   288
17.24   Results   290
17.25   Example 17.6   291
17.26   Results   293
17.27   FIELD8SF   293
17.28   To Calculate the Length of a Boundary
       Formed by (say) nodes 1, 2, and 3   294
17.29   Data   296
17.30   Results   298
17.31   Example 17.7   298
17.32   Results   300
17.33   Example 17.8   302
17.34   Results   304
17.35   Example 17.9   305
17.36   Input Data File   305
17.37   Results   307
17.38   Example 17.10   310
17.39   Results   312
17.40   Screen Dumps   312

Chapter 18   Solution of Helmholtz's Equation   314
18.1   Introduction   314
18.2   Data   316
18.3   Example 18.1   317
18.4   Input Data File for Example 18.1   318
18.5   Results   318

REFERENCES   322

APPENDICES   324

Appendix 1. Computer Program for Plane Pin-Jointed Trusses "TRUSSF" 324

Appendix 2. Computer Program for Beams "BEAMSF" 332

Appendix 3. Computer Program for Beams on an Elastic Foundation "BEAMERF" 340

Appendix 4. Computer Program for Rigid-Jointed Frames "FRAME2DF" 348

Appendix 5. Computer Program for Rigid/Pin-Jointed Plane Frames
"FRAMERP" 361

Appendix 6. Computer Program for Pin-Jointed Space Trusses "STRUSSF" 379

Appendix 7. Computer Program for Three-Dimensional Rigid-Jointed
Frames "3DFRAMEF" 388

Appendix 8. Computer Program for Vibration of Rigid-Jointed Space
Frames "VIB3DSFN" 400

Appendix 9. Computer Program for Bending Moments in Grillages "GRIDSG" 413

Appendix 10. Computer Program for the Vibration of Grillages "VIBGRIDG" 423

Appendix 11. Computer Program for a Slab on an Elastic Foundation
"SLABELAG" 433

Appendix 12. Computer Program for In-Plane Stresses in Plates "STRESS3N" 451

Appendix 13. Computer Program for In-Plane Stresses in Plates "STRESS4N" 461

Appendix 14. Computer Program for In-Plane Stresses in Plates "STRESS8N" 473

Appendix 15. Computer Program for Bending Stresses in Plates "PLATEBEF" 490

Appendix 16. Computer Program for Stresses in Doubly-Curved Shells
"SHELLSTF" 502

Appendix 17. Computer Program for Vibration of In-Plane Plates
"VIBPLANF" 520

Appendix 18. Computer Program for Lateral Vibrations of Flat Plates
"VIBPLATF" 531

Appendix 19. Computer Program for the Vibration of Doubly-Curved Shells
"VIBSHELF" 545

Appendix 20. Computer Program for Stresses in Solids "TETRAF"   562

Appendix 21. Computer Program for Field Problems "FIELD3SF"   572

Appendix 22. Computer Program for Field Problems "FIELD4SF"   581

Appendix 23. Computer Program for Field Problems "FIELD8SF"   591

Appendix 24. Computer Program for the Solution of Helmholtz's Equation "ACOUSTIC"
604

AUTHOR'S PREFACE

This book consists of 18 chapters and 24 computer programs, the latter of which are listed in the appendices of this book.

The computer programs cover quite a wide range of problems in engineering science, from the static analysis of two and three dimensional structures to the stress analysis of thick slabs on elastic foundations, and from two and three dimensional stress analysis problems to two dimensional field problems, including heat transfer and acoustic vibrations.

The book attempts to bridge the gap between theoretical texts on finite element methods, with little or no software support and the giant finite element software packages, which require considerable skill to use. The book is intended to be used as a modern day handbook. It should prove useful to students and graduates in civil, mechanical, structural, aeronautical and electrical engineering and to naval architects. The book should also prove useful to consultant engineers, who cannot afford to purchase the giant software packages, and to casual users of finite element methods, who will find it easier to use the enclosed software than the giant complex finite element packages.

The book commences with an introductory chapter, which gives many useful hints and formulae intended for structural design. Most of the chapters describe how to use a computer program for a particular problem.

Chapter 1 describes a computer program for the static analysis of plane pin-jointed trusses and Chapter 2 describes computer programs for analysing continuous beams on rigid and elastic supports. Chapter 3 describes computer programs for the static analysis of rigid-jointed plane frames and frames with a mixture of rigid and pin joints. Chapter 4 describes a computer program for the static analysis of a pin-jointed space truss, and Chapter 5 describes a computer program for the static analysis of a rigid-jointed space frame. Chapter 6 describes a computer program for the vibration of a rigid-jointed space frame and Chapter 7 describes a computer progam for the static analysis of orthogonal and skew grids. Chapter 8 describes a computer program for the free vibration of orthogonal and skew grids.

Chapter 9 describes a computer program for the stress analysis of a thick slab on an elastic foundation and Chapter 10 describes computer programs for the stress analysis of in-plane plates, using 3 node triangular elements, and 4 and 8 node isoparametric quadrilateral elements.

Chapter 11 describes a computer program for the stress analysis of the out-of-plane bending of plates, based on small deflection elastic theory, and Chapter 12 describes a computer program for the stress analysis of a doubly curved shell. Chapter 13 describes a computer program for the in-plane vibration of plates, and Chapter 14 describes a computer program for the out-of-plane vibration of plates.

Chapter 15 describes a computer program for the free vibration of thin-walled doubly curved shells, and Chapter 16 describes a computer program for the stress analysis of solids.

Chapter 17 describes three computer programs for two dimensional field problems. These programs use 3 node triangular elements, and 4 and 8 node isoparametric quadrilateral elements.

These programs are capable of analysing problems in heat transfer, fluid flow, seepage through porous media, electrostatics, magnetostatics, etc.

Chapter 18 describes a computer program which can solve the two dimensional version of Helmholt'z equation. This program can analyse problems involving acoustic vibration, the oscillations of water in a lake or enclosed harbour, oscillations in electromagnetic waveguides, etc, etc.

As the finite element method is used in all these programs, the problems for analysis can be of quite complex shape and with quite complex boundary conditions. All the chapters include worked examples, and some chapters include screendumps of problems solved.

Acknowledgements The Author would like to thank TASS (AUEW) for permission to extract sections from the seven booklets he wrote for them. His thanks are extended to Miss Sharon Snook for the considerable care and devotion she showed in typing the book.

Warning Although these programs have been written in good faith and after careful testing, neither the author nor the publisher can accept any liability, whatsoever, for the reliability of the computer programs nor for any statements made, whether implied or otherwise, that might arise through the use of these computer programs.

Floppy disks containing the programs in .BAS and .EXE forms can be purchased from:-

Professor C T F Ross
6 Hurstville Drive
Waterlooville
Hampshire
PO7 7NB

Tel: 01705 259304

C T F Ross
University of Portsmouth
December 1995

INTRODUCTORY CHAPTER

The finite element method is one of the most powerful methods of solving partial differential equations which apply over complex shapes and which are governed by complex boundary conditions.

It is based on subdividing a complex shape into many simpler shapes, which are much more mathematically manageable, as shown by Figure I.1.

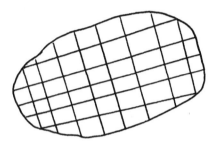

Figure I.1 - Complex shape, divided into simpler shapes or finite elements

A digital computer is used, together with a suitable computer program, to solve the differential equation over each simpler shape or finite element. Then by considering equilibrium and compatibility at the inter-element boundaries, a large number of simultaneous equations results. Solution of these equations leads to the required values for the function at the nodes. The nodes are used to describe each element and also, the entire domain, as shown in Figure I.1.

The finite element method is based on the matrix displacement method [1 to 3], which first appeared in the early 1940's. At that time, the matrix displacement method was used to design aircraft structures, so that they would have a better strength:weight ratio. During that period and for a decade later, it was difficult to apply the method to practical problems, largely because computational power was poor and expensive, and in many cases, it was necessary to employ teams of operators of desktop electro-mechanical calculators to implement the method.

The true finite element method appeared in 1956 [4], although a similar method, based on variational finite differences, was presented by Courant [5], a decade earlier. Since then, an enormous effort has been made at developing the finite element method [6 to 12], so that today, in addition to it being applied to structures and vibrations, the method can be used for heat transfer, fluid flow, acoustics, magneto-statics, electro-statics, electromagnetic waveguides, medicine, weather forecasting etc.

To use the finite element method correctly, it is necessary to adopt the appropriate finite element to mathematically model the domain. For example, if a skeletal structure were being analysed, it would be necessary to use a one dimensional line element, such as that shown in Figure I.2. From Figure I.2(a), it can be seen that the element is straight and it is governed by end nodes. Similarly, from Figure I.2(b), it can be seen that the line element is curved and it has an additional mid-side node.

(a)   Straight element   (b)   Curved element

Figure I.2 - One dimensional line elements

For two dimensional problems, it will be necessary to use straight and curved triangles and quadrilaterals, as shown by Figure I.3, and for solids, the elements can take the form of tetrahedra, hexahedra and curved cubes, as shown in Figure I.4.

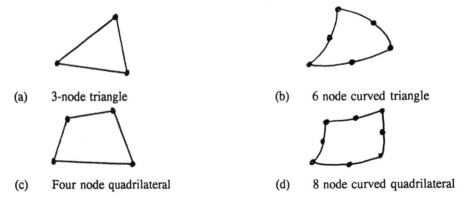

(a)   3-node triangle   (b)   6 node curved triangle

(c)   Four node quadrilateral   (d)   8 node curved quadrilateral

Figure I.3 Two dimensional elements

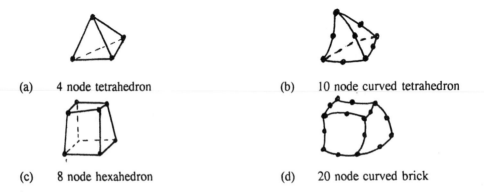

(a)   4 node tetrahedron   (b)   10 node curved tetrahedron

(c)   8 node hexahedron   (d)   20 node curved brick

Figure I.4 - Solid elements

2

Similarly, to represent axisymmetric shells and doubly-curved shells, elements such as those shown in Figure I.5, are often used. In Figure I.5(a), it can be seen that the end circles of the elements are used as nodal circles.

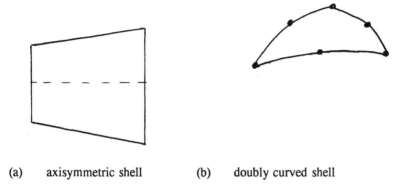

(a)   axisymmetric shell   (b)   doubly curved shell

Figure I.5 - Shell elements

In this book, not all the above elements will be used, although a comprehensive coverage of problems will be made.

All the programs will require the creation of an input data file to contain the input data for the appropriate problem. The input data file must be a text or ASCII file, and it can usually be created quite easily with an editor or with most word processors. The various items of data can be separated by commas, or better still by two or more spaces. It must be emphasised that the user should be careful when using commas to separate data, because if only one full stop were accidentally typed instead of a comma, the results will be faulty. DO NOT USE THE TAB KEY TO INSERT SPACES!

Fortunately, however, error trapping can be quite painless, as most programs include a graphics display prior to processing, and in any case, the user can examine the output file.

Be careful not to include blank lines in the data, as they may be interpreted as zeroes, especially in the case of the first line.

Be careful not to attach a short stiff element to a long flexible one, or numerical instability may result. Numerical instability can also occur for two dimensional problems if an angle is less than 30° or greater than 150°, as shown below:

(a)   angle < 30°   (b)   angle > 150°

Figure I.5 - Unsuitable elements

In general, if a large number of elements is used, the analysis is more likely to be satisfactory than if a small number of elements is used, but care should be taken not to use

3

too many elements, or elements that are so small, that the numerical precision of the computer is exceeded. In particular, it should be ensured that in areas of discontinuity or stress concentration, the elements are made smaller near the discontinuity or stress concentration. This latter feature is called mesh refinement, and an example of a plate with a hole in it is shown in Figure I.6. In this figure, it can be seen that the mesh is finer near the hole, and the reason for this mesh requirement can be seen in Figure I.7, where the stress contours are gathered closer together near the hole. The plate is subjected to a uniform longitudinal in-plane tensile stress, acting perpendicularly to the short edges.

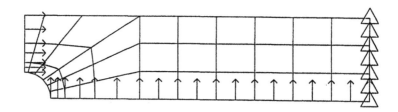

Figure I.6 - Meshed quadrant of a plate with a hole in its centre

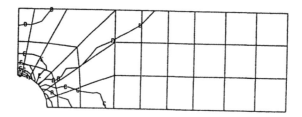

Figure I.7 - Stress contours in a quadrant of a plate with a hole in its centre

For static problem, the finite element method is based on solving the following matrix equation:-

$$\{q_i\} = [K] \{u_i\} \qquad\qquad \text{I.1}$$

where

$\{q_i\}$ = a vector of "known" externally applied loads

$[K]$ = a known system stiffness matrix

4

$\{u_i\}$ = a vector of "unknown" nodal displacements.

Solution of equation I.1 can be carried out by a number of different methods, but two of the best methods are through Gaussian elimination or Choleski decomposition [7, 13].

Most of the static programs take advantage of any sparsity in the stiffness matrix [K] and solve through a skyline technique [14, 15].

For free vibration without damping, the matrix equation to be solved is shown by equation I.2

$$|[K] - \omega^2 [M]| = 0 \qquad \qquad I.2$$

where

$[M]$ = a known system mass matrix

ω = radian frequency

Equation I.2 can be solved by a number of different methods, but the method chosen here is the power method, together with Aitken's acceleration [14, 15]. If a structure is vibrating in air or water at low frequencies, the effects of <u>damping</u> on the magnitude of the resonant frequencies is small.

Loads

Uniformly distributed loads appear in a number of different forms, including as wind loads and as snow loads, and hydrostatics loads are usually due to water, or some other liquid pressure, where the pressure exerted by the liquid increases linearly with depth.

The pressure caused by wind blowing perpendicularly to a <u>very large wide surface</u> is usually calculated by formulae I.3.

$$\text{pressure} = \rho v^2 \qquad \qquad I.3$$

ρ = density of air $\leftrightharpoons$ 1.2 kg/m^3

v = wind velocity in m/s

The pressure caused by wind blowing perpendicularly on the surface of a tall <u>slender building</u> is usually calculated by formulae I.4.

$$\text{pressure} = \rho v^2/2 \qquad \qquad I.4$$

The density of fresh <u>virgin show</u> is usually taken to be about one twelfth of the density of water, but this figure is only very approximate and it depends on the nature of the snow flakes.

Hydrostatic pressure due to water is given by formulae I.5.

Pressure = $\rho g h$ I.5

where ρ = density in kg/m^3 = 1000 kg/m^3 for fresh water

                  g = acceleration due to gravity in m/s^2

                  h = depth of water in metres.

If pressure acts on a flat rectangular surface of length L and width B, the equivalent uniformly distributed load "w" is given by:

Uniformly distributed load = w (load/unit length) = pressure x B

Total load = w x L

Two versions of each program are supplied, namely a Quickbasic version, (.BAS), and an executable version (.EXE), where the latter can be run directly from DOS.

Throughout the text, the following shorthand notation is used:

⌈   For i = 1 to N
│
→   Input x_i, y_i

This notation means that the symbol "i" is cycled from 1 to N, in increments of 1, and for each value of "i" the variables x_i and y_i are inputted.

Screen Dumping

The figures on the screen during processing can be dumped onto a printer, by adopting the following procedure:-

While in DOS, and before running the program, type

GRAPHICS <Enter>

and later, when you see the following prompt on the screen:-

"To continue, type Y"

Press the Shift and Printscreen buttons together.

Make special considerations for structural members in compression, because they can fail by buckling or instability at loads which may only be a small fraction of that to cause yield. For such structures, it will be necessary to carry out an inelastic instability analysis [16, 17] to ensure that the member or structure does not buckle.

If the strut is initially straight, a suitable formulae that can be used is the Rankine-Gordon formulae, shown by equation I.6.

P_R = Ranking-Gordon buckling load

= $A\sigma_c/[1 + a(L/k)^2]$

where

σ_c = compressive yield stress (see Table I.1)

a = a constant, depending on material properties and boundary conditions. (for a pinned-ended strut a = $\sigma_c/[\pi^2 E]$; see Table I.1 for some typical values for a pinned-ended straight strut)

L = strut length

k = least radius of gyration of the strut cross-section

Table I.1 Typical values of Rankine-Gordon constants

Material	1/a	σ_c (MPa)
Mild Steel	7000	300
Wood	750	35
Wrought Iron	9000	250
Duralumin	4300	155

An alternative formula that could be used is the Perry-Robertson formulae of equation I.7.

σ_{cr} = crippling stress
= $\alpha - (\alpha^2 - \beta^2)^{1/2}$

where

α = ½ [σ_c + (1 + η)σ_e]

β = $\sigma_c \sigma_e$

For steel, η (empirical) = 0.003 (L_o/k)

L_o = effective length of strut (see BS449)

= L for a pinned-ended strut

7

CHAPTER 1

FORCES IN PLANE PIN-JOINTED TRUSSES

1.1   Introduction

This computer program which is called TRUSSF is capable of calculating the nodal displacements and elemental forces in plane pin-jointed trusses, subjected to point loads at the joints of the framework, as shown in Figure 1.1.   The trusses can be statically determinate or statically indeterminate, and its members can have different cross-sectional and material properties.

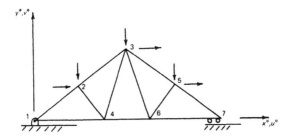

Figure 1.1 - Plane pin-jointed truss

The nodal points of the members are usually taken at the joints. The joints, are assumed to consist of smooth frictionless ball joints, so that it is not possible to subject any member of the framework to a bending moment. Thus, each element or member has only got axial stiffness, and cannot withstand flexure. That is, this program should not be used for frameworks with welded joints!

A typical element is shown in Figure 1.2. This element is called a <u>rod</u> element.

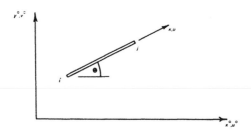

Figure 1.2 - A two dimensional rod element

The element has two end nodes, each with two degrees of freedom, namely, u^o and v^o as shown in Figure 1.2

Thus, for the truss of Figure 1.1, there are a total of 14 degrees of freedom, and three of these are zero; both displacements at node 1, ($u_1°$ and $v_1°$) and one at node 7. ($v_7°$).

The displacement $u_1°$ corresponds to the displacement position 1 and $v_1°$ corresponds to the displacement position 2. The suppressed displacement

$v_7°$ corresponds to the displacement position 14.

These displacement positions are calculated as follows:-

$u_1° = 2 \times 1 - 1 \qquad = 1$

$v_1° = 2 \times 1 \qquad\qquad = 2$

$v_7° = 2 \times 7 \qquad\qquad = 14$

Thus, for the ith node, the displacement position corresponding to

$u_i° = 2 \times i - 1$

& $v_i° = 2 \times i$

As the element has two end nodes, each with two degrees of freedom, the global elemental stiffness matrix $[k°]$ will be a 4 x 4 matrix, as shown by equation 1.1.

$$[k°] = \frac{AE}{l} \begin{array}{c} \\ \\ \\ \\ \end{array} \begin{array}{cccc} u_1° & v_1° & u_2° & v_2° \\ \end{array}$$
$$[k°] = \frac{AE}{l} \begin{bmatrix} c^2 & cs & -c^2 & -cs \\ cs & s^2 & -cs & -s^2 \\ -c^2 & -cs & c^2 & cs \\ -cs & -s^2 & cs & s^2 \end{bmatrix} \begin{array}{c} u_1° \\ v_1° \\ u_2° \\ v_2° \end{array} \qquad 1.1$$

where

$c = \cos \alpha$
$s = \sin \alpha$
A = cross-sectional area
l - elemental length
E = Young's modulus

1.2 Input Data

The input data file should be created in the following sequence.

1. Number of nodal points or pin-joints = NN

9

2. Number of one-dimensional structural members = ES

3. Number of nodes that have zero displacements = NF

4. <u>Details of Zero Displacements</u>

    [FOR i = 1 to NF   )   Input the nodal position of each
    [   )   "Suppressed" node and whether or
    [Input NS_i)   not the appropriate displacements
    [Type 1 if the $u°$)   are zero at these nodes
    [displacement is zero at)
    [this node or 0 if it is not)
    [After this, type 1 if the)
    [$v°$ displacement is zero at)
    [this node, or 0 if it is not)

    where, NS_i = a node with one or more zero displacements.

5. <u>Nodal Co-ordinates</u>

    [FOR i = 1 to NN)   Input the $x°$ and $y°$ global
    [   )   co-ordinates of each node
    [Input $x_i°$, $y_i°$)

6. <u>Member Details</u>

    [FOR EL = 1 to ES
    [
    [Input i - Input "i" node for the member
    [
    [Input j - Input "j" node for the member
    [
    [Input A - Input cross-sectional area for the member
    [
    [Input E - Input Young's modulus of elasticity for the member

7. <u>Details of Nodal Forces</u>

    NC = number of nodes with nodal forces

    [FOR i = 1 to NC
    [
    [Input node i
    [
    [Input loads in $x°$ and $y°$ directions, respectively, at node "i"

10

A typical input data file is shown in Section 1.5; this must be created prior to running the program.

1.3 Results

The results are output onto a data file which must not have the same name as the input data file or the name of any other file that you do not wish to overwrite.

The results are output in the following sequence:

1. Vector of Nodal Displacements

    [FOR i = 1 to NN
    [
    [PRINT $u_i°$, $v_i°$

2. Nodal Forces in Each Member

    [FOR i = 1 to ES
    [
    [PRINT Nodes defining each member and the force in each member (tensile + ve)

3. Reactions in the u° and v° directions, respectively at each of the "suppressed" nodes.

1.4 Example 1.1

Determine the forces in the plane pin-jointed truss shown in Figure 1.3. All members are of constant "E", and all members are of constant "A", except for member 1-3, which has a cross-sectional area of "2A".

As only the forces are required, it will be convenient to assume that A = E = 1. If, however, the nodal displacements are required, then it will be necessary to feed in the true values of A and E.

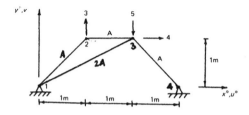

Figure 1.3

11

Number of nodes = 4

Number of members = 4

Number of nodes with zero displacement = 2

Nodal positions of nodes with Zero Displacements, etc

Nodal Position 1 = 1

$u_1^\circ = 0,$ $\therefore$ Type 1

$v_1^\circ = 0,$ $\therefore$ Type 1

Nodal Position 2 = 4

$u_4^\circ = 0,$ $\therefore$ Type 1

$v_4^\circ = 0,$ $\therefore$ Type 1

Nodal Co-ordinates

x_i°	y_i°
0	0
1	1
2	1
3	0

Details of Members

i	j	A	E
1	2	1	1
2	3	1	1
1	3	2	1
3	4	1	1

Nodal Forces

NC = 2

Nodal position 1 = 2,   loads = 0 and 3

Nodal position 2 = 3,   loads = 4 and -5

1.5   Input Data File

A typical input data file for this program namely TRUSS.DAT, is shown below.

This data file was used to solve the problem of Example 1.1.

TRUSS.DAT

```
4
4
2
1,1,1
4,1,1
0,0
1,1
2,1
3,0
1,2,1,1
2,3,1,1
1,3,2,1
3,4,1,1
2
2,0,3
3,4,-5
EOF
```

NB   It should be noted that instead of using commas to separate the various items of data, two or more spaces could have been used. Do not use the TAB key to insert spaces!

1.6   Output

The results for Example 1.1 are given below

Displacements x AE

Node	u°	v°
1	0	0
2	-2.028	10.513
3	0.972	-9.398
4	0	0

Forces in Each Member

Member	Forces	
1	$F_{1\text{-}2} = 4.243$	
2	$F_{2\text{-}3} = 3.000$	in tension
3	$F_{1\text{-}3} = -2.981$	
4	$F_{3\text{-}4} = -5.185$	in compression

where the subscripts i - j represent the nodes defining the member, and the <u>negative</u> <u>sign denotes compression</u>.

Reactions

$H_{x1} = =0.333$ $V_{y1} = -1.667$ (Node 1)   } See figure below
$H_{x3} = -3.667$ $V_{y3} = 3.667$ (Node 4)   }

1.7 Screen Dump

A screen dump of the truss and its deflected form is shown in Figure 1.4.

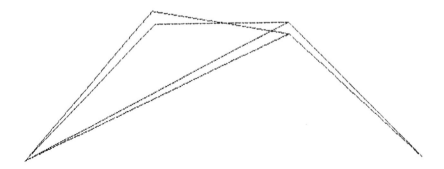

Figure 1.4 - Deflected form of truss

1.8   Example 1.2

Determine the member forces in the plane pin-jointed truss shown in Fig 1.5. For all members

          A = 1E-3m^2
and   E = 2E8 kN/m^2

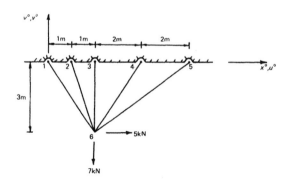

Fig 1.5

The data should be prepared in the following sequence.

6   5   5

1	1	1	}
2	1	1	}
3	1	1	} Details of zero displacement
4	1	1	}
5	1	1	}

15

```
0    0    }
1    0    }
2    0    }        Nodal coordinates
4    0    }
6    0    }
2   -3    }
```

```
1    6    1E-3   2E8   }
2    6    1E-3   2E8   }
3    6    1E-3   2E8   }        Element topology
6    4    1E-3   2E8   }
6    5    1E-3   2E8   }
```

```
1                     }        Details of the concentrated loads
6    5    -7          }
```

1.9   Results

The results were as follows:-

Nodal displacements

Node	u°(m)	v°(m)
1	0	0
2	0	0
3	0	0
4	0	0
5	0	0
6	7.58E-5	-3.27E-5

Member Forces

Member	Force (kN)
1-6	3.84
2-6	3.48
3-6	2.18
6-4	-0.82
6-5	-1.64

Reactions

Node	H_x (kN)	V_y (kN)
1	-2.130	3.195
2	-1.099	3.298
3	0	2.178
4	-0.457	-0.686
5	-1.313	-0.985

CHAPTER 2

BENDING MOMENT IN BEAMS

2.1   Introduction

This chapter describes the use of two programs, namely, BEAMSF and BEAMERF. Both programs are capable of analysing statically determinate or statically indeterminate beams, such as that shown in Figure 2.1. Both programs allow the beams to have a stepped variation in cross-sectional properties, and for the lateral loads to be a complex combination of point, uniformly distributed and hydrostatic loads. (See the Introductory Chapter).

The program BEAMSF is intended for analysing laterally loaded beams resting on rigid supports, and the program BEAMERF is intended for analysing laterally loaded beams resting on rigid or elastic supports.

Both programs calculate the nodal displacements, namely v and θ, and then the nodal bending moments and "reactions".

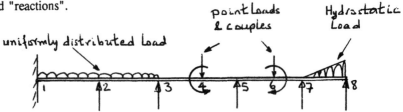

Figure 2.1 - Continuous beam under lateral bending

The beam element has two degrees of freedom at each node, namely v and θ, as shown by Figure 2.2.

Figure 2.2 - Beam element

The displacement positions at the ith node are calculated as follows:-

$$v_i \equiv 2 \times i - 1$$

$$\theta \equiv 2 \times i$$

Each element has two end nodes, as shown by Figure 2.2, and as each node has two degrees of freedom, the elemental stiffness matrix = [k] will be of order 4 x 4, as shown by equation 2.1.

[k] = the elemental stiffness matrix for a beam

$$= EI \begin{bmatrix} & v_1 & \theta_1 & v_2 & \theta_2 & \\ 12/\ell^2 & -6/\ell^2 & -12/\ell^3 & -6/\ell^2 & v_1 \\ -6/\ell^2 & 4/\ell & 6/\ell^2 & 2/\ell & \theta_1 \\ -12/\ell^3 & 6/\ell^2 & 12/\ell^3 & 6/\ell^2 & v_2 \\ -6/\ell^2 & 2/\ell & 6/\ell^2 & 4/\ell & \theta_2 \end{bmatrix}$$

It must be ensured that the beam lies horizontally, and that the nodes are numbered in ascending order from the left of the beam to its right.

2.1   Data for BEAMSF

The input data file should be created in the following sequence:-

1.   Number of elements = LS

2.   Number of nodes with zero displacements = NF

3.   Number of nodes with concentrated loads and couples = NC

4.   Nodal Positions etc, of Zero Displacements

   FOR i = 1 to NF

   Input NS$_i$ (nodal position of suppressed node and whether or not the displacements are zero at this node)

   Type 1 if v = 0 at this node, and 0 if its is not zero; immediately after this, type 1 if θ = 0 at this node, and 0 if it is not zero.

5.   Elastic Modulus = E

6.   If there is any hydrostatic load feed 1, otherwise feed 0,
   (ie NL = 1 or NL = 0, respectively).

7.   Nodal Positions and Value of Concentrated Loads and Couples

   NC = number of nodes with concentrated loads and couples

   FOR i = 1 to NC

   Input "nodal" position of load and couple

   Input value of load and couple at this node

19

8.   If NL = 1, ignore (8).

    <u>Member Details</u> (no hydrostatic load)

    FOR I = 1 to LS

    (Type in the details of the members/elements from left to right)

    Input SA - Input 2nd moment of area for member

    Input XL - Input elemental length

    Input UD - Input value of distributed load/length (+ve if in "y" direction)

9.   If NL = 0, ignore (9)

    <u>Member Details</u> (hydrostatic load)

    FOR I = 1 to LS

    (Type in the details of the members/elements from left to right)

    Input SA - Input 2nd moment of area for element (I)

    Input XL - input elemental length (ℓ)

    Input WA - Value of distributed load on left   )

    Input WB - Value of distributed load on right   )   see Figure 2.3

    Input AS - The distance of WA from the left node   )

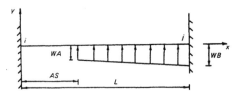

Figure 2.3 - Element in Local Co-ordinates

2.2   Output

    The Nodal Displacements $v_1, \theta_1, v_2 \ \ldots \ v_n, \theta_n$

    where n = LS + 1

The Nodal Bending Moments

Element	Nodes Defining Each Element		Nodal Bending Moments
1	i	j	BM_i
	i	j	BM_j
2	j	k	BM_j
	j	k	
	k		BM_k
	k		BM_k
LS			

2.3 Example 2.1

Determine the nodal bending moments for the beam shown in Figure 2.4. The value of the second moment of area for element 2-3 is twice the value for element 1-2.

As only the moments are required, it will be convenient to assume that $I = E = 1$. If, however, the displacements are required, it will be necessary to feed in the true values for "E" and "I".

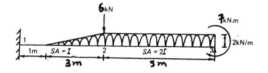

Figure 2.4

The data is as follow:

LS = 2

NF = 2

NC = 2

Details of Zero Displacements

Nodal Position 1 = 1

$v_1 = 0$, $\therefore$ Type 1; $\theta_1 = 0$, $\therefore$ Type 1

Nodal Position 2 = 3

$v_3 = 0$, $\therefore$ Type 1; $\theta_3 \neq 0$ $\therefore$ Type 0

21

E = 1

NL = 1

Nodal Positions and Values of Concentrated Loads and Couples

Nodal Position 1 = 2

Value of load = -6

Value of couple = 0

Nodal Position 2 = 3

Value of load = 0

Value of couple = 7 (clockwise + ve)

Member Details (hydrostatic load data for this example)

SA	XL	WA	WB	AS
1	4	0	-2	1
2	5	-2	-2	0

A typical input data file for BEAMSF is shown below.

The file was used to solve the problem of Example 2.1.

Beams.DAT

```
2
2
2
1,1,1
3,1,0
1
1
2,-6,0
3,0,7
1,4
0,-2,1
2,5
-2,-2,0
EOF
```

The various items of data in this input file can be separated by two spaces instead of commas. Do not use the TAB key for spaces.

2.4 Results

The results are given below:-

Nodal Displacements x EI

Node	v	θ
1	0	0
2	-55.68	4.096
3	0	-16.98

Nodal bending moments (kNm) and reactions (kN)

Node	Moments	Reactions
1	19.45	9.494
2(left)	-15.53	0
2(right)	-15.53	0
3	7.00	9.506

Hogging bending moments are positive.

2.5 Screen Dumps

Screendumps of the bending moment and shearing force diagrams for Example 2.1 are shown in Figure 2.5.

BENDING MOMENT DIAGRAM

SHEARING FORCE DIAGRAM

Figure 2.5 - Screendumps of the BM & SF diagrams for Example 2.1

2.6   Example 2.2

Determine the nodal displacements and bending moments for the continuous beam shown in Figure 2.6

$E = 2E8kN/m^2$ $I = 1E-7 \ m^4$

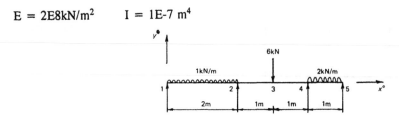

Figure 2.6 - Continuous beam

The input data file is as follows:-

```
4
4
1
1,1,0
2,1,0
4,1,0
5,1,0
2E8
0
3,-6,0
1E-7,2,-1
1E-7,1,0
1E-7,1,0
1E-7,1,-2
EOF
```

2.7   Results

The results are given below:-

Nodal Displacements

Node	v(m)	θ (rads)
1	0	-1.136E-3
2	0	1.894E-4
3	-2.131E-2	-6.629E-4
4	0	-1.629E-2
5	0	6.061E-3

24

Bending Moments and Reactions

Node	Moment (kN/m)	Reactions (kN)
1	0	0.466
2 (left)	1.068	4.455
2 (right)	1.068	
3 (left)	-1.852	0
3 (right)	-1.852	
4 (left)	1.227	5.307
4 (right)	1.227	
5	0	-0.227

Hogging moments are positive.

NB It should be noted that the beam tends to "lift off" node 5. Hence, if node 5 is not firmly secured, another analysis should be carried out after removing the support at node 5.

2.8 Screendumps

A screendump of the bending moment and shearing force diagrams is shown in Figure 2.7.

BENDING MOMENT DIAGRAM

SHEARING FORCE DIAGRAM

Figure 2.7 - BM & SF Diagrams for Example 2.2

25

2.9   Data For BEAMERF

The input data file should be created in the following sequence:-

1.   Number of elements = LS

2.   Number of nodes with elastic or rigid supports = NF

3.   Number of nodes with concentrated loads and couples = NC

4.   Nodal Positions of Elastic or Rigid Supports and Stiffnesses

   FOR i = 1 to NF

   Input NS_i (node); vertical stiffness and rotational stiffness at node NS_i

5.   Elastic modulus = E

6.   If there is any hydrostatic load feed 1, otherwise feed 0, (ie NL = 1 or NL = 0, respectively).

7.   Nodal Positions and Values of Concentrated Loads and Couples

   NC = number of nodes with concentrated loads and couples

   FOR i = 1 to NC

   Input "Nodal" position of load or couple

   Input value of load and couple at this node

8.   If NL = 1, ignore (8)

   Member Details (no hydrostatic load)

   FOR 1 = 1 to LS

   (Type in the details of the members/elements from left to right)

   Input SA - Input 2nd moment of area for member

   Input XL - Input elemental length

   Input UD - Input value of distributed load/length (+ve if in "y" direction)

9.   If NL = 0, ignore (9)

Member Details (hydrostatic load)

FOR I = 1 to Ls

(Type in the details of the members/elements from left to right)

Input SA - Input 2nd moment of area for element

Input XL - Input elemental length

Input WA - Value of distributed load on left)
)
Input WB - Value of distributed load on right)   See Figure 2.3
)
Input AS - The distance of WA from the left node)

2.10   Output

The Nodal Displacement $v_1, \theta_1, v_2, ... v_n, \theta_n$

where n = LS + 1

The Nodal Bending Moments

Element	Nodes Defining Each Element		Nodal Bending Moments
1	i	j	BM_i
	i	j	BM_j
2	j	k	BM_j
	j	k	BM_k
	k		BM_k
	k		
LS			

Reactions (upwards positive)

$V_1, V_2, ... V_n$ (some of which are zero)

2.11   Example 2.3

A cantilever of length 2 m is fixed firmly at node 3, and initially, has its "free" end propped to the same level as the fixed end, as shown in Figure 2.8.

A uniformly distributed load of 10000 N/m is then placed on the cantilever. Determine the deflection of the propped end and the nodal bending moments assuming the following apply:-

Beam

I = 4.91E-6m^4 =   2nd MOA of beam section

E = 2E11 N/m^2 =   Elastic modulus of beam

Prop

A = 2E-4m^2 =   Cross-sectional area of prop

L = 0.3 m   =   length of prop

E = 6.6E10 N/m^2 =   Elastic modulus of prop

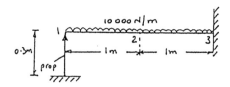

Figure 2.8

$$\text{Stiffness of prop} = \frac{AE}{L} = \frac{2E\text{-}4 \times 6.6E10}{0.3} = 44E6 \text{ Nm}$$

The data is as follows:-

LS = 2

NF = 2

NC = 0

"Support" Displacement Positions with Stiffnesses

Nodal Position 1 = 1;   Vertical stiffness = 4.4E7

                        Rotational Stiffness = 0   --   Free

Nodal Position 2 = 3;   Vertical Stiffness = 1E 30   )
) Rigid
                        Rotational Stiffness = 1E 30   )

28

E = 2E11

NL = 0

Element Details

4.91E-6	1	- 10000
4.91E-6	1	- 10000

A typical input data file for BEAMERF is shown below

This file was used to solve the statically indeterminate beam of Example 2.3

BEAMERF.DAT

```
2
2
0
1,44E6,0
2,1E30,1E30
2E11
0
4.91E-6,1,-10000
4.91E-6,1,-10000
EOF
```

The various items of data in this input data file can be separated by two spaces, instead of commas. Do not use the TAB key for spaces!

2.12 Output

Nodal Displacement

Node	v_i	θ_i
1	-1.69-E-4	1.57E-3
2	-9.01E-4	-5.19E-4
3	0	0

Nodal Moments

M_{1-2} = 0

M_{2-1} = 2438 Nm

M_{2-3} = 2438 Nm

M_{3-2} = 5124 Nm

"Exact" Calculation

v_1 = -1.69E-4

M_1 = 0

M_2 = -2437 Nm

M_3 = 5127 Nm

Reactions (some are zero)

(Node 1) V_1 = 7438 N   )
) vertically upwards (+ve)
(Node 3) V_3 = 12562 N   )

CHAPTER 3

BENDING MOMENTS IN RIGID-JOINTED PLANE FRAMES

3.1   Introduction

This chapter describes two computer programs, namely FRAME2DF & FRAMERP. For both programs, the applied loads can be a complex combination of point, uniformly distributed and hydrostatic loads, and the boundary conditions can be quite complex as well. Additionally, each element can have different cross-sectional properties.

In the case of the computer program FRAME2DF, the frames must be rigid-jointed, but in the case of the computer program FRAMERP, the joints can be a complex combination of rigid or pin-joints or even partial pin-joints.

The computer program FRAMERP allows a multiple combination of loads to be placed between adjacent joints, and for Young's modulus to be different for each element.

For both programs, each element has two end nodes, with three degrees of freedom per node, namely, $u°$, $v°$ and θ as shown in Figure 3.1.

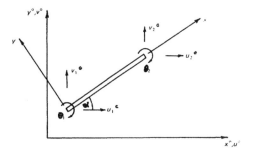

Figure 3.1

Displacement positions are calculated as follows:-

Displacement position corresponding to $u_i°$ $\equiv$ 3 x i - 2

Displacement position corresponding to $v_i°$ $\equiv$ 3 x i - 1

displacement position corresponding to θ_i $\equiv$ 3 x i

As there are six degrees of freedom per element, the elemental stiffness matrix $[k°]$ will be of order 6 x 6. This elemental stiffness matrix consists of an axial component $[k_r°]$ and a flexural component $[k_b°]$, as shown by equation 3.1.

$$[k°] = [k_b°] + [k_r°] \qquad\qquad 3.1$$

where

$$
[k]_b{}^\circ = EI
\begin{array}{cccccc}
u_1{}^\circ & v_1{}^\circ & \theta_1 & u_2{}^\circ & v_2{}^\circ & \theta_2 \\
\end{array}
$$

$$
[k]_b{}^\circ = EI
\left[
\begin{array}{cccccc}
\dfrac{12}{\ell^3}s^2 & & & & & \\[3ex]
-\dfrac{12}{\ell^3}cs & \dfrac{12}{\ell^3}c^2 & & & & \\[3ex]
\dfrac{6}{\ell^2}s & -\dfrac{6}{\ell^2}c & \dfrac{4}{\ell} & & & \\[3ex]
-\dfrac{12}{\ell^3}s^2 & \dfrac{12}{\ell^3}cs & -\dfrac{6}{\ell^2}s & \dfrac{12}{\ell^3}s^2 & & \\[3ex]
\dfrac{12}{\ell^3}cs & -\dfrac{12}{\ell^3}c^2 & \dfrac{6}{\ell^2}c & -\dfrac{12}{\ell^3}cs & \dfrac{12}{\ell^3}c^2 & \\[3ex]
\dfrac{6}{\ell^2}s & -\dfrac{6}{\ell^2}c & \dfrac{2}{\ell} & -\dfrac{6}{\ell^2}s & \dfrac{6}{\ell^2}c & \dfrac{4}{\ell}
\end{array}
\right]
\begin{array}{c}
u_1{}^\circ \\[3ex] v_1{}^\circ \\[3ex] \theta_1 \\[3ex] u_2{}^\circ \\[3ex] v_2{}^\circ \\[3ex] \theta_2
\end{array}
$$

= stiffness matrix for an inclined beam element.

$$[k_r^\circ] = \frac{AE}{\ell}
\begin{array}{cccccc}
u_1^{\;\circ} & v_1^{\;\circ} & \theta_1 & u_2^{\;\circ} & v_2^{\;\circ} & \theta_2 \\
\end{array}
\left[
\begin{array}{cccccc}
c^2 & cs & 0 & -c^2 & -cs & 0 \\
cs & s^2 & 0 & -cs & -s^2 & 0 \\
0 & 0 & 0 & 0 & 0 & 0 \\
-c^2 & -cs & 0 & c^2 & cs & 0 \\
-cs & -s^2 & 0 & cs & s^2 & 0 \\
0 & 0 & 0 & 0 & 0 & 0 \\
\end{array}
\right]
\begin{array}{l}
u_1^{\;\circ} \\
v_1^{\;\circ} \\
\theta_1 \\
u_2^{\;\circ} \\
v_2^{\;\circ} \\
\theta_2 \\
\end{array}$$

3.3

= stiffness matrix for an inclined rod element.

3.2   Data for FRAME2DF

The input data file for FRAME2DF should be created in the following sequence.

1.   Number of nodal points =   NJ

2.   Number of elements = MS

3.   Number of nodes with zero displacements = NF

4.   Number of nodes with concentrated loads and couples = NC

5.   Nodal co-ordinates

    FOR i = 1 to NJ

    Input x_i°, y_i°

6.   Positions and details of zero displacements, etc.

    FOR i = 1 to NF

    Input NS_i and state whether or not the appropriate displacements are zero at this node.

    Type 1 if u° = 0 or 0 if it is not, then   } For
    Type 1 if v° = 0 or 0 if it not, and then   } node
    Type 1 if θ = 0 or 0 if it is not   } NS_i

7.   Elastic modulus = E

33

8.   If there is any hydrostatic load, type 1, otherwise type 0, (ie HY = 1 or HY = 0, respectively).

9.   If HY = 1, ignore (9).

FOR I = 1 to MS

Input i - input "i" node of element

Input j - Input "j" node of element

Input SA - input 2nd moment of area of element

Input A - input cross-sectional area of element

Input UD - input of uniformly distributed load/length (+ve if in "y" direction)

10.   If HY = 0, ignore (10)

FOR I = 1 to MS

Input i - input "i" node of element

Input j - input "j" node of element

Input SA - input 2nd moment of area of element

Input A - Input cross-sectional area of element

Input WA - Input value of distributed load on "left"   )
)
Input WB - Input value of distributed load on "right"   )   see Figure 2.3
)
Input AS - Distance of WA from the left node   )

11.   Nodal Positions and Value of Concentrated Loads and Couples

FOR i = 1 to NC

Input Nodal position of load

Value of load in x° direction   )
)
Value of load in y° direction   )   at this node
)
Value of couple (clockwise +ve)   )

34

3.3 Output

The Nodal Displacements u_1°, v_1°, θ_1, u_2°, v_2° .... u°_{NJ}, v°_{NJ}, θ_{NJ}

Nodal Moments and Axial Forces

Element	Nodes Defining Element	Axial Force	Moment
1	i-j	F_1	M_{i-j}
	i-j	F_1	M_{j-i}
2	k-	F_2	M_{j-k}
	k-		
Ms			

Reactions in x° and y° directions at the suppressed nodes

3.4 Example 3.1

Determine the nodal bending moments for the rigid-jointed skew frame shown in Figure 3.2.

As only the moments are required, it will be convenient to assume $\dfrac{A}{1000} = I = E = 1$.

If, however, the nodal displacements are required, it will be necessary to feed in the true values of A, I and E, where "I" is the second moment of area for the inclined members.

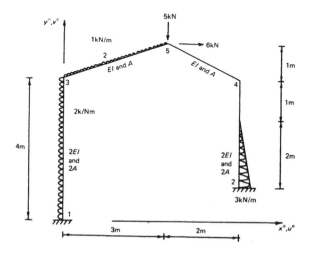

Figure 3.2 - Skew Frame

35

As element 2.4 has a hydrostatic load, the variable HY must be made equal to one.

The data file should be created in the following sequence.

NJ = 5

MS = 4

NF = 2

NC = 1

Nodal Co-ordinates

$x_i{}^\circ$	$y_i{}^\circ$
0	0
5	1
0	4
5	4
3	5

Nodal details of zero displacements

Nodal Position 1 = 1

$u_1{}^\circ = 0$ $\therefore$ Type 1

$v_1{}^\circ = 0$ $\therefore$ Type 1

$\theta = 0$ $\therefore$ Type 1

Nodal Position 2 = 2

$u_2{}^\circ = 0$ $\therefore$ Type 1

$v_1{}^\circ = 0$ $\therefore$ Type 1

$\theta_2 = 0$ $\therefore$ Type 1

E = 1

HY = 1

Member Details

i	j	SA	A	WA	WB	AS
1	3	2	2000	-2	-2	0
3	5	1	1000	-1	-1	0
5	4	1	1000	0	0	0
4	2	2	2000	0	-3	1

Nodal Positions and Values of Concentrated Loads

Nodal Position 1 = 5

Load in x° direction = 6

Load in y° direction = -5

Value of couple = 0

3.5 Output

The Displacements $u_1{}^\circ$, $v_1{}^\circ$, θ_1, $u_2{}^\circ$, $v_2{}^\circ$, θ_2, $u_3{}^\circ$, $v_3{}^\circ$, θ_3, $u_4{}^\circ$, $v_4{}^\circ$, θ_4, $u_5{}^\circ$, $v_5{}^\circ$ and θ_5.

Moments and Forces

Member	Nodes Defining Member	Axial Force	Nodal Moments (kN, m)	
1	1-3	-0.87	$M_{1\text{-}3}$ =	11.60
	1-3	-0.87	$M_{3\text{-}1}$ =	-2.57
2	3-5	-0.71	$M_{3\text{-}5}$ =	-2.57
	3-5	-0.71	$M_{5\text{-}3}$ =	0.28
3	5-4	-9.86	$M_{5\text{-}4}$ =	0.28
	5-4	-9.86	$M_{4\text{-}5}$ =	7.08
4	4-2	-7.13	$M_{4\text{-}2}$ =	7.08
	4-2	-7.13	$M_{2\text{-}4}$ =	13.3

Reactions

$H_{x1} = -7.54 \; V_{y1} = 0.87$ - Node 1

$H_{x2} = -4.28 \; V_{y2} = 7.13$ - Node 2

3.6 "FRAME2DF"

A typical data file for FRAME2DF is given below.

The data file "FRAME2D.INP" was used to analyse the problem of Example 3.1, which was for the <u>hydrostatic load case</u>.

FRAME2DF.INP

```
5
4
2
1
0,0
5,1
0,4
5,4
3,5
1,1,1,1
2,1,1,1
1
1
1,3,2,2000
-2,-2,0
3,5,1,1000
-1,-1,0
5,4,1,1000
0,0,0
4,2,2,2000
0,-3,1
5,6,-5,0
EOF
```

The various items in the above data file can be separated by two or more spaces, instead of commas. <u>Do not use the TAB key</u> to create spaces!

3.7   Example 3.2

Determine the nodal bending moments for the rigid-jointed plane frame shown in Figure 3.3, which is firmly fixed at nodes 1 and 4, and firmly pinned at node 7.

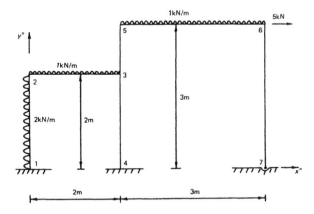

Figure 3.3 - Two bay frame

The input data file for this program, namely FRAMEF.DAT is shown below. This data file is for the <u>uniformly distributed load case</u>, as there is no hydrostatic loading on this framework.

FRAMEF.DAT

```
7
6
3
1
0,0
0,2
2,2
2,0
2,3
5,3
5,0
1,1,1,1
4,1,1,1
7,1,1,0
2E8
0
1,2,1E-7,1E-3
-2
2,3,1E-7,1E-3
-1
```

```
[       3,4,1E-7,1E-3
[       0

[       5,3,1E-7,1E-3
[       0

[       5,6,1E-7,1E-3
[       -1

[       6,7,1E-7,1E-3
        0
        6,5,0,0
        EOF
```

3.8   Output

The results are shown below.

Answers

Member	Node	Axial Force (kN)	Moment (kN m)
1-2	1	2.532	4.86
1-2	2	2.532	-2.773
2-3	2	1.817	-2.773
2-3	3	1.817	4.290
3-4	3	-4.788	1.741
3-4	4	-4.788	-3.344
5-3	5	-0.256	1.810
5-3	3	-0.256	-2.549
5-6	5	4.359	-1.810
5-6	6	4.359	1.923
6-7	6	-2.744	1.923
6-7	7	-2.744	0

Reactions

Node	H_x (kN)	Vy (kN)
1	-5.816	-2.531
4	-2.542	4.787
7	-0.641	2.744

3.9   Screen Dumps

Screen dumps of the bending moment and shearing force diagrams for Example 3.1 are shown in Figure 3.4 and 3.5, and for Example 3.2 are shown in Figures 3.6 and 3.7.

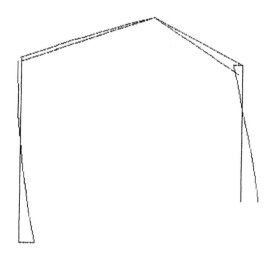

BENDING MOMENT DIAGRAM

Figure 3.4

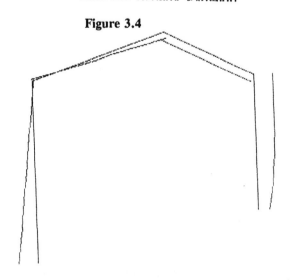

SHEARING FORCE DIAGRAM
Figure 3.5

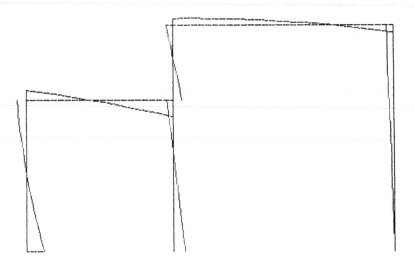

BENDING MOMENT DIAGRAM

Figure 3.6

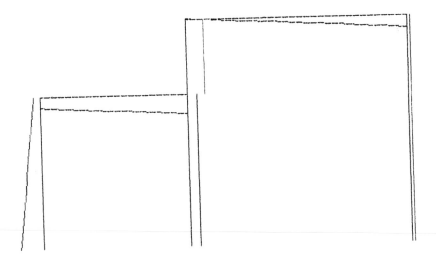

SHEARING FORCE DIAGRAM

Figure 3.7

42

3.10 Frames with a Combination of Rigid and Pin-Joints

This program, which is called FRAMERP can analyse frames which are a combination of rigid and pin-joints and joints which are partially pinned. This program must not be used for continuous beams!

The bending stiffness matrix $[k_b{}^\circ]$ for an inclined beam, that is rigid at one end and pinned at the other, will be different to that of equation 3.2. For the interest of researchers and for readers who may require a deeper understanding of this topic, this new stiffness matrix will be derived for the beam of Figure 3.8.

3.11 Derivation of Stiffness Matrix for the Beam of Figure 3.8

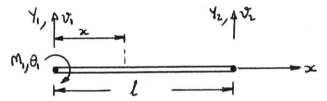

Figure 3.8 - Element with a rigid-joint at node 1 and a pin-joint at node 2

Moments about Node 2

$$Y_1 = - M_1/l \qquad\qquad 3.4$$

$$\text{hence, } Y_2 = M_1/l \qquad\qquad 3.5$$

Now

$$EI \frac{d^2v}{dx^2} = M = Y_1 x + M_1$$

$$EI \frac{dv}{dx} = \frac{Y_1 x^2}{2} + M_1 + A$$

$$EI\, v = \frac{Y_1 x^3}{6} + \frac{M_1 x^2}{2} + Ax + B$$

@ $x = 0$, $\frac{dv}{dx} = -\theta_1$ and $v = v_1$

$$\therefore A = -EI\theta_1 \qquad (3.6)$$

$$\&\ B = EIv_1 \qquad (3.7)$$

@ $x = \ell$, $v = v_2$

$$\therefore EIv_2 = \frac{Y_1 \ell^3}{6} + \frac{M_1 \ell^2}{2} - EI\theta_1 \ell + EIv_1 \qquad (3.8)$$

Substituting (3.4) into (3.8),

$$M_1 = \frac{3EI}{\ell^2} (v_2 - v_1) + \frac{3EI\theta_1}{\ell} \qquad (3.9)$$

Hence, from equations (3.4), (3.5) and (3.9)

$$\begin{Bmatrix} Y_1 \\ M_1 \\ Y_2 \end{Bmatrix} = EI \begin{bmatrix} 3/\ell^3 & -3/\ell^2 & -3/\ell^3 \\ -3/\ell^2 & 3/\ell & 3/\ell^2 \\ -3/\ell^3 & 3/\ell^2 & 3/\ell^3 \end{bmatrix} \begin{Bmatrix} v_1 \\ \theta_1 \\ v_2 \end{Bmatrix} \qquad (3.10)$$

that is the bending stiffness matrix for this beam element $[k_b]$ is given by equation 3.11.

$$[k_b] = EI \begin{array}{c} v_1 \theta_1 v_2 \\ \begin{bmatrix} 3/\ell^3 & -3/\ell^2 & -3/\ell^3 \\ -3/\ell^2 & 3/\ell & 3/\ell^2 \\ -3/\ell^3 & 3/\ell^2 & 3/\ell^3 \end{bmatrix} \begin{array}{c} v_1 \\ \theta_1 \\ v_2 \end{array} \end{array} \qquad 3.11$$

Similarly, for a beam that is pinned at its left end and rigid at its right end, $[k_b]$ takes the form of equation 3.12.

$$[k_b] = EI \begin{array}{c} v_1 v_2 \theta_2 \\ \begin{bmatrix} 3/\ell^3 & -3/\ell^3 & -3/\ell^2 \\ -3/\ell^3 & 3/\ell^3 & 3/\ell^2 \\ -3/\ell^2 & 3/\ell^2 & 3/\ell \end{bmatrix} \begin{array}{c} v_1 \\ v_2 \\ \theta_2 \end{array} \end{array} \qquad 3.12$$

The global bending stiffness matrix $[k_b{}^\circ]$ for a beam inclined as shown in Figure 3.1 is given by:

$$[k_b{}^\circ] = [DC]^T \, [k_b] \, [DC] \qquad 3.13$$

where $[DC] =$

$$\zeta = \begin{bmatrix} c & s & 0 \\ -s & c & 0 \\ 0 & 0 & 1 \end{bmatrix}$$

$\qquad 3.14$

O_3 = a null matrix of order 3 x 3

It should be ensured that the three columns and rows of zeros should be inserted into equations 3.11 and 3.12, prior to carrying out the triple matrix product of equation 3.13, the result which is as follows for the beam of Figure 3.8

$$[k_b{}^\circ] = 3EI \begin{bmatrix} 0 & -3s/\ell^3 & -3s/\ell^2 & 0 & -3s/\ell^3 & 0 \\ 0 & 3c/\ell^3 & 3c/\ell^2 & 0 & 3c/\ell^3 & 0 \\ 0 & -3/\ell^2 & 3/\ell & 0 & 3/\ell^2 & 0 \\ 0 & 3s/\ell^3 & 3s/\ell^2 & 0 & 3s/\ell^3 & 0 \\ 0 & -3c/\ell^3 & 3c/\ell^2 & 0 & 3c/\ell^3 & 0 \\ 0 & 0 & 0 & 0 & 0 & 0 \end{bmatrix} \begin{matrix} u_1{}^\circ \\ v_1{}^\circ \\ \theta_1 \\ u_2{}^\circ \\ v_2{}^\circ \\ \theta_2 \end{matrix}$$

with column headings $u_1{}^\circ \quad v_1{}^\circ \quad \theta_1 \quad u_2{}^\circ \quad v_2{}^\circ \quad \theta_2$

where
$c = \cos \alpha$
$s = \sin \alpha$

3.12 Loading

The end forces and bending moments, for non-standard cases, which are required to determine the equivalent nodal forces for distributed and intermediate point loads are given below.

1. **Case (a) left end "rigid" and right end pinned**

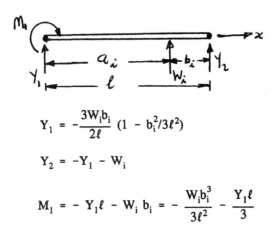

$$Y_1 = -\frac{3W_i b_i}{2\ell}(1 - b_i^2/3\ell^2)$$

$$Y_2 = -Y_1 - W_i$$

$$M_1 = -Y_1\ell - W_i b_i = -\frac{W_i b_i^3}{3\ell^2} - \frac{Y_1\ell}{3}$$

2.   Case (b) left end pinned and right end "rigid"

$$Y_1 = -\frac{3W_i b_i^2}{\ell^3}\left[\frac{\ell}{2} - \frac{b_i}{6}\right]$$

$$Y_2 = -Y_1 - W_i$$

$$M_2 = -Y_1\ell - W_i b_i$$

3.   Case(c) Trapezoidal load - right and pinned and left end rigid

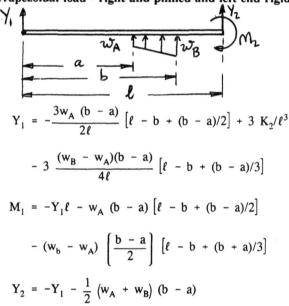

$$Y_1 = -\frac{3w_A(b-a)}{2\ell}\left[\ell - b + (b-a)/2\right] + 3\,K_2/\ell^3$$

$$- 3\,\frac{(w_B - w_A)(b-a)}{4\ell}\left[\ell - b + (b-a)/3\right]$$

$$M_1 = -Y_1\ell - w_A(b-a)\left[\ell - b + (b-a)/2\right]$$

$$- (w_b - w_A)\left[\frac{b-a}{2}\right]\left[\ell - b + (b+a)/3\right]$$

$$Y_2 = -Y_1 - \frac{1}{2}(w_A + w_B)(b-a)$$

4. **Case (d) Trapezoidal load - left end pinned and right end rigid**

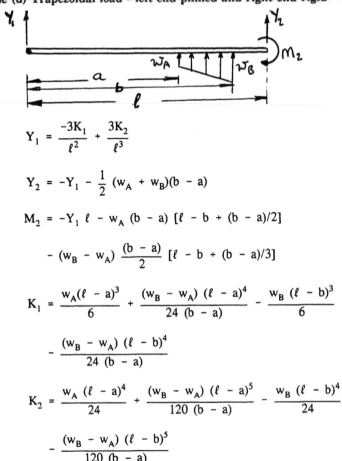

$$Y_1 = \frac{-3K_1}{\ell^2} + \frac{3K_2}{\ell^3}$$

$$Y_2 = -Y_1 - \frac{1}{2} (w_A + w_B)(b - a)$$

$$M_2 = -Y_1 \ell - w_A (b - a) [\ell - b + (b - a)/2]$$

$$- (w_B - w_A) \frac{(b - a)}{2} [\ell - b + (b - a)/3]$$

$$K_1 = \frac{w_A(\ell - a)^3}{6} + \frac{(w_B - w_A)(\ell - a)^4}{24 (b - a)} - \frac{w_B (\ell - b)^3}{6}$$

$$- \frac{(w_B - w_A)(\ell - b)^4}{24 (b - a)}$$

$$K_2 = \frac{w_A (\ell - a)^4}{24} + \frac{(w_B - w_A)(\ell - a)^5}{120 (b - a)} - \frac{w_B (\ell - b)^4}{24}$$

$$- \frac{(w_B - w_A)(\ell - b)^5}{120 (b - a)}$$

3.13 Moment-nodal dispalcements relationships

To obtain the matrix relating <u>bending moment to nodal dispalcements</u>, an expressions for

$\frac{d^2v}{dx^2}$ will now be obtained for a beam which is <u>"rigid at the left end and pinned at the right</u>

<u>end</u>.

Let

$$v = \alpha_1 + \alpha_2 x + \alpha_3 x^2 + \alpha_4 x^3 \qquad\qquad 3.16$$

where, α_1, α_2, α_3 and α_4 = arbitrary constants.

$$\frac{dv}{dx} = \alpha_2 + 2\alpha_3 x + 3\alpha_4 x^2 \tag{3.17}$$

$$\frac{d^2v}{dx^2} = 2\alpha_3 + 6\alpha_4 x \tag{3.18}$$

The boundary conditions are:

$$@ \quad x = 0, \quad v = v_1, \quad \& \quad \theta = \theta_1 = - \frac{dv}{dx}$$

Substituting these two boundary conditions into equations (3.16 and (3.17), the following are obtained for α_1 and α_2.

$$\alpha_1 = v_1 \tag{3.19}$$

$$\alpha_2 = -\theta_1 \tag{)3.20}$$

The third boundary condition is:

$$@ \quad x = \ell, \quad \frac{d^2v}{dx^2} = 0 \tag{3.21}$$

$$\therefore \quad \alpha_3 = - 3\alpha_4 \ell$$

The fourth and last boundary condition is:

$@ \quad x = \ell, \quad v = v_2$

$$\therefore \quad v_2 = \alpha_1 + \alpha_1 \ell + \alpha_3 \ell^2 + \alpha_4 \ell^3 \tag{3.22}$$

Substituting (3.19) and (3.20) into (3.22), and solving (3.21) and (3.22).

$$\alpha_3 = \frac{3(v_2 - v_1)}{2\ell^2} + \frac{3\theta_1}{2\ell} \tag{3.23}$$

$$\alpha_4 = \frac{-(v_2 - v_1)}{2\ell^3} - \frac{\theta_1}{2\ell^2} \tag{3.24}$$

Substituting (3.23) and (3.24) into (3.18).

$$\frac{d^2v}{dx^2} = [3\xi/\ell^2 \quad - 3\xi/\ell^2 \quad - 3\xi/\ell \quad \begin{Bmatrix} v_1 \\ v_2 \\ \theta_2 \end{Bmatrix} \tag{3.25}$$

where
$\xi = x/l$

Hence M can be calculated from the expression

$$M = EI \left[\frac{d^2v}{dx^2} \right] \tag{3.26}$$

3.14   Deflection distributions

As this program allows a multiple combination of loads to act between adjacent joints, it is necessary to obtain expressions for deflection distributions for various combinations of rigid and pinned joints. A typical loaded beam, which has a rigid joint at its left end and a pinned joint at its rigid end.

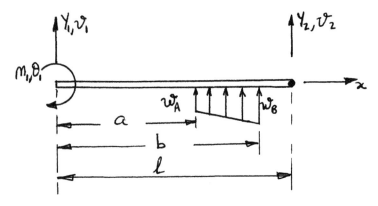

$$EI \frac{d^2v}{dx^2} = Y_1 x + M_1 \qquad\qquad + \frac{w_A (x - a)^2}{2} + \frac{(w_B - w_A)(x - a)^3}{(b - a)\ 6} \qquad\qquad - \frac{w_B (x - b)^2}{2} - \frac{(w_B - w_A)(x - b)^3}{(b - a)\ 6}$$

$$EI \frac{dv}{dx} = \frac{Y_1 x^2}{2} + M_1 x + A \qquad + \frac{w_A(x - a)^3}{6} + \frac{(w_B - w_A)(x - a)^4}{24(b - a)} \qquad - \frac{w_B (x - b)^3}{6} - \frac{(w_B - w_A)(x - b)^4}{24(b - a)}$$

$$EI\ v = \frac{Y_1 x^3}{6} + \frac{M_1 x^2}{2} + Ax + B \qquad + \frac{w_A (x - a)^4}{24} + \frac{(w_B - w_A)(x - a)^5}{120(b - a)} \qquad - \frac{w_B (x - b)^4}{24} - \frac{(w_B - w_A)(x - b)^5}{120(b - a)}$$

@ x = 0   v = 0   ∴ $\underline{B = 0}$

@ x = 0   dv/dx = 0   ∴ $\underline{A = 0}$

@ x = ℓ v = 0

$$0 = \frac{Y_1 \ell^3}{6} + \frac{M_1 \ell^2}{2} + K_2 \qquad\qquad\qquad (3.27)$$

Moms abt node (2)

$$M_1 + Y_1\ \ell + w_A\ (b - a)\ [\ell - b + (b - a)/2] +$$

$$(w_B - w_A)\ \frac{(b - a)}{2}\ [\ell - b + (b - a)/3] \qquad\qquad (3.28)$$

Multiply (3.28) by $\ell^2/_2$

$$\frac{Y_1 \ell^3}{2} + \frac{M_1 \ell^2}{2} + \frac{w_A \ell^2 (b - a)}{2}\ [\ell - b + (b - a)/2]$$

$$+ (w_B - w_A)\ell^2\ \frac{(b - a)}{4}\ [\ell - b + (b - a)/3] \qquad\qquad (3.28)$$

Take (3.27) from (3.28)

$$0 = Y_1 \ell^3 \left[\frac{1}{2} - \frac{1}{6} \right] + \frac{w_A \ell^2 (b - a)}{2}\ [\ell - b + (b - a)/2] - K_2$$

$$+ (w_B - w_A)\ell^2 \frac{(b - a)}{4}\ [\ell - b + (b - a)/3]$$

$$Y_1 = \frac{-3w_A\,(b - a)}{2\ell}\,[\ell - b + (b - a)/2] + 3K_2/\ell^3$$

From (2)

$$M_1 = -Y_1\ell - w_A\,(b - a)\,[\ell - b + (b - a)/2]$$

$$\frac{-(w_B - w_A)\,(b - a)}{2}\,[\ell - b + (b - a)/3]$$

$v = \dfrac{1}{EI}$	$\left(\dfrac{Y_1 x^3}{6} + \dfrac{M_1 x^2}{2}\right.$	$+\dfrac{w_A(x - a)^4}{24} + \dfrac{(w_B - w_A)(x - a)^5}{120(b - a)}$	$-\dfrac{w_B(x - b)^4}{24} - \dfrac{(w_B - w_A)\,(x - b)^5}{120\,(b - a)}\left.\right)$	

RV

$$Y_1 + Y_2 + 1/2\,(w_A + w_B)\,(b - a) = 0$$

$$Y_2 = -Y_1 - 1/2\,(w_A + w_B)\,(b - a)$$

3.15   Description of how to use the program FRAMERP

A description of the capabilities and use of this program will now be given.

This program calculates and plots the bending moments, shearing forces and deflections in rigid-jointed/pin-jointed frameworks, subjected to various combinations of distributed and point loads.

The joints can be rigid, but allowances are made for internal pin-joints, where the latter can be partial or full.

For a framework, the nodal points need only be taken at the joints or supports, as allowances are made for the input of multiple point and hydrostatic loads between any pair of adjacent nodes, in addition to point loads and couples at nodes themselves.

The geometric properties of the elements can be different.

Calculations are also made of the support reactions.

NB if an element has intermediate load/s, and at the same time, one of its ends is firmly *fixed or pinned to a rigid-support*, then the support node should be designated the "i" or start node, while the remaining node should be taken as the "j" or finish mode

52

The element is governed by two end nodes, namely an "i" (or start) node and a "j" (or finish) node, where each node has three global degrees of freedom $u°$, $v°$ and θ together with global axes $x°$ and $y°$, as shown in Figure 3.10.

The local degrees of freedom are u, v and θ and the local axes are x and y.

Positive loads are shown in Figure 3.9 and 3.10.

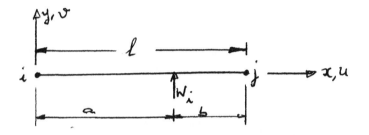

Figure 3.9 - Positive point load

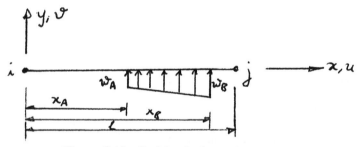

Figure 3.10 - Positive hydrostatic load

3.16   Computer Program

In the computer program the following symbols were used to represent the quantities of Figures 3.9 and 3.10.

ME = an element

NUD (ME) = number of hydrostatic loads on element ME

NWLOAD (ME) - number of point loads on element ME

UDA = w_A

UDB = w_B

53

XA = x_A

XB = x_B

WLOAD = W_i

XAW = a

BL = b = ℓ - a

NB If an <u>internal joint is pinned</u>, then a <u>negative sign should be placed before the nodal number</u> of that joint, in the element topology section of the input, as demonstrated in the given examples.

3.17 Input Data File

The input data file should be created in the following sequence:

NJ = number of nodes of joints

MS = number of elements

NF = number of nodes with zero displacements

NC = number of nodes with additional point loads and couples

Co-ordinates

FOR I = 1 to NJ

Input x_i y_I global coordinates of the nodes

NEXT I

Details of zero displacements

FOR I = 1 TO NJ

Type 1 if u_i° = 0, else type 0

Type 1 if v_i° = 0, else type 0

Type 1 if θ_i = 0, else type 0

Element Topology

For I = 1 TO MS

54

i   j   SA   A   E   NUDL(I)   NWLOAD(I)

If NUDL (I) > 0 then type in UDA, UDB, XA and XB for NUDL(I)

If NWLOAD (I) > 0 Then type in WLOAD and XAW for NWLOAD (I)

NEXT I

where,

```
i = start node                           }
j = finish node                          }
SA = 2nd moment of area                  }        For element I
A = cross-sectional area                 }
E = Young's modulus                      }
NUDL (I) = number of hydrostatic loads   }
NWLOAD (I) - number of point loads       }
```

Details of additional point loads and couples at nodes

FOR I = 1 to NC

Type in node number

Type in value of point load in $x°$ direction

Type in value of point load in $y°$ direction

Type in value of couple in θ direction

NEXT I

3.18   Example 3.3

To demonstrate the input of data, consider the frame of Figure 3.11, where element 2-3 has an internal pin at node 3.

$E = 2E8 \text{ kN/m}^2$
$I = IE\text{-}7 \text{ m}^4$
$A = 1E\text{-}3 \text{ m}^2$

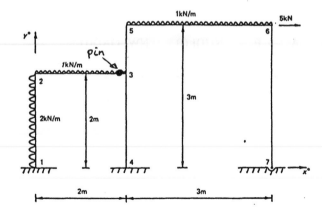

Figure 3.11 - Internal pin at node 3 for element 2-3

3.19 The input data file FRAMERP.DAT

```
7, 6, 3, 1              }
0,0  0,2  2,2  2,0      }        co ordinates
2,3  5,3  5,0           }

1, 1, 1, 1              }
4, 1, 1, 1              }        Details of zero displacements
7, 1, 1, 0              }
```

Element Topology

i	j	SA	A	E	NUDL	NWLOAD
1, -2, -2,	2, 0, 2	1E-7,	1E-3,	2E-8,	1	0
2, -1, -1,	-3 0, 2	1E=7,	1E-3,	2E8,	1,	0
3,	4,	1E-7	1E-3,	2E-8,	0,	0
5,	3,	1E-7,	1E-3,	2E8,	0,	0
5, -1, -1,	6, 0, 3,	1E-7,	1E-3,	2E8	1,	0
6, 6, 5,	7, 0, 0	1E-7	1E-3,	2E8,	0,	0

Details of additional nodal point loads

56

3.20 Results

The results for Example 3.3 are given below

Displacements

Node	$u°(m)$	$v°(m)$	θ (rads)
1	0	0	0
2	0.263	4.044E-6	0.110
3	0.263	-1.819E-5	0.215
4	0	0	0
5	0.462	-1.526E-5	0.149
6	0.462	-5.378E-5	7.286E-3
7	0	0	0.227

Bending Moments and Axial Forces

Element	Node	Moment (kN m)	Axial Force (kN)
1-2	1	6.348	0.404
	2	-2.809	
2-3	2	-2.809	2.578
	3	0	
3-4	3	-0.701	-1.819
	4	-3.594	
5-3	5	3.324	0.585
	3	-0.701	
5-6	5	-3.324	4.022
	6	2.932	
6-7	6	2.932	-3.585
	7	0	

3.21 Example 3.4

Similarly for the frame of Figure 3.12, which has an internal pin-joint at node 2, the data for the element topology is:-

1, -2, 1E-7, 1E-3, 2E8, 1, 0, -2, -2, 0, 2

-2, 3, 1E-7, 1E-3, 2E8, 1, 0, -1, -1, 0, 2

3, 4, 1E-7, 1E-3, 2E8, 0, 0

5, 3, 1E-7, 1E-3, 2E8, 0, 0

5, 6, 1E-7, 1E-3, 2E8, 1, 0, -1, -1, 3

6, 7, 1E-7, 1E-3, 2E8, 0, 0

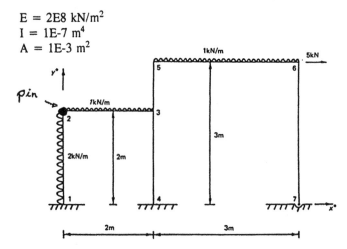

$E = 2E8 \ kN/m^2$
$I = 1E-7 \ m^4$
$A = 1E-3 \ m^2$

Figure 3.12 - Internal pin at node 2

3.22 Examples 3.5

Similarly, for the frame of Figure 6, where there are two internal pins at the ends of one element, namely, 2-3, the data for element topology is:-

1, 2, 1E-7, 1E-3, 2E8, 1, 0, -2, -2, 0, 2

-2, -3, 1E-7, 1E-3, 2E8, 1, 0, -1, -1, 0, 2

3, 4, 1E-7, 1E-3, 2E8, 0, 0

5, 3, 1E-7, 1E-3, 2E8, 0, 0

5, 6, 1E-7, 1E-3, 2E8, 1, 0, -1, -1, 3

6, 7, 1E-7, 1E-3, 2E8, 0, 0

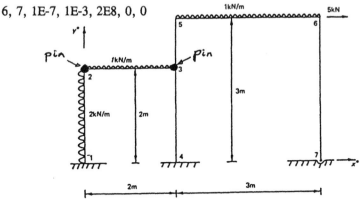

Figure 3.13 - Internal pinned joints at both ends of an element

3.23 Example 3.6

The data for the problem of Figure 3.14, is as follows:-

5, 4, 2, 1

0, 0,   5, 1,   0, 4,   5, 4,   3, 5

1, 1, 1, 1   2, 1, 1, 1

1, 3, 2, 2000, 1, 1, 0, -2, -2, 0, 4   }

3, 5, 1, 1000, 1, 1, 0, -1, -1, 0, 3.162   }   Element topology

5, 4, 1, 1000, 1, 0, 0   }

2, 4, 2, 2000, 1, 1, 0, 3, 0, 0, 2   }

5, 6, -5, 0 - Details of point loads at node 5.

NB   Element 2-4 must be fed in as 2, 4 and NOT as 4,2 for the reasons given in Section
       3.15.

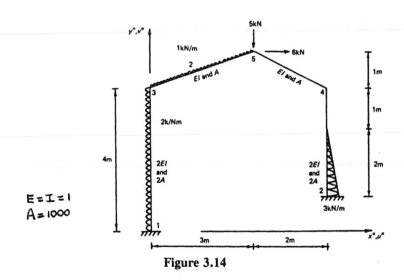

Figure 3.14

3.24 Screen Dumps

Screen dumps of the bending moment and shearing force diagrams for Example 3.3 are given in Figure 3.15 and 3.16.

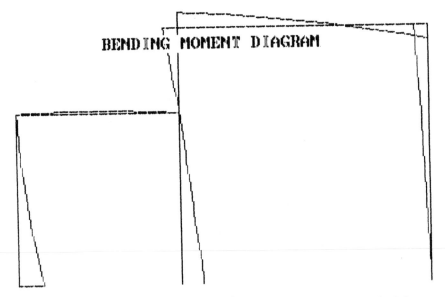

BENDING MOMENT DIAGRAM

Figure 3.15 - Screen dump of the BM diagram for Example 3.3

Figure 3.16 Screen dump of the SF diagram for Example 3.3

CHAPTER 4

FORCES IN PIN-JOINTED SPACE TRUSSES

4.1   Introduction

This program which is called STRUSSF, is meant for determining nodal displacements and member forces in a three-dimensional pin-jointed space truss subjected to concentrated loads at its nodes (or pin-joints).

All the joints are assumed to be in the form of smooth frictionless ball joints, which cannot transfer couples. That is, the computer program should not be applied to welded or rigid-jointed frameworks! Additionally, the computer program should not be applied to frameworks which carry loads between nodal points. For such frameworks, see Chapter 5.

Each member of the truss possesses only axial stiffness and has end nodes with three degrees of freedom per node, as shown in Figure 4.1.

The members of the truss can have different values of cross-sectional area and elastic modulus.

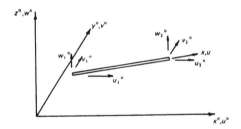

Figure 4.1 - Three-dimensional Rod

The displacement positions can be obtained as follows:-

Displacement position $u_i^\circ \equiv 3 \times i - 2$

Displacement position $v_i^\circ \equiv 3 \times i - 1$

Displacement position $w_i^\circ \equiv 3 \times i$

As each element of the framework has two nodes, and as each node has three degrees of freedom, namely, u°, v° and w°, as shown in Figure 4.1, the elemental stiffness matrix $[k^\circ]$ will be of order 6 x 6, as shown by equation (4.1).

$$[k^\circ] = \frac{AE}{\ell}
\begin{array}{c}
\begin{array}{cccccc}
u_1^\circ & v_1^\circ & w_1^\circ & u_2^\circ & v_2^\circ & w_2^\circ
\end{array} \\
\left[
\begin{array}{cccccc}
C^2x,x^\circ & & & & & \\
Cx,x^\circ,Cx,y^\circ & C^2x,y^\circ & & & & \\
Cx,x^\circ Cx,z^\circ & Cx,y^\circ Cx,z^\circ & C^2x,z^\circ & & & \\
-C^2x,x^\circ & -Cx,x^\circ Cx,y^\circ & -Cx,x^\circ Cx,z^\circ & C^2x,x^\circ & & \\
-Cx,x^\circ Cx,y^\circ & -C^2x,y^\circ & -Cx,y^\circ Cx,z^\circ & Cx,x^\circ Cx,y^\circ & C^2x,y^\circ & \\
Cx,x^\circ Cx,z^\circ & Cx,y^\circ Cx,z^\circ & -C^2x,z^\circ & Cx,x^\circ Cx,z^\circ & Cx,y^\circ Cx,z^\circ & C^2x,z^\circ
\end{array}
\right]
\begin{array}{c}
u_1^\circ \\ v_1^\circ \\ w_1^\circ \\ u_2^\circ \\ v_2^\circ \\ w_2^\circ
\end{array}
\end{array}$$

(4.1)

which can be written in the form:

$$[k^\circ] =
\left[
\begin{array}{c|c}
a & -a \\
\hline
-a & a
\end{array}
\right]$$

(4.2)

where

$$[a] = \frac{AE}{\ell}
\left[
\begin{array}{ccc}
C^2x,x^\circ & & \text{Symmetrical} \\
Cx,x^\circ Cx,y^\circ & C^2x,y^\circ & \\
Cx,x^\circ Cx,z^\circ & Cx,y^\circ Cx,z^\circ & C^2x,z^\circ
\end{array}
\right]$$

(4.3)

$$\left.
\begin{array}{l}
\ell = \sqrt{[(x_j^\circ - x_i^\circ)^2 + (y_j^\circ - y_i^\circ)^2 + (z_j^\circ - z_i^\circ)^2]} \\
Cx,x^\circ = (x_j^\circ - x_i^\circ)/\ell \\
Cx,y^\circ = (y_j^\circ - y_i^\circ)/\ell \\
Cx,z^\circ = (z_j^\circ - z_i^\circ)/\ell
\end{array}
\right\}$$

(4.4)

4.2   Input Data

The input data file should be created in the following sequence:-

1.   Number of pin-joints or nodes = NJ

2.   Number of members = MS

3.   Number of nodes with zero displacements = NF

4.   Nodal Co-ordinates

   For I = 1 to NJ

   Input $x_i^°$, $y_i^°$

5.   Details of Zero Displacements

   FOR i = 1 to NJ

   Input NS_i

   Type 1, if $u^° = 0$ at node NS_i, else type 0, and

   after this,

   Type 1, if $v^° = 0$ at node NS_i, else type 0, and

   after this,

   Type 1, if $w^° = 0$ at node NS_i, else type 0.

6.   Members Details

   FOR I = 1 to MS

   Input i - Input "i" node for member

   Input j - input "j" node for member

   Input A - Input cross-sectional area for member

   Input E - Input elastic modulus for member

7.   Externally Applied Loads

   NC = number of nodes with concentrated loads

FOR i = 1 to NC

Input Node i

Input loads in $x°$, $y°$ and $z°$ directions, respectively, at node i

4.3 Output

The Nodal <u>displacements</u> $u_1°$, $v_1°$, $w_1°$, $u_2°$, $v_2°$, $w_2°$, ... $u_{NJ}°$, $v_{NJ}°$, $w_{NJ}°$

<u>Axial Forces in Members</u>

Member	Nodes Defining Member	Force
1	i-j	$F_{i\text{-}j}$
2		
MS		

NB   Tensile forces are positive

<u>Reactions</u> in $x°$, $y°$ and $z°$ directions

4.4 Example 4.1

Determine the forces in the three-dimensional pin-jointed truss shown in Figure 4.2. All members are of constant AE.

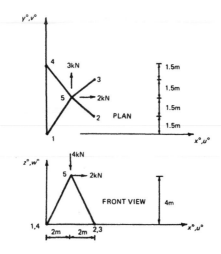

Figure 4.2 - Three-dimensional Truss

The input data file for this problem should be created in the following manner.

NJ = 5

MS = 4

NF = 4

Nodal Co-ordinates

$x_i{}^\circ$	$y_i{}^\circ$	$z_i{}^\circ$
0	0	0
4	1.5	0
4	4.5	0
0	6	0
2	3	4

Details of Zero Displacements

Nodal Position 1 = 1

$u_1{}^\circ = 0$ $\therefore$ Type 1

$v_1{}^\circ = 0$ $\therefore$ Type 1

$w_1{}^\circ = 0$ $\therefore$ Type 1

66

Nodal Position 2 = 2

$u_2{}^\circ = 0$ $\therefore$ Type 1

$v_2{}^\circ = 0$ $\therefore$ Type 1

$w_2{}^\circ = 0$ $\therefore$ Type 1

Nodal Position 3 = 3

$u_3{}^\circ = 0$ $\therefore$ Type 1

$v_3{}^\circ = 0$ $\therefore$ Type 1

$w_3{}^\circ = 0$ $\therefore$ Type 1

Nodal Position 4 = 4

$u_4{}^\circ = 0$ $\therefore$ Type 1

$v_4{}^\circ = 0$ $\therefore$ Type 1

$w_4{}^\circ = 0$ $\therefore$ Type 1

Member Details

i	j	A	E
1	5	1	1
2	5	1	1
3	5	1	1
4	5	1	1

Externally Applied Loads

NC = 1

Nodal position = 5

Value of load in x° direction = 2 }

Value of load in y° direction = 3 }   at node 5

Value of load in z° direction = -4 }

4.5 Output

The Nodal Displacements u_1°, v_1°, w_1°, u_2°, v_2°, w_2°, u_3°, v_3°, w_3°, u_4°, v_4°, w_4°, u_5° and w_5°

Forces in Members

Member	Nodes defining Member	Force (kN)
1	1-5	1.96
2	2-5	-1.08
3	3-5	-3.64
4	4-5	-1.96

The -ve sign denotes compression

Reactions (kN)

$X_1 = -0.73$ $Y_1 = -1.09$ $Z_1 = -1.46$ (Node 1)

$X_2 = -0.46$ $Y_2 = 0.34$ $Z_2 = 0.92$ (Node 2)

$X_3 = -1.54$ $Y_3 = -1.16$ $Z_3 = 3.08$ (Node 3)

$X_4 = 0.73$ $Y_4 = -1.09$ $Z_4 = 1.46$ (Node 4)

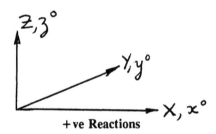

+ve Reactions

4.6 Input Data File

A typical input data file for STRUSSF is shown below.

This data file was used to solve the problem of Example 4.1.

STRUSS.DAT

```
5
4
4
0,0,0
4,1.5,0
4,4.5,0
0,6,0
2,3,4
1,1,1,1
2,1,1,1
3,1,1,1
4,1,1,1
1,5,1,1
2,5,1,1
2,5,1,1
3,5,1,1
4,5,1,1
1
5,2,3,-4
EOF
```

The various items in the above data file can be separated by two or more spaces, instead of commas. Do not use the TAB key to create spaces!

4.7   Example 4.2

Determine the member forces in the pin-jointed space truss of Fig 4.3, which is firmly pinned at nodes 1 to 4. It may be assumed that $AE = 1$.

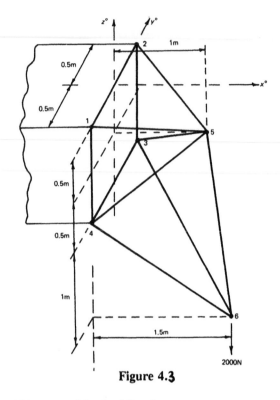

Figure 4.3

The input data should be created in the following sequence.

6	7	4		
$x_i^{\,o}$	$y_i^{\,o}$	$z_i^{\,o}$		
0	-0.5	0		
0	0.5	0		
0	0.5	-1		Global co-ordinates
0	-0.5	-1		
1	0	-0.5		
1.5	0	-2		

1	1	1	1	
2	1	1	1	
3	1	1	1	Details of zero displacements
4	1	1	1	

Element Topology

i	j	A	E
1	5	1	1
2	5		
3	5		
5	4		
5	6		
6	3		
4	6	↓	↓

6   0   0   -2000 ← Details of the concentrated load

4.8   Results

The results are given below:-

Displacement x AE

Node	u°	v°	w°
1	0	0	0
2	0	0	0
3	0	0	0
4	0	0	0
5	393.7	0	-4724
6	-9637	0	-12585

Member	Force (N)
1-5	1837
2-5	1837
3-5	-1312
5-4	-1312
5-6	2711
6-3	-535
4-6	-535

Reactions (N)

Node	X	Y	Z
1	-1500	-750	750
2	-1500	750	750
3	1500	-679	250
4	1500	679	250

CHAPTER 5

STATIC ANALYSIS OF THREE DIMENSIONAL RIGID-JOINTED FRAMES

5.1   Introduction

This program, which is called 3DFRAMEF, is intended for analysing three-dimensional rigid-jointed space frames, under distributed and concentrated loads.

Each node has six load degree of freedom, (u, v, w, θ_x, θ_y and θ_z), and these are shown with the six global degrees of freedom (u°, v°, w°, θ_x°, θ_y° and θ_z°), in Figure 5.1.

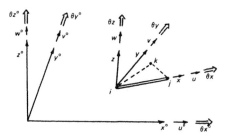

Figure 5.1 - General Case of the One dimensional Member

The element is shown in its global axes, namely x°, y° and z°, in Figure 5.1 and in its local axes, namely x, y and z, in Figure 5.2.

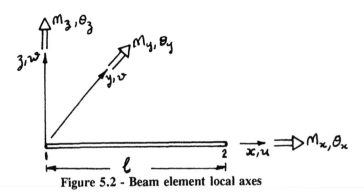

Figure 5.2 - Beam element local axes

The node "k" is in the local x-y plane, so that it defines one of the principal planes of bending, as shown in Figure 5.3. The other principal plane is in the local x-z plane and it is perpendicular to the x-y principal plane, as shown in Figure 5.3.

The directions of the couples and the rotations are shown according to the right-hand screw rule.

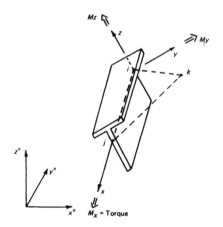

Figure 5.3 - Principal Planes x-y and x-z

The <u>right hand screw rule</u> defines the direction of rotation by pointing the right hand in the direction of the double tailed arrow and rotating the right hand in a clockwise direction.

As each element has two nodes and there are six degrees of freedom per node, the elemental stiffness matrix in local coordinates is of order 12 x 12, as shown by Table 5.1.

Table 5.1

$[k] =$

	u_1	v_1	w_1	θ_{x1}	θ_{y1}	θ_{z1}	u_2	v_2	w_2	θ_{x2}	θ_{y2}	θ_{z2}	
	AE/ℓ												u_1
	0	$12EI_z/\ell^3$											v_1
	0	0	$12EI_y/\ell^3$						symmetrical				w_1
	0	0	0	GJ/ℓ									θ_{x1}
	0	0	$-6EI_y/\ell^2$	0	$4EI_y/\ell$								θ_{y1}
	0	$6EI_z/\ell^2$	0	0	0	$4EI_z/\ell$							θ_{z1}
	$-AE/\ell$	0	0	0	0	0	AE/ℓ						u_2
	0	$-12EI_z/\ell^3$	0	0	0	$-6EI_z/\ell^2$	0	$12EI_z/\ell^3$					v_2
	0	0	$-12EI_y/\ell^3$	0	$6EI_y/\ell^2$	0	0	0	$12EI_y/\ell^3$				w_2
	0	0	0	$-GJ/\ell$	0	0	0	0	0	GJ/ℓ			θ_{x2}
	0	0	$-6EI_y/\ell^2$	0	$2EI_y/\ell$	0	0	0	$6EI_y/\ell^2$	0	$4EI_y/\ell$		θ_{y2}
	0	$6EI_z/\ell^2$	0	0	0	$2EI_z/\ell$	0	$-6EI_z/\ell^2$	0	0	0	$4EI_z/\ell$	θ_{z2}

The elemental stiffness matrix in global coordinates [k°] is given by equation (5.1).

$$[k°] = [Ξ]^T [k] [Ξ] \qquad (5.1)$$

where

$$(5.2)$$

$$[\zeta] = \begin{bmatrix} C_{x,x°} & C_{x,y°} & C_{x,z°} \\ C_{y,x°} & C_{y,y°} & C_{y,z°} \\ C_{z,x°} & C_{z,y°} & C_{z,z°} \end{bmatrix} \qquad (5.3)$$

= a matrix of directional cosines

The directional cosines $C_{x,x°}$, $C_{x,y°}$ and $C_{x,z°}$ are calculated according to equation (4.4), whilst the directional cosines $C_{y,x°}$, $C_{y,y°}$ and $C_{y,z°}$, are calculated by constructing a triangular plane through the centroid of the section in the x-y principal plane of bending, as shown in Figures 5.4 and 5.5. These directional cosines have been derived by Argyris [18] and Petyt [19].

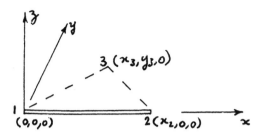

Figure 5.4 - A principal plane of bending

75

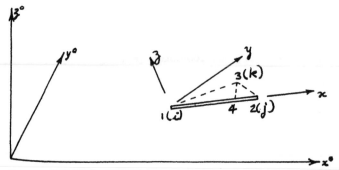

Figure 5.5 - Triangular plane in local and global coordinates

Let

$$\{\Psi_i\} = \begin{Bmatrix} x_i^\circ \\ y_i^\circ \\ z_i^\circ \end{Bmatrix}; \quad \{\Psi_j\} = \begin{Bmatrix} x_j^\circ \\ y_j^\circ \\ z_j^\circ \end{Bmatrix} \tag{5.4}$$

So that

$$\{\Psi_{ji}^\circ\} = \begin{Bmatrix} x_j^\circ - x_i^\circ \\ y_j^\circ - y_i^\circ \\ z_j - z_i^\circ \end{Bmatrix}; \quad \{\Psi_{ji}\} = \begin{Bmatrix} x_j - x_i \\ y_j - y_i \\ z_j - z_i \end{Bmatrix} \tag{5.5}$$

By constructing a line 3-4, as in Figure 5.5, Petyt has shown that

$$[C_{y,x^\circ} \; C_{y,y^\circ} \; C_{y,z^\circ}] = [\Psi^\circ{}_{34}]/\ell_{34^\circ} \tag{5.6}$$

where

$$(\ell_{34^\circ})^1 = [\Psi_{34^\circ}] \{\Psi_{34^\circ}\}$$

$$[\Psi_{44^\circ}] = [[I] - \{\varsigma_1[\varsigma_1]]\{\Psi_{31^\circ}\}$$

$$[\varsigma_1] = 1/\ell_{21} \{\Psi_{21^\circ}\}^T = [C_{x,x^\circ} C_{x,y^\circ} \; C_{x,z^\circ}]$$

$$(\ell_{21^\circ})^2 = \{\Psi_{21^\circ}\}T \{\Psi_{21^\circ}\} \tag{5.7}$$

The x-z principal plane is perpendicular to the x-y principal plane and the o-z axis, and by using a cross-product of the two previously obtained vectors, Petyt has shown the directional cosines of the x-z plane to be given by:-

$$[C_{z,x^\circ}, \; C_{z,y^\circ}, \; C_{z,z^\circ}] = 1/\Delta \; [\Delta_{y^\circ,z^\circ} \; \Delta_z{}^\circ, \; x^\circ \; \Delta_{x^\circ,y^\circ}] \tag{5.8}$$

where

Δ = area of triangle 123

76

$$= \sqrt{[\Delta_{y^\circ,z^\circ}{}^2 + \Delta_{z^\circ,x^\circ}{}^2 + \Delta_{x^\circ,y^\circ}{}^2]} \qquad (5.9)$$

Δ_{x°,y° = projected area of Δ_{123} on x° - y° plane

$$= \frac{1}{2} \begin{vmatrix} x_1^\circ & y_1^\circ & 1 \\ x_2^\circ & y_2^\circ & 1 \\ x_3^\circ & y_3^\circ & 1 \end{vmatrix} \qquad (5.10)$$

Δ_{y°,z° = project area of Δ_{123} on to plane y° - z°

$$= \frac{1}{2} \begin{vmatrix} y_1^\circ & z_1^\circ & 1 \\ y_2^\circ & z_2^\circ & 1 \\ y_3^\circ & z_3^\circ & 1 \end{vmatrix} \qquad (5.11)$$

Δ_{x°,z° = projected area of Δ_{123} on to the plane x° - z°

$$(5.12\_)$$

The relationship between local $[P_i\}$ and global $\{P_i\}$ elemental nodal forces is given by:

$$\{P_i\} = [\Xi] \{P_i^\circ\} \qquad (5.13)$$

and the relationship between local $\{u_i\}$ and global $\{u_i^\circ\}$ elemental nodal displacements given by:

$$\{u_i\} = [\Xi] \{u_i^\circ\} \qquad (5.14)$$

where,

$\{P_i\}^T = [X_1 Y_1 Z_1 M_{x1} (= T_1) M_{y1} M_{z1} X_2 Y_2 Z_2 M_{x2} M_{y2} M_{z2}]$

$\{P_i^\circ\}^T = [X_1^\circ Y_1^\circ Z_1^\circ M_{x1}^\circ M_{y1}^\circ M_{z1}^\circ X_2^\circ Y_2^\circ Z_2^\circ M_{x2}^\circ M_{y2}^\circ M_{z2}^\circ]$

$\{u_i\}^T = [u_1 v_1 w_1 \theta_{x1} \theta_{y1} \theta_{z1} u_2 v_2 w_2 \theta_{x2} \theta_{y2} \theta_{z2}]$

$\{u_i^\circ\}^T = [u_1^\circ v_1^\circ w_1^\circ \theta_{x1}^\circ \theta_{y1}^\circ \theta_{z1}^\circ u_2^\circ v_2^\circ w_2^\circ \theta_{x2}^\circ \theta_{y2}^\circ \theta_{z2}^\circ]$

5.2   Input Data

The input data file should be created in the following sequence:

NJ = number of actual joints or nodes

NF - number of nodes with zero displacements

IM - number of imaginary nodes. These are required in addition to the actual nodes if it's is not possible to define the principal x-y planes of the members with the actual nodes (see later).

ML - number of different types of member. Each particular member type has the same sectional and material properties.

MS - number of members.

Global Coordinates of Nodes

For i = 1 to NJ + IM

Input x_i° y_i° z_i°

Details of Member Types

For i = 1 to ML

Input E_i G_i AR_i IX_i IY_i IZ_i UDY_i UDZ_i

where

E = elastic modulus

G = rigidity modulus

AR = cross-sectional area = A

IX = torsional constant = J

IY = second moment of area about the (local) principal plane x-y = Iy

IZ = second moment of area about the (local) principal plane x-z - I_z

UDY = distributed load in local "y" direction (see Figure 5.6)

UDZ = distributed load in local "z" direction (see Figure 5.7)

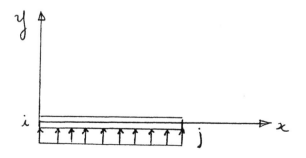

Figure 5.6 - Positive "UDY" distributed load

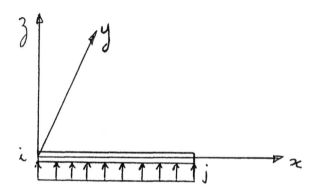

Figure 5.7 - Positive "UDZ" distributed load

Nodes Defining Each Member

For ME = 1 to Ms

Input i_{ME} j_{ME} k_{ME}

where,

i = the node which is of the origin of the local co-ordinate system, or start node

j = the node which is at the end of the member, or finish node, so that i→j represents the "x" direction.

k = a node in the (local) principal plane x-y, as shown in Figure 5.3. In many cases, "k" can be an existing node in the structure, which may conveniently define the x-y principal plane of the member. If there is no such convenient node, then an imaginary node must be used. Imaginary nodes are used for defining principal planes of bending and do not increase the size of the stiffness matrix, nor the size of the half bandwidth.

79

If there is more than one type of member (ie ML > 1), then, immediately after feeding in i, j and k, the member type should be fed in, as follows:-

FOR ME = 1 to MS

Input i j k

Member type (a number 1, 2, 3, etc)

NC - number of nodes with concentrated loads

For i = 1 to NC

PW_i - nodal position

Force in x° direction

Force in y° direction

Force in z° direction

Couple in x° direction   }

Couple in y° direction   }   according to the right-hand screw rule

Couple in z° direction   }

Details of Zero Displacements

FOR i = 1 to NF

NS_i - "Nodal" position of the zero displacements

Type 1 if u° is zero, else 0

Type 1 if v° is zero, else 0

Type 1 if w° is zero, else 0

Type 1 if θ_x° is zero; else 0

Type 1 if θ_y° is zero; else 0

Type 1 if θ_z° is zero; else 0

5.3 Output

The output is as follows:

Global Displacements

$u_1° \ v_1° \ w_1° \ \theta_{x1}° \ \theta_{y1}° \ \theta_{z1}° \ u_2° \ v_2° \ ..... \ \theta_{xNJ}° \ \theta_{yNJ}° \ \theta_{zNJ}°$

Axial Forces and Moments

FOR ME = 1 to MS

Axial force T BM_y BM_z (for member i-j-k)

where,

T = torque   }

BM_y = bending moment about x-y plane }   local axes - see Figure 5.8 and 5.9

BM_z = bending moment about x-z plane }

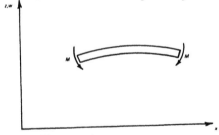

Figure 5.8 - Positive Bending Moment About x-y Plane

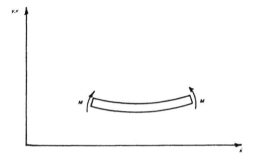

Figure 5.9 - positive Bending Moment About x-z Plane

81

5.4   Example 5.1

Determine the nodal displacements, moments and forces in the rigid-jointed space frame shown in Figure 5.10. The frame is similar to that of Reference [18], (page 35), but the loading is not the same. All members are of circular cross-sectional with the following properties.

$A = 25.13 \text{ in}^2 \qquad I_{xx} = J = 125.7 \text{ in}^4 \; I_{xy} = I_{xz} = 62.83 \text{ in}^4$

$E = 13,500 \text{ tonf/in}^2 \; G = 5,500 \text{ tonf/in}^2$

Length of each member 120 ins

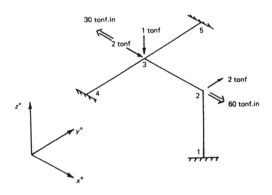

Figure 5.10 - Rigid-jointed Space Frame

The data should be created in the following sequence:

NJ = 5

NF = 3 (there are three fixed nodes with 6 degrees of freedom per node)

IM = 0   As the members are of circular cross-section, each cross-section has an infinite number of principal axes of bending, thus, for this particular case, no imaginary nodes are required to define the x-y axes. (Almost any node, apart from i and j will be suitable to be used as a k node).

ML = 1   There is only one "type" of member.

MS = 4   There are four members.

82

Global Co-ordinates of Nodes

x°	y°	z°		
120	0	0	}	Any suitable point can be taken
120	0	120	}	as the origin, but the system
0	0	120	}	must be the normal right-handed
0	-120	120	}	system for x°, y° and z°
0	120	120	}	

E = 13,500   }

G = 5,500   }

AR = 25.13   }

IX = 125.7   }

IY = 62.83   }   Member properties

IZ = 62.83   }

UDY = 0   }

UDZ = 0   }

Nodes Defining Each Member

i	j	k	
1	2	3	(x-y is a vertical plane)
3	2	1	(x-y is a vertical plane)
4	3	2	(x-y is a horizontal plane)
5	3	2	(x-y is a horizontal plane)

NC = 2 (There are two nodes which have externally applied concentrated loads).

Nodal Positions and "Values" of Concentrated Loads

Nodal position (1) = 2

| 0 | 2 | 0 | | 60 | 0 | 0 |

Forces   Couples

Nodal Position (2) = 3

| 2 | 0 | -1 | | -3 | 0 | 0 |

Forces   Couples

Displacement Positions of Suppressed Displacements

Nodal position 1 =   1, 1 1 1 1 1 1

Nodal position 2 =   4, 1 1 1 1 1 1

Nodal position 3 =   5, 1 1 1 1 1 1

5.5   Input Data File

A typical input data file namely 3D.Dat for this computer program is given below. This data file was used to solve Example 5.1.

3D.DAT

```
5
3
0
1
4
120,0,0
120,0,120
0,0,120
0,-120,120
0,120,120
13500,5500,25.13,125.7,62.83,62.83,0,0
1,2,3
3,2,1
4,3,2
5,3,2
2
2,0,2,0,60,0,0
3,2,0,-1,-30,0,0
1,1,1,1,1,1,1
```

4,1,1,1,1,1,1
5,1,1,1,1,1,1
EOF

The various items in the above data file may be separated by two or more spaces, instead of by commas. <u>Do not use the TAB key</u> to create spaces.

5.6 Results

Displacements are zero at nodes (1), (4) and (5). Values for deflections at nodes (2) and (3) are:-

$u_2{}^\circ = 0.125$ ins, $v_2{}^\circ = 0.364$ ins, $w_2{}^\circ = -1.06E-4$ ins

$u_3{}^\circ = 0.125$ ins, $v_3{}^\circ = 1.08E-4$ ins, $w_3{}^\circ = -0.595$ ins

<u>Moments and Forces in Units of tonf and ins</u>

Member	Node	Axial Force	Torque	BM_{x-y}	BM_{x-z}
1-2-3	1	- 0.299	19.46	98.41	35.76
1-2-3	2	- 0.299	19.46	- 68.35	27.35
3-2-1	3	- 0.526	- 835	- 53.78	- 8.51
3-2-1	2	- 0.526	- 8.35	19.46	27.35
4-3-2	4	+ 0.305	- 4.256	-11.44	57.67
4-3-2	3	+ 0.305	- 4.256	1.861	- 71.11
5-3-2	5	- 0.305	4.256	30.62	30.78
5-3-2	3	- 0.305	4.256	- 40.21	- 17.33

5.7 Screen Dump

Convenient viewing angles are $\alpha = \beta = 30^\circ$. A screen dump of the deflected form of the frame is given in Figure 5.11.

85

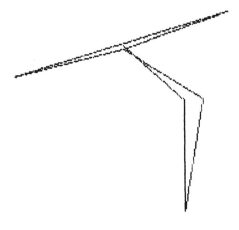

Figure 5.11 - Deflected form of frame of Example 5.1

5.8   Example 5.2

Determine the nodal displacements moments and forces in the rigid-jointed space frame of Figure 5.12, which apart from the loading, has the same properties as that of Example 5.1. The distributed load on member 1-2 is of magnitude 3.33×10^{-2} Tonf/in in the $y°$ direction and the distributed load on member 3-5 is of magnitude 1.667×10^{-2} Tonf/in in the $(-z°)$ direction (see Figures 5.13 and 5.14).

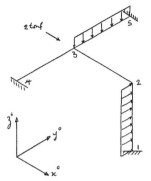

Figure 5.12

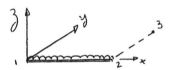

Figure 5.13 Load downwards and "z" upwards

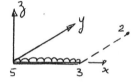

Figure 5.14 load downwards and "z" upwards

5.9 Input Data File

The input data file for Example 5.2 is as follows:

```
5       3    0    3    4

120    0    0
120    0    120
0      0    120
0    -120   120
0     120   120

13500 5500  25.13  125.7  62.83  62.83
0       -3.33E-2
13500 5500  25.13  125.7  62.83  62.83
0     0
13500 5500  25.13  125.7  62.83  62.83

0       -1.667E-2
```

```
1    2    3    1
3    2    1    2
4    3    2    2
5    3    2    3
1
3    2    0    0    0    0    0
1    1    1    1    1    1    1
4    1    1    1    1    1    1
5    1    1    1    1    1    1
EOF
```

5.10   Results

The results are as follows

Deflections (ins) and Rotations (rads)

Node	u°	v°	w°	$\theta_n{}^\circ$	$\theta_y{}^\circ$	$\theta_z{}^\circ$
1	0	0	0	0	0	0
2	0.125	0.415	-1.0576-4	-3.246E-3	5.942E-	3.862E-3
3	0.125	1.234E-4	-5.952E-2	-6.211E-4	4	1.087e-3
4	0	0	0	0	-7.387E-	0
5	0	0	0	0	4	0
					0	
					0	

Forces (tonf) 7 Moments & Torques (tonf.in)

Node	Axial Force	Torque	$M_{x\text{-}y}$	$M_{x\text{-}3}$
1	-0.299	22.25	141.0	-35.76
2	-0.299	22.25	-15.13	27.36
3	-0.526	-15.13	-61.52	-8.51
2	-0.526	-15.13	22.26	27.36
4	0.349	-4.26	-12.25	59.60
3	0.349	-4.26	3.47	-75.00
5	-0.349	4.26	49.83	28.84
3	-0.349	4.26	-18.60	-13.46

88

CHAPTER 6

VIBRATION OF RIGID-JOINTED SPACE FRAMES

6.1   Introduction

This computer program, which is called VIB3DSF, can determine the natural frequencies of rigid-jointed space frames with different element types.

For the free vibration of continuous beams, pin-jointed space trusses and two dimensional pin-jointed and rigid-jointed frames, the reader is referred to references [7] and [21]. In reference [21], the computer programs are written in FORTRAN 77, so that these programs are readily transportable between different machines, regardless of whether they are micros or minis or mainframes. Additionally, the FORTRAN versions of these programs run much faster than their QUICKBASIC counterparts.

In the case of VIB3DSF, the elements can have different cross-sectional sizes, material properties, and the program is capable of analysing large, small or medium sized structures.

To enable the solution of large structures on microcomputers, the continuous reduction techniques of Irons [22] is employed. It must be emphasised that for the vibration analysis of very large structures, it is very necessary to use some form of reduction technique.

The process of Irons consists of defining the nodal displacements as either master displacements or slave displacements. As the particulars of each element are fed in, "completed" slave displacements can be eliminated, so long as lower slave displacements have been previously eliminated.

It should be noted that for smaller restructures, it may not be necessary to eliminate any displacements, but it must be remembered that there are six degrees of freedom per node, as shown in Figure 5.1.

Three nodes are used for the topological description of each element, namely i, j and k, as described in Chapter 5, where the node k is used to define the principal axes of bending, so that the i j k plane is in fact the oxy plane of the element. The second moment of area of the section, I_y, corresponds to the oxy plane of the element.

It should be noted that for many problems, the node k can be made an existing node on the structure, but this depends on the directions of the principal axes of bending of the element's cross-section. For circular cross-sections, where there are an infinite number of principal planes, the node k can be made almost any node on the structure, except for the i and j nodes of the element itself.

The stiffness matrix [k] for the element is given in Chapter 5 and the elemental mass matrix, in global co-ordinates, namely [m°] is given by equation (6.1)

$$[m°] = [\Xi]^T[m][\Xi] \tag{6.1}$$

where

[Ξ] =   a matrix of directional cosines, which is given in Chapter 5.

[m] =   elemental mass matrix (see Table 6.1)

ℓ =   L = elemental length

Table 6.1 - Elemental mass matrix [m] for the generalized case of the one-dimensional member

	u_1	v_1	w_1	θ_{x1}	θ_{y1}	θ_{z1}	u_2	v_2	w_2	θ_{x2}	θ_{y2}	θ_{z2}	
	1/3												u_1
	0	$13/15 +$ $6I_z/5AL^2$											v_1
	0	0	$13/15 +$ $6I_y/5AL^2$										w_1
	0	0	0	$I_p/3A$									θ_{x1}
	0	0	$-11L/210 -$ $I_y/10AL$	0	$L^2/105 +$ $2I_y/15A$								θ_{y1}
	0	$11L/210 +$ $I_z/10AL$	0	0	0	$L^2/105 +$ $2I_z/15A$							θ_{z1}
	1/6	0	0	0	0	0	1/3						u_2
	0	$9/70 -$ $6I_z/5AL^2$	0	0	0	$13L/420 -$ $I_z/10AL$	0	$13/35 +$ $6I_z/5AL^2$					v_2
	0	0	$9/70 -$ $6I_y/5AL^2$	0	$-13L/420 +$ $I_y/10AL$	0	0	0	$13/35 +$ $6I_y/5AL^2$				w_2
	0	0	0	$I_p/6A$	0	0	0	0	0	$I_p/3A$			θ_{x2}
	0	0	$13L/420 -$ $I_y/10AL$	0	$L^2/140 -$ $I_y/30A$	0	0	0	$11L/210 +$ $I_y/10AL$	0	$L^2/105 +$ $2I_y/15A$		θ_{y2}
	0	$13L/420 +$ $I_z/10AL$	0	0	0	$-L^2/140$ $-I_z/30A$	0	$-11L/210 -$ $I_z/10AL$	0	0	0	$L^2/105 +$ $2I_z/15A$	θ_{z2}

6.2 Continuous Reduction Technique

As the elemental stiffness and mass matrices are of order 12 x 12, it is necessary to reduce the sizes of the system matrices for large structures, or numerical instability might result, apart form the fact that the computer's random access memory may not be large enough to store the system matrices.

The process of Iron's is described in references [7] and [22], and will not be covered in detail in the present text. It is based on representing the dynamical behaviour of the structure with a similar one, which has a smaller number of degrees of freedom. Iron's process ensures that the energy of the reduced system is the same as the whole unreduced system, by eliminating irrelevant displacements, but ensuring that any energy possessed or created by these slave or irrelevant displacements is not neglected.

Obvious slaves are rotations as distinct from translations, or relatively small displacements, such as those that occur near the root of a cantilever. Care should be taken not to eliminate key displacements, or important eigenmodes can be missed out altogether!

It must be ensured that if very large problems are being analysed, that sensible reduction is carried out, otherwise the problem can become so big, that the numerical precision of the machine can be exceeded. The reduction process is now described.

If all the displacements at a node are to be completely eliminated, then the nodal number is followed by three zeroes, when it is last fed in; eg if the node to be completely eliminated is 19, then in place of 19, type 19000. If all the displacements are to be eliminated, except for the u°, v° and w° displacements at a node, then the nodal number should be made negative, when it is last fed in' eg if only u_{19}°, v_{19}° and w_{19}° are the only required displacements to remain at node 19, type-19.

It should be noted that it is a simple matter to alter the program so that only one displacement remains or displacements apart from u°, v° and w° are left.

6.3 Zero nodes and other details

If a node is completely fixed, so that all its displacements are zero, it should be defined as a zero node. For such a node as all its "displacements" are the coefficients corresponding to these zero displacements, in the stiffness and mass matrices, it will not appear in these matrices.

If there is more than one member type, then the number of the member type should be fed in. Additional member types occur, if the sectional properties of different members have different cross-sectional or material properties. If i = 0 or/and j = 0 or/and k = 0, then the co-ordinates of the zero nodes should be fed in order.

6.4 The input data file

The input data file should be created in the following sequence.

NJ =   Number of non-zero joints

NF =   Number of fixed (or zero) nodes

NSUP =   Number of zero displacements at pin-jointed supports, etc.

NIMK =   Number of imaginary nodes. (These are required in addition to the actual nodes to define the oxy plane of an element - see Figures 5.1 to 5.3).

N =   Number of free displacements remaining. (These are the displacements that are left after the reduction of the slave displacements and the elimination of the displacements corresponding to a zero or fixed nodal point, but not a pin-joint).

M1 =   Number of frequencies

NMATL =   Number of member types

MEMS =   Number of members or elements

Nodal Co-ordinates

For i =   1 to NJ + NIMK   }
               Input $x_i°$, $y_i°$ and $z_i°$ }   See Chapter 5
               NEXT i

Member Types

i = 1 to NMATL

ρ = RHO$_i$ = density

E = E$_i$ = Young's modulus

G = G$_i$ = rigidity modulus

A = CSA$_i$ = cross-sectional area

J = TC$_i$ = torsional constant

I$_y$ = SMAY$_i$ = 2nd moment of area about local x-y plane   }

I$_z$ = SMAZ$_i$ = 2nd moment about local x-z plane   } see Figure 5.2

I$_p$ = POLAR$_i$ = Polar 2nd moment of area

NEXT i

Element Topology

For i = 1 to MEMS

Input i, j and k nodes for each element.

(The node k is used to define one of the principal axes of bending of the element, so that the i-j-k plane is the oxy plane of the element. The 2nd moment of area I_y, (SMAY), corresponds to the oxy plane - see Figure 5.3).

If NMATL > 1, then type in the member type for this member.

NEXT i

Pinned Nodes, etc

For i - 1 to NSUP

NS_i- Type in the finished displacement position of the zero displacement. This displacement must be a u, v or w displacement and it should be ensured that this node is not completely eliminated.

NEXT i

Concentrated Masses

NCONC = number of concentrated masses

For i = 1 to NCONC

POS_i = "u°" displacement position of the mass - the nodes must not be completely eliminated and the mass is added to the mass matrix in the u°, v° and w° directions at the particular node.

$MASS_i$ = Value of mass at POS_i

NEXT i

Zero Nodes' Co-ordinates

Type in the x°, y° and z° co-ordinates of the zero nodes, as appropriate and in order.

6.5   Example 6.1

Determine the three lowest natural frequencies of vibration for the rigid-jointed space frame of Figure 6.1. It may be assumed that the frame is pinned at its base and that it is composed of symmetrical section members with the following properties:-

ρ = 7.35 x 10^{-4} lbfs^2/in^4

E = 30 x 10^6 lbf/in^2

G = 1.15 x 10^7 lbf/in^2

CSA = 6.3 x 10^{-2} in^2

TC = 1.3 x 10^{-3} in^4

SMAY = 3.26 x 10^{-4} in^4

SMAZ = 3.26 x 10^{-4} in^4

POLAR = 6.51 x 10^{-4} in^4

As the structure is symmetrical, it will be necessary to make the structure slightly unsymmetrical, by altering the co-ordinates of some of the FREE NODES. This process is necessary to determine frequencies and eigenmodes, apart from the first, for symmetrical structures only. The reason for this is that the computer program adopts the power method for determining the eigenvalues, and the power method cannot determine equal eigenvalues.

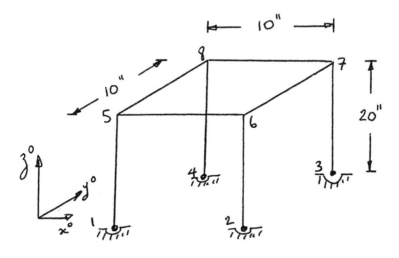

Figure 6.1 - Rigid-jointed Space Frame

6.6   Input

Data

The input data file should be created in the following sequence.

NJ =   8

NF =   0

NSUP =   12 (The displacements $u_1{}^\circ$, $v_1{}^\circ$, $w_1{}^\circ$, $u_2{}^\circ \rightarrow w_4{}^\circ = 0$).

NIMK =   0 (As the sections are symmetrical)

N =   24 (These correspond to $u_1{}^\circ$, $v_1{}^\circ$, $w_1{}^\circ$, $u_2{}^\circ$, $v_2{}^\circ$, $w_2{}^\circ$, to $u_8{}^\circ$, $v_8{}^\circ$, $w_8{}^\circ$).

M1 =   3

NMATL =   1 (There is only one member type).

MEMS =   8

Nodal Co-ordinates

$x_i{}^\circ$	$y_i{}^\circ$	$z_i{}^\circ$
0	0	0
10	0	0
10	10	0
0	10	0
1E-2	1E-2	20.01
10.01	1E-2	19.99
10.01	10.01	20.01
1E-2	10.01	19.99

Material Properties

RHO = 7.35E-4   E = 3E7   G = 1.15E7
CSA = 6.3E-2   TC = 1.3E-3   SMAY = 3.26E-4
SMAZ = 3.26E-4   POLAR = 6.51E-4

Element Topology

i	j	k
-1	5	8
-2	6	7
-3	7	6

-4	8	5
5	6	2
-5	8	7
-6	7	8
-7	-8	5

Zero Displacements at pin-jointed supports, etc

1	2	3 (@ node 1)
4	5	6 (@ node 2)
7	8	9 (@ node 3)
10	11	12 (@ node 4)

NCONC = 0

6.7 Computer Output

The results are as follows:-

Mode 1 $n_1 = 9.178$ Hz

Node	u°	v°	w°
1	0	0	0
2	0	0	0
3	0	0	0
4	0	0	0
5	1	1	-7.68E-4
6	1	1	-1.00E-3
7	1	1	-1.23E-3
8	1	1	-1.00E-3

Mode 2 $n_2 = 9.192$Hz

Node	u°	v°	w°
1	0	0	0
2	0	0	0
3	0	0	0
4	0	0	0
5	-0.613	0.983	-1.42E-4
6	-0.613	1	-7.48E-6
7	-0.629	1	-2.28E-4
8	-0.629	0.983	-3.63E-4

Mode 3 $n_3 = 10.387$ Hz

Node	u°	v°	w°
1	0	0	0
2	0	0	0
3	0	0	0
4	0	0	0
5	0.974	-0.968	-2.27E-6
6	0.974	1	-9.84E-4
7	-0.994	1	-3.64E-6
8	-0.994	-0.968	9.78E-4

6.8   Example 6.2

Determine the three lowest natural; frequencies of vibration for the rigid-jointed space frame of Example 6.1, assuming that it is fixed at its base.

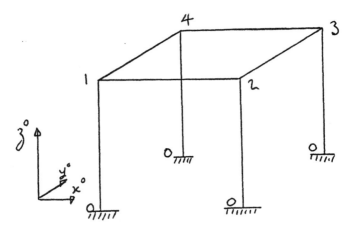

Figure 6.2

6.9   Data

The input data file should be prepared in the sequence shown below.

The data should be fed in as follows:-

NJ = 4
NF = 4
NSUP = 0
NIMK = 0
N = 12 (ie u_1°, v_1°, w_1°, $u_2^\circ \rightarrow u_4^\circ$, v_4°, w_4°)

M1 = 3
NMATL = 1
MEMS = 8

$x_i^°$ $y_i^°$ $z_i^°$

1E-2	1E-2	20.01
10.01	1E-2	19.99
10.01	10.01	20.01
1E-2	10.01	19.99

RHO = 7.35E-4   E = 3E7   G = 1.15E7
CSA = 6.3E-2   TC = 1.3E-3   SMAY = 3.26E-4
SMAZ = 3.26E-4   POLAR = 6.15E-4

i	j	k
0	1	4
0	2	3
0	3	2
0	4	1
1	2	3
-1	4	3
-2	3	4
-3	-4	1

NCONC = 0

Co-ordinates for Zero Nodes

$x^°$	$y^°$	$z^°$
0	0	0 (Element 1)
10	0	0 (Element 2)
10	10	0 (Element 3)
0	10	0 (Element 4)

6.10   Computer Output

The results are as follows:-

Mode 1 N_1 = 19.774 Hz

Node	$u^°$	$v^°$	$w^°$
1	1	1	-5.79E-4
2	1	1	-1.00E-3
3	1	1	-1.42E-3
4	1	1	-1.00E-3

Mode 2 $n_2 = 19.778$ Hz

Node	u°	v°	w°
1	-0.168	1	-2.41E-4
2	-0.168	1	-1.71E-4
3	-0.168	1	-5.91E-4
4	-0.168	1	-6.61E-4

Mode 3 $n_3 = 32.805$ Hz

Node	u°	v°	w°
1	1	-1	-3.48E-7
2	1	1	-1.00E-3
3	-1	1	-7.28E-7
4	-1	-1	1.00E-3

6.11   Example 6.3

Determine the three lowest natural frequencies of vibration for the rigid-jointed space frame for Example 6.2, assuming that there are additional concentrated masses of 5×10^{-4} lbfs^2/in and 1×10^{-3} lbfs^2/in at nodes 1 and 3 respectively, and that the sectional properties of the vertical members have twice the magnitude of the horizontal members.

6.12   Input Data

The input data file should be prepared in the following sequence.

NJ = 4	NF = 4	NSUP = 0
NIMK = 0	N = 12	M1 = 3
NMATL = 1	MEMS = 8	

x_i° y_i° z_i°

1E-2	1E-2	20.01
10.01	1E-2	19.99
10.01	10.01	20.01
1E-2	10.01	19.99

Details of Member Types

RHO = 7.35E-4	E = 3E7	G = 1.15E7	CSA = 6.3E-2
TC = 1.3E-3	SMAY = 3.26E-4	SMAZ = 3.26E-4	POLAR = 6.51E-4
RHO = 7.35E-4	E= 3E7	G = 1.15E7	CSA = 0.126
TC = 2.6E-3	SMAY = 6.52E-4	SMAZ = 6.52E-4	POLAR = 1.302E-3

i	j	k	Member Type
0	1	4	2
0	2	3	2
0	3	2	2
0	4	1	2
1	2	3	1
-1	4	3	1
-2	3	4	1
-3	-4	1	1

NCONC = 2

POS_1 = 1 (u° displacement position for 1st concentrated mass)

$MASS_1$ = 5E-4

POS_2 = 7 (u° displacement position for 2nd concentrated mass)

$MASS_2$ = 1E-3

Co-ordinates for Zero Nodes

x°	y°	z°
0	0	0
10	0	0
10	10	0
0	10	0

6.13 Computer Output

The results are as follows:-

Mode 1 n_1 = 19.158 Hz

Node	u°	v°	w°
1	0.985	0.985	-6.48E-4
2	0.985	1	-9.91E-4
3	1	1	-1.34E-3
4	1	0.985	-9.91E-4

Mode 2 $n_2 = 19.119$ Hz

Node	u°	v°	w°
1	-0.754	0.895	-4.64E-5
2	-0.754	1	1.75E-4
3	-0.857	1	-9.59E-5
4	-0.857	0.895	-3.17E-4

Mode 3 $n_3 = 70.532$ Hz

Node	y°	v°	w°
1	0.999	1	-1.06E-3
2	0.999	-0.825	-3.49E-5
3	-0.825	-0.826	7.80E-4
4	-0.825	1	-3.58E-5

Comparing the solution for Example 6.3 with the solution for Example 6.2, it appears that the computer solution has missed a mode for Example 6.3.

6.13   Example 6.4

Determine the three lowest resonant frequencies of vibration for the model tower shown in Figure 6.3 and 6.4, and whose input data file is shown in Section 6.15.

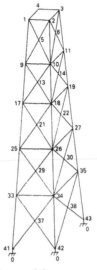

Figure 6.3

102

Figure 6.4

6.15 Input data file for Example 6.4

44	4	0	0	42	3	3	124

1.25001	1.25001	29.7501
4.125	1.25	29.75
4.125	4.125	29.75
1.25	4.125	29.75
2.688	1.146	26.938
4.229	2.688	26.938
2.688	4.229	26.938
1.146	2.688	26.938
1.027	1.027	24.125
4.348	1.027	24.125
4.348	4.348	24.125
1.027	4.348	24.125
2.688	0.915	21.5
4.46	2.688	21.5
2.688	4.460	21.5
0.915	2.688	21.5
0.803	0.803	18.875
4.572	0.803	18.875
4.572	4.572	18.875

0.803	4.572	18.875
2.688	0.68	16.0
4.694	2.688	16.0
2.688	4.694	16.0
0.68	2.688	16.0
0.559	0.559	13.125
4.816	0.559	13.125
4.816	4.816	13.125
0.559	4.816	13.125
2.688	0.426	10.0
4.95	2.688	10.0
2.688	4.95	10.0
0.426	2.688	10.0
0.293	0.293	6.875
5.082	0.293	6.875
5.082	5.082	6.875
0.293	5.082	6.875
2.688	0.178	4.188
5.197	2.688	4.188
2.688	5.197	4.188
0.178	2.688	4.188
0.032	0.032	0.75
5.343	0.032	0.75
5.343	5.343	0.75
0.032	5.343	0.75

0.00079	14000000	5190000	0.0103	0.00008746	0.00004373
0.00004373	8.746E-5				

0.00079	14000000	5190000	0.0159	0.0002854	0.0001427
0.0001427	2.85E-4				

0.0875	14000000	5190000	0.0103	0.00008746	0.00004373
0.00004373	8.74E-5				

1	2	3
3		
1	4	3
3		
2	3	4
3		
3	4	1
3		
1	9	12
2		
1	5	9
1		
-1	8	4
1		

2	10	11
2		
2	6	10
1		
-2	5	1
1		
3	7	11
1		
3	6	11
1		
-3	11	12
2		
4	12	9
2		
4	8	1
1		
-4	7	3
1		
5	10	9
1		
5000	9	10
1		
6	11	10
1		
6000	10	11
1		
7	11	12
1		
7000	12	11
1		
8	12	9
1		
8000	9	12
1		
9	12	11
1		
9	10	11
1		
9	17	18
2		
9	13	17
1		
-9	16	12
1		
10	11	12
1		
10	18	11
2		

10	14	11
1		
-10	13	9
1		
11	12	20
1		
11	14	10
1		
11	15	12
1		
12	20	17
2		
-11	19	18
2		
12	15	11
1		
-12	16	9
1		
13	17	18
1		
13000	18	17
1		
14	18	19
1		
14000	19	18
1		
15	20	19
1		
15000	19	20
1		
16	17	20
1		
16000	20	17
1		
17	25	26
2		
17	18	26
1		
17	20	28
1		
17	21	25
1		
-17	24	25
1		
18	19	20
1		
18	26	27
2		

18	22	19	
1			
18000	21	17	
1			
19	27	28	
2			
19	20	28	
1			
19	22	18	
1			
-19	23	20	
1			
20	28	27	
2			
20	23	19	
1			
20000	24	17	
1			
21	25	26	
1			
21000	26	25	
1			
22	26	27	
1			
22000	27	26	
1			
23	27	28	
1			
23000	28	27	
1			
24	28	25	
1			
24000	25	28	
1			
25	33	34	
2			
25	26	27	
1			
25	28	27	
1			
25	29	33	
1			
25000	32	28	
1			
26	34	35	
2			
26	27	35	
1			

26	30	27
1		
-26	29	25
1		
27	35	36
2		
27	28	36
1		
27	30	26
1		
27000	31	28
1		
28	36	35
2		
28	31	27
1		
-28	32	25
1		
29	34	33
1		
29000	33	34
1		
30	35	34
1		
30000	34	35
1		
31	35	36
1		
31000	36	35
1		
32	36	33
1		
32000	33	36
1		
33	41	42
2		
33	34	35
1		
33	36	35
1		
33	37	41
1		
-33	40	36
1		
34	35	43
1		
34	42	43
2		

34	27	42
1		
34000	38	35
1		
35	43	44
2		
35	36	44
1		
35	38	34
1		
-35	39	43
1		
36	44	43
2		
36	39	35
1		
36000	40	33
1		
37	42	41
1		
37000	41	42
1		
38	43	42
1		
38000	42	43
1		
39	43	44
1		
39000	44	43
1		
40	44	41
1		
40000	41	44
1		
41000	0	0
2		
42000	0	0
2		
43000	0	0
2		
44000	0	0
2		
0		
0	0	0
5.375	0	0
5.375	0	0
0	0	0

```
5.375  5.375  0
5.375  0      0
0      5.375  0
0      0      0
```

6.16 Results

The result are shown in Figure 6.5, together with plan views of the eigenmodes.

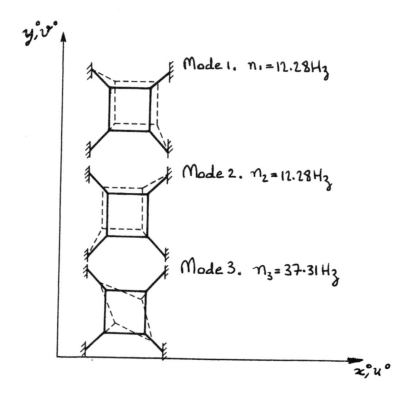

Figure 6.5 - Plan view of the Eigenmodes of model tower

CHAPTER 7

BENDING MOMENTS IN GRILLAGES

7.1   Introduction

This chapter describes the computer program GRIDSG which is suitable for the analysis of flat, horizontal cross-stiffened grids under vertical loads. The grids can be skew or orthogonal (Figure 7.3), and their internal and external boundaries can be irregular.

The boundary conditions can be quite complex, so that various combinations of elastic, clamped and simply-supported edges can be catered for, together with intermediate discrete rigid or elastic supports.

The vertical loading can be uniformly distributed along the length of any element, and this loading can be in addition to concentrated vertical loads applied at the nodes.

Similarly, the sectional and material properties can have different values for various elements.

The method of analysis is the matrix displacement method, and the element in local co-ordinates is shown in Figure 7.1.

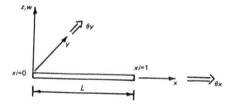

Figure 7.1 - Beam Element in Local Co-ordinates

The element is also shown in Figure 7.2, with respect to both local axes (x, y and z), and global axes (x, y° and z°).

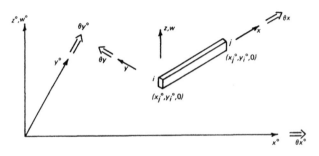

Figure 7.2 - Beam Element

111

As the beam is a horizontal member, its local z axis is parallel to its global z° axis, so that $w = w^\circ$. The rotations $\theta_x{}^\circ$, $\theta_y{}^\circ$, etc are according to the right-hand screw rule.

For elastic supports type in the stiffness of the support, translational or rotations, (see Chapter 2); for a zero displacement, type 1E30 for the value of the elastic restraint.

The elemental stiffness matrix $[k^\circ]$, in global coordinates is given by:-

$$[k^\circ] = [DC]^T[k][DC] \tag{7.1}$$

where

$$[k] = \begin{array}{ccccccc} w_1 & \theta_{x1} & \theta_{y1} & w_2 & \theta_{x2} & \theta_{y2} & \\ 12EI/\ell^3 & & & & & & w_1 \\ 0 & GJ/\ell & & & & & \theta_{x1} \\ -6EI/\ell^2 & 0 & 4EI/\ell & & & & \theta_{y1} \\ -12EI/\ell^3 & 0 & 6EI/\ell^2 & 12EI/\ell^3 & & & w_2 \\ 0 & -GJ/\ell & 0 & 0 & GJ/\ell & & \theta_{x2} \\ -6EI/\ell^2 & 0 & 2EI/\ell & 6EI/\ell^2 & 0 & 4EI/\ell & \theta_{y2} \end{array} \tag{7.2}$$

$$[DC] = \left[\begin{array}{c|c} \varsigma & O_3 \\ \hline O_3 & \varsigma \end{array} \right] \tag{7.3}$$

$$[\varsigma] = \begin{bmatrix} 1 & 0 & 0 \\ 0 & c & s \\ 0 & -s & c \end{bmatrix} \tag{7.4}$$

O_3 = a null matrix of order 3
$c = \cos \alpha$
$s = \sin \alpha$
α is angle between the x and x° axes

$L = \ell = $ length of element

Solution of the simultaneous equations takes account of the sparsity of the system stiffness matrix, so that it is possible to analyse quite large grillages.

Figure 7.3 shows the grid elements for an orthogonal grid and a skew grid, where the nodes governing each element, are usually taken at the grid intersections.

112

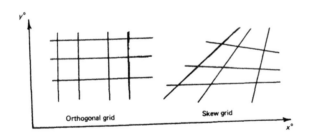

Orthogonal grid

Skew grid

Figure 7.3 - Orthogonal and Skew Grillages

7.2 The Input Data

The data should be prepared in the following manner.

NJ = Number of nodes or joints
NF = Number of nodes that have zero displacements or elastic restraints
NC = Number of nodes with out-of-plane concentrated loads (in $z°$ direction)

Global Nodal Co-ordinates

For i = 1 to NJ

Input $x_i°$ $y_i°$

NEXT i

MS =   Number of Members
NTYPE =   Number of member types

Details of Each Member Type

For i   = 1 to NTYPE
E = E_i = elastic modulus of member "i"
G = G_i = rigidity modulus of member "i"
I = SMA_i = 2nd moment of area about a horizontal plane of member "i"
J = TC_i = torsional constant of member "i"
UDL_i = uniformly distributed load acting on member 2i"
NEXT i

Element Topology

For i = 1 to MS
i = "i" node
j = "j" node
If NTYPE < > 1 then type in member type
NEXT i

113

Concentrated Loads

For i = 1 to NC
POS$_i$ = nodal position of concentrated load
CONC$_i$ = value of out-of-plane concentrated load (in "z" direction)
NEXT i

Nodal Positions, etc, of Zero Displacements

For i = 1 to NF
SUN = Number of the node with the elastic or rigid restraints
Type in elastic restrain in w° direction (N/m)
type in elastic restraint in θ_x° direction (Nm/rad)
Type in elastic restraint in θ_y° direction (Nm/rad)

NEXT i

7.3   Output

The nodal displacements in global co-ordinates are output in ascending order, as follows:-

$$w_1°, \theta_{x1}°, \theta_{y1}°, w_2°, \theta_{x2}° \ldots\ldots w_{NJ}°, \theta_{xNJ}°, \theta_{yNJ}°$$

Next, the torques and bending moments on individual elements are output in local co-ordinates.

Member	ξ	Torque	Bending Moment
i-j	0		
i-j	1		
ℓ-m	0		
ℓ-m	1		

ξ = 0 is at the 'origin' of the member (that is at x = 0)
ξ = 1 is at the other end of the member (that is at x = L)

7.4   Example 7.1

Determine the nodal displacements, torques and bending moments for the skew grillage shown in Figure 6.4.. The grillage may be assumed to be simply supported at nodes, 1, 4, 5, 8, 9 and 12, and the following apply:-

SMA of x° direction member = 1.25E-5 m^4 SMA of other members = 1.25E-5 m^4

TC of x° direction members = 2.5E-5 m^4 TC of other members = 2.5E-5 m^4

E = 2E11 N/m^2 G = 7.69E10 N/m^2

UDL (x° direction members) = -0.5 kN/m   UDL (other members) = -1kN/m

There is an additional downward load of 5 kN at node 7 (that is W_7 = -5000 N).

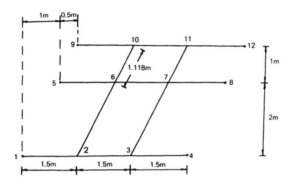

Figure 7.4 - Skew Grillage

7.5   Data for Example 7.1

The data is as follows:-

NJ = 12
NF = 6
NC = 1

$x_i°$	$y_i°$
0	0
1.5	0
3	0
4.5	0
1	2
2.5	2
4	2
5.5	2
1.5	3
3	3
4.5	3
6	3

MS = 13
NTYPE = 2

E   G   SMA   TC   UDL

115

2E11	7.69E10	1.25E-5	2.5E-5	-500
2E11	7.69E10	1.25E-5	2.5E-5	-1000

Element Topology

i	j	Member Type
1	2	1
2	3	1
3	4	1
5	6	1
6	7	1
7	8	1
9	10	1
11	12	1
2	6	2
3	7	2
6	10	2
7	11	2

Concentrated Loads

Node = 7

Load = -5000

Details of Restraints

Node	w direction	$\theta_x °$ direction	$\theta_y °$ direction
1	1E30	0	0
4	1E30	0	0
5	1E30	0	0
8	1E30	0	0
9	1E30	0	0
12	1E30	0	0

7.6   Results

The results are as follows:-

Displacements

Node	w° (m)	θ_x° (rads)	θ_y° (rads)
1	0	-2.108E-4	2.038E-3
2	-2.578E-3	-2.108E-4	1.109E-3
3	-2.764E-3	-9.871E-4	-9.351E-4
4	0	-9.871E-4	-2.311E-3
5	0	6.910E-4	2.509E-3
6	-3.131E-3	6.910E-4	1.273E-3
7	-3.434E-3	-1.262E-4	-1.114E-3
8	0	-1.262E-4	-2.891E-3
9	0	8.400E-4	2.344E-3
10	-2.864E-3	8.400E-4	1.069E-3
11	-2.751E-3	3.216E-4	-1.151E-3
12	0	3.216E-4	-2.189E-3

Moments and Torques (Nm)

Node	ξ	Torque	Bending Moment
1	0	0	0
2	1	0	-2910
2	0	-995	-2941
3	1	-995	-3684
3	0	0	-4398
4	1	0	0
5	0	0	0
6	1	0	-3933
6	0	-1047	-2659
7	1	-1047	-5108
7	0	0	-5736
8	1	0	0
9	0	0	0
10	1	0	-4063
10	0	-664	-3953
11	1	-664	-3261
11	0	0	-3272
12	1	0	0
2	0	473	-876
6	1	473	70
3	0	194	1209
7	1	194	-2276
6	0	-199	-1437
10	1	-199	643
7	0	287	-1059
11	1	287	-599

7.7   Example 7.2

Determine the bending moment diagrams for the grillage of Figure 7.5, which may be assumed to be simply-supported round its boundary, and compare the results with that of the Vedeler solution [23]. The grillage is subjected to an upward uniform pressure of 1 lbf/in^2, and this may be assumed to be equivalent to uniformly distributed loads of 15.944 lbf/in in both the $x°$ and $y°$ directions. The other details of the grillage are given in Section 7.8.

118

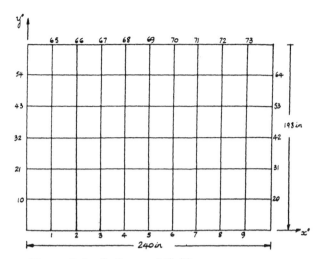

Figure 7.5 - Orthogonal Grillage

7.8   Input Data File for Example 7

73	28	0

24	0
48	0
72	0
96	0
120	0
144	0
168	0
192	0
216	0
0	33
24	33
48	33
72	33
96	33
120	33
144	33
168	33
192	33
216	33
240	33
0	66
24	66
48	66
72	66
96	66
120	66
144	66

119

168	44
192	66
216	66
240	66
0	99
24	99
48	99
72	99
96	99
120	99
144	99
168	99
192	99
216	99
240	99
0	132
24	132
48	132
72	132
96	132
120	132
144	132
168	132
192	132
216	132
240	132
0	165
24	165
48	165
72	165
96	165
120	165
144	165
168	165
192	165
216	165
240	165
24	198
48	198
72	198
96	198
120	198
144	198
168	198
192	198
216	198

```
104    2

1E7    3.846E6      2.88   0.2    15.944
1E7    3.846E6      1.45   0.1    15.944

10     11     1
11     12     1
12     13     1
↓      ↓      ↓
↓      ↓      ↓
62     63     1
63     64     1
1      11     2
2      12     2
3      13     2
↓      ↓      ↓
↓      ↓      ↓
62     72     2
63     73     2
1      1      0      0
2      1      0      0
3      1      0      0
4      1      0      0
5      1      0      0
6      1      0      0
7      1      0      0
8      1      0      0
9      1      0      0
10     1      0      0
20     1      0      0
21     1      0      0
31     1      0      0
32     1      0      0
42     1      0      0
43     1      0      0
53     1      0      0
54     1      0      0
64     1      0      0
65     1      0      0
66     1      0      0
67     1      0      0
68     1      0      0
69     1      0      0
70     1      0      0
71     1      0      0
72     1      0      0
73     1      0      0
```

7.9   Results for Examples 7.2

The bending moment diagrams for some members are shown in Figures 7.6 and 7.7.

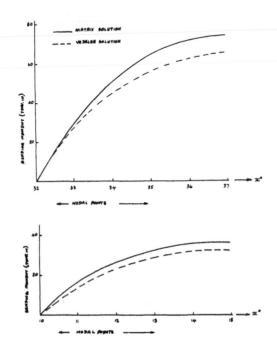

Figure 7.6 - Bending moment diagrams for some members in "x" direction

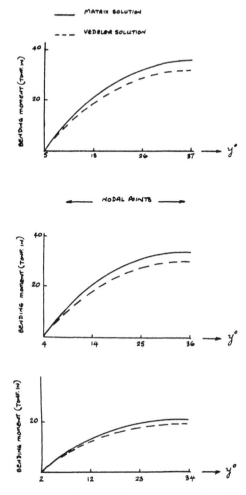

Figure 7.7 - Bending moment diagrams for some members in "y" direction

The main reason for the difference between the two sets of results is that in the Vedeler solution, the 1 lbf/in^2 pressure was assumed to act all over the grillage, so that some of the pressure was directly transmitted to the boundary, but in the matrix solution, the entire pressure loading was carried by the grid members.

7.10   Example 7.3

Determine the bending moment diagrams for the ship's deck of Figure 7.8. The x° direction members are subjected to a downward uniformly distributed load of 500 N/m and there are two additional downward concentrated loads of 4000 N and 3000 N acting at nodes 15 and 20 respectively. All edges may be assumed to be simply-supported, and the following may be assumed to apply:

Uniformly distributed load in y° direction members = 0

A_x = 1.2E-3 m²
A_y = 3.2E-3 m²
I_x = 8.6 x 10^{-7}m⁴
I_y = 3.6 x 10^{-6} m⁴
J_x = 9 x 10^{-8} m⁴
J_y = 2.2 x 10^{-7} m⁴
E = 2 x 10^{11} N/m²
G = 7.69 x 10^{10}N/m²
ρ = 7860 kg/m³

where the suffix "x" refers to x° direction members and the suffix "y" refers to y° direction members

Other details are given in Section 7.11, in units of N and m.

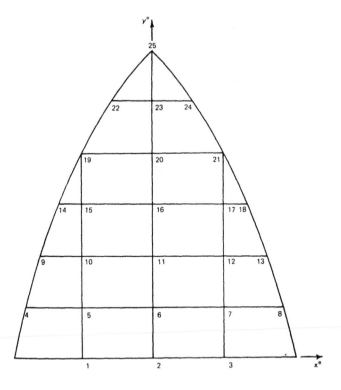

Figure 7.8 - Ship Grillage

7.1   Input Data File for Example 7.3

25	14	2
-1.4	0	
0	0	
1.4	0	
-2.71	1.0	
-1.4	1.0	
0	1.0	
1.4	1.0	
2.71	1.0	
-2.38	2.0	
-1.4	2.0	
0	2.0	
1.4	2.0	
2.38	2.0	
-1.93	3.0	
-1.4	3.0	
0	3.0	
1.4	3.0	
1.93	3.0	
-1.4	4.0	
0	4.0	
1.4	4.0	
-0.77	5.0	
0	5.0	
0.77	5.0	
0	6.0	
30	2	

2E11	7.69E10	8.6-E7	1E-7	-500
2E11	7.69E10	3.6E-6	3E-7	0
4	5	1		
5	6	1		
6	7	1		
7	8	1		
9	10	1		
10	11	1		
11	12	1		
12	13	1		
14	15	1		
15	16	1		
16	17	1		
17	18	1		
19	20	1		
20	21	1		
22	23	1		
23	24	1		
1	5	2		

2	6	2
3	7	2
5	10	2
6	11	2
7	12	2
10	15	2
11	16	2
12	17	2
15	19	2
16	20	2
17	21	2
20	23	2
23	25	2
15	-4000	
20	-3000	

1	1E30	0	0
2	1E30	0	0
3	1E30	0	0
4	1E30	0	0
8	1E30	0	0
9	1E30	0	0
13	1E30	0	0
14	1E30	0	0
18	1E30	0	0
19	1E30	0	0
21	1E30	0	0
22	1E30	0	0
24	1E30	0	0
25	1E30	0	0

7.12 Results for Example 7.3

The bending moment diagrams for some members are given in Figure 7.9

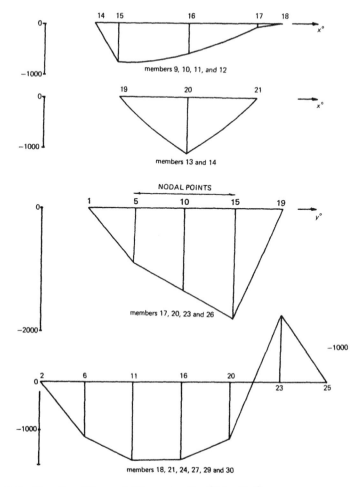

Figure 7.9 - Bending Moment Diagrams for Ship Grillage

7.13 Screendumps

A screendump of the deflected form of the grid is shown in Figure 7.10, and the bending moment diagram in Figure 7.11.

127

DEFLECTED FORM OF GRILLAGE

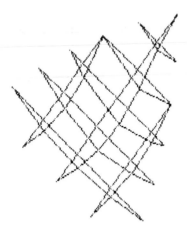

Figure 7.10 - Deflected Form of Ship Deck

BENDING MOMENT DIAGRAM

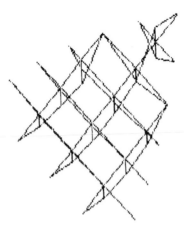

Figure 7.11 - Bending Moment Diagram for Ship Deck

128

CHAPTER 8

VIBRATION OF GRILLAGES

8.1   Introduction

This program, which is called VIBGRIDG, can determine the resonant frequencies and eigenmodes of orthogonal and skew grids, as shown in Figure 8.1. A <u>FORTRAN</u> version of the program is published in reference [21].

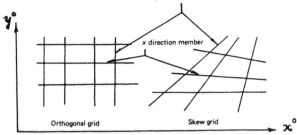

Figure 8.1 - Orthogonal and Skew Grids

The grids can have two different element types, namely "x" elements and "y" elements. These elements can have different geometrical and material properties, and are not necessarily orthogonal.

The grids can be simply-supported or fixed or can have a complex combination of simply-supported and fixed supports, together with discrete intermediate supports.

The element is shown in local and global co-ordinates in Figure 7.1 and 7.2, and its mass matrix in global co-ordinates $[m°]$ is given by:-

$$[m°] = [DC]^T [m] [DC] \tag{8.1}$$

where $[DC]$ is given by equation (7.3).

$$[m] = \rho AL$$

	w_1	θ_{x1}	θ_{y1}	w_2	θ_{x2}	θ_{y2}	
	$13/35 + 6I/5A\ell^2$						w_1
	0	$I_p/3A$					θ_{x1}
	$-11\ell/210 - I/10A\ell$	0	$\ell^2/105 + 2I/15A$				θ_{y1}
	$9/70 - 6I/5A\ell^2$	0	$-13\ell/420 + I/10A\ell$	$13/35 + 6I/5A\ell^2$			w_2
	0	$I_p/6A$	0	0	$I_p/3A$		θ_{x2}
	$13\ell/420 - I/10A\ell$	0	$\ell^2/140 - I/30A$	$11\ell/210 + I/10A\ell$	0	$\ell^2/105 + 2I/15A$	θ_{y2}

129

where

ℓ = length of element
I = 2nd moment of area of cross-section about the x-y plane
J = torsional constant
I_p = 2nd polar moment of area of cross-section about the "x" axis
A = cross-sectional area
E = elastic modulus
G = rigidity modulus
ρ = density

8.2   Input Data File

The input data file should be prepared in the following sequence:-

NJ =   Number of non-zero nodes.
NF =   Number of nodes, with zero displacements
M1 =   Number of frequencies (must be $< = $ NJ x 3)
MEMX =   Number of "x" members
MEMY =   Number of "y" members

Material Properties, etc

For   i = 1 to NMATL

| RHO_i = material density

| E_i = elastic modulus

| G_i = rigidity modulus

| $CSAX_i$ = cross-sectional area of an "x" member

| $CSAY_i$ = cross-sectional area of a "y" member

| TCX_i = torsional constant of an "x" member

| TCY_i = torsional constant of a "y" member

| $POLX_i$ = polar 2nd moment of area of an "x" member

| $POLY_i$ = polar 2nd moment of area of a "y" member

→   NEXT i

130

Element Topology

For   MEM = 1 to MEMS
|
| Type in "i" node followed by "j" node
|
→   NEXT MEM

where MEMS = MEMX + MEMY

Details of Zero Displacements

For   i = 1 to NF
|
| NS_i = the number of the node with the zero displacements
|
| If $w\,(NS_i) = 0$, type 1; else type 0
|
| If $\theta_x(NS_i) = 0$, type 1; else type 0
|
| If $\theta_y\,(NS_i) = 0$, type 1; else type 0
|
→   NEXT i

Additional Concentrated Masses

NCONC = Number of concentrated masses.

For i = 1 to NCONC
|
| POS_i = the number of the node with the concentrated mass
|
| $MASS_i$ = value of concentrated mass at the above position.
|
→   NEXT i

8.3   Example 8.1

Determine to the lowest natural frequencies for the grillage of Figure 8.2, which is simply-supported at nodes 1 to 6, [24].

It may be assumed that no additional concentrated masses or forces act on this grid.

131

The material properties, etc, are as follows:-

RHO $= 7860$ kg/m^3 E $= 2 \times 10^{11}$ N/m^2

G $= 7.69 \times 10^{10}$ N/m^2

CSA $= 0.004$ m^2

SMA $= 1.25 \times 10^{-5}$ m^4

TC $=$ POL $= 2.5 \times 10^{-5}$ m^4

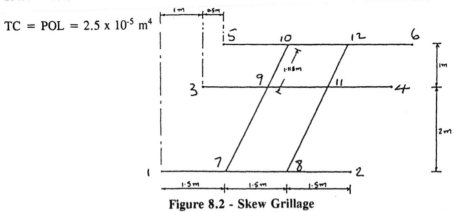

Figure 8.2 - Skew Grillage

8.4   Input Data File for Example 8.1

The input data file is as follows:-

```
12    6    3    9    4
0     0    4.5  0    1     2     5.5   2
1.5   3    6    3    1.5   0     3     0
2.5   2    3    3    4     2     4.5   3
2E11  7.69E10    7860
4E-3  4E-3 1.25E-5    1.25E-5
2.5E-5      2.5E-5     2.5E-5      2.5E-5
1     7    7    8    8     2     3     9
9     11   11   4    5     10    10    12
12    6    7    9    8     11    9     10
11    12
1     1    0    0
2     1    0    0
3     1    0    0
4     1    0    0
5     1    0    0
6     1    0    0
0
```

8.5   Results for Example 8.1

The results are shown in Figure 8.3 to 8.5.

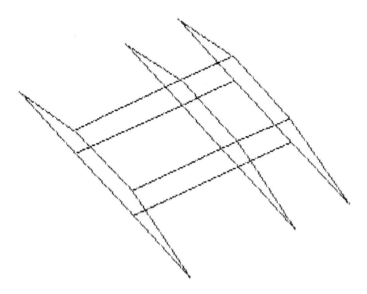

Figure 8.3 - Mode 1 (17.01 Hz) (Simple Supports)

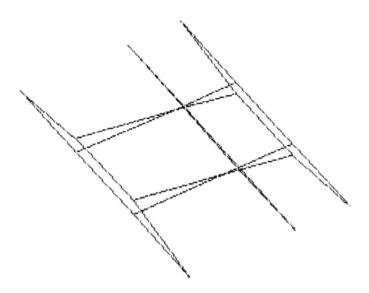

Figure 8.4 - Mode 2 (19.44 Hz) (Simple Supports)

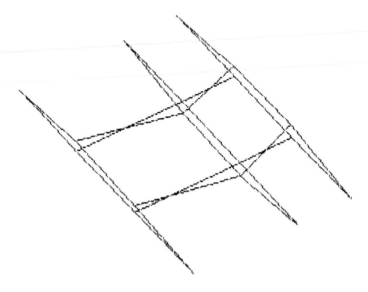

Figure 8.5 - Mode 3(54.58 Hz) (Simple Supports)

8.6   Example 8.2

Determine the resonant frequencies for the skew grid of Figure 8.2, assuming that it is firmly fixed at nodes 1 to 6.

The <u>input data file</u> is identical to Section 8.4, except for the details of the zero displacements, which are as follows:-

1	1	1	1
2	1	1	1
3	1	1	1
4	1	1	1
5	1	1	1
6	1	1	1

8.7   Results for Example 8.2

The results are shown in Figure 8.6 to 8.8.

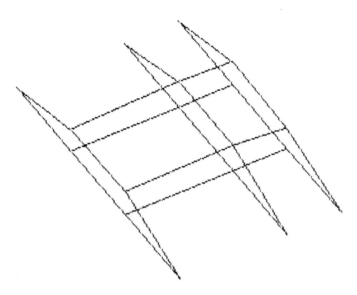

Figure 8.6 - Mode 1 (38.17 Hz) (Clamped Edges)

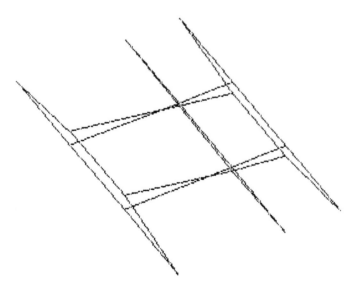

Figure 8.7 - Modes (45.38 Hz) (Clamped Edges)

135

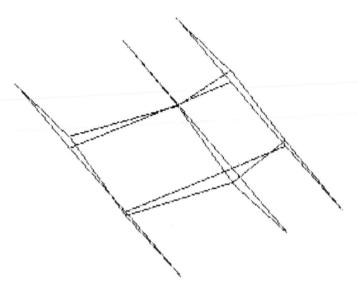

Figure 8.8 - Mode 3 (83.36 Hz) (Clamped Edges)

8.8   Example 8.3

Determine the resonant frequencies for the ship's deck of Figure 7.3, assuming there are no additional masses, other than the self mass of the structure.

The details of the structure are given in Section 8.9.

8.9   Input Data File for Example 8.3

```
25  14   3  16   14
-1.4   0
0   0
1.4   0
-2.71   1
-1.4   1
0   1
1.4   1
2.71   1
-2.38   2
-1.4   2
0   2
1.4   2
2.38   2
-1.93   3
-1.4   3
0   3
1.4   3
1.93   3
-1.4   4
0   4
1.4   4
-0.77   5
0   5
0.77  5
0   6
2E11   7.69E10   7860   1.2E-3   3.2E-3   8.6E-7   3.6E-6   1E-7   3E-7   9E-8   2.2E-7
4   5
5   6
```

```
6   7
7   8
9   10
10  11
11  12
12  13
14  15
15  16
16  17
17  18
19  20
20  21
22  23
23  24
1   5
2   6
3   7
5   10
6   11
7   12
10  15
11  16
12  17
15  19
16  20
17  21
20  23
23  25
1   1   0   0
2   1   0   0
3   1   0   0
4   1   0   0
8   1   0   0
9   1   0   0
13  1   0   0
14  1   0   0
18  1   0   0
19  1   0   0
21  1   0   0
22  1   0   0
24  1   0   0
25  1   0   0
0
```

8.10   Results for Example 8.3

The results are shown in Figure 8.9 to 8.11.

FREQUENCY= 13.67023 Hz. (MODE= 1)
TO CONTINUE TYPE Y

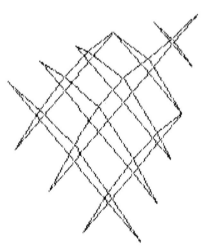

Figure 8.9 - Mode 1 for Ship's Deck

137

FREQUENCY= 27.7919 Hz. (MODE= 2)
TO CONTINUE TYPE Y

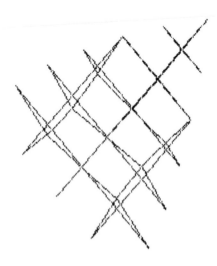

Figure 8.10 - Mode 2 for Ship's Deck

FREQUENCY= 33.0702 Hz. (MODE= 3)

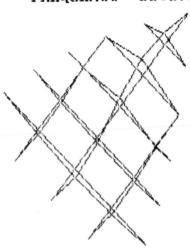

Figure 8.11 - Mode 3 for Ship's Deck

CHAPTER 9

SLAB ON AN ELASTIC FOUNDATION

9.1   Introduction

This computer program, which is called SLABELAG, is suitable for determining the nodal displacements and stresses in a slab on an elastic foundation. The slab can be of simple or of irregular shape, with cut-outs and other discontinuities, and its supports can be simple or fixed or elastic, or almost any other combination. The size of the problem that can be tackled, can be quite large, depending on the amount of RAM available.

The element is the flat eight node isoparametric element shown in Figure 9.1. This element can be used to represent a rectangle or a quadrilateral with straight or curved sides, but care must be taken to ensure that the element is not too thin or too distorted, as numerical instability can occur. If the element represents a rectangle, then the aspect ratio of the rectangle should be less than 4:1. For further details on numerical instability, the reader should consult references, [ ] and [ ].

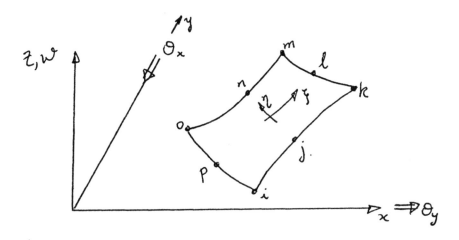

Figure 9.1 - Eight Node Flat Quadrilateral in the x-y Plane

The element is intended to analyse thick slabs, [9], but can be used for medium thickness plates, so long at the out-of-plane deflection "w" does not exceed half the plate thickness. The thickness of the plate can be uniform or vary linearly or parabolically.

The program can cater for concentrated nodal loads in addition to pressure loading, where the latter is uniform over an element.

To guard against locking, three Gauss points were used in the ξ direction and three in the η direction. The ξ direction points from the "p" node to the "ℓ" node, and the η direction points from the "j" node to the "n" node. Values of ξ and η at the nodes are as shown in Table 9.1.

Table 9.1 - Values of ϵ and n at the elemental nodes

Node	ξ	η
i	-1	-1
j	0	-1
k	1	-1
ℓ	1	0
m	1	1
n	0	1
o	-1	1
p	-1	0

The rotational displacements θ_x and θ_y are shown in Figure 9.1, where it can also be seen that the deflection "w" is perpendicular to the plane of the plate. The plate is flat and in the x-y plane of Figure 9.1.

9.2   Input

The input data file should be fed in the following sequence:-

(1)   NNODE = number of nodes

(2)   ELEMS = number of elements.

(3)   ALL$ = "Y" or ALL$ = "N". (If ALL$ = "Y", then all the supports are elastic and have equal stiffnesses.

        If ALL$ = "Y" then ignore (3a)

(3a)   NFIX = Number of nodes with zero or elastic displacements.

        (If ALL$ = "Y" then NFIX will be made equal to NNODE in the program).

(4)   If there is a pressure load, PRESS$ = "Y"; else PRESS$ = "N".

(5)   NCONC = number of nodes with out-of-plane concentrated loads

        (+ve if in the "Z" direction).

(6)   E = Young's modulus of elasticity.

(7)   NU = Poissons' ratio

140

(8)   If the plate thickness is constant, then THICKNESS$ = "Y"; else THICKNESS$ = "N".

(9)   If THICKNESS$ = "Y", then type in plate thickness

(10)   If THICKNESS$ = "N" then

For I = 1 to NNODE

Type in the thickness at each node, starting from node 1 and ending at NNODE (ie TH (I))

NEXT I

(11)   **Global co-ordinates at nodes**

For I = 1 to NNODE

Type in x(I), y(I)

NEXT I

(12)   If ALL$ - "Y", type in vertical stiffness, followed by the rotational stiffnesses in the θ_x and θ_y direction.

If ALL$ = "Y", ignore (12a)

(12a)   **The nodal positions and values of the vertical and rotational stiffnesses.**

For I = 1 to NFIX

POS (I) = node number of support

SUPPSTIFF = "vertical" stiffness at the above node.

ROTX = rotational stiffness at the above node in the "θ_x" direction.

ROTY = rotational stiffness at the above node in the "θ_y" direction.

NB   If a displacement is zero, type in 1E20 for its stiffness in the appropriate position. Similarly, if there is no stiffness at a displacement position, type in zero.

DO NOT TYPE IN NEGATIVE STIFFNESSES!

NEXT I

(13)　If NCONC ≠ 0 then

For I = 1 to NCONC

Type in the node number corresponding to the out-of-plane load.

Type in the value of the out-of-plane concentrated load, corresponding to this node.

(14)　**Element Topology**

For I = 1 to ELEMS

Type in the elemental node numbers i, j, k, ℓ, m, n, o, p.

If there is pressure loading (ie PRESS$ = "Y"), the n type in the lateral pressure that is acting on this element.

If PRESS$ = "N", then <u>do not type anything</u> after the element topology for the Ith element.

The above details should be typed in for each element, starting from element 1, and ending at ELEMS.

NEXT I.

9.3　Output

The output includes details of the input, together with the following:-

For I = 1 to NNODE

PRINT w, θ_x, θ_y at NODE I

NEXT I

For I = 1 to ELEMS

For XI = -1 or XI = 1

For ETA = -1 or ETA - 1

PRINT XI, ETA, σ_x, σ_y, τ_{xy}, τ_{yz} and τ_{xz}

NEXT ETA

NEXT XI

NEXT I

where σ_x, σ_y etc are shown in Figure 9.2.

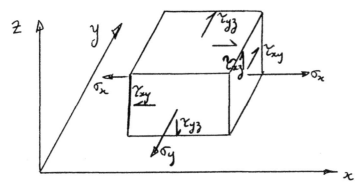

Figure 9.2

9.4   Example 9.1

Determine the nodal displacements and stresses for the square plate of Figure 9.3, which can be seen to be a quarter of a square plate, fixed along its edges. The following may be assumed:-

E = 3E7 lbf/in^2 v = 0.3

Thickness = 0.25 ins   Pressure = 2 lbf/in^2

NCONC = 0

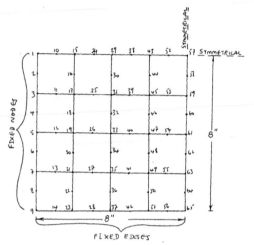

Figure 9.3 - Square Plate Under Uniform Lateral Pressure

The data is shown in Section 9.5, and the output in Section 9.6.

9.5 Input Data File for Example 9.1

SLABELAG.DAT

```
65,16,"N",32,"Y",0
3E7,.3
"Y"
.25
0,8,  0,7,  0,6,  0,5,  0,4,  0,3,  0,2,  0,1,  0,0
1,8,  1,6,  1,4,  1,2,  1,0
2,8,  2,7,  2,6,  2,5,  2,4,  2,3,  2,2,  2,1,  2,0
3,8,  3,6,  3,4,  3,2,  3,0
4,8,  4,7,  4,6,  4,5,  4,4,  4,3,  4,2,  4,1,  4,0
5,8,  5,6,  5,4,  5,2,  5,0
6,8,  6,7,  6,6,  6,5,  6,4,  6,3,  6,2,  6,1,  6,0
7,8,  7,6,  7,4,  7,2,  7,0
8,8,  8,7,  8,6,  8,5,  8,4,  8,3,  8,2,  8,1,  8,0
1,  1E20,1E20,1E20,  2,  1E20,1E20,1E20,  3,  1E20,1E20,1E20
4,  1E20,1E20,1E20,  5,  1E20,1E20,1E20,  6,  1E20,1E20,1E20
7,  1E20,1E20,1E20,  8,  1E20,1E20,1E20,  9,  1E20,1E20,1E20
10,  0,0,1E20,  14,  1E20,1E20,1E20,  15,  0,0,1E20
23,  1E20,1E20,1E20,  24,  0,0,1E20,  28,  1E20,1E20,1E20
29,  0,0,1E20,  37,  1E20,1E20,1E20,  38,  0,0,1E20
42,  1E20,1E20,1E20,  43,  0,0,1E20,  51,  1E20,1E20,1E20
52,  0,0,1E20,  56,  1E20,1E20,1E20,  57,  0,1E20,1E20
58,  0,1E20,0,  59,  0,1E20,0,  60,  0,1E20,0
61,  0,1E20,0,  62,  0,1E20,0,  63,  0,1E20,0
64,  0,1E20,0,  65,  1E20,1E20,1E20
3,11,17,16,15,10,1,2,  2
5,12,19,18,17,11,3,4,  2
7,13,21,20,19,12,5,6,  2
9,14,23,22,21,13,7,8,  2
17,25,31,30,29,24,15,16,  2
19,26,33,32,31,25,17,18,  2
21,27,35,34,33,26,19,20,  2
23,28,37,36,35,27,21,22,  2
31,39,45,44,43,38,29,30,  2
33,40,47,46,45,39,31,32,  2
35,41,49,48,47,40,33,34,  2
37,42,51,50,49,41,35,36,  2
45,53,59,58,57,52,43,44,  2
47,54,61,60,59,53,45,46,  2
49,55,63,62,61,54,47,48,  2
51,56,65,64,63,55,49,50,  2
EOF
```

9.6 Results for Example 9.1

Comparisons are made in Table 9.2 with the computer results and the result obtained from classical small deflection elastic theory.

Table 9.2 - Computer and Classical Theory Results

	δ (in)	σ(lbf/in²) at centre of fixed edge	σ (lbf/in²) (centre)
Computer	3.69E-3	1653.7	1102.7
Theory	3.86E-3	2521	1135

δ = maximum deflection at node 57

Screendumps of various stress contours are shown in Figure 9.4 to 9.6.

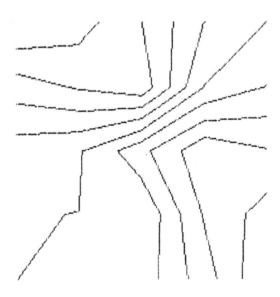

Figure 9.4 - Lines of Constant Maximum Shear-Stress

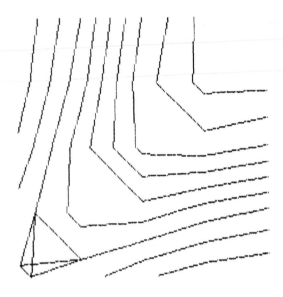

Figure 9.5 - Lines of Constant Maximum Principal Stress

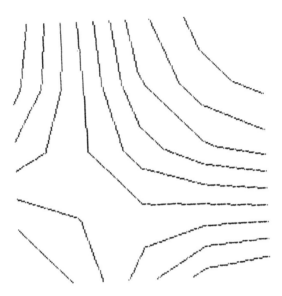

Figure 9.6 - Lines of Constant Minimum Principal Stress

9.7 Example 9.2

Determine the nodal displacements and stresses for the plate of Figure 9.3, assuming that there are simple-support along nodes 1 to 9 and nodes 14, 23, 28, 37, 42, 51, 56 and 65.

The <u>data</u> is the same as for Example 9.1, except for the boundary conditions, which are of the following form:-

Node   1E30 0   0

9.8 Results

Comparison between traditional small deflection elastic theory and the computer output is shown in Table 9.3.

Table 9.3 - Comparison Between Computer Results and Classical Theory

	δ (in)	σ (lbf/in²) at centre
Computer	1.24E-2	2354
Theory	1.24E-2	2354

From Tables 9.2 and 9.3, it can be seen that the computer results are very good for the simply-supported case and that, a more refined mesh should have been taken for the fixed edges case.

9.9   Example 9.3

Determine the nodal displacement and stresses for the flat skew plate of Figure 9.7, which is subjected to six downward concentrated loads, each of magnitude 2000 lbf. The following may be assumed to apply:-

E = 3E6 lbf/in²   NU = 0.15

Thickness = 9 ins

All nodes are supported on elastic props, each of stiffness 10 000 lbf/in.

The computer data is shown in Section 9.10.

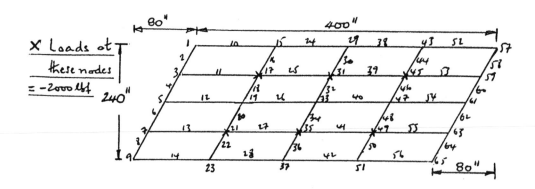

Figure 9.7 - Skew Slab on Elastic Foundations

9.10 Input Data File for Example 9.3

The input data file for Example 9.3 is shown below.

SKEWSLAB.DAT

```
65,16,"Y","N",6
3E6,.15
"Y"
9
80   240   70  210   60  180   50  150   40  120   30   90   20
10    30    0    0
130   240  110  180   90  120   70   60   50    0
180   240  170  210  160  180  150  150  140  120
130    90  120   60  110   30  100    0
230   240  210  180  190  120  170   60  150    0
280   240  270  210  260  180  250  150  240  120
230    90  220   60  210   30  200    0
330   240  310  180  290  120  270   60  250    0
380   240  370  210  360  180  350  150  340  120
330    90  320   60  310   30  300    0
430   240  410  180  390  120  370   60  350    0
480   240  470  210  460  180  450  150  440  120
430    90  420   60  410   30  400    0
10000   0   0
17  -2000    0    0   21  -2000    0    0   31  -2000    0    0
35  -2000    0    0   45  -2000    0    0   49  -2000    0    0
3,11,17,16,15,10,1,2
5,12,19,18,17,11,3,4
7,13,21,20,19,12,5,6
9,14,23,22,21,13,7,8
17,25,31,30,29,24,15,16
19,26,33,32,31,25,17,18
21,27,35,34,33,26,19,20
23,28,37,36,35,27,21,22
31,39,45,44,43,38,29,30
33,40,47,46,45,39,31,32
35,41,49,48,47,40,33,34
37,42,51,50,49,41,35,36
45,53,59,58,57,52,43,44
47,54,61,60,59,53,45,46
49,55,63,62,61,54,47,48
51,56,65,64,63,55,49,50
EOF
```

9.11 Results

Some displacements are shown in Table 9.4.

Table 9.4 - Some results for the skew slab

Node	w° (m)	Element	ξ	η	$\sigma_x^{\,\circ}$ (lbf/in²)
1	-8.215E-3		-1	-1	-19.8
9	3.774E-4		-1	1	-20.8
33	-3.151E-2	19-26-33-32-31-25-17-18	1	-1	-20.3
57	3.774E-4		1	1	-20.9
65	-8.215E-3		-	-	-

Screendumps of some of the stress contours are shown in Figure 9.8 to 9.10.

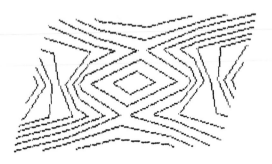

Figure 9.8 Lines of constant maximum shear stress

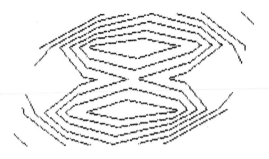

Figure 9.9 Lines of constant maximum principal stress

150

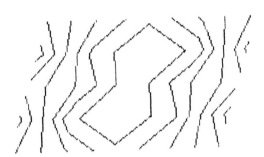

Figure 9.10 - Lines of Constant Minimum Principal Stress

9.12 Conclusions

1.   A sufficient number of elements should be used to describe the plate, and this can only be assessed through trial and error. (See the examples in this chapter for guidance).

2.   Do not distort the element too much, as numerical instability can occur. The ideal element is of a square shape.

3.   The maximum deflection of a plate should not exceed half its thickness.

4.   Extreme values of dimensions should not be used.

5.   The program can easily be modified to analyse orthotropic slabs or sandwich structures [9]

CHAPTER 10

IN-PLANE STRESSES IN PLATES

10.1 Introduction

This chapter describes three programs, namely, STRESS3N, STRESS4N & STRESS8N; all three computer programs calculate nodal deflections and elemental stresses in flat plates under in-planes forces. The plates can be of quite complex shape, with complex boundary conditions, as shown in Figure 10.1. The computer programs can cater for either plane stress or plane strain [16].

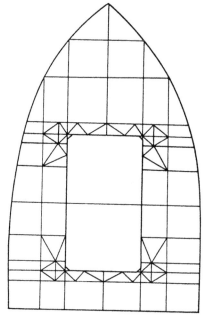

Figure 10.1 Ship's deck

In the case of the computer program STRESS3N, the program uses the 3 node constant strain triangular element of Turner et al [4], while in the case of the computer program STRESS4N, the program uses the 4 node quadrilateral element of Ergatoudis et al [25]; a similar element was produced by Taig [26].

In the case of the computer program STRESS8N, the program uses the eight node isoparametric curved quadrilateral element of Ergatoudis et al [25]. All three programs have graphical displays, which include the deflected form of the plate, together with various stress contours.

A brief description of each element, together with a description of how to use each program will now be given.

10.2   The Computer Program STRESS3N

This computer program adopts the 3 node constant strain triangular element (CST) of Turner et al [4], as shown in Figure 10.2

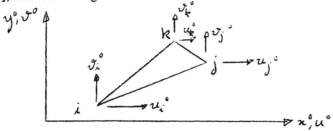

Figure 10.2 - 3 node triangular element

As each node has two degrees of freedom, namely $u°$ and $v°$, the element has a total of six degrees of freedom, as shown in Figure 10.2.

The elemental stiffness matrix was derived on the basis of the following assumptions for the displacement functions $u°$ and $v°$.

$$u° = \alpha_1 + \alpha_2 x° + \alpha_3 y°$$
$$v° = \alpha_4 + \alpha_5 x° + \alpha_6 y°$$
                                                                            10.1

where

α_1 to α_6 = arbitrary constants

The stiffness matrices for both plane stress and plane strain has been derived in reference [7], and will not be rederived here.

10.3   Data for STRESSN

The input data file should be created in the following sequence.

MS = number of elements
NN = number of nodes
NF = number of nodes with zero displacements
E = Elastic modulus
NU = Poisson's ratio
T = Plate thickness

If a plane stress analysis is required, type 1, but if a plane strain analysis is required, type zero. (ie SS = 1 or SS = 0).

Nodes describing each element

-     ME = 1(1)MS

|

→ i_{ME} j_{ME} k_{ME} - type these in an anticlockwise order

Nodal coordinates

-     i = 1(1)NN

|

→ x_i° y_i°

NC = no of nodes with concentrated loads

Nodal positions and values of concentrated loads

-     i = 1(1)NC

|

| Nodal position of concentrated loads

|

| Component of load in x° direction

|

→   Component of load in y° direction

Zero displacement details

-     i = 1(1)NF

|

| NS_i - no of node with zero displacement

|

| Type 1 if the u° displacement is zero at this node; else type 0

|

→   Type 1 if the v° displacement is zero at this node; else type 0

The <u>output</u> for STRESS3F is as follows:-

-     i = 1(1)NN

|

→ u_i° v_i°

-     i = 1(1)MS

|

| σ_x°, σ_y, τ_{xy}°

|

→ σ_1, σ_2

154

where

$\sigma_x^{\,\circ}$ = direct stress in the x° direction
$\sigma_y^{\,\circ}$ = direct stress in the y° direction

$\tau_{x^{\circ}y^{\circ}}$ = shear stress in the x°-y° plane

σ_1 = maximum principal stress
σ_2 = minimum principal stress

10.4 Example 10.1

Using the computer program STRESS3N, determine the nodal displacements and elemental
stresses for the cantilever plate of Figure 10.3, which has a square hole in its centre. The
following may be assumed:-

1E11 lbf/in²; ν = 0.32; t = 2E-2 in

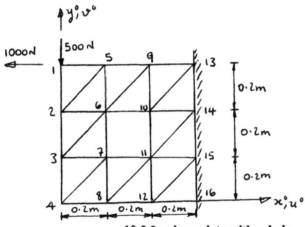

10.3 In-plane plate with a hole

As the plate is <u>thin</u>, the condition of <u>plane stress</u> should be assumed.

10.4 Input data file for Example 10.1

The input data file should be prepared in the following sequence:-

MS = 16
NN = 16
NF = 4
E = 1E11
NU = 0.32
T = 2E-2
SS = 1 (ie plane stress)

i	j	k	
5	1	2	}
5	2	6	}
6	2	3	}
6	3	7	}
7	3	4	}
7	4	8	}
9	5	6	}
9	6	10	} counter-clockwise order
11	7	8	}
11	8	12	}
13	9	10	}
13	10	14	}
14	10	11	}
14	11	15	}
15	11	12	}
15	12	16	}

x	y
0	0.6
0	0.4
0	0.2
0	0.0
0.2	0.6
0.2	0.4
0.2	0.2
0.2	0.0
0.4	0.6
0.4	0.4
0.4	0.2
0.4	0.0
0.6	0.6
0.6	0.4
0.6	0.2
0.6	0.0

NC = 1

1   -1000   -500

13,1,1   14,1,1   15,1,1   16,1,1

10.5 Results

Some nodal displacement (m)

Node	u°	v°
1	-3.03E-6	-3.87E-6
2	-8.29E-7	-3.01E-6
3	1.89E-8	-2.66E-6
4	8.38E-7	-2.59E-6
13	0	0
14	0	0
15	0	0
16	0	0

Some elemental stress values (Pa)

Element	σ_x°	σ_y°	$\tau_{x^\circ y^\circ}$	σ_1	σ_2
5-1-2	472,207	-277,793	-27,793	473,236	-278,822
5-2-6	79,205	-58,373	16,918	81,255	-60,422
6-2-3	45,377	-164,086	96,789	83,252	-201,962
6-3-7	-24,902	-26,063	79,339	53,859	-104,824
13-9-10	517,537	43,848	-31,438	519,614	41,771
13-10-14	102,820	32,902	78,202	153,521	-17,799
14-10-11	103,318	34,459	47,546	127,592	10,186
14-11-15	12,656	4,050	78,731	87,202	-70,496
15-11-12	24,359	40,622	14,207	48,859	16,122
15-12-16	260,690	-83,421	91,166	-44,905	-299,206

10.6 Input Data File STRESS3.DAT

A typical data file for STRESS3N.EXE, namely STRESS3.DAT is given below. This data file is for the solution of Example 10.1.

```
16
16
4
1E11,.32
2E-2
1
5,1,2
5,2,6
6,2,3
6,3,7
7,3,4
7,4,8
9,5,6
9,6,10
11,7,8
11,8,12
13,9,10
13,10,14
14,10,11
14,11,15
15,11,12
15,12,16
0,0.6
0,0.4
0,0.2
0,0.0
0.2,0.6
0.2,0.4
0.2,0.2
0.2,0.0
0.4,0.6
0.4,0.4
0.4,0.2
0.4,0.0
0.6,0.6
0.6,0.4
0.6,0.2
0.6,0.0
1
1,-1000,-500
13,1,1
14,1,1
15,1,1
16,1,1
EOF
```

10.7   Stress Contours

Various stress contours for this plate are shown in Figures 10.4 to 10.6.

Figure 10.4 Maximum Shear Stress Contours

10.5 Maximum Principal Stress Contours

159

Figure 10.6 Minimum Principal Stress Contours

10.8 Example 10.2

Determine the nodal displacements and the elemental stresses for the plate shown in Fig 10.7, assuming

(a)   plane strain
(b)   plane stress

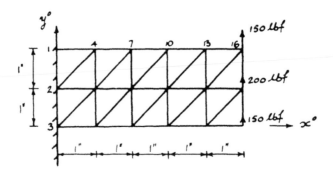

Figure 10.7

160

It may be assumed that:

Elastic modulus, $E = 30 \times 10^6$ lbf/in^2
Poisson's ratio $\nu = 0.3$
Thickness, $t = 0.1$ in

Data

MS = 20
NN = 18
NF = 3

(a)   SS = 0 for plane strain

or

(b)   SS = 1 for plane stress

E = 3E7
NU = 0.3
T = 0.1

Nodal Points for Each Element

i	j	k
1	2	4
2	5	4
$\downarrow$	$\downarrow$	$\downarrow$
15	18	17

Coordinates

x_i°	y_i°
0.0	2.0
0.0	1.0
$\downarrow$	$\downarrow$
5.0	0.0

NC = 3

16	0	150
17	0	200
18	0	150
1	1	1
2	1	1
3	1	1

Results

The maximum tensile and compressive stresses are given in Table 10.1 for (a), (b) and the predictions of simple beam theory

Table 10.1 Stresses in cantilever plate (lbf/in^2)

Element	$x°$(in)	$y°$(in)	(a)	(b)	Simple beam theory
1	0.333	1.667	-22273	-21576	-23333
4	0.667	0.333	19491	19188	21667

The deflection distributions for the bottom edge for the two cases are compared with the predictions for simple bema theory in Table 10.2.

Table 10.2 Deflection of cantilever plate (ins)

$x°$(in)	0	1	2	3	4	5
(a)	0.00	0.00054	0.00147	0.00275	0.00426	0.00591
(b)	0.00	0.00057	0.00162	0.00306	0.00477	0.00663
Beam theory	0.00	0.00058	0.00217	0.0045	0.00733	0.0104

From the above, it can be seen that whereas the stress calculations "appear" to be reasonable, but the deflection calculations are underestimated; that is the finite element mathematical models are too stiff. A more refined mesh should give better results.

10.9 The computer program STRESS4N

This program can determine the nodal displacements and stresses for an in-plane plate in either a plane stress or a plane strain condition.

The element has four nodes, (at its corners), and each node has two degrees of freedom, making a total of 8 degrees of freedom per element, as shown in Figure 10.8.

162

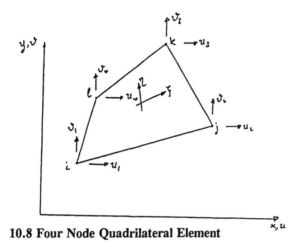

10.8 Four Node Quadrilateral Element

The direct stresses in the x and y directions and the principal stresses are calculated at a number of points for each element, and also the shear stresses in the x-y plane.

The non-dimensional co-ordinates ξ and η point in the direction 1 to 2 and 2 to 3 respectively. The values of ξ and η at the nodes are as follows:-

Node 1 - $\xi = -1$ and $\eta = -1$

Node 2 - $\xi = 1$ and $\eta = -1$

Node 3 - $\xi = 1$ and $\eta = 1$

Node 4 - $\xi = -1$ and $\eta = 1$

The displacement positions are calculated as follows:-

Displacement position corresponding to $u_i \equiv 2^*i\text{-}1$

Displacement position corresponding to $v_i = 2^*i$

where i is the nodal number.

The assumed displacement functions are:-

$u = \alpha_1 + \alpha_2 x + \alpha_3 y + \alpha_4 xy$

$v = \alpha_5 + \alpha_6 x + \alpha_7 y + \alpha_8 xy$

which can be put in the following matrix form:

$= [N] \{u_i\}$

$$\left\{ \begin{array}{c} u \\ v \end{array} \right\} = \left[\begin{array}{cccccccc} N_1 & 0 & N_2 & 0 & N_3 & 0 & N_4 & 0 \\ 0 & N_1 & 0 & N_2 & 0 & N_3 & 0 & N_4 \end{array} \right] \left\{ \begin{array}{c} u_1 \\ v_1 \\ u_2 \\ v_2 \\ u_3 \\ v_3 \\ u_4 \\ v_4 \end{array} \right\}$$

where

α_i = arbitrary constant

[N] = a matrix of shape functions

$\{u_i\}$ = a vector of nodal displacements

$N_1 = \frac{1}{4}(1 - \xi)(1 - \eta)$

$N_2 = \frac{1}{4}(1 + \xi)(1 - \eta)$

$N_3 = \frac{1}{4}(1 + \xi)(1 + \eta)$

$N_4 = \frac{1}{4}(1 - \xi)(1 + \eta)$

The stiffness matrix is given by:

$[k] = \int \int [B]^T [D] [B] \, dx \, dy$

$= \int_{-1}^{1} \int_{-1}^{1} [B]^T [D] [B] \det | J | \, d\xi \, d\eta$ \hfill 10.1

where,

$[B] = [B_1 \ B_2 \ B_3 \ B_4]$

$$B_i = \begin{bmatrix} \dfrac{\partial N_i}{\partial x} & 0 \\[2ex] 0 & \dfrac{\partial N_i}{\partial y} \\[2ex] \dfrac{\partial N_i}{\partial y} & \dfrac{\partial N_i}{\partial x} \end{bmatrix} \qquad\qquad 10.2$$

$$\left\{ \begin{array}{c} \dfrac{\partial N_i}{\partial x} \\[2ex] \dfrac{\partial N_i}{\partial y} \end{array} \right\} = [J^{-1}] \left\{ \begin{array}{c} \dfrac{\partial N_i}{\partial \xi} \\[2ex] \dfrac{\partial N_i}{\partial \eta} \end{array} \right\} \qquad\qquad 10.3$$

$$[J] = \text{a Jacobian} = \begin{bmatrix} \dfrac{\partial N_1}{\partial \xi} & \dfrac{\partial N_2}{\partial \xi} & \dfrac{\partial N_3}{\partial \xi} & \dfrac{\partial N_4}{\partial \xi} \\[2ex] \dfrac{\partial N_1}{\partial \eta} & \dfrac{\partial N_2}{\partial \eta} & \dfrac{\partial N_3}{\partial \eta} & \dfrac{\partial N_4}{\partial \eta} \end{bmatrix} \begin{bmatrix} x_1 & y_1 \\ x_2 & y_2 \\ x_3 & y_3 \\ x_4 & y_4 \end{bmatrix} \qquad 10.4$$

x_i and y_i are the co-ordinates of the ith node.

$$\frac{\partial N_1}{\partial \xi} = -\tfrac{1}{4}(1-\eta)$$

$$\frac{\partial N_2}{\partial \xi} = \tfrac{1}{4}(1-\eta)$$

$$\frac{\partial N_3}{\partial \xi} = \tfrac{1}{4}(1-\eta)$$

$$\frac{\partial N_4}{\partial \xi} = -\tfrac{1}{4}(1+\eta)$$

$$\tfrac{1}{4} \qquad\qquad 10.5$$

$$\frac{\partial N_1}{\partial \eta} = -\tfrac{1}{4}(1-\xi)$$

$$\frac{\partial N_2}{\partial \eta} = -\tfrac{1}{4}(1+\xi)$$

$$\frac{\partial N_3}{\partial \eta} = \tfrac{1}{4}(1+\xi)$$

$$\frac{\partial N_4}{\partial \eta} = \tfrac{1}{4}(1-\xi)$$

$$[D] = E^1 \begin{vmatrix} 1 & \zeta & 0 \\ \zeta & 1 & 0 \\ 0 & 0 & \gamma \end{vmatrix} \qquad\qquad 10.6$$

where,

for plane stress

$E^1 = E/(1 - \nu^2)$

$\varsigma = \nu$

$\gamma = (1 - \nu)/2$

and,

for plane strain

$E^1 = E(1 - \nu/[(1 + \nu)(1 - 2\,\nu)]$

166

$\varsigma = \nu/(1 - \nu)$

$\gamma - (1 - 2\nu)/[2(1 - \nu)]$

E = Young's modulus of elasticity

ν = Poisson's ratio

By substitution of 10.2, 10.3, 10.4, 10.5 and 10.6 into 10.1, the stiffness matrix [k] can be determined. In the present program, it was found sufficient to use two Gauss points in the ξ direction and two Gauss points in the η direction, making a total of four Gauss points per element.

10.10 Data

The input data should be prepared in the following sequence:-

(1)   NJ = number of nodes.

(2)   ES = number of elements.

(3)   NF = number of nodes with zero displacements.

(4)   E = elastic modulus

(5)   NU = Poisson's ratio

(6)   T = plate thickness

(7)   If plane stress, type in 1, but if plane strain, type in 0

(8)   Nodal co-ordinates

  For i = 1 to NJ

  Input x_i, y_i

  NEXT i

(9)   Suppressed displacement positions

  For i = 1 to NF

  Input NS_i - the ith "suppressed" nodal position

  Type I/O as appropriate

  NEXT i

167

(10)   Element details

For EL = 1 to ES

Input i_{EL} j_{EL} k_{EL} ℓ_{EL}

(The nodes describing element EL, fed in an anti-clockwise direction).

NEXT EL

Note The ξ direction will be from i_{EL} to j_{EL} and the η direction from j_{EL} to k_{EL}.

(11)   NC = number of nodes with concentrated loads.

For i = 1 to NC

Input Nodal position of load

Input value of load in x direction

Input value of load in y direction

NEXT i

RESULTS

Nodal Displacements

For i = 1 to NJ

Print u_i, v_i

NEXT i

Element Stresses

For EL = 1 to ES

For XI = - 1 to 1

For ETA = -1 to 1

Print σ_{xEL}, σ_{yEL}, τ_{xyEL}, σ_1, σ_2

NEXT ETA

NEXT XI

NEXT EL

10.11 Example 10.3

Determine the nodal displacements and stresses for the square plate with a hole in its centre, as shown in Figure 10.9. The plate is firmly fixed at nodes 13, 14, 15 and 16 and the following apply:-

$E = 1 \times 10^{11}$ N/m^2

$\nu = 0.3$

$T = 2 \times 10^{-2}$ m

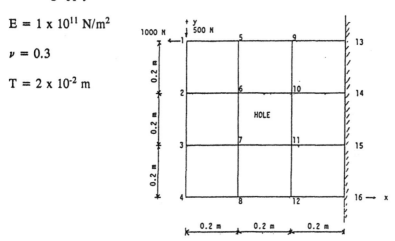

Figure 10.9 - Plate with Hole

The data should be prepared in the following sequence:-

NJ = 16

ES = 8

NF = 4

E = 1E11

NU = 0.32

T = 2E-2

PS = 1 (Plane Stress)

169

Nodal Co-ordinates

0	0.6
0	0.4
0	0.2
0	0
0.2	0.6
0.2	0.4
0.2	0.2
0.2	0
0.4	0.6
0.4	0.4
0.4	0.2
0.4	0
0.6	0.6
0.6	0.4
0.6	0.2
0.6	0

Zero displacement details

13	1	1	14	1	1	15	1	1	16	1	1

Element details

i	j	k	ℓ
1	2	6	5
2	3	7	6
3	4	8	7
5	6	10	9
7	8	12	11
9	10	14	13
10	11	15	14
11	12	16	15

NC = 1

Nodal loads

Nodal position = 1

Load in x direction = -1000

Load in y direction = - 500

170

10.12 Input data file for STRESS4N

The input data file for Example 10.3, namely STRESS4.DAT, is as follows:-

STRESS4.DAT

```
16
8
4
1E11,.32
2E-2
1
0,.6
0,.4
0,.2
0,0
.2,.6
.2,.4
.2,.2
.2,0
.4,.6
.4,.4
.4,.2
.4,0
.6,.6
.6,.4
.6,.2
.6,0
13,1,1
14,1,1
15,1,1
16,1,1
1,2,6,5
2,3,7,6
3,4,8,7
5,6,10,9
7,8,12,11
9,10,14,13
10,11,15,14
11,12,16,15
1
1,-1000,-500
EOF
```

10.13 Results

Node i	Nodal Displacements(m)	
	u_i	v_i
1	-3.68E-6	-4.82E-6
2	-6.74E-7	-3.81E-6
3	1.90E-7	-3.33E-6
4	1.19E-6	-3.32E-6
5	-2.32E-6	-2.09E-6
6	-5.69E-7	-2.08E-6
7	9.67E-8	-2.19E-6
8	1.18E-6	-2.11E-6
9	-1.23E-6	-7.49E-7
10	-1.92E-7	-4.21E-7
11	-4.45E-8	-5.12E-7
12	8.10E-7	-7.49E-7
13	0	0
14	0	0
15	0	0
16	0	0

Some of the stresses (N/m^2) are as follows:-

Element	ξ	η	σ_x	σ_y	τ_{xy}	σ_1	σ_2
1	-1	-1	576359	-320105	-51407	579297	-323043
1	-1	0	665557	-41361	67239	671896	-47670
1	-1	1	754755	237382	185885	814616	177523
1	0	-1	227399	-431772	-146180	258363	-462735
1	0	0	316598	-153028	-27534	318207	-154637
1	0	1	405796	125716	91112	432827	98685
1	1	-1	-121560	-543439	-240953	-12259	-652740
1	1	0	-32362	-264695	-122307	20154	-317211
1	1	1	56837	14049	-3661	57147	13738
8	-1	-1	67048	140063	-64889	178017	29093
8	-1	0	45905	73993	16060	81284	38615
8	-1	1	24763	7924	97018	113726	-81039
8	0	-1	-171065	63867	-42435	71297	-178495
8	0	0	-192207	-2203	38523	5311	-199721
8	0	1	-213349	-68272	119482	-1033	-280588
8	1	-1	-409178	-12329	-19971	-11327	-410180
8	1	0	-430320	-78399	60987	-68129	-440589
8	1	1	-451462	-144468	141945	-88896	-507034

These results show the simplex triangular element of reference [4] to be considerably in error.

10.14 Screendumps

Screendumps for Example 10.3 are shown in Figure 10.10 and 10.12.

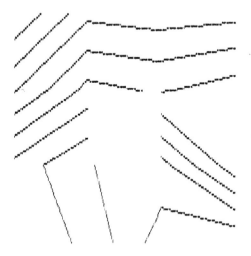

Figure 10.10 - Maximum shear stress contours

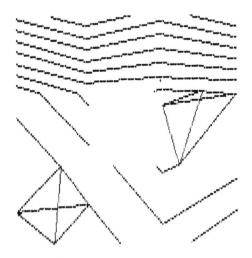

Figure 10.11 - Maximum principal stress contours

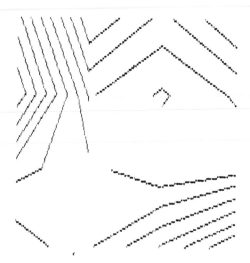

10.12 - Minimum principal stress contours

10.15 Example 10.4

Determine the nodal displacements and stresses for the plane stress problem of Figure 10.13. The elements have been distorted to demonstrate their distorting capabilities, but caution must be exercised when carrying out this feature [6,8].

E = 3*10^7 lbf/in²   ν = 0.3   T 0.1 in

Figure 10.13 - Cantilever Plate

The data should be fed in as follows:-

NJ = 18

ES = 10

NF = 3

174

E = 3E7

NU = 0.3

T = 0.1

PS = 1

Nodal co-ordinates

x_i	y_i
0	3
0	2
0	1
0	1
1	3
1	2
1	1
2	3
2	2
2	1
3	3
3	2
3	1
4.25	3
4	2
3.75	1
5	3
5	2
5	1

Suppressed displacement positions

1	1	1		2	1	1		3	1	1

Element Details

i	j	k	ℓ
1	2	5	4
2	3	6	5
4	5	8	7
5	6	9	8
7	8	11	10
8	9	12	11
10	11	14	13
11	12	15	14
13	14	17	16
14	15	18	17

Nodal loads

NC = 3

Nodal position 1 = 16

Load in x direction = 0

Load in y direction = 150

Nodal position 2 = 17

load in x direction = 0

Load in y direction = 200

Nodal position 3 = 18

Load in x direction = 0

Load in y direction = 150

10.16 Results

| Node | Displacements (ins) | |
	u	v
1	0	0
2	0	0
3	0	0
4	-9.75E-4	7.86E-4
5	-8.50E-8	6.22E-4
6	9.75E-4	7.87E-4
7	-1.76E-3	2.33E-3
8	7.89E-7	2.24E-3
9	1.76E-3	2.34E-3
10	-2.32E-3	4.58E-3
11	5.61E-6	4.51E-3
12	2.31E-3	4.58E-3
13	-2.69E-3	7.97E-3
14	1.09E-6	7.23E-3
15	2.59E-3	6.57E-3
16	-2.76E-3	0.0102
17	-3.43E-6	0.0101
18	2.76E-3	0.0102

Some of the stress values are as follows:-

Element	ξ	η	σ_x	σ_y	τ_{xy}
			Stress (ℓbf/in^2)		
1	-1	-1	-32151	-9645	9073
1	-1	0	-31339	-6940	3447
1	-1	1	-30528	-4236	-2179
1	0	-1	-16077	-4823	8127
1	0	0	-15265	-2118	2500
1	0	1	-14454	586	-3125
1	1	-1	-2.8	-0.8	7180
1	1	0	809	2704	1554
1	1	1	1620	5409	-4072
2	-1	-1	-2.8	-0.8	7180
2	-1	0	-815	-2708	1551
2	-1	1	1627	-5415	-4076
2	0	-1	16078	4823	8127
2	0	0	15265	2116	2499
2	0	1	14453	-592	-3128
2	1	-1	32158	947	9075
2	1	0	31346	6940	3447
2	1	1	30533	4233	-2181
10	-1	-1	-717	-1940	3631
10	-1	0	-584	-1495	2626
10	-1	1	-450	-1050	1621
10	0	-1	1865	-1075	3269
10	0	0	1984	-680	2378
10	0	1	2102	-284	1483
10	1	-1	3931	-384	2980
10	1	0	4038	-28	2176
10	1	1	4144	329	1372

According to elementary theory

$\sigma_x \text{ (max}^m) = + 37500 \text{ lbf/in}^2$

$\sigma_x \text{ (min}^m) = -37500 \text{ lbf/in}^2$

$v_{16} = u_{17} = v_{18} = 0.0104 \text{ ins}$

This seems to compare favourably with the results from the computer program and are considerably better than the computer results for the simplex triangular element of reference[4].

10.17 The Computer Program STRESS8

This program adopts an eight node isoparametric element and can determine nodal displacements and stresses for an in-plane plate under plane stress or plane strain.

The element has four of its nodes and its corners and the other four nodes at its "mid-sides", as shown in Figure 10.14.

Each node has two degrees of freedom per node, making a total of sixteen degrees of freedom per element. The "mid-side" nodes need not be exactly at the mid-sides of the element.

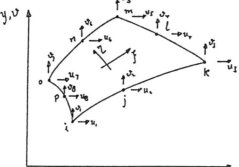

Figure 10.14 - Eight Node Isoparametric Element

The nodal displacements are u and v, and these are in the x and y direction, respectively. The curvilinear co-ordinates ϵ and n point from p to ℓ and j to n respectively. The values for ϵ and n at the nodes are as follows:-

Node i

$\xi = -1$ $\qquad$ $\eta = -1$

Node j

$\xi = 0$ $\qquad$ $\eta = -1$

Node k

$\xi = 1$ $\qquad$ $\eta = -1$

Node ℓ

$\xi = 1$ $\qquad$ $\eta = 0$

Node m

$\xi = 1$ $\qquad$ $\eta = 1$

179

Node n

$\xi = 0$ $\qquad \eta = 1$

Node 0

$\xi = -1$ $\qquad \eta = 1$

Node p

$\xi = -1$ $\qquad \eta = 0$

The displacement positions are calculated as follows:-

Displacement position corresponding to $u_i = 2^*i\text{-}1$

Displacement position corresponding to $v_i = 2^*i$

where, i = nodal number.

The assumed displacement functions are:-

$u = \alpha_1 + \alpha_2 x + \alpha_3 y + \alpha_4 xy + \alpha_5 x^2 + \alpha_6 y^2 + \alpha_7 x^2y + \alpha_8 y^2 x$

$v = \alpha_9 + \alpha_{10} x + \alpha_{11} y + a_{12} xy + a_{13} x^2 + \alpha_{14} y^2 + \alpha_{15} x^2y + \alpha_{16} y^2 x$

which can be put in the following matrix form:

$$
\begin{Bmatrix} u \\ v \end{Bmatrix} = \begin{bmatrix} N_1 & 0 & N_2 & 0 & N_3 & 0 & N_4 & \dots & N_8 & 0 \\ 0 & N_1 & 0 & N_2 & 0 & N_3 & 0 \dots & & 0 & N_8 \end{bmatrix} \begin{Bmatrix} u_1 \\ v_1 \\ u_2 \\ v_2 \\ u_3 \\ v_3 \\ \downarrow \\ u_8 \\ v_8 \end{Bmatrix}
$$

$= [N] \{u_i\}$

where

α_i = arbitrary constants

[N] = a matrix of nodal displacement

$N_1 = -\frac{1}{4}(1 - \xi)(1 - \eta)(\xi + \eta + 1)$

$N_2 = \frac{1}{2}(1 - \xi^2)(1 - \eta)$

$N_3 = \frac{1}{4}(1 + \xi)(1 - \eta)(\xi - \eta - 1)$

$N_4 = \frac{1}{2}(1 - \eta^2)(1 + \xi)$

$N_5 = \frac{1}{4}(1 + \xi)(1 = \eta)(\xi + \eta - 1)$

$N_6 = \frac{1}{2}(1 - \xi^2)(1 + \eta)$

$N_7 = -\frac{1}{4}(1 - \xi)(1 + \eta)(\xi - \eta + 1)$

$N_8 = \frac{1}{2}(1 - \eta^2)(1 - \xi)$

The stiffness matrix is obtained from equation 10.2, where

[B] = [B_1 B_2 B_3 B_4 B_5 B_6 B_7 B_8]

and [B_i] is as in the equation 10.2

and, $\dfrac{\partial N_i}{\partial x}$ and $\dfrac{\partial N_i}{\partial y}$ are as in equation 10.3, where

$$J] = \begin{bmatrix} \dfrac{\partial N_1}{\partial \xi} & \dfrac{\partial N_2}{\partial \xi} & \dfrac{\partial N_3}{\partial \xi} & \cdots & \dfrac{\partial N_8}{\partial \xi} \\[4mm] \dfrac{\partial N_1}{\partial \eta} & \dfrac{\partial N_2}{\partial \eta} & \dfrac{\partial N_3}{\partial \eta} & \cdots & \dfrac{\partial N_8}{\partial \eta} \end{bmatrix} \begin{bmatrix} x_1 & y_1 \\ x_2 & y_2 \\ x_3 & y_3 \\ | & | \\ | & | \\ x_8 & y_8 \end{bmatrix} \qquad 10.7$$

$$\frac{\partial N_1}{\partial \xi} = 0.25 \, (1 - \eta)(2\xi + \eta)$$

$$\frac{\partial N_1}{\partial \eta} = 0.25(1 - \xi)(\xi + 2\eta)$$

$$\frac{\partial N_2}{\partial \xi} = -\xi(1 - \eta)$$

$$\frac{\partial N_2}{\partial \eta} = - \, 0.5(1 - \xi^2)$$

$$\frac{\partial N_3}{\partial \xi} = 0.25(1 - \eta(2\xi - \eta)$$

$$\frac{\partial N_3}{\partial \eta} = 0.25(1 + \xi)(- \xi + 2\eta)$$

$$\frac{\partial N_4}{\partial \xi} = 0.5(1 \quad \eta^2)$$

$$\frac{\partial N_4}{\partial \eta} = - \, \eta(1 + \xi)$$

10.8

$$\frac{\partial N_5}{\partial \xi} = 0.25 \, (1 + \eta)(2\xi + \eta)$$

$$\frac{\partial N5}{\partial \eta} = 0.25(1 + \xi)(\xi + 2\eta)$$

$$\frac{\partial N_6}{\partial \xi} = -\xi(1 + \eta)$$

$$\frac{\partial N_6}{\partial \eta} = \tfrac{1}{2}(1 + \xi^2)$$

$$\frac{\partial N_7}{\partial \eta} = 0.25(1 + \eta(2\xi - \eta)$$

$$\frac{\partial N_7}{\partial \eta} = 0.25(1 - \xi)(- \xi + 2\eta)$$

$$\frac{\partial N_8}{\partial \xi} = - \, 0.5(1 - \eta^2)$$

$$\frac{\partial N_8}{\partial \eta} = - \, \eta(1 - \xi)$$

The matrix [D] is given in equation 10.6 and use of this matrix, together with the other relevant matrices, gives an expression for the stiffness matrix [k].

Integration was carried out using three Gauss points in the ξ direction and three in the η direction, making a total of nine Gauss points per element.

10.18 Data

The data should be typed in the following sequence:-

(1)   NJ = number of nodes

(2)   ES = number of elements

(3)   NF = number of nodes with zero displacements

(4)   E = elastic modulus

(5)   NU = Poisson's ratio

(6)   T = plate thickness

(7)   If plane stress, type in 1; else if plane strain, type in 0

(8)   Nodal co-ordinates

   For i = 1 to NJ

   Input x_i, y_i

   NEXT i

(9)   Zero displacement details

   For i = 1 to NF

   Input NS_i - nodal position of zero displacements

   Type 1 if $u^\circ = 0$; else type 0

   Type 1 if $v^\circ = 0$; else type 0

   NEXT i

(10)   Element details

183

For EL = 1 to ES

input i_{EL}, j_{EL}, k_{EL} ℓ_{EL}, m_{EL}, n_{EL}, o_{EL} p_{EL}

(The nodes defining element EL, fed in an anti-clockwise direction).

NEXT EL

Note ξ direction will be from p_{EL} to ℓ_{EL} and the η direction from j_{EL} to n_{EL}

(11) NC = number of nodes with concentrated loads

(12) For i = 1 to NC

Input nodal position of load

Input value of load in x direction

Input value of load in y direction

NEXT i

Results

Nodal Displacements

For i = 1 to NJ

Print u_i, v_i

NEXT i

Element Stresses

For EL = 1 to ES

For XI = -1 to 1

For ETA = -1 to 1

Print σ_{xEL}, σ_{yEL}, τ_{xyEL}, σ_1, σ_2

NEXT ETA

NEXT XI

NEXT EL

10.19 Example 10.5

Determine the nodal displacements and stresses for the problem of Example 10.3.

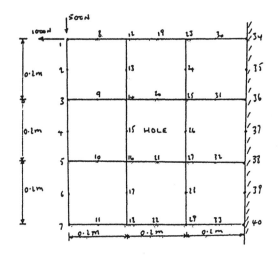

Figure 10.15 - Square Plate with Square Hole

NJ = 40

ES = 8

NF = 7

E = 1E11

NU = 0.32

T = 2E-2

PS = 1 (plane stress)

Nodal co-ordinates

0	0.6
0	0.5
0	0.4
0	0.3
0	0.2
0	0.1
0	0
0.1	0.6
0.1	0.4
0.1	0.2
0.1	0
0.2	0.6
0.2	0.5
0.2	0.4
0.2	0.3
0.2	0.2
0.2	0.1
0.2	0
0.3	0.6
0.3	0.4
0.3	0.2
0.3	0
0.4	0.6
0.4	0.5
0.4	0.4
0.4	0.3
0.4	0.2
0.4	0.1
0.4	0
0.5	0.6
0.5	0.4
0.5	0.2
0.5	0
0.6	0.6
0.6	0.5
0.6	0.4
0.6	0.3
0.6	0.2
0.6	0.1
0.6	0

Suppressed displacement positions

34 1 1	35 1 1	36 1 1	37 1 1
38 1 1	39 1 1	40 1 1	

Element details

1, 2, 3, 9, 14, 13, 12, 8

3, 4, 5, 10, 16, 15, 14, 9

5, 6, 7, 11, 18, 17, 16, 10

12, 13, 14, 20, 25, 24, 23, 19

16, 17, 18, 22, 29, 28, 27, 21

23, 24, 25, 31, 36, 35, 34, 30

25, 26, 27, 32, 38, 37, 36, 31

27, 28, 29, 33, 40, 39, 38, 32

NC = 1

Nodal position of load = 1

Load in xi direction = - 1000

Load in y direction = - 500

10.20 Input data file for STRESS8N.EXE

The input data file for Example 10.5 is as follows:-

STRESS8.DAT

```
40
8
7
1E11,0.32
2E-2
1
0,.6
0,.5
0,.4
0,.3
0,.2
0,.1
0,0
.1,.6
.1,.4
.1,.2
.1,0
.2,.6
.2,.5
.2,.4
.2,.3
.2,.2
.2,.1
.2,0
.3,.6
.3,.4
.3,.2
.3,0
.4,.6
.4,.5
.4,.4
.4,.3
.4,.2
.4,.1
.4,0
.5,.6
.5,.4
.5,.2
.5,0
.6,.6
.6,.5
.6,.4
.6,.3
.6,.2
.6,.1
.6,0
34,1,1
35,1,1
36,1,1
37,1,1
38,1,1
39,1,1
40,1,1
1,2,3,9,14,13,12,8
3,4,5,10,16,15,14,9
5,6,7,11,18,17,16,10
12,13,14,20,25,24,23,19
16,17,18,22,29,28,27,21
23,24,25,31,36,35,34,30
25,26,27,32,38,37,36,31
27,28,29,33,40,39,38,32
1
1,-1000,-500
EOF
```

10.21 Results

Node i	Nodal Displacements (m)	
	u_i	v_i
1	-5.23E-6	-6.66E-6
2	-1.80E-6	-5.40E-6
3	-6.05E-7	-4.36E-6
4	-1.16E-7	-3.89E-6
5	2.22E-7	-3.77E-6
6	6.92E-7	-3.77E-6
7	1.27E-6	-3.77E-6
8	-3.72E-6	-3.57E-6
9	-5.00E-7	-3.44E-6
10	1.90E-7	-3.22E-6
11	1.28E-6	-3.20E-6
12	-2.52E-6	-2.34E-6
13	-1.52E-6	-2.28E-6
14	-4.02E-7	-2.38E-6
15	-4.21E-8	-2.63E-6
16	6.58E-8	-2.60E-6
17	6.40E-7	-2.53E-6
18	1.31E-7	-2.5E-6
19	-1.95E-6	-1.49E-6
20	-3.13E-7	-1.23E-6
21	1.20E-7	-1.55E-6
22	1.26E-7	-1.64E-6
23	-1.30E-6	-7.63E-7
24	-7.42E-7	-5.72E-7
25	-2.36E-7	-4.72E-7
26	-6.46E-8	-4.67E-7
27	-6.39E-8	-5.76E-7
28	3.79E-7	-6.82E-7
29	9.20E-7	-8.33E-7
30	-6.32E-7	-3.10E-7
31	-1.37E-7	-1.50E-7
32	-2.91E-8	-1.94E-7
33	4.66E-7	-2.92E-7
34	0	0
35	0	0
36	0	0
37	0	0
38	0	0
39	0	0
40	0	0

Stresses (N/m^2)

Element	ξ	η	σ_x	σ_y	τ_{xy}	σ_1	σ_2
1	-1	-1	1.37E6	-919334	-203896	1391443	-937325
1	-1	0	1.42E6	205149	-195243	1450772	174545
1	-1	1	1.11E6	220434	-244485	1174587	157790
1	0	-1	-75496	-1.17E6	40281	-74020	-1174772
1	0	0	131652	-25099	-19755	134103	-27551
1	0	1	-16144	13900	-137685	137379	-139624
1	1	-1	-213221	-1.01E6	294923	-115707	-1105180
1	1	0	154355	164244	166197	325571	-6972
1	1	1	166988	226956	-20421	233249	160697
8	-1	-1	72180	108080	30827	125804	54459
8	-1	0	45165	41263	33893	77164	9264
8	-1	1	29331	9386	38268	58905	-20188
8	0	-1	-155042	78919	25502	81667	-157790
8	0	0	-193668	-13389	35401	-6687	-200372
8	0	1	-221114	-70756	46607	-57481	-234391
8	1	-1	-437299	32146	28425	37861	-439017
8	1	0	-487536	-85652	45156	-80641	-492551
8	1	1	-526593	-168510	63195	-157685	-537423

The results for this plate show that the mesh for the 4 node quadrilateral analysis of Example 10.3 was unsatisfactory and that of the simplex triangular element of reference [4] was totally inadequate. Even for the present case, a much more refined mesh was preferable.

10.22 Screendumps

Screendumps for Example 10.5 are shown in Figure 10.16 to 10.18

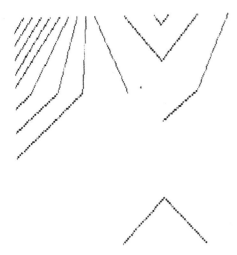

Figure 10.16 - Maximum shear stress contours

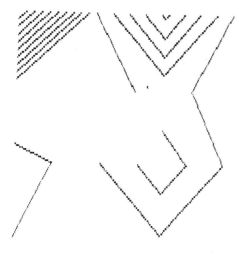

Figure 10.17 - Maximum principal stress contours

Figure 10.18 - Minimum principal stress contours

10.23 Example 2.3

Determine the nodal displacements and stresses for the cantilever plate of Example 10.2. The isoparametric elements of Figure 10.19 have deliberately been distorted to test their distorting capabilities.

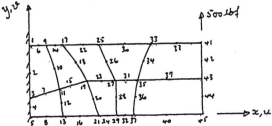

Figure 10.19 - Isoparametric Mesh for Cantilever

The data should be fed in as follows:-

NJ = 45

ES = 10

NF = 5

E = 3E7

NU = 0.3

T = 0.1

PS = 1 (plane stress)

192

Nodal co-ordinates

0	2
0	1.25
0	0.5
0	0.25
0	0
0.25	2
0.5	0.65
0.5	0
0.5	2.0
0.7	1.5
0.9	0.75
0.95	0.4
1	0
0.75	2
1.3	0.9
1.5	0
1	2
1.4	1.5
1.65	1
1.85	0.5
2	0
1.5	2
2.1	1
2.25	0
2	2
2.25	1.5
2.5	1
2.5	0.5
2.5	0
2.75	2
2.85	1
2.75	0
3.5	2
3.25	1.5
3.15	1
3.05	0.5
3	0
4.25	2
4	1
4	0
5	2
5	1.5
5	1
5	0.5
5	0

Zero displacement details

1	1	1	2	1	1
3	1	1	4	1	1

Element details

1, 2, 3, 7, 11, 10, 9, 6

3, 4, 5, 8, 13, 12, 11, 7

9, 10, 11, 15, 19, 18, 17, 14

11, 12, 13, 16, 21, 20, 19, 15

17, 18, 19, 23, 27, 26, 25, 22

19, 20, 21, 24, 29, 28, 27, 23

25, 26, 27, 31, 35, 34, 33, 30

27, 28, 29, 32, 37, 36, 35, 31

33, 34, 35, 39, 43, 42, 41, 38

35, 36, 37, 40, 45, 44, 43, 39

Concentrated Loads

NC = 5

Nodal position 1 = 41

Load in x direction = 0   load in y direction = 100

Nodal position 2 = 42

Load in xi direction = 0   load in y direction = 100

Nodal position 3 = 43

Load in x direction = 0   load in y direction = 100

Nodal position 4 = 44

Load in x direction = 0   load in y direction = 100

Nodal position 5 = 45

Load in x direction = 0   load in y direction = 100

10.24 Results

Node i	Nodal Displacements (ins)	
	u_i	v_i
1	0	0
2	0	0
3	0	0
4	0	0
5	0	0
6	-2.97E-4	1.62E-4
7	1.80E-4	2.22E-4
8	6.09E-4	3.69E-4
9	-5.99E-4	3.70E-4
10	-3.77E-4	4.27E-4
11	2.29E-4	6.51E-4
12	6.13E-4	7.45E-4
13	1.13E-3	9.07E-4
14	-9.04E-4	5.95E-4
15	1.24E-4	1.21E-3
16	1.60E-3	1.70E-3
17	-1.14E-3	9.00E-4
18	-7.20E-4	1.42E-3
19	8.87E-6	1.85E-3
20	9.09E-4	2.29E-3
21	2.00E-3	2.69E-3
22	-1.60E-3	1.70E-3
23	4.20E-6	2.81E-3
24	2.18E-3	3.27E-3
25	-2.00E-3	2.69E-3
26	-1.05E-3	3.19E-3
27	-4.14E-6	3.78E-3
28	1.14E-3	3.81E-3
29	2.34E-3	3.88E-3
30	-2.48E-3	4.54E-3
31	3.79E-6	4.73E-3
32	2.49E-3	4.53E-3
33	-2.84E-3	6.68E-3
34	-1.33E-3	5.89E-3
35	8.50E-7	5.57E-3
36	1.29E-3	5.30E-3
37	2.62E-3	5.22E-3
38	-3.06E-3	9.03E-3
39	-2.09E-6	8.20E-3
40	3.00E-3	8.23E-3
41	-3.14E-3	0.0116
42	-1.53E-3	0.0115
43	3.58E-6	0.0115
44	1.52E-6	0.0115
45	3.16E-3	0.0116

Stresses (lbf/in²)

Element	ξ	η	σ_x	σ_y	τ_{xy}
1	-1	-1	-38832	-11650	6418
1	-1	0	-36874	-3077	4420
1	-1	1	-34588	6590	2230
1	0	-1	-9900	-2967	2503
1	0	0	-12094	-1691	1867
1	0	1	-17654	-1538	2839
1	1	-1	14747	4424	2477
1	1	0	12503	2925	2921
1	1	1	4614	1288	7420
2	-1	-1	14747	4424	2477
2	-1	0	11565	-469	3142
2	-1	1	5607	3146	6510
2	0	-1	26042	7812	4046
2	0	0	22481	-55	1841
2	0	1	18163	542	2445
2	1	-1	42974	12892	6564
2	1	0	34341	1051	1140
2	1	1	28677	-892	-314
10	-1	-1	-587	-1022	5163
10	-1	0	220	587	4140
10	-1	1	-103	-1271	4970
10	0	-1	7222	-339	2842
10	0	0	3728	320	2176
10	0	1	-335	-2679	3165
10	1	-1	14314	-301	1216
10	1	0	8135	291	375
10	1	1	566	-3747	1491

This problem shows the two quadrilateral element, and in particular the eight node quadrilateral element. The problem also shows that the elements can be distorted to some extent. Care, however [6,8], must be exercised in not distorting the elements too much, as numerical errors can result.

10.25 Example

Calculate and plot the maximum deflection distribution for the plate of Figure 10.20 [27], using STRESS3F and STRESS8F, for the plane stress condition.

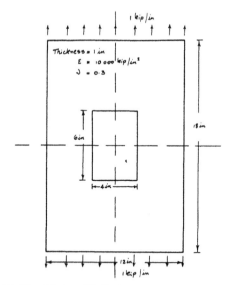

Figure 10.20 - Plate with hole

The meshes for the two computer programs are shown in Figures 10.21 and 10.22 and the results are shown in Figure 10.23.

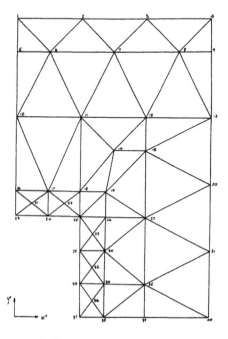

Figure 10.21 - Mesh for plate with hole

197

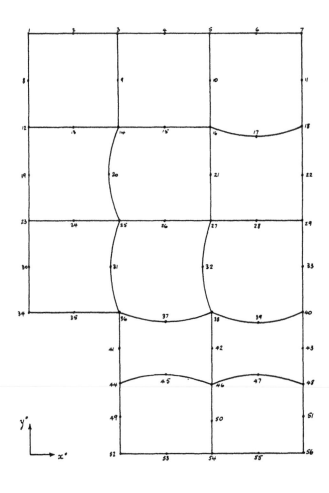

Figure 10.22 - Isoparametric mesh for plate with hole

10.26 Results

3 node triangular element - $v_1° = 0.00132$ in, $v_2° = 0.00125$ in
$v_3° = 0.00113$ in, $v_4° = 0.00101$ in

8 node quadrilateral element - $v_1° = 0.00138$ in $v_5 = 0.00116$ in
$v_2° = 0.00132$ in $v_6° = 0.00105$ in
$v_3° = 0.00131$ in $v_7° = 0.00105$ in
$v_4° = 0.0012$ in

A comparison is made in Figure 10.23 of the edge displacement of Clough's plate using the triangular element program and also the isoparametric element program and this can be seen to be quite good.

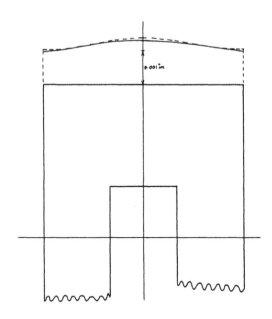

Constant stress triangular element (plane stress) ____

8 node isoparametric element (plane stress) ----

Figure 10.23 - Edge displacement of plate with hole

199

CHAPTER 11

BENDING STRESSES IN FLAT PLATES

11.1 Introduction

This chapter describes the computer program PLATEBEF, which can calculate nodal displacements and elemental stresses in flat plates, subjected to out-of-plane bending. The stiffness matrix is the same as that used by Mohr and Milner [7, 28], and it is based on small deflection elastic theory; that is, the maximum deflection of the plate must not exceed half of the plate thickness. The element is the three node triangular element shown in Figure 11.1. Each node has three degrees of freedom, namely w, θ_x and θ_y, so that there are a total of nine degrees of freedom per element;

where $\theta_x = \partial w / \partial y$

$\theta_y = -\partial w / dx$

w = lateral deflection

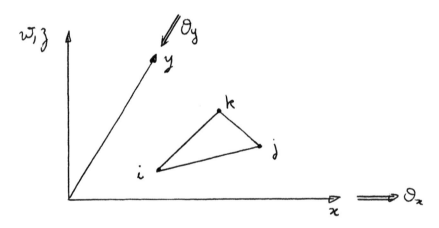

Figure 11.1 - Flat plate bending element

In Figure 11.1, the directions of rotation of θ_x and θ_y are shown according to the right hand screw rule.

NB It should be noted that the topological description of each element, namely the nodes, i j and k, must be made in a <u>counter-clockwise direction</u>. Numerical integration, over each triangular element, was carried out using 3 Gauss-Radau points [6,8].

11.2 Input Data File

The data should be prepared in the following sequence:-

NN = number of nodes

MS = number of elements

NF = number of nodes which have zero displacements

Details of zero displacements

FOR i = 1 to NF

SU_i = number of the node with the zero displacements

IF w = 0 at this node, type 1; else type 1 (ie SUW = 1 or 0)

If θ_y = 0 at this node, type 1; else type 1 (ie SUX = 1 or 0)

If θ_x = 0 at this node, type 1; else type 0 (ie SUY = 1 or 0)

NEXT i

Nodal co-ordinates

FOR i = 1 to NN

Type x_i, y_i

NEXT i

TH = the plate thickness = t

E = Young's modulus

NU = Poisons' ratio = ν

Element Topology

FOR ME = 1 to MS

Type in i_{ME}, j_{ME} and k_{ME} (for each element in a <u>counterclockwise order)</u>

NEXT ME

Details of point loads

NC = number of points loads

IF NC = 0, ignore the following loop

FOR i = 1 to NC

Type in the nodal number where load occurs

Type in WC - the point load in the direction of z - positive upwards

NEXT i

Pressure details

Type in the pressure which acts perpendicularly to the plate surface - <u>positive upwards</u>

11.3   Results

The output which can include screendumps is as follows:-

FOR i = 1 to NN

w_i, θ_{yi}, θ_{xi}

NEXT i

FOR ME = 1 to MS

$M_x = D(\phi_x + \nu \phi_y)$ }

$M_y = D(\phi_y + \nu \phi_x)$ }   For each element, at its centroid

$M_{x-y} = D (\phi_{xy}) (1 - \nu)/2$

NEXT ME

where

$$D = \frac{E \times TH^3}{12(1 - NU * NU)}$$

$$\nu = N$$

$$\phi_x = \frac{\partial^2 w}{\partial x^2}$$

$$\phi_y = \frac{\partial^2 w}{\partial y^2}$$

$$\phi_{xy} = \frac{2\partial^2 w}{\partial x.dy}$$

Also output, for each element, at its centroid, are membrane stresses, shear stresses and principal stresses.

11.4 Example 11.1

Determine the nodal displacements and elemental values of M_x, M_y, M_{xy} and stresses for the plate shown in Figure 11.2, which is subjected to a uniform lateral pressure of 1 kN/m². The plate is simply-supported along its left and bottom edges, and the top and right edges are axes of symmetry.

The other details of the plate are:-

$E = 2 \times 10^{11}$ N/m² $\nu = 0.3$

Plate thickness = 1m

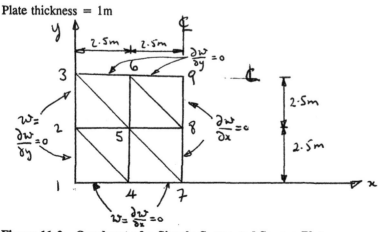

Figure 11.2 - Quadrant of a Simply-Supported Square Plate

11.5 Data

The data is as follows, and an input data file for the problem is shown in Section 11.7.

NN = 9
MS = 8
NF = 8
TH = 1
E = 2E8
NU = 0.3

Element Topology

i	j	k	
			}
			}
1	4	2	}
2	4	5	}
2	5	3	} Counter-clockwise order
3	5	6	}
4	7	5	}
5	7	8	}
5	8	6	}
6	8	9	}

Nodal co-ordinates

x	y
0	0
0	2.5
0	5
2.5	0
2.5	2.5
2.5	5
5	0
5	2.5
5	5

External loads

NC = 0

Pressure = 1

Details of zero displacements

Node	w	∂w/∂x	∂w/∂y
1	1	1	1
2	1	0	1
3	1	0	1
4	1	1	0
6	0	0	1
7	1	1	0
8	0	1	0
9	0	1	1

11.6 Results

Maximum $w = w_9 = 2.163 \times 10^{-6}$ m

Exact value [29] $\hat{w}$ = $\dfrac{406 \times 12 \ (1-0.3^2) \times 1 \times 10^4}{10^5 \times 2 \times 10^8 \times 1^3}$

= $\underline{2.217 \times 10^{-6} \text{ m}}$

Maximum value for the bending moment from the program = $\underline{3.86 \text{ kN m/m}}$ at the centroid of element 8

Exact value for the bending moment [29]

= $\underline{4.789 \text{ kNm/m}}$

The slightly lower value obtained form the computer program is partly due to the fact that this value was not calculated at the centre of the plate, ie at node 9.

11.7 Input data file

The input data file for Example 11.1 is given below.

PLATEBEN.DAT

```
9
8
8
1,1,1,1
2,1,0,1
3,1,0,1
4,1,1,0
6,0,0,1
7,1,1,0
8,0,1,0
9,0,1,1
0,0
0,2.5
0,5
2.5,0
2.5,2.5
2.5,5
5,0
5,2.5
5,5
1
2e9,0.3
1,4,2
2,4,5
2,5,3
3,5,6
4,7,5
5,7,8
5,8,6
6,8,9
0
1
```

11.8 Screendumps

Screendumps for Example 11.1 are given in Figure 11.3 to 11.5

LINES OF CONSTANT MAXIMUM SHEAR STRESS

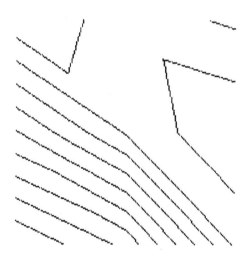

Figure 11.3 - Maximum shear Stress Contours

LINES OF CONSTANT MAXIMUM PRINCIPAL STRESS

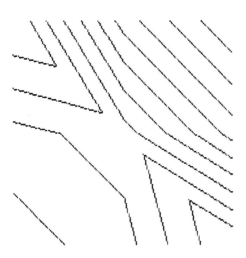

Figure 11.4 - Maximum principal stress contours

LINES OF CONSTANT MINIMUM PRINCIPAL STRESS

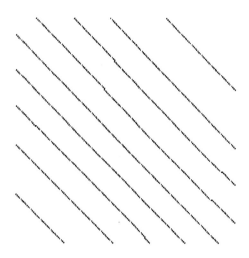

Figure 11.5 - Minimum principal stress contours

11.9   Example 11.2

Using the mesh of Figure 11.6, determine the nodal displacements and elemental stresses for the flat plate of this figure, which is firmly clamped along its edges.

Plate thickness = 0.013 ins $E = 10 \times 10^6$ lbf/in^{12}

$\nu = 0.33$ pressure = q = 1 lbf/in^2

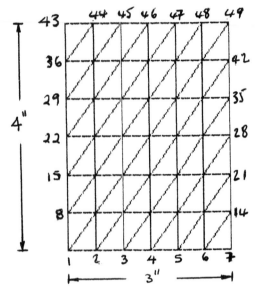

Figure 11.6 - Plate clamped along edges

11.10   Input data file

The input data file for Example 11.2 is as follows:-

210

PLATEB2.INP

```
        49
        72
        24
        1,  1,1,1
        2,  1,1,1
        3,  1,1,1
        4,  1,1,1
        5,  1,1,1
        6,  1,1,1
        7,  1,1,1
        8,  1,1,1
       14,  1,1,1
       15,  1,1,1
       21,  1,1,1
       22,  1,1,1
       28,  1,1,1
       29,  1,1,1
       35,  1,1,1
       36,  1,1,1
       42,  1,1,1
       43,  1,1,1
       44,  1,1,1
       45,  1,1,1
       46,  1,1,1
       47,  1,1,1
       48,  1,1,1
       49,  1,1,1
       0,0
       .5,0
       1,0
       1.5,0
       2,0
       2.5,0
       3,0
       0,.667
       .5,.667
       1,.667
       1.5,.667
       2,.667
       2.5,.667
       3,.667
       0,1.333
       .5,1.333
       1,1.333
       1.5,1.333
       2,1.333
       2.5,1.333
       3,1.333
       0,2
       .5,2
       1,2
       1.5,2
       2,2
       2.5,2
       3,2
       0,2.667
       .5,2.667
       1,2.667
       1.5,2.667
       2,2.667
       2.5,2.667
       3,2.667
       0,3.333
       .5,3.333
       1,3.333
       1.5,3.333
       2,3.333
       2.5,3.333
       3,3.333
       0,4
       .5,4
       1,4
       1.5,4
       2,4
       2.5,4
       3,4
```

```
.013
10E6,.33
1,9,8
1,2,9
2,10,9
2,3,10
3,11,10
3,4,11
4,12,11
4,5,12
5,13,12
5,6,13
6,14,13
6,7,14
8,16,15
8,9,16
9,17,16
9,10,17
10,18,17
10,11,18
11,19,18
11,12,19
12,20,19
12,13,20
13,21,20
13,14,21
15,23,22
15,16,23
16,24,23
16,17,24
17,25,24
17,18,25
18,26,25
18,19,26
19,27,26
19,20,27
20,28,27
20,21,28
22,30,29
22,23,30
23,31,30
23,24,31
24,32,31
24,25,32
25,33,32
25,26,33
26,34,33
26,27,34
27,35,34
27,28,35
29,37,36
29,30,37
30,38,37
30,31,38
31,39,38
31,32,39
32,40,39
32,33,40
33,41,40
33,34,41
34,42,41
34,35,42
36,44,43
36,37,44
37,45,44
37,38,45
38,46,45
38,39,46
39,47,46
39,40,47
40,48,47
40,41,48
41,49,48
41,42,49
0
1: REM PRESSURE LOAD ONLY
EOF
```

11.10 Results

From computer program,

Maximum deflection = w_{25} = 8.344E-2 ins

Maximum bending stress (on element 48) - ± 12880 lbf/in²

From Roark and Young [29]

$a/b = \dfrac{4}{3} = 1.333 \therefore \alpha = 0.0203$ and $\beta_1 = 0.411$

$$\text{Maximum deflection} = \frac{0.0208qb^4}{Et^3} = \underline{0.075 \text{ ins}}$$

$$\text{Maximum stress (at centre of long edge)} = \frac{\beta_1 qb^2}{t^2}$$

$$= \frac{0.411 \times 1 \times 3^2}{0.013^2} = \pm 21890 \text{ lbf/in}^2$$

A better value for bending stress could have been obtained from the computer program, if a more refined mesh were taken near the centre of the longer edges. This is because the computer program calculates the bending stress at the centroid of the element.

11.12 Screendumps

Screendumps of various stress contours are shown in Figures 11.7 to 11.9.

LINES OF CONSTANT MAXIMUM SHEAR STRESS

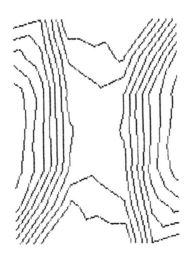

Figure 11.7 - Maximum shear stress contours

LINES OF CONSTANT MAXIMUM PRINCIPAL STRESS

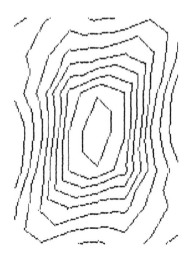

Figure 11.8 - Maximum principal stress contours

215

LINES OF CONSTANT MINIMUM PRINCIPAL STRESS

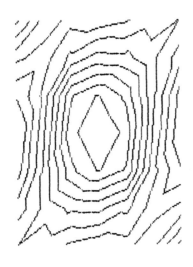

Figure 11.9 - Minimum principal stress contours

CHAPTER 12

STRESSES IN DOUBLY-CURVED SHELLS

12.1   Introduction

This chapter describes the computer program SHELLSTF, which is capable of calculating nodal displacements and elemental stresses in doubly-curved shells of complex shape and with complex boundary conditions.

The element used in this program is a flat shell element which is obtained by superimposing the constant strain triangular element of Turner et al [4] with the out-of-plane bending element of Mohr & Milner [7, 28].

In local coordinates the elemental stiffness matrix is an 18 x 18 matrix, [7], which can be formed from 3 x 3 submatrices, namely [k_{ij}] of the following form:-

$$[k_s] = \begin{bmatrix} k_{11} & k_{12} & k_{13} \\ k_{21} & k_{22} & k_{23} \\ k_{31} & k_{32} & k_{33} \end{bmatrix}$$

(12.1)

where the elements of equation (12.1) are:

$$[k_{ij}] = \begin{array}{cccccc} & u & v & w & \theta_x & \theta_y & \theta_z \\ \begin{bmatrix} k_p^{xx} & k_p^{xy} & 0 & 0 & 0 & 0 \\ k_p^{yx} & k_p^{yy} & 0 & 0 & 0 & 0 \\ 0 & 0 & k_b^{zz} & k_b^{zx} & k_b^{zy} & 0 \\ 0 & 0 & k_b^{xz} & k_b^{xx} & k_b^{xy} & 0 \\ 0 & 0 & k_b^{yz} & k_b^{yx} & k_b^{yy} & 0 \\ 0 & 0 & 0 & 0 & 0 & 0 \end{bmatrix} & \begin{array}{c} u \\ v \\ w \\ \theta_x \\ \theta_y \\ \theta_z \end{array} \end{array}$$

The element is shown in global co-ordinates in Figure 12.1, and the elemental stiffness matrix, in global co-ordinates is given by:-

$$[k^\circ]_{18x18} = [\Xi]^T_{18x18} [k]_{18x18} [\Xi]_{18x18}$$

12.2

where

$$[\bar{Z}] = \begin{bmatrix} \zeta & & & & & \\ & \zeta & & & & \\ & & \zeta & & & \\ & & & \zeta & & \\ & & & & \zeta & \\ & & & & & \zeta \end{bmatrix}$$

12.3

$[\bar{Z}]$ is given by equation (5.3)

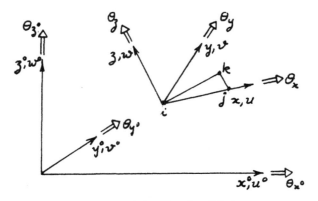

Figure 12.1 - Flat "shell" element

The element can be used for analysis of doubly curved shells and folded slabs of various shape, see Figures 12.2 and 12.3.

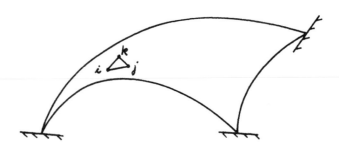

Figure 12.2 - Doubly curved shell

218

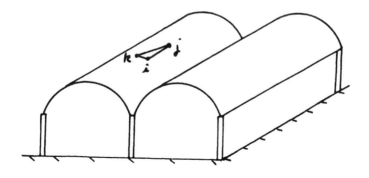

Figure 12.3 - Cylindrical Shells

If a coplanar section is met, the stiffness matrix of equation 12.2 can become singular, because of the row and column of zeros in equation 12.1. However, in the computer program SHELLST, this deficiency was avoided when a coplanar section was set, by placing a small number, instead of the zero, on the bottom diagonal element of the matrix of equation 12.1. The magnitude of this number is related to the magnitude of the non-zero diagonal elements of the matrix of equation 12.1.

12.2   Input Data File

The input data file should be prepared in the following sequence:-

NN = number of nodes
MS = number of elements
NF = number of nodes with zero displacement

For i = 1 to NF
Type in the number of the node with the zero displacement
If u° = 0, type 1, else type 0
If v° = 0, type 1 else type 0
If w° = 0, type 1, else type 0
If θ_x° = 0, type 1, else type 0
If θ_y° = 0, type 1, else type 0
If θ_z° = 0, type 1, else type 0
NEXT i

Coordinates

For i = 1 to NN
Type in x_i° y_i° z_i°
NEXT i

T = wall thickness of shell
E = Young's modulus

NU = ν = Poisson's ratio

Element topology

For I = 1 to MS
Type in i_I, j_I and k_I (in a counter-clockwise direction)
NEXT I

External Loads

NC = number of concentrated loads

IF NC = 0, ignore the next loop

For i = 1 to NC
Type in the node number with the loads
Type in x° component of load of this node
Type in y° component of load of this node
Type in z° component of load of this node
Type in θ_x° component of couple at this node   }
Type in θ_y° component of couple at this node   }   according to the RH screw rule
Type in θ_z° component of couple of this node
NEXT i

Type in the pressure p (+ve if on inside surface).

12.3   Example 12.1

Determine the nodal displacements and elemental stress for the cylindrical shell of Figure 12.4, which may be assumed to be firmly clamped along its edges, and subjected to an internal pressure of 1 lbf/in². Details of the shell are given in Section 12.1, where it can be seen that the cylindrical shell rises to a height of 0.067 ins along its centre line.

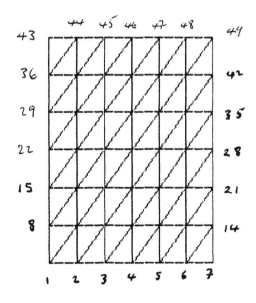

Figure 12.4

12.4 Input data file

The input data file for example 12.1 is as follows:-

SHELLST.DAT

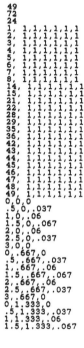

```
49
72
24
1,  1,1,1,1,1,1
2,  1,1,1,1,1,1
3,  1,1,1,1,1,1
4,  1,1,1,1,1,1
5,  1,1,1,1,1,1
6,  1,1,1,1,1,1
7,  1,1,1,1,1,1
8,  1,1,1,1,1,1
14,  1,1,1,1,1,1
15,  1,1,1,1,1,1
21,  1,1,1,1,1,1
22,  1,1,1,1,1,1
28,  1,1,1,1,1,1
29,  1,1,1,1,1,1
35,  1,1,1,1,1,1
36,  1,1,1,1,1,1
42,  1,1,1,1,1,1
43,  1,1,1,1,1,1
44,  1,1,1,1,1,1
45,  1,1,1,1,1,1
46,  1,1,1,1,1,1
47,  1,1,1,1,1,1
48,  1,1,1,1,1,1
49,  1,1,1,1,1,1
0,0,0
.5,0,.037
1,0,.06
1.5,0,.067
2,0,.06
2.5,0,.037
3,0,0
0,.667,0
.5,.667,.037
1,.667,.06
1.5,.667,.067
2,.667,.06
2.5,.667,.037
3,.667,0
0,1.333,0
.5,1.333,.037
1,1.333,.06
1.5,1.333,.067
```

221

```
2,1.333,.06
2.5,1.333,.037
3,1.333,0
0,2,0
.5,2,.037
1,2,.06
1.5,2,.067
2,2,.06
2.5,2,.037
3,2,0
0,2.667,0
.5,2.667,.037
1,2.667,.06
1.5,2.667,.067
2,2.667,.06
2.5,2.667,.037
3,2.667,0
0,3.333,0
.5,3.333,.037
1,3.333,.06
1.5,3.333,.067
2,3.333,.06
2.5,3.333,.037
3,3.333,0
0,4,0
.5,4,.037
1,4,.06
1.5,4,.067
2,4,.06
2.5,4,.037
3,4,0
.013:
10E6,.33
1,9,8
1,2,9
2,10,9
2,3,10
3,11,10
3,4,11
4,12,11
4,5,12
5,13,12
5,6,13
6,14,13
6,7,14
8,16,15
8,9,16
9,17,16
9,10,17
10,18,17
10,11,18
11,19,18
11,12,19
12,20,19
12,13,20
13,21,20
13,14,21
15,23,22
15,16,23
16,24,23
16,17,24
17,25,24
17,18,25
18,26,25
18,19,26
19,27,26
19,20,27
20,28,27
20,21,28
22,30,29
22,23,30
23,31,30
23,24,31
24,32,31
24,25,32
25,33,32
25,26,33
26,34,33
26,27,34
27,35,34
27,28,35
29,37,36
29,30,37
30,38,37
30,31,38
31,39,38
31,32,39
32,40,39
32,33,40
33,41,40
33,34,41
34,42,41
34,35,42
36,44,43
36,37,44
37,45,44
37,38,45
38,46,45
38,39,46
39,47,46
39,40,47
40,48,47
40,41,48
41,49,48
41,42,49
0
1
```

12.5 Results

Some nodal deflections, namely $w°$, are given in Table 12.1, and some elemental stresses are given in Table 12.2.

Table 12.1 - Deflections [$w°$ (ins)] along the centre line

Node	4	11	18	25	32	39	46
$w°$	0	2.26E-3	3.37E-3	3.4E-3	3.37E-3	2.26E-3	0

Table 12.2 - Elemental Stresses (lbf/in²)

Element No	Internal Surface		Membrane	
	σ_1	σ_2	σ_x	σ_y
3-4-11(6)	606.3	-284.6	23.0	69.8
15-23-22(25)	1853.2	404.1	701.7	928.7
17-25-24(29)	897.2	266.2	691.8	915.8
17-18-25(30)	955.8	311	1295.9	397.7

NB External surface stresses should also be considered

Approximate checks (for membrane stress and radial deflection)

Radius of cylindrical shell = R = $\dfrac{1.5^2}{2 \times 0.067}$ = 16.79 ins

Membrane Stress

$$\sigma_{(membrane)(maxm)} = \frac{pR}{t} = \frac{1 \times 16.79}{0.013} = \underline{1291.5 \text{ lbf/in}^2}$$

This compares favourably with σ_x (membranae) for element No 30.

Radial deflection

$$\frac{Ew}{R} = \frac{pR}{t} \quad \therefore \quad w = \frac{pR^2}{Et} = \frac{1 \times 16.79^2}{10 \times 10^6 \times 0.013} = 2.17E{-}3 \text{ ins}$$

Comparing the stress values of the **cylindrical shell** with those obtained for the **flat plate of** Figure 11.6, it can be seen that the cylindrical shell, although shallow, resists the lateral pressure much more efficiently than the flat plate.

12.6   Screendumps

Screendumps of various stress contours are shown in Figure 12.5 to 12.7.

LINES OF CONSTANT MAXIMUM SHEAR STRESS

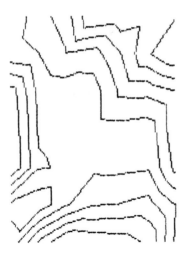

Figure 12.5 - Maximum shear stress contours

LINES OF CONSTANT MAXIMUM PRINCIPAL STRESS

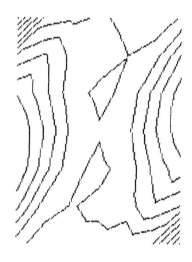

Figure 12.6 Maximum principal stress contours

225

LINES OF CONSTANT MINIMUM PRINCIPAL STRESS

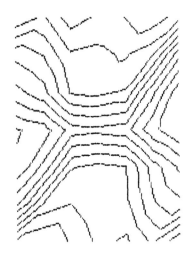

Figure 12.7 Minimum principal stress contours

CHAPTER 13

IN-PLANE VIBRATION OF PLATES

13.1 Introduction

This chapter describes the computer program VIBPLANF, which can be used for determining the resonant frequencies and eigenmodes of a plate, vibrating in its own plane.

The element used in this program, in the constant strain triangular element of Turner et al [4], which has three corner nodes and two degrees of freedom, namely u° and v°, per node, as shown in Figure 10.2.

The mass matrix of the element is derived in reference [7] and will not be rederived here.

13.2 input data file

The input data file should be created in the following sequence:-

NN = number of nodes
LS = number of elements
NF = number of nodes with zero displacement

Details of zero displacement

For i = 1 to NF
Type in the node number of the node with the zero displacements
If u_o = 0 at this node, type 1; else type 0
If v_o = 0 at this node, type 1; else type 0
NEXT i

Nodal co-ordinates

For i = 1 to NN
Type in $x_i{}^\circ$ and $y_i{}^\circ$
NEXT i

M_1 = number of frequencies required (this must not exceed 2 x NN)

TH = plate thickness
E = Young's modulus
NU = ν = Poisson's ratio
RH = density of plate material

Element Topology
For LE = 1 to LS
Type in i_{LE}, j_{LE} and k_{LE} - the nodes describing each element in a counter-clockwise order
NEXT Le

Concentrated masses

NC = number of additional concentrated masses

IF NC = 0, ignore the next loop

For i = 1 to NC
Type in the node with the concentrated mass
Type in the concentrated mass at this node
NEXt i

13.3   Example 13.1

Determine the resonant frequencies and eigenmodes for the shear wall [30], shown in Figure 13.1, which can be assumed to be firmly fixed at its base. The following may be assumed to apply to the shear wall:-

$E = 3.4474E10 \text{ N/m}^2$ $\nu = 0.1$
$\rho = 568.2 \text{ kg/m}^3$
$TH = 0.2289 \text{ m}$

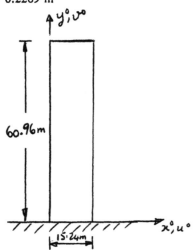
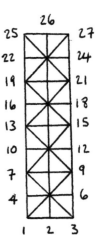

Figure 13.1 Cantilevered shear wall and mesh

13.4   Input data file

The input data file should be created in the following sequence.

NN = 27
LS = 32
NF = 3

Details of Zero Displacements

Node	$u^\circ = 0?$	$v^\circ = 0?$
1	1	1
2	1	1
3	1	1

Nodal co-ordinates

x_i°	y_i°
0	0
7.62	0
15.24	0
0	7.62
7.62	7.62
15.24	7.62
0	15.24
7.62	15.24
15.24	15.24
0	22.86
7.62	22.86
15.24	22.86
0	30.48
7.62	30.48
15.24	30.48
0	38.02
7.62	38.02
15.24	38.02
0	45.72
7.62	45.72
15.24	45.72
0	53.34
7.62	53.34
15.24	53.34
0	60.96
7.62	60.96
15.24	60.96

M1 = 5
TH = 0.2289
E = 3.447E10
NU = 0.11 = ν
TH = 568.2

Element Topology

i	j	k
1	2	5
1	5	4
2	3	5
3	6	5
4	5	7
5	8	7
5	9	8
5	6	9
7	8	11
7	11	10
8	9	11
9	12	11
10	11	13
11	14	13
11	12	15
11	15	14
13	14	17
13	17	16
14	15	17
15	18	17
16	17	19
17	20	19
17	18	21
21	20	17
19	20	23
19	23	22
20	21	23
21	24	23
22	23	25
23	26	25
23	24	27
23	27	26

NC = 0

The input data file for Example 13.1 is given in Section 13.5.

13.5   Input data file for Example 13.1

VIBPLANF.INP

```
27
32
3
1,  1,1
2,  1,1
3,  1,1
0,0
7.62,0
15.24,0
0,7.62
7.62,7.62
15.24,7.62
0,15.24
7.62,15.24
15.24,15.24
0,22.86
7.62,22.86
15.24,22.86
0,30.48
7.62,30.48
15.24,30.48
0,38.02
7.62,38.02
15.24,38.02
0,45.72
7.62,45.72
15.24,45.72
0,53.34
7.62,53.34
15.24,53.34
0,60.96
7.62,60.96
15.24,60.96
5
.2289
34.474E9,  .11,  568.2
1,2,5
1,5,4
2,3,5
3,6,5
4,5,7
5,8,7
5,9,8
5,6,9
7,8,11
7,11,10
8,9,11
9,12,11
10,11,13
11,14,13
11,12,15
11,15,14
13,14,17
13,17,16
14,15,17
15,18,17
16,17,19
17,20,19
17,18,21
21,20,17
19,20,23
19,23,22
20,21,23
21,24,23
22,23,25
23,26,25
23,24,27
23,27,26
0
EOF
```

13.6 Results

The results are shown in Table 13.1 where they are compared with the analytical solution of Carr [30,31], where he treated the shear wall as a deep beam.

Table 13.1 Resonant Frequencies (Hz) of the cantilevered shear wall

Mode	VIBLANF	Analytical
1	6.39	4.97
2	32.04	26.39
3	32.16	31.94
4	74.90	62.07
5	96.83	95.83

13.7 Screendumps

Figures 13.2 to 13.6 present screen dumps of the first five eigenmodes.

EIGENMODE= 1   FREQUENCY= 6.391249 Hz

Figure 13.2

232

EIGENMODE= 2   FREQUENCY= 32.04435 Hz

Figure 13.3

EIGENMODE= 3   FREQUENCY= 32.1638 Hz

Figure 13.4

EIGENMODE= 4   FREQUENCY= 74.90165 Hz

Figure 13.5

EIGENMODE: 5   FREQUENCY= 96.83264 Hz

Figure 13.6

CHAPTER 14

LATERAL VIBRATION OF FLAT PLATES

1.   Introduction

This chapter describes the computer program VIBPLATF, which can determine the resonant frequencies and eigenmodes of a flat plate undergoing lateral vibrations. These plates can be of complex shape with complex boundary conditions.

The element used in this program in the three node triangular element described in Chapter 10. Each of the three corner nodes of the element, has three degrees of freedom, namely w, $\partial w/\partial y$ and $\partial w/\partial x$, so that each element has nine degrees of freedom.

The elemental mass matrix, in global co-ordinates namely $[m°]$ is given by:-

$$[m°] = \int_{vol} [T]^T \, [\bar{N}]^T \, [DC]^T \, \rho \, [DC] \, [\bar{N}] \, [T] \, d \, (Vol)$$

where

ρ = density of plate material

$$[DC] = \begin{bmatrix} \zeta_{ax}^2 & \zeta_{ay}^2 & \sqrt{2} \, \zeta_{ax} \times \zeta_{ay} \\ \zeta_{ax}^2 & \zeta_{by}^2 & \sqrt{2} \, \zeta_{bx} \times \zeta_{by} \\ \zeta_{cx}^2 & \zeta_{cy}^2 & \sqrt{2} \, \zeta_{cx} \times \zeta_{cy} \end{bmatrix}^{-1}$$

$\zeta_{ax} = x_{21}/L_a;$ $\qquad$ $\zeta_{ay} = y_{21}/L_a$

$\zeta_{bx} = x_{32}/L_b;$ $\qquad$ $\zeta_{by} = y_{32}/L_b$

$\zeta_{cx} = x_{13}/L_c;$ $\qquad$ $\zeta_{cy} = y_{13}/L_c$

$x_{21} = x_2° - x_1°;$ $\qquad$ $y_{21} = y_2° - y_1°$, etc etc

$w = [\bar{N}] \, \{w_i\}$

where

$[w_i] = [w_1 \, w_2 \, w_3 \, ... \, w_{10}]$

$\bar{N}_1 = (9 \, L_1{}^3 - 9L_1{}^2 + 2L_1)/2$

$\bar{N}_2 = (9 \, L_2{}^3 - 9L_2{}^2 + 2L_2)/2$

$\bar{N}_3 = (9 L_3{}^3 - 9L_3{}^2 + 2L_3)/2$

$\bar{N}_4 = 13.5 L_1{}^2L_2 - 4.5 L_1 L_2$

$\bar{N}_5 = 13.5 L_1 L_2{}^2 - 4.5 L_1 L_2$

$\bar{N}_6 = 13.5 L_2{}^2 L_3 - 4.5 L_2 L_3$

$\bar{N}_7 = 13.5 L_2 L_3{}^2 - 4.5 L_2 L_3$

$\bar{N}_8 = 13.5 L_3{}^2 L_1 - 4.5 L_3 L_1$

$\bar{N}_9 = 13.5 L_3 L_1{}^2 - 4.5 L_3 L_1$

$\bar{N}_{10} = 27 L_1 L_2 L_3$

$[T] = (1/27)$

27	0	0	0	0	0	0	0	0
0	0	0	27	0	0	0	0	0
0	0	0	0	0	0	27	0	0
20	$2a_1$	$2b_1$	7	$-a_1$	$-b_1$	0	0	0
7	a_1	b_1	20	$-2a_1$	$-2b_1$	0	0	0
0	0	0	20	$2c_1$	$2d_1$	7	$-c_1$	$-d_1$
0	0	0	7	c_1	d_1	20	$-2c_1$	$-2d_1$
7	$-e_1$	$-f_1$	0	0	0	20	$2e_1$	$2f_1$
20	$-2e_1$	$-2f_1$	0	0	0	7	e_1	f_1

Σ rows (4 to 9)/4 - Σ row (1 to 3)/6

Numerical integration was carried out over each triangular element, by using 3 Gauss-Radau points [6,8]

14.2 Input data file

The input data file should be prepared in the following sequence:-

NN = number of nodes
LS = number of elements
NF = number of nodes with zero displacements

Details of zero displacements

For i = 1 to NF
Type in the node with the zero displacement
If w = 0, at this node, type 1; else type 0
If $\partial w/\partial y = 0$ at this node, type 1; else type 0
If $\partial w/\partial x = 0$, at this node, type 1; else type 0
NEXT i

Nodal co-ordinates

For i = 1 to NN
Type in x_i and y_i
NEXT i

M1 = number of frequencies
TH = plate thickness
E - Young's modulus
ν = Poisson's ratio
RH = density

Element topology

For LE = 1 to LS
Type in i_{LE}, j_{LE} and k_{LE} (in a counter clockwise order)
NEXT LE

NC = number of concentrated masses

IF NC = 0, ignore the next loop.

For i = 1 to NC
Type in the node number with the concentrated mass
Type in the value of the mass at this node
NEXT i

14.3   Example 14.1

Determine the resonant frequencies for the triangular plate [30] of Figure 14.1, which is clamped firmly along its right edge. The following may be assumed to apply:-

$E = 2.067 \times 10^{11}$ N/m² $\nu = 0.3$ $\rho = 7890$ kg/m³

Plate thickness = 1.55 mm

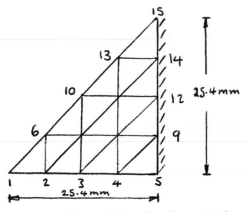

Figure 14.1 - Cantilevered triangular plate

14.4   Input data

The input data should be fed in the following sequence:-

NN = 15
LS = 16
NF = 5

Details of Zero Displacements

Node	w = 0?	∂w/dy = 0?	∂w/dx = 0?
5	1	1	1
9	1	1	1
12	1	1	1
14	1	1	1
15	1	1	1

Nodal co-ordinates

x	y
0	0
0.0635	0
0.127	0
0.1905	0
0.254	0
0.0635	0.0635
0.127	0.0635
0.1905	0.0635
0.254	0.0635
0.1905	0.1905
0.254	0.1905
0.254	0.254

M1 = 6
TH = 1.55E-3
NU = 0.3
RH = 7890

Element Topology

i	j	k
1	2	6
2	7	6
2	3	7
3	8	7
3	4	8
4	9	8
4	5	9
6	7	10
7	11	10
7	8	11
8	12	11
8	9	12
10	11	13
11	14	13
11	12	14
13	14	15

NC = 0

The input data file for Example 14.1 is given in Section 14.5.

241

14.5   Input data file for Example 14.1

VIBPLATF.INP

```
15
16
5
5,  1,1,1
9,  1,1,1
12,  1,1,1
14,  1,1,1
15,  1,1,1
0,0
.0635,0
.127,0
.1905,0
.254,0
.0635,.0635
.127,.0635
.1905,.0635
.254,.0635
.127,.127
.1905,.127
.254,.127
.1905,.1905
.254,.1905
.254,.254
6
1.55E-3
2.067E11,.3,7890
1,2,6
2,7,6
2,3,7
3,8,7
3,4,8
4,9,8
4,5,9
6,7,10
7,11,10
7,8,11
8,12,11
8,9,12
10,11,13
11,14,13
11,12,14
13,14,15
0:REM NO CONCENTRATED MASSES
EOF
```

14.6   Results

The first six resonant frequencies, calculated by VIBPALTF are compared with the experimentally obtained values from reference [30], in Table 14.1.

Table 14.1 - Resonant Frequencies (Hz) for a cantilevered plate

Mode	VIBPLATF	Experiment
1	36.34	34.5
2	137.5	136
3	193.2	190
4	341.2	325
5	444.4	441
6	583.9	578

14.7 Screendumps

Screendumps of the first six eigenmodes of the cantilevered plate of Example 14.1 are shown in Figure 14.2 to 14.7.

```
TO CONTINUE, PRESS Y
EIGENMODE= 1
FREQUENCY= 36.33796 Hz
```

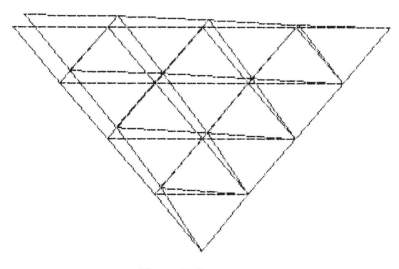

Figure 14.2

TO CONTINUE, PRESS Y
EIGENMODE= 2
FREQUENCY= 137.4987 Hz

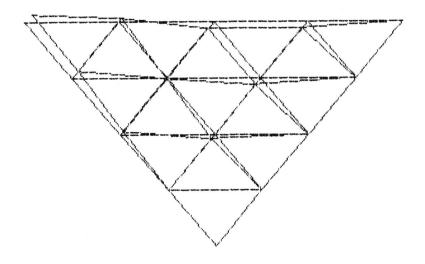

Figure 14.3

TO CONTINUE, PRESS Y
EIGENMODE= 3
FREQUENCY= 193.1846 Hz

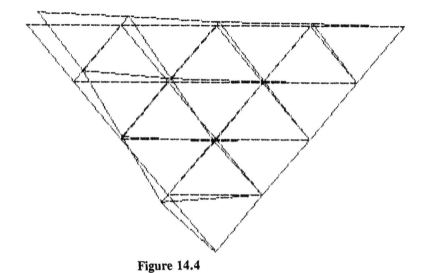

Figure 14.4

TO CONTINUE, PRESS Y
EIGENMODE= 4
FREQUENCY= 341.1743 Hz

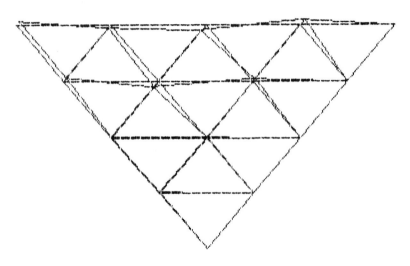

Figure 14.5

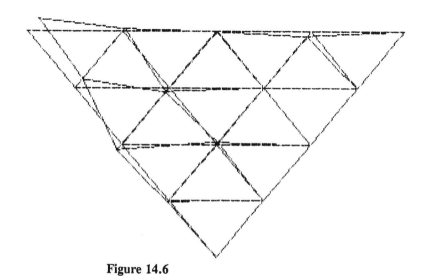

Figure 14.6

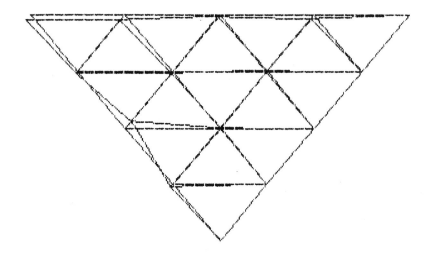

Figure 14.7

CHAPTER 15

VIBRATION OF THIN-WALLED DOUBLY-CURVED SHELLS

15.1   Introduction

This chapter describes the computer program VIBSHELF, which can calculate the resonant frequencies and eigenmodes of a doubly-curved shell.

The computer program uses the element described in Chapter 10, where the in-plane triangular element of Turner et al [6] is superimposed with the out-of-plane bending element at Mohr and Milner [28, 7]. The elemental mass matrix, in local co-ordinates is of order 18 x 18, and can be considered to be composed of 3 x 3 supermatrices, of the form shown by equation 15.1.

$$[m_s] = \begin{bmatrix} m_{11} & m_{12} & m_{13} \\ m_{21} & m_{22} & m_{23} \\ m_{31} & m_{32} & m_{33} \end{bmatrix}$$

15.1

$$[m_{ij}] = \begin{bmatrix} m_p^{xx} & m_p^{xy} & 0 & 0 & 0 & 0 \\ m_p^{yx} & m_p^{yy} & 0 & 0 & 0 & 0 \\ 0 & 0 & m_b^{zz} & m_b^{zx} & m_b^{zy} & 0 \\ 0 & 0 & m_b^{xz} & m_b^{xx} & m_b^{xy} & 0 \\ 0 & 0 & m_b^{yz} & m_b^{yx} & m_b^{yy} & 0 \\ 0 & 0 & 0 & 0 & 0 & 0 \end{bmatrix} \begin{matrix} u \\ v \\ w \\ \theta_x \\ \theta_y \\ \theta_z \end{matrix}$$

where m_p refers to in-plane masses (see Chapter 13)
where m_b refers to out-of-plane plate bending masses (see Chapter 14)

The elemental mass matrix in global co-ordinates $[m°]$ is given by:-

$$[m°] = [\Xi]^T [m] [\Xi]$$

15.2

where

249

[Ξ] is given by equation 5.2.

The eigenvalue economiser technique of Irons [ ] is used to reduce the displacements $\theta_x°$, $\theta_y°$ and $\theta_z°$, so that the eigenmodes are described only by the $u°$, $v°$ and $w°$ displacement, for more details of this technique see Chapter 6.

15.2   Input Data

The input data file should be prepared in the following sequence:-

NN = number of non-zero nodal points
N = number of free displacements remaining after reduction and after removing the zero displacements corresponding to the fixed nodes, but **not** the simply-supported nodes.

LS = number of elements

NF = number of nodes in the pin-joints

IF NF = 0, ignore the next loop
For i = 1 to NF

Type in the **finished** displacement position of the zero displacement
Next i

Nodal co-ordinates of non-zero nodes

For i = 1 to NN
Type in $x_i°$, $y_i°$ and $z_i°$
Next i

M1 = number of frequencies ($M_1 < = N$)
TH = wall thickness of shell
E = Young's modulus
NU = ν = Poisson's ratio
RH = ρ = material density

Element topology

For LE = 1 to LS

Type in i_{LE}, j_{LE} and k_{LE} (in a counter-clockwise order)

If i_{LE} = 0 (ie i_{LE} is fixed), type in the co-ordinates of the i node

If j_{LE} = 0 (ie j_{LE} is fixed), type in the co-ordinates of the j node

If k_{Le} = 0 (ie k_{LE} is fixed), type in the co-ordinates of the k node.

NB

When i or j or k are fed in the for the last time, and they are **not zero**, they should be preceded by a **negative sign**, so that the θ_x°, θ_y° and θ_z° displacements are reduced by the techniques of Bruce Irons [22] NEXT LE.

15.3 Example 15.1

Determine the first six resonant frequencies for the plate of Example 14.1 using VIBSHELF.

15.4 Input data file for Example 15.1

The input data file for Example 15.1, namely VIBSHELL INP, is as follows:

VIBSHELL.INP

```
10
30
16
0
0,0,0
.0635,0,0
.127,0,0
.1905,0,0
.0635,.0635,0
.127,.0635,0
.1905,.0635,0
.127,.127,0
.1905,.127,0
.1905,.1905,0
6
1.55E-3,2.067E11,.3,7890
-1,2,5
2,6,5
-2,3,6
3,7,6
-3,4,7
4,0,7
-4,0,0
-5,6,8
6,9,8
-6,7,9
7,0,9
-7,0,0
-8,9,10
9,0,10
-9,0,0
-10,0,0
.254,.0635,0
.254,0,0
.254,.0635,0
.254,.127,0
.254,.0635,0
.254,.127,0
.254,.1905,0
.254,.127,0
.254,.1905,0
.254,.1905,0
.254,.254,0
0
EOF
```

251

15.5 Results

The results are as follows:-

Frequencies (Hz) for the cantilevered plate

Mode	VIBSHELF	Experiment
1	36.35	34.5
2	137.80	136.0
3	195.10	190.0
4	349.20	325.0
5	474.40	441.0
6	631.70	578.0

It appears that the reduction process caused VIBSHELF to miss the real mode 6.

15.6   Example 15.2

Using a slightly modified version of VIBSHELF on an ATARI Mega 4, the resonant frequencies of the clamped cylindrical shell of reference [30] page 312, and reference [32] was analysed; see Table 15.2 and screendumps of Figures 15.1 to 15.4.

Table 15.2 Resonant frequencies (Hz) of clamped cylindrical shell

Mode	m,n	Ross [32]	VIBSHELF	Analytical
1	1, 2	830	854	870
2	1, 3	944	966	958
3	1, 3	1288	1290	1288
4	2, 1	1343	1381	1364

From Table 15.2, it can be seen that in general, VIBSHELF gives better results than the partially conforming element of Ross [33],

where m = number of half waves in x° direction
where n = number of half waves in y° direction

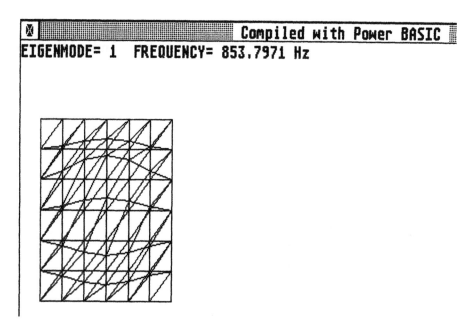

Figure 15.1

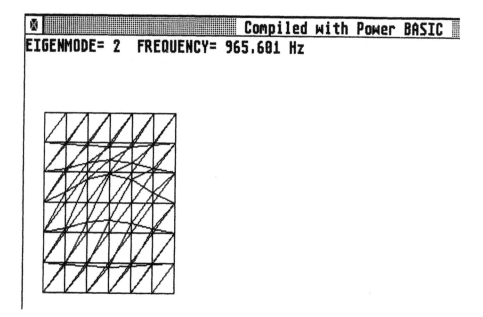

Figure 15.2

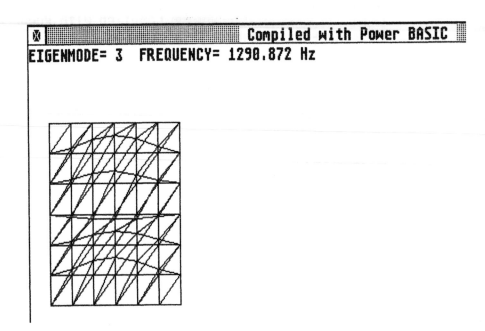

Figure 15.3

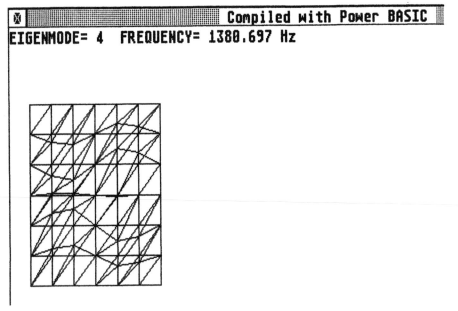

Figure 15.4

CHAPTER 16

STRESSES IN SOLIDS

16.1 Introduction

This chapter describes the computer program TETRAF, which is capable of calculating nodal displacements and elemental stresses in solids of complex shape and with complex boundary conditions.

The program adopts the four node tetrahedron shown in Figure 16.1.

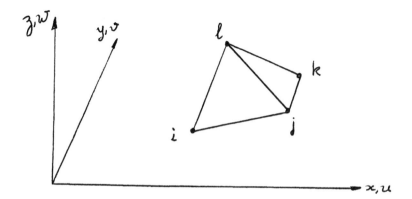

Figure 16.1 - Four node tetrahedron

Each node has 3 degrees of freedom, namely, u, v and w, so that there are a total of 12 degrees of freedom per element. Hence, the elemental stiffness matrix is of order 12 x 12.

The elemental stiffness matrix is derived in reference [7], and it will not be rederived here. Its derivation however, is based on the following assumptions for displacements.

$$u = \alpha_1 + \alpha_2 x + \alpha_3 y + \alpha_z y$$

$$v = \alpha_5 + \alpha_6 x + \alpha_7 y + \alpha_8 z \qquad\qquad 16.1$$

$$w = \alpha_9 + \alpha_{10} x + \alpha_{11} y + \alpha_{12} z$$

From equation 16.1, it can be seen that the element is a constant strain element.

In order to simplify the input of data, the topology is input with the aid of the irregular cubic element, shown in Figure 16.2.

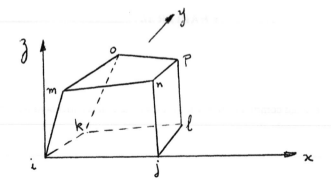

Figure 16.2 - Irregular cubic element

The computer program automatically sub-divides this cube into five tetrahedral elements, [34] with the topology shown in Figure 16.3 and Table 16.1.

Table 16.1 - Topology of the Five tetrahedra

Element	Nodes			
	1	2	3	4
1	i	j	l	n
2	i	l	k	o
3	n	o	m	i
4	n	p	o	l
5	i	l	n	o

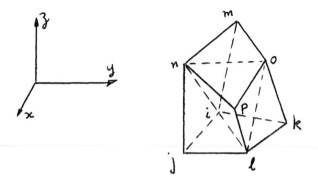

Figure 16.3 Five Tetrahedra

256

16.2   Data

The input data file should be prepared in the following manner:-

NN = number of nodes
ES = number of **cubic** elements
NF = number of nodes with zero degrees of freedom
E = Young's modulus
NU = ν = Poisson's ratio

Nodal co-ordinates

For i = 1 to NN
Type in x_i ,y_i and z_i (the nodal co-ordinates)
NEXT i

Details of zero displacements

For i = 1 to NF
Type in the node number of the node with the zero displacements
If u = 0 at the node, type 1; else type 0
If v = 0 at this node, type 1; else type 0
If w = 0 at this node, type 1; else type 0
NEXT i

Topology of Cubic Elements

For I = 1 to ES
Type in i, j, k, l, m, n, o and p for the cubic element I, in the **order shown in Figures 16.2 and 16.3**
NEXT I

Nodal Loads

NC = number of nodes with concentrated loads

For i = 1 to NC
Type in the concentrated load in the x direction
Type in the concentrated load in the y direction
Type in the concentrated load in the z direction
NEXT i

16.3   Output

The following results are outputs:-

The **Nodal displacements** u, v and w.

The elemental stresses

$$\sigma_x, \sigma_y, \sigma_z, \tau_{yz}, \tau_{zx} \text{ and } \tau_{xy}$$

where σ_x, σ_y abd σ_z are normal stresses in x, y and z direction.
τ_{yz}, τ_{zx} and τ_{xy} are shear stresses in yz, xz and xy planes.

16.4 Example 16.1

Determine the nodal displacements and elemental stresses for the solid cantilever of
Chandrupatla and Belegundu [33], shown in Figure 16.4.

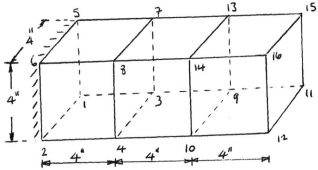

$E = 3E7 \text{ lbf/in}^2 \ \nu = 0.3$

Figure 16.4 - "Solid" cantilever

16.5 Input data file

The input data file for Example 16.1, is shown below.

TETRAF.DAT

```
16,3,4
3E7,.3
0,4,0,  0,0,0,  10,4,0,  10,0,0
0,4,4,  0,0,4,  10,4,4,  10,0,4
20,4,0,  20,0,0,  30,4,0,  30,0,0
20,4,4,  20,0,4,  30,4,4,  30,0,4
1,1,1,1,  2,1,1,1
5,1,1,1,  6,1,1,1
1,2,3,4,5,6,7,8
3,4,9,10,7,8,13,14
9,10,11,12,13,14,15,16
2
15,0,0,-20000,  16,0,0,-20000
EOF
```

16.6 Results

The output file for Example 16.1 is given below:-

TETRAF.OUT

```
3D STRESS USING A 4 NODE TETRAHEDRON

NUMBER OF NODES= 16
NUMBER OF ELEMENTS= 3
NUMBER OF NODES WITH ZERO DISPLACEMENTS= 4
ELASTIC MODULUS= 3E+07
POISSON'S RATIO= .3

THE NODAL COORDINATES (X,Y & Z) IN GLOBAL COORDINATES ARE:

NODE= 1        X( 1 )= 0       Y( 1 )= 4       Z( 1 )= 0
NODE= 2        X( 2 )= 0       Y( 2 )= 0       Z( 2 )= 0
NODE= 3        X( 3 )= 10      Y( 3 )= 4       Z( 3 )= 0
NODE= 4        X( 4 )= 10      Y( 4 )= 0       Z( 4 )= 0
NODE= 5        X( 5 )= 0       Y( 5 )= 4       Z( 5 )= 4
NODE= 6        X( 6 )= 0       Y( 6 )= 0       Z( 6 )= 4
NODE= 7        X( 7 )= 10      Y( 7 )= 4       Z( 7 )= 4
NODE= 8        X( 8 )= 10      Y( 8 )= 0       Z( 8 )= 4
NODE= 9        X( 9 )= 20      Y( 9 )= 4       Z( 9 )= 0
NODE= 10       X( 10 )= 20     Y( 10 )= 0      Z( 10 )= 0
NODE= 11       X( 11 )= 30     Y( 11 )= 4      Z( 11 )= 0
NODE= 12       X( 12 )= 30     Y( 12 )= 0      Z( 12 )= 0
NODE= 13       X( 13 )= 20     Y( 13 )= 4      Z( 13 )= 4
NODE= 14       X( 14 )= 20     Y( 14 )= 0      Z( 14 )= 4
NODE= 15       X( 15 )= 30     Y( 15 )= 4      Z( 15 )= 4
NODE= 16       X( 16 )= 30     Y( 16 )= 0      Z( 16 )= 4

THE NODAL POSITIONS, ETC. OF THE ZERO DISPLACEMENTS ARE:

NODE= 1
DISPLACEMENT IN X DIRECTION IS ZERO AT NODE   1
DISPLACEMENT IN Y DIRECTION IS ZERO AT NODE   1
DISPLACEMENT IN Z DIRECTION IS ZERO AT NODE   1
NODE= 2
DISPLACEMENT IN X DIRECTION IS ZERO AT NODE   2
DISPLACEMENT IN Y DIRECTION IS ZERO AT NODE   2
DISPLACEMENT IN Z DIRECTION IS ZERO AT NODE   2
NODE= 5
DISPLACEMENT IN X DIRECTION IS ZERO AT NODE   5
DISPLACEMENT IN Y DIRECTION IS ZERO AT NODE   5
DISPLACEMENT IN Z DIRECTION IS ZERO AT NODE   5
NODE= 6
DISPLACEMENT IN X DIRECTION IS ZERO AT NODE   6
DISPLACEMENT IN Y DIRECTION IS ZERO AT NODE   6
DISPLACEMENT IN Z DIRECTION IS ZERO AT NODE   6
```

ELEMENT TOPOLOGY

ELEMENT NO. 1   NODES= 1 - 2 - 3 - 4 - 5 - 6 - 7 - 8
ELEMENT NO. 2   NODES= 3 - 4 - 9 - 10 - 7 - 8 - 13 - 14
ELEMENT NO. 3   NODES= 9 - 10 - 11 - 12 - 13 - 14 - 15 - 16

NODAL POSITIONS & VALUES OF THE CONCENTRATED LOADS
NC=3 = 3
NODE= 15
LOAD IN X DIRECTION= 0
LOAD IN Y DIRECTION= 0
LOAD IN Z DIRECTION=-20000
NODE= 16
LOAD IN X DIRECTION= 0
LOAD IN Y DIRECTION= 0
LOAD IN Z DIRECTION=-20000
TERAHEDRON NO. 1 - 4 - 2 - 6
VOLUME= 26.66667
TERAHEDRON NO. 1 - 4 - 3 - 7
VOLUME= 26.66667
TERAHEDRON NO. 6 - 7 - 5 - 1
VOLUME= 26.66667
TERAHEDRON NO. 6 - 7 - 8 - 4
VOLUME= 26.66667
TERAHEDRON NO. 1 - 4 - 6 - 7
VOLUME= 53.33333
TERAHEDRON NO. 3 - 10 - 4 - 8
VOLUME= 26.66667
TERAHEDRON NO. 3 - 10 - 9 - 13
VOLUME= 26.66667
TERAHEDRON NO. 8 - 13 - 7 - 3
VOLUME= 26.66667
TERAHEDRON NO. 8 - 13 - 14 - 10
VOLUME= 26.66667
TERAHEDRON NO. 3 - 10 - 8 - 13
VOLUME= 53.33333
TERAHEDRON NO. 9 - 12 - 10 - 14
VOLUME= 26.66667
TERAHEDRON NO. 9 - 12 - 11 - 15
VOLUME= 26.66667
TERAHEDRON NO. 14 - 15 - 13 - 9
VOLUME= 26.66667
TERAHEDRON NO. 14 - 15 - 16 - 12
VOLUME= 26.66667
TERAHEDRON NO. 9 - 12 - 14 - 15
VOLUME= 53.33333

THE NODAL VALUES OF THE DISPLACEMENTS ARE:-

NODE= 1   U(1)=-2.030434E-13   V(1)= 2.207637E-14
W(1)= 3.761135E-14
NODE= 2   U(2)=-3.048412E-13   V(2)=-5.871623E-14
W(2)=-8.454651E-14
NODE= 3   U(3)=-3.475938E-03   V(3)=-3.040355E-03
W(3)=-1.143339E-02
NODE= 4   U(4)=-3.641875E-03   V(4)=-3.44407E-03
W(4)=-1.096714E-02

260

```
NODE= 5       U( 5 )= 3.045654E-13        V( 5 )=-5.775599E-14
W( 5 )=-8.536443E-14
NODE= 6       U( 6 )= 2.032267E-13        V( 6 )= 5.557179E-14
W( 6 )= 4.590458E-14
NODE= 7       U( 7 )= 3.637974E-03        V( 7 )=-3.17903E-03
W( 7 )=-1.121999E-02
NODE= 8       U( 8 )= 3.518077E-03        V( 8 )=-2.627677E-03
W( 8 )=-1.107717E-02
NODE= 9       U( 9 )=-5.42732E-03         V( 9 )=-2.928398E-03
W( 9 )=-.0372521
NODE= 10      U( 10 )=-6.058523E-03       V( 10 )=-3.151382E-03
W( 10 )=-3.667532E-02
'CUBE' NO. 1 - 2 - 3 - 4 - 5 - 6 - 7 - 8
THE STRESSES IN TETRAHEDRON NO.  1 - 4 - 2 - 6  ARE:-
SIGMAX=-14707.57              SIGMAY=-6303.245              SIGMAZ=-6303.245
TAU Y-Z= 6.820554E-07         TAU X-Z=-12654.39             TAU X-Y=-3973.927
THE STRESSES IN TETRAHEDRON NO.  1 - 4 - 3 - 7  ARE:-
SIGMAX=-11367.22              SIGMAY=-1016.704              SIGMAZ=-2114.656
TAU Y-Z=-1744.988             TAU X-Z= 7328.522             TAU X-Y=-3029.439
THE STRESSES IN TETRAHEDRON NO.  6 - 7 - 5 - 1  ARE:-
SIGMAX= 14691.82              SIGMAY= 6296.494              SIGMAZ= 6296.494
TAU Y-Z=-6.089462E-07         TAU X-Z=-12946.14             TAU X-Y=-3668.112
THE STRESSES IN TETRAHEDRON NO.  6 - 7 - 8 - 4  ARE:-
SIGMAX= 11345.85              SIGMAY= 46.32761              SIGMAZ= 2592.395
TAU Y-Z= 1943.007             TAU X-Z= 7872.352             TAU X-Y=-2686.079
THE STRESSES IN TETRAHEDRON NO.  1 - 4 - 6 - 7  ARE:-
SIGMAX= 18.49163              SIGMAY= 787.5307              SIGMAZ=-706.3827
TAU Y-Z= 17.58042             TAU X-Z=-2300.485             TAU X-Y= 6678.762
'CUBE' NO. 3 - 4 - 9 - 10 - 7 - 8 - 13 - 14
THE STRESSES IN TETRAHEDRON NO.  3 - 10 - 4 - 8  ARE:-
SIGMAX=-8488.807              SIGMAY=-582.803               SIGMAZ=-3546.736
TAU Y-Z= 1010.018             TAU X-Z=-9009.575             TAU X-Y=-2691.722
THE STRESSES IN TETRAHEDRON NO.  3 - 10 - 9 - 13  ARE:-
SIGMAX=-6191.469              SIGMAY=-401.828               SIGMAZ=-722.5731
TAU Y-Z=-1118.174             TAU X-Z= 3480.688             TAU X-Y=-1558.146
THE STRESSES IN TETRAHEDRON NO.  8 - 13 - 7 - 3  ARE:-
SIGMAX= 8507.982              SIGMAY=-370.1935              SIGMAZ= 4041.865
TAU Y-Z=-811.9935             TAU X-Z=-9323.016             TAU X-Y=-2178.631
THE STRESSES IN TETRAHEDRON NO.  8 - 13 - 14 - 10  ARE:-
SIGMAX= 6379.831              SIGMAY= 295.2995              SIGMAZ= 829.9443
TAU Y-Z= 1177.222             TAU X-Z= 3717.958             TAU X-Y=-1215.731
THE STRESSES IN TETRAHEDRON NO.  3 - 10 - 8 - 13  ARE:-
SIGMAX=-103.8281              SIGMAY=-112.4531              SIGMAZ=-264.3022
TAU Y-Z= 85.38406             TAU X-Z=-1933.337             TAU X-Y= 3822.098
'CUBE' NO. 9 - 10 - 11 - 12 - 13 - 14 - 15 - 16
THE STRESSES IN TETRAHEDRON NO.  9 - 12 - 10 - 14  ARE:-
SIGMAX=-3460.956              SIGMAY=-32.04973              SIGMAZ=-2220.507
TAU Y-Z= 243.3718             TAU X-Z=-4832.303             TAU X-Y=-1268.969
THE STRESSES IN TETRAHEDRON NO.  9 - 12 - 11 - 15  ARE:-
SIGMAX=-1528.855              SIGMAY=-283.3555              SIGMAZ=-548.6563
TAU Y-Z=-293.804              TAU X-Z=-637.4532             TAU X-Y=-25.93614
THE STRESSES IN TETRAHEDRON NO.  14 - 15 - 13 - 9  ARE:-
SIGMAX= 3810.511              SIGMAY=-5.394853              SIGMAZ= 2396.959
TAU Y-Z=-184.3216             TAU X-Z=-5035.489             TAU X-Y=-955.7006
THE STRESSES IN TETRAHEDRON NO.  14 - 15 - 16 - 12  ARE:-
SIGMAX= 1317.493              SIGMAY= 527.7435              SIGMAZ=-2160.063
TAU Y-Z= 578.487              TAU X-Z=-653.6618             TAU X-Y=-126.6868
THE STRESSES IN TETRAHEDRON NO.  9 - 12 - 14 - 15  ARE:-
SIGMAX=-69.11523              SIGMAY=-270.2402              SIGMAZ=-190.4219
TAU Y-Z=-56.92677             TAU X-Z=-1921.053             TAU X-Y= 1188.636
```

16.7 Approximate Checks

From simple beam theory

$$I = \frac{4 \times 4^3}{12} = 21.33 \text{ in}^4$$

Deflections

$$\text{End deflection} = \frac{-40\ 000 \times 12^3}{3 \times 3E7 \times 21.33} = \underline{-0.036 \text{ ins}}$$

This compares favourably with the computed value of - 0.0367 ins

Bending stresses

Maximum bending moment = 40 000 x 12 = 480 000 lbf in

$$\text{Maximum bending stress} = \pm \frac{480\ 000 \times 2}{21.33} = \pm \underline{45\ 007 \text{ lbf/in}^2}$$

This does not compare favourably with the maximum computed value for stress, and may be partly due to the fact that the computed stress is calculated at the centroid of the element.

16.8 Screendumps

Screendumps of the mesh for the structure and its deflected form are shown in Figure 16.5 and 16.6

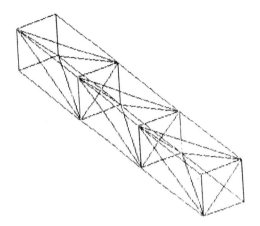

Figure 16.5 - Mesh for "solid" cantilever

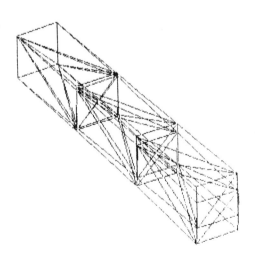

Figure 16.6 - Deflected form of cantilever

CHAPTER 17

TWO DIMENSIONAL FIELD PROBLEMS

17.1   Introduction

This chapter describes three computer programs, namely FIELD3SF, FIELD4SF and FIELD8SF. These programs solve steady state field problems which are described by equation 17.1; this equation is known as **Poisson's equation** when $Q \neq 0$ and as **Llaplace's equation** when $Q = 0$.

$$\left(\frac{\partial^2 \phi}{\partial x^2} + \frac{\partial^2 \phi}{\partial y^2} \right) + Q = 0 \qquad\qquad 17.1$$

where
ϕ = the function to be determined
k = a constant
Q = an input or an output

Equation (17.1) is usually associated with the boundary conditions of equations (17.2).

(a) $\phi = \phi_B$ on the boundary

(b)   and/or K $\left[l_x \dfrac{\partial \phi}{\partial x} + l_y \dfrac{\partial \phi}{\partial y} \right]$ $= q_G + h_L(\phi - \phi_\infty) = 0$ $\qquad$ 17.2

where,

l_x and l_y are the direction cosines of a vector that is normal to the surface.

17.2   Torsion

For the torsion of non-circular sections, equation 17.1, becomes

$$\frac{\partial^2 \phi}{\partial x^2} + \frac{\partial^2 \phi}{\partial y^2} + 2 = 0 \qquad\qquad 17.3$$

with the boundary conditions,

(a) $\phi = \phi_B$ on the boundary

(b)   and/or $\dfrac{\partial \phi}{\partial n} = l_x \dfrac{\partial \phi}{\partial x} + l_y \dfrac{\partial \phi}{\partial y} = 0$

The condition (b) is naturally achieved on a 'free' boundary,

where,

ϕ = shear stress function.

From torsion theory the following apply:

$$J = \text{torsional constant}$$

$$= 2 \text{ x volume under } \phi$$

$$= 2 \int_{\text{Area}} \phi \ d(A)$$

Now,

$$T = G\theta J$$

$$= 2G\theta \int_{\text{Area}} \phi \ d(A)$$

where,

θ = angle of twist/unit length.

The shearing stresses are given by:

$$\tau_{xz} = G\theta \ \dfrac{\partial \phi}{\partial y}$$

$$\tau_{yz} = -G\theta \ \dfrac{\partial \phi}{\partial x}$$

$\}$ See Figure 17.1

$$\tau = \sqrt{\tau_{xz}^2 + \tau_{yz}^2} = \text{resultant shear stress}$$

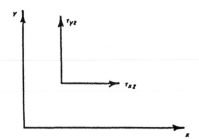

Fig 17.1 - Shear Stresses

17.3   Heat Transfer

For steady-state heat transfer, 17.1 becomes

$$K\left[\frac{\partial^2 T}{\partial x^2} + \frac{\partial^2 T}{\partial y^2}\right] + Q = 0 \qquad\qquad 17.4$$

with boundary conditions,

(a) $T = T_B$, where T_B = boundary temperature

(b)   and/or

$$K\left[l_x \frac{\partial T}{\partial x} + l_y \frac{\partial T}{\partial y}\right] + h_L (T - T_\infty) + q_G = 0$$

where,

T = Temperature (°K)
T_∞ = known fluid temperature (°K)
K = Conductivity (kW/m °K)
Q = Heat generated with the body (kW/m³
($+^{ve}$ if heat is put into the body)
h_L = Convection coefficient (kw/m²°K)
q_G = Heat flux (kW/m²) ($+^{ve}$ when heat is moving out of the body).

NB The heat flux q_G and the convection loss h_L $(T - T_\infty)$ must not take place over the same boundary.

266

17.4   Stream Functions

For the irrotational flow of an ideal fluid, equation 17.1 becomes,

$$\frac{\partial^2 \psi}{\partial x^2} + \frac{\partial^2 \psi}{\partial y^2} = 0 \qquad\qquad 17.5$$

where
ψ = stream function

The flow velocities are obtained from

$$V_x = \frac{\partial \psi}{\partial y} \text{ and } V_y = -\frac{\partial \psi}{\partial x}$$

and the volume rate,
$Q = \psi_j - \psi_i$

where,

ψ_j and ψ_i are adjacent stream functions, and,
Q = flow rate/depth in z direction.

An alternative formulation

$$\frac{\partial^2 \phi}{\partial x^2} + \frac{\partial^2 \phi}{\partial y^2} = 0$$

$$V_x = \frac{\partial \phi}{\partial y} \text{ and } V_y = -\frac{\partial \phi}{\partial x}$$

Also

$$V_n = l_x \frac{\partial \phi}{\partial x} + l_y \frac{\partial \phi}{\partial y} = \frac{\partial \phi}{\partial n} = 0$$

where,

ϕ = velocity potential function.

It should be noted that the condition $\partial \phi / \partial n = 0$ occurs naturally over a 'free' boundary.

17.5   Ground water flow

For ground water flow in the horizontal x - y plane of a confined aquifer, the symbols in equation 17.1 have the following definitions:

K = coefficient of permeability (m^3/day/m^2)
ϕ = piezometric pressure head (m), measured from the bottom of the aquifer
Q = recharge (m^3/day) - when pumping, Q becomes -^{ve}.

The boundary conditions are:

(a) $\phi = \phi_B$ (boundary values of piezometric pressure head)

(b)   and/or K $\dfrac{\partial \phi}{\partial n}$ + q = 0

where,

q = seepage term representing water moving out of the boundary (m^3/day).

17.6   Electrical Field Problems

In this case, Poisson's equation becomes:-

$$\epsilon \left[\frac{\partial^2 u}{dx^2} + \frac{\partial^2 u}{\partial y^2} \right] = - \rho \qquad\qquad 17.6$$

where u = a on a boundary S_1

and u = b on a boundary S_2

where for an isotropic dielectric medium,

ϵ = k = permittivity (F/m)
ρ = Q = volume charge density (C/m^3)
u = ϕ = the electric potential to be determined (V).

It should be noted that the permittivity "ϵ" is defined in terms of the relative permittivity "ϵ_R" and the permittivity of free space "ϵ_o",

that is, $\epsilon = \epsilon_R \epsilon_o$

where ϵ_o = 8 .854 x 10^{-12} F/m

For **rubber**, $\epsilon_R = 2.5$ to 3.0

For a typical co-axial cable, assumed $\rho = 0$

For an **Electrostatic field**

$u = \phi =$- Electric force field intensity, to be determined.

$\epsilon = k =$ permittivities

17.7 Magnetic field problems

In this case, Poisson's equation takes the form:-

$$\mu \left[\frac{\partial^2 u}{\partial x^2} + \frac{\partial^2 u}{\partial y^2} \right] = 0 \qquad\qquad 17.7$$

where,

$\mu = k =$ permeability (H/m)

$u = \phi$ - the magnetic field potential, to be determined (A).

It should be noted that the permeability "μ" is defined in terms of the relative permeability "μ_R" and the permeability of free space "μ_o"

$\mu = \mu_R \, \mu_o$

where

$\mu_o = 4\pi \times 10^{-7}$ H/m

For pure iron, $\mu_R = 4000$

For Al or Cu, $\mu_R = 1$

Each of the three programs will now be described

17.8 The computer program FIELD3SF

This computer program adopts a simplex triangular element with corner nodes, with one degree of freedom per node, as shown in Figure 17.2, that is, the value of the unknown ϕ should be plotted perpendicularly to the plane of the triangle, as shown in Figure 17.3.

A linear distribution was assumed for ϕ over a triangular element, so that, the volume under any element $= (\phi_i + \phi_j + \phi_k) A/3$, where A = area of triangle.

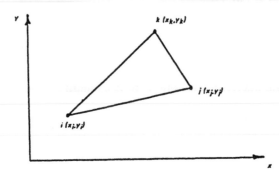

Figure 17.2 - Triangular element

The program calculates nodal values of ϕ and, (if required), slopes of ϕ in both x and y directions at the centroid of each element.

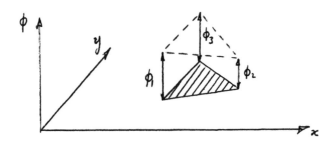

Figure 17.3 - Variation of ϕ over a triangular element

17.9 Data

The data should be prepared in the following sequence:-

MS = number of elements

NN = number of nodes

NF = number of known nodal values of the function ϕ on the boundaries

ZK = number of different values of K

If Q = constant over the entire surface, type 1; else type zero.
If Q = constant over the entire surface, type in the value of Q
K = a constant (see appropriate equation)

270

If h_L has a value on any boundary, type 1; else type zero
if q_G has a value on any boundary, type 1; else type zero

Nodal points describing each element

-      ME = 1(1)MS
|
→ i_{ME} j_{ME} k_{ME} - anti-clockwise order

Nodal co-ordinates

-      i = 1(1)NN
|
→ x_i y_i

Boundary conditions

-      i = 1(1) NF
| NS_i - nodal position of boundary
→ ϕ_B - value of function at the above node

Other element details

For ME = 1 to MS

If Q is not constant over the entire surface, Type in the value of Q for this element

If there is **more** than one value of "K", type in the KTYPE (ie 1, 2, 3 etc).

If h_L or q_G have a value on any boundary for this element; type Y; else type any other key; unless both h_L and q_G are zero throughout.

Then type,

either, LH = Number of boundaries on which h_L has a value

-      I = 1(1)LH
| i   j - nodes defining boundary
| h_L
→   ∞

-      I = 1(1)GQ
| i   j - nodes defining boundary
| q_G
→   Next ME

Do you require the **slopes** of the Function? Type "Y" or "N"

17.10 Output

$\phi_1 \quad \phi_2 \quad \phi_3 \quad \phi_N$

$\dfrac{\partial \phi}{\partial x}$ and $\dfrac{\partial \phi}{\partial y}$ (optional) at element centroids

NB

For problems involving the distribution of **electric potential**, "q_G" is the externally applied boundary current.

For problems, involving **magnetostatics**, "q_G" is the externally applied magnetic field intensity.

17.11 Example 17.1

Determine the torsional constant for the rectangular cross-section shown in Fig 17.4. The 8 element mesh should be used.

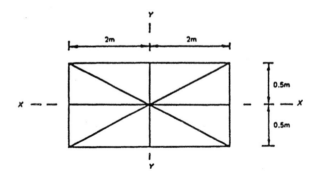

Figure 17.4 - Rectangular section with 8 element

For this case, ϕ = zero on the boundaries, and the problem is symmetrical about the XX and YY axes. Hence, analysis can be carried out by considering only one quarter of the section, as shown in Fig 17.5.

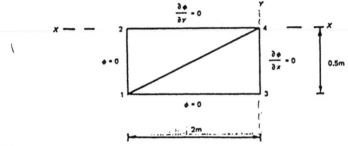

Fig 17.5

The data should be fed in as follows:

MS = 2
NN = 4
NF = 3
ZK = 1   - There is only one element type

1   Q = 2

K = 1
          0   0

i	j	1
4	2	1
4	1	3

Nodal co-ordinates

x_i	y_i
0	0
0	0.5
2	0
2	0.5

Boundary conditions

Nodes Values of functions

1	0
2	0
3	0

"N"

17.12 Results

$$\phi_q = \phi_2 = \phi_3 = 0$$
$$\phi_4 = 0.3137$$

Using numerical integration (that is, Simpson's rule),

J = 2 x volume under function
J = 1.115 m^4

From 'exact' solution, J = 1.12 m^4

17.13 Example 17.2

Determine the nodal temperatures for the heat transfer problem of Figure 17.6..

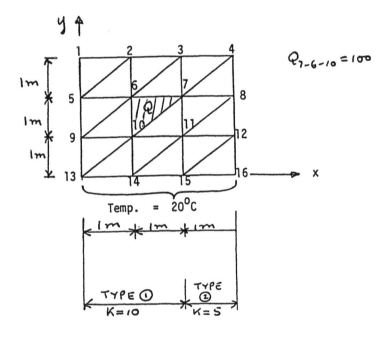

Figure 17.6 - Heat transfer problem

The input data should be prepared as follows:-

MS = Number of elements = 18

NN = Number of nodes = 16

NF = Number of known boundary values = 4

ZK = Number of different values of "K" = 2

0 - "Q" not constant

$K = 10$ - Type (1)

$K = 5$ - Type (2)

0 - No h_L on boundary

0 - No q_G on boundary

Nodes defining each element (anti-clockwise)

i	j	k
2	1	5
2	5	6
3	2	6
3	6	7
4	3	7
4	7	8
6	5	9
6	9	10
7	6	10
7	10	11
8	7	11
8	11	12
10	9	13
10	13	14
11	10	14
11	14	15
12	11	15
12	15	16

Nodal co-ordinates

x	y
0	3
1	3
2	3
3	3
0	2
1	2
2	2
3	2
0	1
1	1
2	1
3	1
0	0
1	0
2	0
3	0

Boundary Values

Node	Value
13	20
14	20
15	20
16	20

Element details

K-Type	Q
1	0
1	0
1	0
1	0
2	0
2	0
1	0
1	0
1	100
1	0
2	0
2	0
1	0
1	0
1	0
1	0
2	0
2	0
"N"	

17.14 Results

Node	Temperature °C
1	23.18
2	23.37
3	23.43
4	23.21
5	22.98
6	23.44
7	23.54
8	23.00
9	21.85
10	22.21
11	21.91
12	21.70
13	20
14	20
15	20
16	20

17.15 FIELD4SF

This computer program solves two-dimensional steady state field problems, governed by equations of the Laplace and Poisson type. It adopts a four node isoparametric quadrilateral element, which has one degree of freedom per node, making a total of four degrees of freedom per element, as shown in Figures 17.7 and 17.8.

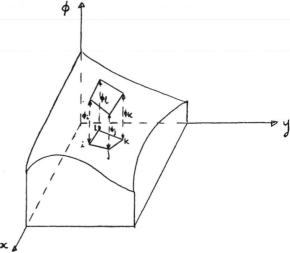

Figure 17.7 - Field Problem

The differential equation solved by this program and also by FIELD8SF is shown by equation 17.8.

$$\frac{\partial}{\partial x} (K_x \frac{\partial \phi}{\partial x}) + \frac{\partial}{\partial y} (K_y \frac{\partial \phi}{\partial y}) + Q = 0 \qquad 17.8$$

with the boundary conditions

$$\phi = \phi_B \text{ on boundary } S_1 \qquad\qquad\qquad\qquad 17.9$$

and/or

$$K_x \frac{\partial \phi}{\partial x} \ell_x + K_y \frac{\partial \phi}{\partial y} \ell_y + q_G + h_L (\phi - \phi_\infty) = 0 \qquad 17.10$$

on boundaries S_2, S_3, etc.

where ℓ_x and ℓ_y are directional cosines of a vector that is normal to the boundary, and the other symbols represent different quantities, depending on the problem.

These equations can be seem to be of similar form to equations 17.1 and 17.2.

To use these quadrilateral elements, it is convenient to obtain the relationship between the rectangular co-ordinates x and y and the non-dimensional co-ordinates ξ and η, as shown in Figure 17.8, where ξ points in the direction form i to j and η from i to ℓ.

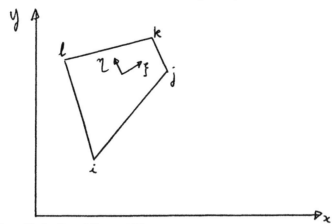

Figure 17.8 - Four Node Quadrilateral Element

The values of ξ and η at the nodes are given below.

Node	ξ	η
i	-1	-1
j	1	-1
k	1	1
ℓ	-1	1

Integration is carried out in each of the two directions, using two gauss points per direction.

17.16 Data

The data should be prepared in the following manner:-

(1)   NN = number of nodes.

(2)   ES = number of elements.

(3)   NF = number of known values of "ϕ" at the boundary nodes.

(4)   KZ = number of K-Types, ie the total number of different values for K_x or K_y.

(5)   If KZ = 1 then Input K_x and K_y

279

(6)   If KZ > 1 then

For I - 1 to KZ

Input K_x (I) and K_y (I)

NEXT I

(7)   **Nodal co-ordinates**

For I = 1 to NN

Input x (I) y(I)

Next I

(8)   **Boundary Conditions**

If NF = 0, ignore (8)

For I = 1 to NF

Input NS(I) - Boundary node

Input W(I) - value of "ϕ" at this node

NEXT I

(9)   If h_L has a value on any boundary, type 1; else 0

(10)   If Q_G has a value on any boundary, type 1; else 0

(11)   If q = 0 or a constant over the entire surface, type 1; else 0

(12)   If Q = 0 or a constant over the entire surface, type in the **value of Q**.

(13)   **Element details**

For EL = 1 to ES

Input i_{EL}, j_{EL}, k_{EL}, ℓ_{EL}

(Element nodes, fed in an anti-clockwise direction)

NEXT EL

(14)   For EL = 1 to ES

(14a) If KZ > 1, then Input K-Type for **this particular element**

(14b) If Q = 0 or a constant over the entire surface, ignore (14b); else type in the value
of **"Q" for this particular element.**

(14c) If h_L and q_G are both zero, ignore (14c), else if either h_L or q_G has a value on any
boundary for this element, type in "Y" or "N"; then type in:-

either LH = number of boundaries in which h_L has a value, (must be < 4), followed
by

For I = 1 to LH

Input i j

nodes defining the boundary.

Input h_L, ϕ_∞ - values referring to the above boundary

NEXt I

and/or

GQ = number of boundaries on which q_G has a value, followed by

For I = 1 to GQ

Input i j

nodes defining boundary.

Input q_G - value referring to the above boundary

NEXT I

NEXT EL

NB h_L and q_G cannot both have values on the same boundary.

17.17 Results

For I = 1 to NN

Print $\phi(I)$

NEXT I

17.18 Example 17.3

Determine the torsional constant for a torque bar, which has the solid rectangular cross-section shown in Figure 17.9, where four elements are used.

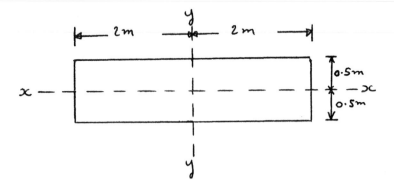

Figure 17.9 - Rectangular Cross-Section

For this problem, the value of the shear stress function ψ will be zero on the external boundaries, and as the problem is symmetrical about both xx and yy, it will be convenient to consider only one quadrant of the section, as shown in Figure 17.10. The boundaries 2-4 and 3-4 are assumed "free" and the condition $\dfrac{\partial \psi}{\partial n} = 0$, occurs naturally along them.

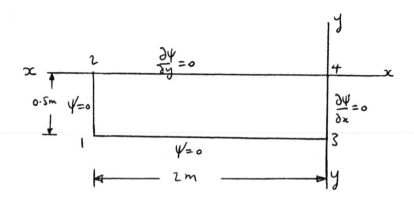

Figure 17.10

The data should be prepared in the following sequence:-

NN = 4

ES = 1

NF = 3

KZ = 1

KX = 1

KY = 1

Nodal co-ordinates

x_i	y_i
0	0
0	0.5
2	0
2	0.5

Boundary conditions

$NS_1 = 1$	$W_1 = 0$
$NS_2 = 2$	$W_2 = 0$
$NS_3 = 3$	$W_3 = 0$

h_L has no value, so type in 0.

q_G has no value, so type in 0.

Q is constant, so type in 1.

Q = 2

Element details

1,3,4,2

Nodes defining the element, fed in an anti-clockwise direction.

ξ will point in direction from 1 to 3, and

η will point in direction from 3 to 4.

17.19 Results

The results for ψ are given below, together with those obtained from the triangular element of Section 17.8.

Node	Quadrilateral ψ_i	Triangular ψ_i
1	0	0
2	0	0
3	0	0
4	0.3529	0.3137

To calculate the torsional constant "J", integration over the surface will be carried out using Simpson's rule.

$$\text{Vol} = \frac{2}{3} (1 * 0 + \frac{4 * 0.5}{3} (1 * 0 + 4 * \psi_4 + 1 * 0) + 1 * 0) = \frac{16 \psi_4}{9}$$

$$J = 2 * \text{ volume under "}\psi\text{"}$$

$$= \frac{32}{9} \psi_4 = 1.255 \text{ m}^4, \text{ which compares favourably with the "exact" value of } 1.12 \text{ m}^4.$$

17.20 Example 17.4

Determine the nodal values of the stream functions for the irrotational fluid flow problem of Figure 17.11. It may be assumed that the value of stream function along the bottom edge are zero, and those along the top edge are 10, and also that the problem is of unit thickness.

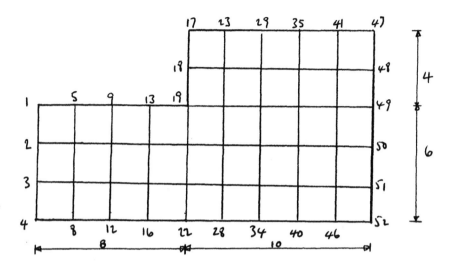

Figure 17.11 - Fluid Flow Problem

17.21 Input data file

The input data file for Example 17.4, namely, FIELD4,INP, is as follows:

FIELD4.INP

```
52
37
22
1,  1,1
0,6
0,4
0,2
0,0
2,6
2,4
2,2
2,0
4,6
4,4
4,2
4,0
6,6
6,4
6,2
6,0
8,10
8,8
8,6
8,4
8,2
8,0
10,10
10,8
10,6
10,4
10,2
10,0
```

285

```
12,10
12,8
12,6
12,4
12,2
12,0
14,10
14,8
14,6
14,4
14,2
14,0
16,10
16,8
16,6
16,4
16,2
16,0
18,10
18,8
18,6
18,4
18,2
18,0

4,0
8,0
12,0
16,0
22,0
28,0
34,0
40,0
46,0
52,0
1,10
5,10
9,10
13,10
19,10
18,10
17,10
23,10
29,10
35,10
41,10
47,10
0,0
1,0
1,2,6,5
2,3,7,6
3,4,8,7
5,6,10,9
6,7,11,10
7,8,12,11
9,10,14,13
10,11,15,14
11,12,16,15
13,14,20,19
14,15,21,20
15,16,22,21
17,18,24,23
18,19,25,24
19,20,26,25
20,21,27,26
21,22,28,27
23,24,30,29
24,25,31,30
25,26,32,31
26,27,33,32
27,28,34,33
29,30,36,35
30,31,37,36
31,32,38,37
32,33,39,38
33,34,40,39
```

```
35,36,42,41
36,37,43,42
37,38,44,43
38,39,45,44
39,40,46,45
41,42,48,47
42,43,49,48
43,44,50,49
44,45,51,50
45,46,52,51
EOF
```

17.22 Results

The nodal values of "ψ" are given below, where they are compared with those obtained form the triangular element.

Node	quadrilateral ψ	Triangular ψ
1	10	10
2	6.657	6.65
3	3.324	3.32
4	0	0
5	10	10
6	6.650	6.64
7	3.317	3.31
8	0	0
9	10	10
10	6.619	6.61
11	3.285	3.28
12	0	0
13	10	10
14	6.515	6.52

15	3.182	3.21
16	0	0
17	10	10
18	10	10
19	10	10
20	6.05	6.25
21	2.984	3.05
22	0	0
23	10	10
24	9.118	9.12
25	7.588	7.88
26	5,421	5.42
27	2.708	2.73
28	0	0
29	10	10
30	8.531	8.61
31	6.822	6.96
32	4.765	4.84
33	2.448	2.47
34	0	0
35	10	10
36	8.279	8.34
37	6.442	6.54
38	4.430	4.50
39	2.259	2.29
40	0	0
41	10	10
42	8.169	8.21
43	6.271	6.54
44	4.268	4.33
45	2.164	2.20
46	0	0
47	10	10
48	8.138	8.18
49	6.222	6.28
50	4.220	4.28
51	2.136	2.17
52	0	0

These two columns of values for "ψ" compare favourably.

17.23 Example 17.5

Determine the temperature distribution for the heat conduction problem of Figure 17.12, where the temperature at node 9 is 50°C. Four equal sized elemental may be used and unit thickness may be assumed.

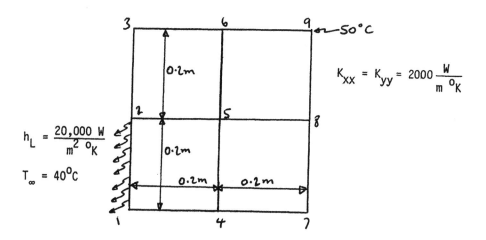

Figure 17.12 - Heat conduction Problem

The data should be typed in as follows:-

NN = 9

ES = 4

NF = 1

KZ = 1

KX = 2000

KY = 2000

x_i	y_i
0	0
0	0.2
0	0.4
0.2	0
0.2	0.2
0.2	0.4
0.4	0
0.4	0.2
0.4	0.4

$NS_1 = 9$

$W_1 = 50$

1   0
1

$Q = 0$

Element Details

1, 4, 5, 2

2, 5, 6, 3

4, 7, 8, 5

5, 8, 9, 6

Y   1

1   2   nodes defining the boundary on which h_L has a value

$h_L = 20,000$

$T_\infty = 40$

N

N

N   Elements 2 to 4

N

17.24 Results

The results are given in Table 17, where they are compared with the results obtained from the triangular element and mesh of Figure 17.13.

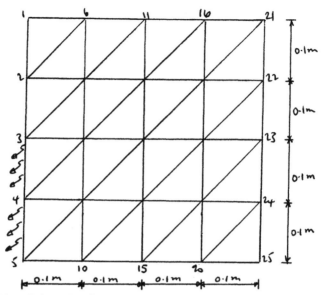

Figure 17.13 - Mesh Using Triangular Element

Table 17.1 - Values of Nodal Temperature

i	Quadrilateral T_i	i	Triangular T_i
1	41.26	1	41.34
2	42.34	3	42.38
3	43.95	5	43.86
4	43.62	11	43.50
5	44.39	13	44.12
6	44.67	15	45.05
7	44.34	21	44.30
8	44.96	23	45.20
9	50	25	50

From Table 17.1, it can be seen that the less refined 4 element quadrilateral mesh compared favourably withe the 32 element triangular mesh. This fact was found to be particularly encouraging, as the input for the 4 node quadrilateral was considerably simpler than that of the triangular element.

17.25 Example 17.6

Determine the temperature distribution for the heat conduction problem of Example 17.5, given that:

$K_{xx} = K_{yy} = 1500$ W/m° K for elements 1, 4, 5, 2 and 4, 7, 8, 4

$K_{xx} = K_{yy} = 2500$ W/m° K for elements 2, 5, 6, 3 and 5, 8, 9, 6

The data should be fed in as follows:-

NN = 9

ES = 4

NF = 1

KZ = 2 - There are two K-Types

K_x = 1500 K_y = 1500 for K-Type 1

K_x = 2500 K_y = 2500 for K-Type 2

x_i	y_i
0	0
0	0.2
0	0.4
0.2	0
0.2	0.2
0.2	0.4
0.4	0
0.4	0.2
0.4	0.4

NS_1 = 9

W_1 = 50

1	0
1	Q = 0

Element details

1, 4, 5, 2

2, 5, 6, 3

4, 7, 8, 5

5, 8, 9, 6

KTYPE =   1 for element 1, 4, 5, 2

Y	1	1	2	20,000	40
KTYPE =	2 for element 2, 5, 6, 3				

N

KTYPE =   1 for element 4, 7, 8, 5

N

KTYPE =   2 for element 5, 8, 9, 6

N

17.26 Results

i	T_i
1	41.04
2	42.85
3	44.60
4	44.20
5	45.15
6	45.26
7	45.06
8	45.76
9	50.00

17.27 FIELD8SF

This computer program solves steady state field problems governed by equations of the Laplace and Poisson type. It adopts an eight node isoparametric quadrilateral element, which has one degree of freedom per node, making a total of eight degrees of freedom per element, as shown in Figure 17.14.

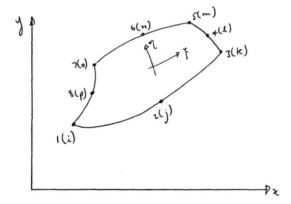

Figure 17.14 - Eight Node Quadrilateral Element

The differential equations and boundary conditions are in equations 17.8 to 17.10.

The value of ξ and η at the nodes are as follows:-

Node	ϵ	n
i	-1	-1
j	0	-1
k	1	-1
l	1	0
m	1	1
n	0	1
o	-1	1
p	-1	0

17.38 To calculate the length of a boundary governed by (say) nodes, 1, 2 and 3

It is necessary to determine an expression for calculating the length of a second order curve, defined by three nodes, as this is required for determining the effects of q_G or h_L (convection) on any boundary of the 8 node isoparametric element.

This is usually difficult, because of the curvature of the boundary, but the length can be calculated numerically by the following method.

This procedure was found to be quite economical, when used for the program described in this chapter.

Consider a second order curve, defined by the nodes 1, 2 and 3, as shown in Figure 17.15, where the co-ordinates are (x_1, y_1), (x_2, y_2) and (x_3, y_3) at nodes 1, 2 and 3, respectively.

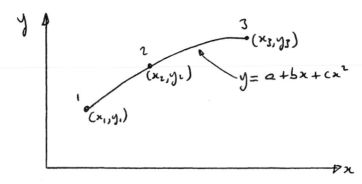

Figure 17.15 - Second Order curve

Let $\quad X_1 = Y_1 = 0$

$\qquad X_2 = x_2 - x_1$

$\qquad Y_2 = y_2 - y_1$

$\qquad X_3 = x_3 - x_1$

$\qquad Y_3 = y_3 - y_1$

By substitution of the above data into the 2nd order equation, the constant a, b and c are determined, as follows:-

$a = 0$

$b = Y_2/X_2 - cX_2$

$c = (Y_3/X_3 - Y_2/X_2)/(X_3 - X_2)$

Now, for this 2nd order equation,

$$L_{123} = \int_0^{X_3} 1[1 + \left(\frac{dy}{dx}\right)^2]^{\frac{1}{2}}dx$$

$$= \int_0^{X_3} [1 + \frac{1}{2}\left(\frac{dy}{dx}\right)^2 +]dx$$

$$L_{123} = (\frac{b^2}{2} + 1)X_3 + bc\ X_3^2 + \frac{2c^2}{3}\ X_3^3$$

This expression is satisfactory for the general case of the 2nd order equation, but breaks down when (a) $X_3 - X_2 = 0$

or   (b) $X_2 = 0$

or   (c) $X_3 = 0$

To overcome these possibilities, the following process was adopted:-

If X2 = 0 and X3= 0 then LL = Y3: GOTO*

If ABS (X3 - X2) = 0 then X2 = 0.99*X3

If X2 = 0 then X2 = 0.001*X3

If X3 = 0 then X3 = 0.001*X2

Rem calculation for LL is now done.

* LL = ABS (LL)

where,

$LL = L_{123}$

The above routine allowed for the possibility that the boundary was vertical, (X2 = X3 = 0), and for three other unusual conditions.

Integration over the surface of the element, was carried out using 3 Gauss points in the ξ direction and 3 in the η direction, making a total of 9 Gauss points per element.

19.29 Data

The data should be prepared in the following manner:

(1)   NN - number of nodes.

(2)   ES = number of elements.

(3)   NF - number of known values of "ϕ" at the boundary nodes.

(4)   KZ - number of different values for K_x or K_y (K-Types).

(5)   If KZ = 1 then Input K_x and K_y - these are constants in the "x" and "y" directions.

(6)   If KZ > 1 then

For I = 1 to KZ

Input K_x (I) and K_y (I)

NEXT I

(7)   **Nodal co-ordinates**

For I = 1 to NN

Input x(I) y (I)

NEXT I

(8)   **Boundary Conditions**

If NF = 0, ignore (8a)

For I = 1 to NF

Input NS(I) - Boundary node

Input W(I) - value of "ϕ" at this node

NEXT I

(9)   If h_L has a value on any boundary, type 1; else 0

(10)   If q_G has a value on any boundary, type 1; else 0

(11)   If Q = 0 or a constant over the entire surface, type 1; else 0

(12)   If Q = 0 or a constant over the entire surface, type in the **value of Q**

(13)   **Element details**

For EL = 1 to ES

Input i_{EL}, j_{EL}, k_{EL}, l_{EL}, m_{EL}, n_{EL}, o_{EL}, p_{EL}

(Element nodes, fed in an anti-clockwise direction, starting from a corner node)

NEXT EL

(14)   For EL = 1 to ES

(14a)   If KZ > 1, then Input K-Type

(14b)   If Q = 0 or a constant over the entire surface, ignore (14b); else type in the value of **"Q" for this particular element**

(14c)   If h_L and q_G are both zero, ignore (14c), else if either h_L or q_g has a value on any boundary for this element; type in "Y" or "N"; then type in:-

either

LH = number of boundaries on which h_L has a value, (must be < - 4), followed by

For I = 1 to LH

Input mid-side node of boundary

Input h_L, ϕ_∞ - values referring to the above boundary

NEXT I

and/or

GQ = number of boundaries in which q_G has a value, followed by

For I = 1 to GQ

Input **mid-side** node of boundary

Input q_G - value on this boundary

NEXT I

NEXT EL

NB h_L and q_G cannot both have values on the same boundary.

17.30 Results

For I = 1 to NN

Print $\phi(I)$

Next I

17.31 Example 17.7

Determine the torsional constant for a torque bar, which has the solid rectangular cross-section shown in Figure 2.3, where a mesh of four symmetrical elements is adopted.

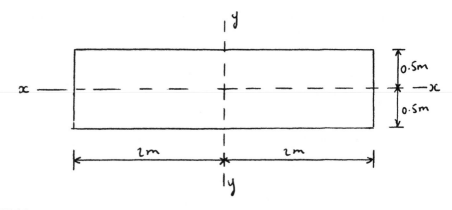

Figure 17.16

298

As the section is symmetrical about xx and yy, only one element need be considered, as shown in Figure 2.4. The condition $\frac{\partial \psi}{\partial n} = 0$ occurs naturally on the "free" boundaries 3-5-8 and 6-7-8.

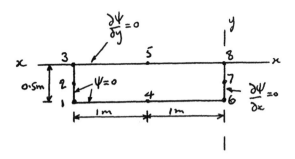

Figure 17.17

The data should be fed in as follows:-

NN = 8

ES = 1

NF = 5

KZ = 1

KX = 1

KY = 1

Nodal co-ordinates

x_i	y_i
0	0
0	0.25
0	0.5
1	0
1	0.5
2	0
2	0.25
2	0.5

Boundary conditions

Nodal position	Value of "ψ"
1	0
2	0
3	0
4	0
6	0

0   0
1

$Q = 2$

Element details

1, 4, 6, 7, 8, 5, 3, 2 - Fed in an **anti-clockwise** order, starting from a **corner node.**

ξ will point in direction from 2 to 7, and

η will point in direction from 4 to 5

NB   If the element nodes were fed in the order: 3, 2, 1, 4, 6, 7, 8, 5, then ξ will have pointed in the direction from 5 to 4 and η from 2 to 7.

17.32 Results

Node	ψ
1	0
2	0
3	0
4	0
5	0.255
6	0
7	0.191
8	0.203

To calculate J, (using Simpson's rule)

	1	2	3	Ordinates 4	5	4^1	6	7	8	7^1	6	
"ψ"	0	0	0	0	.255	0	0	.191	.203	.191	0	
Simpson's multipliers				1	4	1	1	4	2	4	1	

F(Area)	0	1.02	1.92
Area	0	0.17	0.14
SM	1	4	1
F(Vol	0	0.68	0.14

Volume under "ψ" = 1/3 x (0 + 0.68 + 0.14)
= 0.56 m³

J = 2 x volume under "ψ"
= 1.12 m³

which agrees exactly with the analytical value for the torsional constant "J".

It can be seen, however, from the results for "ψ", that there are numerical errors, as ψ_7 was less than ψ_5 and this was not possible.

This deficiency was rectified when the mesh of Figure 17.18 was adopted instead of that of Figure 17.17.

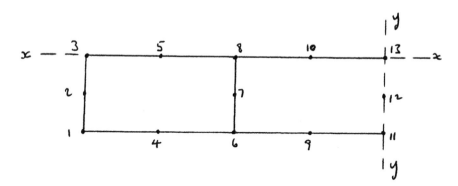

Figure 17.18

The results for "ψ" are as follows:-

| Node | 1 | 2 | 3 | 4 | 5 | 6 | 7 | 8 | 9 | 10 | 11 | 12 | 13 |
| ψ | 0 | 0 | 0 | 0 | .206 | 0 | .184 | .226 | 0 | .246 | 0 | .184 | .251 |

301

17.33 Example 17.8

Determine the temperature distribution for the heat conduction problem of Figure 17.9 given that the temperature at node 21 is 50° and that four equal sized elements are used.

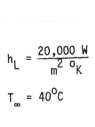

$$h_L = \frac{20,000 \text{ W}}{m^2 \text{ }^\circ K}$$

$$T_\infty = 40^\circ C$$

Figure 17.19

The data should be fed in as follows:-

NN = 21

ES = 4

NF = 1

KZ = 1

KX = 1

KY = 1

x_i	y_i
0	0
0	0.1
0	0.2
0	0.3
0	0.4
0.1	0
0.1	0.2
0.1	0.4
0.2	0
0.2	0.1
0.2	0.2
0.2	0.3
0.2	0.4
0.3	0
0.3	0.2
0.3	0.4
0.4	0
0.4	0.1
0.4	0.2
0.4	0.3
0.4	0.4

$NS_1 = 21$

$W_1 = 50$

1   0

1

$Q = 0$

Element details

5, 4, 3, 7, 11, 12, 13, 8	}
1, 6, 9, 10, 11, 7, 3, 2	} Fed in an anti-clockwise order,
11, 15, 19, 20, 21, 16, 13, 12	} from a corner node **first**
11, 10, 9, 14, 17, 18, 19, 15	}

N
Y   1
2 - mid-side node defining the boundary 1-2-3

303

$T_\infty = 40$
$h_L = 20,000$
N
N

17.34 Results

The results are shown in Table 17.2, where they are compared with those obtained from the simplex triangular element of Section 17.8. The two columns for "ϕ" compare favourably, but it is expected that the 8 node quadrilateral will be superior for the more complex problems.

Table 17.2 - Values of Nodal Temperatures

8 Node Quadrilateral		Simplex Triangle	
NODE i	T_i	NODE i	T_i
1	40.00	1	41.35
2	40.00	2	41.5
3	40.00	3	42.38
4	41.59	4	43.53
5	42.35	5	43.86
6	41.02	6	42.60
7	41.43	8	43.3
8	42.29	10	44.19
9	42.00	11	43.5
10	42.02	12	43.67
11	42.39	13	44.12
12	43.18	14	44.7
13	43.21	15	45.05
14	42.45	16	44.09
15	43.43	18	44.82
16	45.44	20	46.61
17	42.82	21	44.3
18	42.85	22	44.51
19	43.35	23	45.2
20	45.40	24	46.65
21	50.0	25	50

17.35 Example 17.9

Determine the nodal values of the stream functions for the irrotational fluid flow problem of Figure 17.20. It may be assumed that the value of the stream function along the bottom edge from nodes 5 to 45 is zero, and that the value of the stream function along the top edge from nodes 1 to 19 to 17 to 39 is ten.

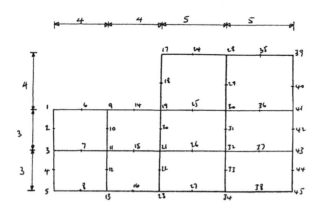

Figure 17.20

17.36 Input Data file

The input data file for Example 17.8, namely, FIELD8,INP, is as follows:-

FIELD8.INP

```
                45
                10
                20
                1,1,1
                0,6
                0,4.5
                0,3
                0,1.5
                0,0
                2,6
                2,3
                2,0
                4,6
                4,4.5
                4,3
                4,1.5
                4,0
                6,6
                6,3
                6,0
                8,10
                8,8
                8,6
                       305
```

```
8,4.5
8,3
8,1.5
8,0
10.5,10
10.5,6
10.5,3
10.5,0
13,10
13,8
13,6
13, 4.5
13,3
13,1.5
13,0
15.5,10
15.5,6
15.5,3
15.5,0
18,10
18,8
18,6
18,4.5
18,3
18,1.5
18,0
5,0
8,0
13,0
16,0
23,0
27,0
34,0
38,0
45,0
1,10
6,10
9,10
14,10
19,10
18,10
17,10
24,10
28,10
35,10
39,10
0,0
1,0
1,2,3,7,11,10,9,6
3,4,5,8,13,12,11,7
9,10,11,15,21,20,19,14
11,12,13,16,23,22,21,15
17,18,19,25,30,29,28,24
19,20,21,26,32,31,30,25
21,22,23,27,34,33,32,26
28,29,30,36,41,40,39,35
30,31,32,37,43,42,41,36
32,33,34,38,45,44,43,37
eof
```

17.37 Results

The nodal values of the stream function are given below, where they can be seen to be of similar magnitude to those predicted by both the simplex triangular element and the four node quadrilateral element. The big advantages of using the 8 node quadrilateral elements are that it is simpler to input the data, and the results are likely to be more precise for the more complex problem.

Node	1	2	3	4	5	6	7	8	9	10
ψ_i	10	7.49	4.98	2.49	0	10	4.97	0	10	7.47

Node	11	12	13	14	15	16	17	18	19	20
ψ_i	4.93	2.45	0	10	4.83	0	10	10	10	7.07
Node	21	22	23	24	25	26	27	28	29	30
ψ_i	4.46	2.21	0	10	7.45	3.98	0	10	8.45	6.58
Node	31	32	33	34	35	36	37	38	39	40
ψ_i	5.15	3.42	1.74	0	10	6.31	3.25	0	10	8.13
Node	41	42	43	44	45					
ψ_i	6.29	4.73	3.22	1.61	0					

Figure 17.21 and 17.22 show the stream functions for the irrotational flow of a fluid through a duct near a sudden contraction.

Two different refinements of mesh were chosen, but the differences between these two appear to be small. The stream functions for four node quadrilateral elements are shown by the dotted lines and those for the eight node quadrilateral, by the full lines.

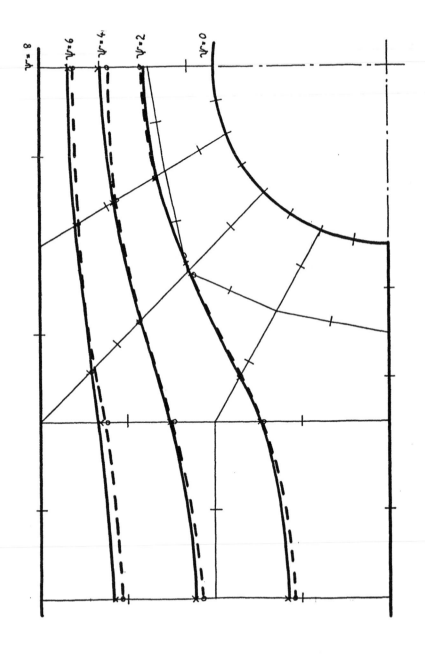

Figure 17.21 - Flow through a duct near a sudden contraction

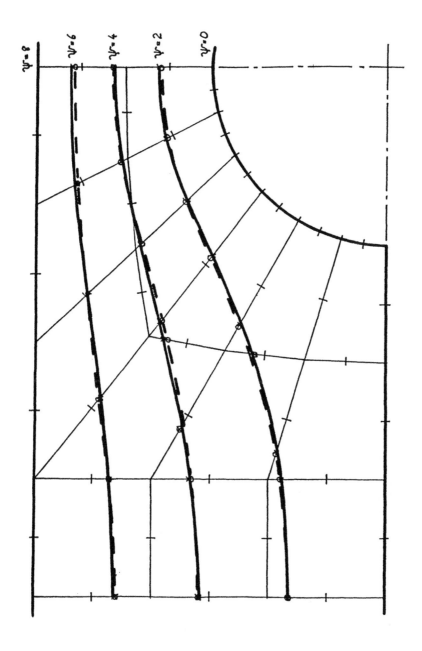

Figure 17.22 - Flow through a duct near a sudden contraction (more refined mesh)

17.38 Example 17.10

Determine the temperature distribution for the heat conduction problem of Example 17.6, using an eight node isoparametric quadrilateral element.

NN = 21

ES = 4

NF = 1

KZ = 2 - There are two K-Types

K_x = 1500 K_y = 1500 for K-TYPE 1

K_x = 2500 K_y = 2500 for K-TYPE 2

x_i	y_i
0	0
0	0.1
0	0.2
0	0.3
0	0.4
0.1	0
0.1	0.2
0.1	0.4
0.2	0
0.2	0.1
0.2	0.2
0.2	0.3
0.2	0.4
0.3	0
0.3	0.2
0.3	0.4
0.4	0
0.4	0.1
0.4	0.2
0.4	0.3
0.4	0.4

$NS_1 = 21$

$W_1 = 50$

1   0

1    $Q = 0$

5, 4, 3, 7, 11, 12, 13, 8
1, 6, 9, 10, 11, 7, 3, 2
11, 15, 19, 20, 21, 16, 13, 12
11, 10, 9, 14, 17, 18, 19, 15

K-TYPE = 2 for element 5, 4, 3, 7, 11, 12, 13, 8

N

K-TYPE = 1 for element 1, 6, 9, 10, 11, 7, 3, 2

Y

1   2

$h_L = 20,000$ $T_\infty = 40$

K-TYPE = 2 for element 11, 15, 19, 20, 21, 16, 13, 12

N

K-TYPE = 1 for element 11, 10, 9, 14, 17, 18, 19, 15

"N"

17.39 Results

i	T_i
1	41.25
2	41.44
3	42.79
4	43.89
5	44.42
6	42.82
7	43.72
8	44.41
9	43.78
10	44.01
11	44.44
12	45.06
13	45.09
14	44.38
15	45.37
16	46.70
17	44.77
18	44.79
19	45.33
20	46.78
21	50.00

17.40 Screendumps

Screendumps of Example 17.4, using the 3 node triangular element of Section 17.8, are shown in Figure 17.23 and 17.24.

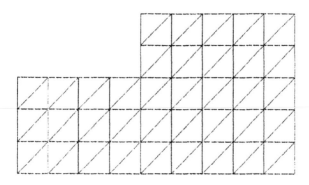

Figure 17.23 - Mesh for flow through a sudden enlargement

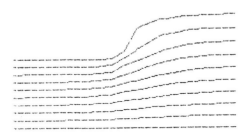

Figure 17.24 - Streamlines for flow through a sudden enlargement

CHAPTER 18

SOLUTION OF HELMHOTZ'S EQUATION

18.1   Introduction

This chapter describes the computer program ACOUSTIC, which is capable of solving Helmholtz's equation in two dimensions. Helmholtz's equation, which is shown in two dimensions, in equation (18.1), describes a number of physical problems, including acoustic vibration, the propagation of electromagnetic waves in waveguides, the vibration of a membrane, the oscillatory motion of a mass of water in a lake or an enclosed harbour, etc, etc.

$$\frac{\partial}{\partial x}(k_x \frac{\partial \phi}{\partial x}) + \frac{\partial}{\partial y}(ky \frac{\partial \phi}{\partial y}) + \omega^2 \phi = 0 \qquad \qquad 18.1$$

where

k_x and k_y are values of physical constants in the x and y directions respectively
ϕ = a function, used to define the eigenmodes.
ω = the eigenvalue to be determined

If $k = k_x = k_y$, the equation 18.1 can be written in the form shown by equation 18.2.

$$k\left(\frac{\partial^2 \phi}{\partial x^2} + \frac{\partial^2 \phi}{\partial y^2}\right) = -\omega^2 \phi \qquad \qquad 18.2$$

For acoustic vibration

$\phi = p$ = acoustic pressure

ω = radian frequency (rads/s)

$k = \overline{c}^2$ for acoustic vibration
$k = 1/(\epsilon_d \mu_o \epsilon_o)$ for electro-magnetic waveguides

$k = T/\rho$ for vibrating membranes

where

ϵ_d = permittivity of the dielectric

μ_o = permeability of free space

ϵ_o = permittivity of free space

T = tension/unit length in membrane

ℓ = mass/unit area

$\overline{C}$ = speed of sound in the fluid.

$$\overline{C} = \sqrt{K/\ell_F}$$

= 331.3 m/s in air at 0°C
= 1498 m/s in H_2O at 25°C

K $\quad$ = bulk modulus of the fluid
$\quad\quad$ = 1.32 x 10^5 N/m^2 air at 0°C
$\quad\quad$ = 2.05 x 10^9 N/m^2 for H_2O

ℓ_f $\quad$ = density of the fluid
$\quad\quad$ = 1.2 kg/m^3 for air
$\quad\quad$ = 1000 kg/m^3 for water

For oscillations of a mass of water in an **enclosed lake,**

K = gh
ω = $2\pi/T$ or n = 1/T
h = depth of water
g = acceleration due to gravity
T = period of oscillation

The equation of motion solved by the computer program, ACOUSTIC, is shown by equation 18.3.

$$|[H] - \lambda [G]| = 0 \qquad\qquad\qquad 18.3$$

For unconstrained vibration, [H] can be singular, hence, [G] was inverted in ACOUSTIC, and premultiplied into equation 18.3, to avoid this situation, to give equation 18.4.

$$|[G]^{-1} [H] \lambda [I]| = 0$$

where
[H] = Σ[h]

[G] = Σ[g]

$\lambda = \omega^2$ = eigenvalue

n = (ω/2π{ Hz = frequency

$$[g] = 1/\bar{C}^2 \int_A [N]^T [N] \, d(A) \qquad 18.5$$

$$[h] = \frac{1}{(4\Delta)} \begin{bmatrix} b_i^2 & & \\ b_i b_j & b_j^2 & \\ b_i b_k & b_j b_k & b_k^2 \end{bmatrix}$$

$$+ \frac{1}{(4\Delta)} \begin{bmatrix} c_i^2 & & \\ c_i c_j & c_j^2 & \\ c_i c_k & c_j c_k & c_k^2 \end{bmatrix}$$

18.6

b_i, c_i Δ etc, are defined in reference [7] page 418. [N] = a matrix of shape function [7].

The element used in this program, is the three node simplex triangular element described in chapters 10 and 17. Each node has one degree of freedom, namely ϕ_i, so that the element has a total of 3 degrees of freedom, hence [h] and [g] are of order 3 x 3.

NB

It should be noted that solving equation 18.4 by the Power method, which is used here, results in calculating the largest eigenvalues first, hence, if small values of ω are required, **all the eigenvalues should be calculated.**

18.2   Data

The data file should be prepared in the following sequence:-

NN = number of nodal points

LS = number of elements

NF = number of nodes with zero values of the function

If NF = 0, ignore the next loop

Details of zero values of the function

For i = 1 to NF

Type in the node number with the zero value of the function

NEXT i

Nodal co-ordinates

For i = 1 to NN

Type in x_i and y_i

NEXT i

M1 = number of frequencies required (this must not be greater than NN; normally M1 = NN)

Type in $\bar{C}$ (ie $\sqrt{k}$)

Element Topology

for LE = 1 to LS

Type in i(LE), j(LE) and k(LE) (Type these nodes in a **counter-clockwise order**)

NEXT LE

18.3 Example 18.1

Determine the acoustic resonant frequencies of vibration for the rectangular room of Figure 18.1. [35] given that $\bar{C}$ = 1.

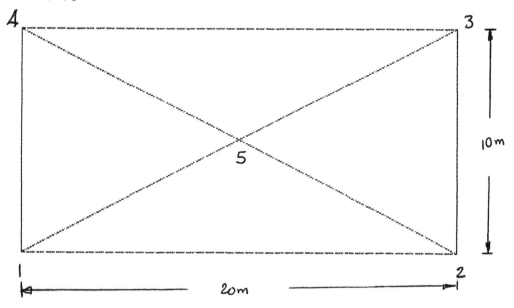

Figure 18.1 - Rectangular room

18.4   Input data file for Example 18.1

The input data file for this five element model, is as follows;-

ACOUSTIC.INP

```
5
4
0
0,0
20,0
20,10
0,10
10,5
5
1
3,4,5
3,5,2
2,5,1
1,5,4
```

18.5   Results

The results for w^2 are given in table 18.1, where they are compared with the results of Segerlind [35] and the "exact" values.

Table 18.1 - Values of "λ" (ie ω^2)

Mode	Segerlind	ACOUSTIC	Exact
1	0.45	0.45	0.395
2	0.15	0.15	0.123
3	0.12	0.12	0.987
4	0.03	0.03	0.0247
5	0.00	0.00	0

18.6   Screendumps

Screendumps of the 5 eigenmodes are given in Figures 18.2 to 18.6.

EIGENMODE= 1   FREQUENCY= .1067645 Hz

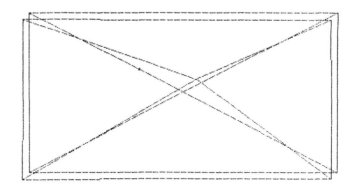

Figure 18.2

EIGENMODE= 2   FREQUENCY= 6.164049E-02 Hz

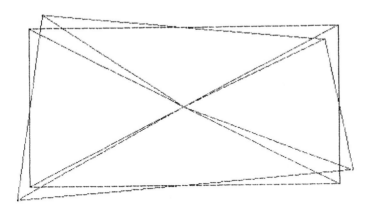

Figure 18.3

EIGENMODE= 3   FREQUENCY= 5.512789E-02 Hz

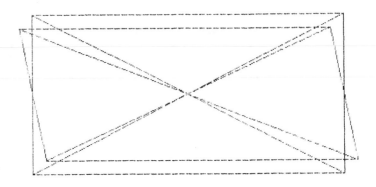

Figure 18.4

EIGENMODE= 4   FREQUENCY= .027569 Hz

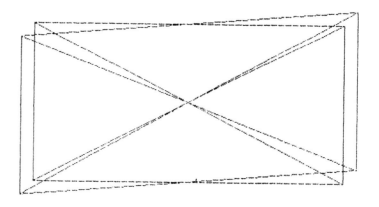

Figure 18.5

EIGENMODE= 5   FREQUENCY= 1.064219E-05 Hz

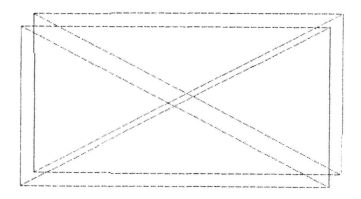

Figure 18.6

References

1. Levy, S, Computation of Influence Coefficients for Aircraft Structures with Discontinuities and Sweepback, J Aero Sci, 14, 547-560, Oct 1947.

2. Levy S, Structural Analysis and Influence coefficients for Delta Wings, J Aero Sci, 20, 449-454, July 1953.

3. Argyris, J H, Energy Theorems and Structural Analysis, Aircraft Eng, Oct-Nov, 1954 and Feb-May 1955.

4. Turner, M J, Clough, R W Martin, H C and Topp, L J, Stiffness and Deflection Analysis of Complex structures, J Aero, Sci, 23, 805-823, 1956.

5. Courant, R, Variational Methods for the Solution of Problems of Equilibrium and vibration, Bull, Am. Math. Soc., 49, 1-23,1943.

6. Zienkiewicz, O C and Taylor, R L, The Finite Element Method", McGraw-Hill, Vols 1 and 2 4th Ed, 1989 and 1991.

7. Ross, C T F, Finite Element Methods in Engineering Science, Horwood, 1990.

8. Irons, B and Ahmad, S, Techniques of finite Elements, Ellis Horwood, 1980.

9. Cook, R. D., Concepts and Applications of Finite Element Analysis, 2nd Ed., Wiley, 1974.

10. Przemieniecki, J S., Theory of Matrix Structural Analysis, McGraw-Hill, 1968.

11. Baker, A J., Finite Element Computational Fluid Mechanics, McGraw-Hill, 1985.

12. Fenner, R T, Finite Element Methods for Engineers, MacMillan, 1975.

13. Collar, A.R and Simpson. A., Matrices and Engineering Dynamics, Ellis Horwood, 1987.

14. Jennings, A., Matrix Algebra for Engineers and Scientists, Wiley, 1977.

15. Kikuchi, J., Finite Element Methods in Mechanics, Cambridge University Press, 1986.

16. Ross, C T F, Advanced Applied Stress analysis, Horwood, 1987.

17. Case, J, Lord Chilver and Ross, C T F, Strength of Materials and Structures, Arnold, 1992.

18. Argyris, J H., Recent Advances in Matrix Methods of Structural Analysis, Pergamon Press, 1964.

19. Petyt, M., The Application of Finite Element Techniques to Plate and Shell Problems, ISVR Report No 120, Feb 1965.

21. Ross, C T F, Finite Element Programs for Structural Vibrations, Springer-Verlage, 1991.

22. Irons, B., Structural Eigenvalue Problems: Elimination of Unwanted Variables, J A I A A, 3, 961, 1965.

23. Vedeler, G., The Distribution of Load in Longitudinal Strength Calculations, Trans, RINA, 89, 16-31, 1947.

24. Venancio-Filho, F and Iguti, F. Vibrations of Grids by the Finite Element Methods, Computers and Structures, 3, 1331-1344, Pergamon Press, 1973.

25. Ergatoudis, I, Irons, BM and Zienkiewicz, P C, Curved isoparametric 'quadrilateral;' elements for finite element analysis, I J solids and Structures, 4, 31-42, 1968.

26. Taig, I C, Structural analysis by the matrix displacement method, Engl Electric Aviation Rept, No 5017, 1961.

27. Clough, R W, The Finite Element Method in Plane Stress Analysis, Proc 2nd Conf Elec Comp, 345-479, 1960.

28. Mohr, G A and Milner, H R, Finite Element Method, Heinemann, 1986.

29. Young, W C, Roarks Formulas for Stress and Strain, McGraw-Hill, 6th Edition, 1989.

30. Petyt, M, Introduction to Finite Element Vibration Analysis, Cambridge University Press, 1990.

31. Carr, J B, The Effect of Shear Flexibility and Rotary Inertia on the Natural frequencies of Uniform Beams, Aeronautical Quarterly 21, 79-90, 1970.

32. Ross, C T F, Free Vibration of Thin Shells, J Sound and Vib, 39, 337-344, 1975.

33. Ross, C T F, Partially Conforming Plate Bending Elements for Static and Dynamic analyses, J Strain Analysis, 8, 260-263, 1973.

34. Chandrupatla, T R and Belegundu, A D, Introduction to Finite Elements in Engineering, Prentice-Hall, 1991.

35. Segerlind, L J, Applied Finite Element Analysis, Wiley, 1984.

Computer Program for Plane Pin-Jointed Trusses "TRUSSF"

```
90 CLS
100 GOSUB 9000
105 PRINT : PRINT "FORCES IN PLANE PIN-JOINTED TRUSSES": PRINT
110 PRINT : PRINT "TYPE IN THE NAME OF YOUR INPUT FILE"
120 INPUT FILENAME$
130 PRINT "IF YOU ARE SATISFIED WITH THIS FILENAME, TYPE Y; ELSE TYPE N"
140 A$ = INKEY$: IF A$ = "" THEN GOTO 140
142 IF A$ = "Y" OR A$ = "y" THEN GOTO 148
144 IF A$ = "N" OR A$ = "n" THEN GOTO 110
146 GOTO 140
148 OPEN FILENAME$ FOR INPUT AS #1
150 PRINT : PRINT "TYPE IN THE NAME OF YOUR OUTPUT FILE"
152 INPUT OUT$
154 PRINT "IF YOU ARE SATISFIED WITH THIS FILENAME, TYPE Y; ELSE TYPE N"
155 B$ = INKEY$: IF B$ = "" THEN GOTO 155
156 IF B$ = "Y" OR B$ = "y" THEN GOTO 159
157 IF B$ = "N" OR B$ = "n" THEN GOTO 150
GOTO 155
159 OPEN OUT$ FOR OUTPUT AS #2
160 NNODE = 2: NDF = 2
170 REM PRINT "TYPE IN NUMBER OF NODES"
180 INPUT #1, NN
182 IF NN < 3 OR NN > 499 THEN GOSUB 9995: PRINT "NO. OF NODES="; NN: STOP
190 REM PRINT "TYPE IN NUMBER OF MEMBERS"
200 INPUT #1, ES
202 IF ES < 2 OR ES > 499 THEN GOSUB 9995: PRINT "NO. OF ELEMENTS="; ES: STOP
210 N2 = NN * 2
220 REM PRINT : PRINT "TYPE IN THE NUMBER OF NODES WITH ZERO
DISPLACEMENTS"
230 INPUT #1, NF
232 IF NF < 2 OR NF > NN THEN GOSUB 9995: PRINT "NO. OF NODES WITH ZERO
DISPLACEMENTS="; NF: STOP
240 DIM SK(4, 4), DC(4, 4)
250 DIM U(N2), SG(4), SL(4), X(NN), Y(NN)
260 DIM XA(ES), XE(ES), XI(ES), XJ(ES)
270 DIM NS(NF * 2), RH(NN), RV(NN), NP(NF), DIAG(N2)
275 REM "TYPE IN THE DETAILS OF THE NODES WITH ZERO 'DISPLACEMENTS'"
280 FOR II = 1 TO NF * 2: NS(II) = 0: NEXT II
282 FOR II = 1 TO NF: NP(II) = 0: NEXT II
283 FOR II = 1 TO NF
285 REM PRINT "TYPE IN THE NODE NUMBER OF THE 'SUPPRESSED' NODE "; II
INPUT #1, SU, SUX, SUY
286 IF SU < 1 OR SU > NN THEN GOSUB 9995: PRINT "NO. OF THE NODE WITH ZERO
DISPLACEMENTS="; NF: STOP
292 REM "IS HORIZONTAL DISPLACEMENT ZERO AT NODE "; SU; " ?
293 IF SUX = 1 THEN Y$ = "Y" ELSE Y$ = "N"
294 IF Y$ = "Y" OR Y$ = "y" OR Y$ = "N" OR Y$ = "n" THEN GOTO 297
296 GOSUB 9995: STOP
297 IF Y$ = "Y" OR Y$ = "y" THEN NS(II * 2 - 1) = SU * 2 - 1
```

```
300 REM "IS VERTICAL DISPLACEMENT ZERO AT NODE "; SU; " ?
302 IF SUY = 1 THEN Y$ = "Y" ELSE Y$ = "N"
304 IF Y$ = "Y" OR Y$ = "N" OR Y$ = "y" OR Y$ = "n" THEN GOTO 310
305 GOSUB 9995: STOP
310 IF Y$ = "Y" OR Y$ = "y" THEN NS(II * 2) = SU * 2
315 NP(II) = SU
320 NEXT II
322 REM DETAILS OF NODAL COORDINATES
380 REM "TYPE IN THE NODAL COORDINATES"
390 FOR II = 1 TO NN
400 REM "X & Y COORDS FOR NODE "; II; "=";
410 INPUT #1, X(II), Y(II)
440 NEXT II
445 MX = 0
447 REM START OF ELEMENT CYCLE
450 FOR EL = 1 TO ES
460 REM "I,J NODES, ETC. FOR MEMBER "; EL
470 INPUT #1, I, J, AR, E
472 IF I < 1 OR I > NN THEN GOSUB 9995: PRINT "I="; I; " FOR ELEMENT"; EL: STOP
480 REM PRINT "J NODE FOR MEMBER "; EL; "=";
492 IF J < 1 OR J > NN THEN GOSUB 9995: PRINT "J="; J; " FOR ELEMENT"; E; : STOP
497 IF I = J THEN GOSUB 9995: PRINT "I = J FOR ELEMENT "; EL: STOP
500 REM "CROSS SECTIONAL AREA FOR MEMBER "; I; "-"; J; "=";AR
515 IF AR < = 0 THEN GOSUB 9995: PRINT "AREA="; AR; " FOR ELEMENT "; EL: STOP
520 REM "ELASTIC MODULUS FOR MEMBER "; I; "-"; J; "=";E
532 IF E < 1 THEN GOSUB 9995: PRINT "YOUNG'S MODULUS="; E; " FOR ELEMENT";
EL: STOP
539 REM END OF INPUT FOR MEMBER EL
540 XI(EL) = I
550 XJ(EL) = J
560 XA(EL) = AR
561 XE(EL) = E
562 FOR II = 1 TO NNODE: IF II = 1 THEN IN = I ELSE IN = J
564 FOR IJ = 1 TO NDF
565 II1 = NDF * (IN - 1) + IJ
566 FOR JJ = 1 TO NNODE: IF JJ = 1 THEN JN = I ELSE JN = J
567 FOR JI = 1 TO NDF
568 JJ1 = NDF * (JN - 1) + JI
569 IIJJ = JJ1 - II1 + 1: IF IIJJ > DIAG(JJ1) THEN DIAG(JJ1) = IIJJ
570 NEXT JI, JJ, IJ, II, EL
572 DIAG(1) = 1: FOR II = 1 TO N2: DIAG(II) = DIAG(II - 1) + DIAG(II): NEXT II
573 NT = DIAG(N2)
574 DIM A(NT), Q(N2), C(N2)
575 GOSUB 2000
576 GOSUB 2600
577 Z9 = 0: GOSUB 10000
578 FOR EL = 1 TO ES
579 I = XI(EL): J = XJ(EL)
580 AR = XA(EL)
581 E = XE(EL)
584 PRINT : PRINT "MEMBER NO."; EL; " UNDER COMPUTATION": PRINT
585 L = SQR((X(J) - X(I)) ^ 2 + (Y(J) - Y(I)) ^ 2)
590 KS = (X(J) - X(I)) / L
```

```
600 SN = (Y(J) - Y(I)) / L
610 SK(1, 1) = KS ^ 2
620 SK(3, 3) = SK(1, 1)
630 SK(2, 1) = KS * SN
640 SK(1, 2) = SK(2, 1)
650 SK(3, 4) = SK(2, 1)
660 SK(4, 3) = SK(3, 4)
670 SK(3, 1) = -SK(1, 1)
680 SK(1, 3) = SK(3, 1)
690 SK(1, 4) = -SK(2, 1)
700 SK(4, 1) = SK(1, 4)
710 SK(2, 3) = SK(1, 4)
720 SK(3, 2) = SK(2, 3)
730 SK(2, 2) = SN ^ 2
740 SK(4, 4) = SK(2, 2)
750 SK(2, 4) = -SK(2, 2)
760 SK(4, 2) = SK(2, 4)
770 CN = AR * E / L
780 FOR II = 1 TO 4
790 FOR JJ = 1 TO 4
800 SK(II, JJ) = SK(II, JJ) * CN
810 NEXT JJ
820 NEXT II
830 I1 = 2 * I - 2
835 J1 = 2 * J - 2
840 FOR II = 1 TO 2
845 FOR JJ = 1 TO 2
850 MM = I1 + II
860 MN = I1 + JJ
870 NM = J1 + II
880 N9 = J1 + JJ
910 IF JJ > II THEN GOTO 922
915 A(DIAG(MM) + MN - MM) = A(DIAG(MM) + MN - MM) + SK(II, JJ)
920 A(DIAG(NM) + N9 - NM) = A(DIAG(NM) + N9 - NM) + SK(II + 2, JJ + 2)
922 IF I1 > J1 THEN GOTO 935
925 A(DIAG(NM) + MN - NM) = A(DIAG(NM) + MN - NM) + SK(II + 2, JJ)
930 GOTO 940
935 A(DIAG(MM) + N9 - MM) = A(DIAG(MM) + N9 - MM) + SK(II, JJ + 2)
940 NEXT JJ, II
970 NEXT EL
980 FOR II = 1 TO NF * 2
990 N9 = NS(II)
995 IF N9 = 0 THEN GOTO 1010
997 DI = DIAG(N9)
1000 A(DI) = A(DI) * 1E+12 + 1E+12
1010 NEXT II
1040 FOR II = 1 TO N2
1065 C(II) = Q(II)
1070 NEXT II
1090 N = N2
1095 PRINT : PRINT "SOLUTION OF THE SIMULTANEOUS EQUATION IS NOW TAKING
PLACE": PRINT
1100 GOSUB 2280
```

```
1110 FOR II = 1 TO N2
1120 U(II) = C(II)
1160 NEXT II
1180 PRINT #2, : PRINT #2, "THE NODAL DISPLACEMENTS U & V ARE:-": PRINT #2,
1190 FOR II = 1 TO NN
1192 PRINT #2, "NODE="; II; "     ";
1195 FOR JJ = 1 TO 2
1200 PRINT #2, U(II * 2 - 2 + JJ); "     ";
1205 NEXT JJ
1210 PRINT #2, : NEXT II
1220 FOR II = 1 TO 4
1230 FOR JJ = 1 TO 4
1240 DC(II, JJ) = 0
1250 NEXT JJ
1260 NEXT II
1290 FOR EL = 1 TO ES
1300 AR = XA(EL)
1310 E = XE(EL)
1320 I = XI(EL)
1330 J = XJ(EL)
1340 L = SQR((X(J) - X(I)) ^ 2 + (Y(J) - Y(I)) ^ 2)
1350 KS = (X(J) - X(I)) / L
1360 SN = (Y(J) - Y(I)) / L
1370 DC(1, 1) = KS
1380 DC(2, 2) = KS
1390 DC(3, 3) = KS
1400 DC(4, 4) = KS
1410 DC(1, 2) = SN
1420 DC(3, 4) = SN
1430 DC(2, 1) = -SN
1440 DC(4, 3) = -SN
1450 I1 = 2 * I - 2
1460 J1 = 2 * J - 2
1470 FOR II = 1 TO 2
1480 MM = I1 + II
1490 NM = J1 + II
1500 SG(II) = U(MM)
1510 SG(II + 2) = U(NM)
1520 NEXT II
1530 FOR II = 1 TO 4
1540 SL(II) = 0
1550 FOR JJ = 1 TO 4
1560 SL(II) = SL(II) + DC(II, JJ) * SG(JJ)
1570 NEXT JJ
1580 NEXT II
1590 FC = AR * E * (SL(3) - SL(1)) / L
1600 PRINT #2, : PRINT #2, "FORCE IN MEMBER "; I; "-"; J; "="; FC
1610 GOSUB 4000
1620 NEXT EL
1622 GOSUB 4500
1629 GOSUB 11000
1630 STOP
2000 REM ***INPUT OF LOADS***
```

```
2010 REM "TYPE IN THE NUMBER OF NODES WITH CONCENTRATED LOADS"
2020 INPUT #1, NC
2030 IF NC < 1 OR NC > NN THEN GOSUB 9995: PRINT "NO. OF NODES WITH
CONCENTRATED LOADS="; NC: STOP
2050 DIM POSW(NC)
2060 FOR II = 1 TO NC
2070 REM "TYPE IN THE NODE NUMBER OF THE NODE, PLUS THE CONCENTRATED
LOADS"
2080 INPUT #1, CN, HW, VW
2090 IF CN < 1 OR CN > NN THEN GOSUB 9995: PRINT "NO. OF NODE WITH THE
CONCENTRATED LOADS="; CN: STOP
2100 POSW(II) = CN
2120 REM "THE HORIZONTAL COMPONENT OF LOAD (+VE TO RIGHT) "; HW
2130 REM "THE VERTICAL COMPONENT OF LOAD (+VE UPWARDS) "; VW
2140 REM
2160 Q(CN * 2 - 1) = HW: Q(CN * 2) = VW
2180 NEXT II
2190 RETURN
REM SOLUTION OF SIMULTANEOUS EQUATIONS
2280 A(1) = SQR(A(1))
2285 FOR II = 2 TO N
2290 DI = DIAG(II) - II
2295 LL = DIAG(II - 1) - DI + 1
2300 FOR JJ = LL TO II
2305 X = A(DI + JJ)
2310 DJ = DIAG(JJ) - JJ
2315 IF JJ = 1 THEN GOTO 2345
2320 LB = DIAG(JJ - 1) - DJ + 1
2325 IF LL > LB THEN LB = LL
2330 FOR KK = LB TO JJ - 1
2335 X = X - A(DI + KK) * A(DJ + KK)
2340 NEXT KK
2345 A(DI + JJ) = X / A(DJ + JJ)
2350 NEXT JJ
2355 A(DI + II) = SQR(X)
2360 NEXT II
2370 C(1) = C(1) / A(1)
2375 FOR II = 2 TO N
2380  DI = DIAG(II) - II
2385 LL = DIAG(II - 1) - DI + 1
2390 X = C(II)
2395 FOR JJ = LL TO II - 1
2400 X = X - A(DI + JJ) * C(JJ)
2405 NEXT JJ
2410 C(II) = X / A(DI + II)
2415 NEXT II
2420 FOR II = N TO 2 STEP -1
2425 DI = DIAG(II) - II
2430 X = C(II) / A(DI + II)
2435 C(II) = X
2440 LL = DIAG(II - 1) - DI + 1
2445 FOR KK = LL TO II - 1
2450 C(KK) = C(KK) - X * A(DI + KK)
```

```
2455 NEXT KK
2460 NEXT II
2465 C(1) = C(1) / A(1)
2500 RETURN
2600 REM OUTPUT 1
2610 PRINT #2, : PRINT #2, "FORCES IN PIN-JOINTED PLANE TRUSSES"
2620 PRINT #2, : PRINT #2, "PROGRAM BY DR.C.T.F.ROSS"
2630 PRINT #2, : PRINT #2, "NUMBER OF NODES="; NN
2640 PRINT #2, "NUMBER OF MEMBERS="; ES
2650 PRINT #2, "NUMBER OF NODES WITH ZERO DISPLACEMENTS="; NF
2660 PRINT #2, "DETAILS OF NODES WITH ZERO DISPLACEMENTS"
2670 FOR II = 1 TO NF
2680 SU = INT((NS(II * 2 - 1) + 1) / 2)
2690 IF SU = 0 THEN SU = INT(NS(II * 2) / 2)
2700 PRINT #2, "NODE="; SU
2710 N9 = NS(II * 2 - 1)
2720 IF N9 > 0 THEN PRINT #2, "HORIZONTAL DISPLACEMENT AT NODE "; SU; " IS
ZERO"
2730 N9 = NS(II * 2)
2740 IF N9 > 0 THEN PRINT #2, "VERTICAL DISPLACEMENT AT NODE "; SU; " IS ZERO"
2750 NEXT II
2760 PRINT #2, : PRINT #2, "NODAL COORDINATES"
2770 FOR II = 1 TO NN
2780 PRINT #2, "X("; II; ")="; X(II); "    Y("; II; ")="; Y(II)
2790 NEXT II
2810 PRINT #2, "ELEMENT DETAILS"
2820 FOR EL = 1 TO ES
2830 PRINT #2, "MEMBER "; XI(EL); "-"; XJ(EL)
2840 PRINT #2, "CROSS-SECTIONAL AREA="; XA(EL)
2850 PRINT #2, "YOUNG'S MODULUS="; XE(EL)
2860 NEXT EL
2890 PRINT #2, : PRINT #2, "NUMBER OF NODES WITH CONCENTRATED LOADS="; NC
2900 FOR II = 1 TO NC
2910 PRINT #2, "NODE="; POSW(II): WPOSN = POSW(II)
2920 PRINT #2, "HORIZONTAL LOAD="; Q(2 * WPOSN - 1); "    VERTICAL LOAD=";
Q(WPOSN * 2)
2950 NEXT II
2960 RETURN
4000 REM CALCULATION OF REACTIONS
4010 RH = 0: RV = 0
4020 FOR II = 1 TO NF
4030 IF XI(EL) = NP(II) THEN SU = NP(II): GOTO 4060
4040 NEXT II
4050 GOTO 4100
4060 RH = -FC * KS: RV = -FC * SN
4070 RH(SU) = RH(SU) + RH
4080 RV(SU) = RV(SU) + RV
4100 FOR II = 1 TO NF
4110 IF XJ(EL) = NP(II) THEN SU = NP(II): GOTO 4230
4120 NEXT II
4200 GOTO 4320
4230 RH = FC * KS: RV = FC * SN
4300 RH(SU) = RH(SU) + RH
```

```
4310 RV(SU) = RV(SU) + RV
4320 RETURN
4500 REM OUTPUT OF REACTIONS
4510 PRINT #2, : PRINT #2, " THE REACTIONS ARE:": PRINT #2,
4520 FOR II = 1 TO NN
4530 IF RH(II) = 0 THEN GOTO 4550
4540 GOTO 4560
4550 IF RV(II) = 0 THEN GOTO 4580
4560 PRINT #2, "NODE="; II; "   HORIZONTAL REACTION="; RH(II); "   VERTICAL
REACTION="; RV(II)
4580 NEXT II
4600 RETURN
9000 PRINT : PRINT "Copyright of DR.C.T.F.ROSS"
9010 PRINT "NOT TO BE COPIED"
9020 PRINT "Have you read the book:-"
9030 PRINT "FINITE ELEMENT PROGRAMS IN STRUCTURAL ENGINEERING &
CONTINUUM MECHANICS"
9040 PRINT "BY"
9050 PRINT "C.T.F.ROSS."
9060 PRINT "TYPE Y/N"
9065 A$ = INKEY$: IF A$ = "" THEN GOTO 9065
9070 IF A$ = "Y" OR A$ = "y" THEN GOTO 9090
9080 PRINT : PRINT "GO BACK & READ IT!": PRINT : STOP
9090 RETURN
9995 REM ***WARNING***
9996 PRINT : PRINT "INCORRECT DATA": PRINT
9997 RETURN
10000 IF Z9 = 0 OR Z9 = 77 THEN SCREEN 9, 0: WIDTH 80: PRINT : PRINT "TO
CONTINUE, TYPE Y"
10004 IF Z9 = 77 OR Z9 = 99 THEN GOTO 10280
10005 XZ = 1E+12
10010 YZ = 1E+12
10020 XM = -1E+12
10030 YM = -1E+12
10040 FOR EL = 1 TO ES
10110 I = XI(EL)
10120 J = XJ(EL)
10130 IF X(I) < XZ THEN XZ = X(I)
10140 IF Y(I) < YZ THEN YZ = Y(I)
10150 IF X(J) < XZ THEN XZ = X(J)
10160 IF Y(J) < YZ THEN YZ = Y(J)
10170 IF X(I) > XM THEN XM = X(I)
10180 IF Y(I) > YM THEN YM = Y(I)
10190 IF X(J) > XM THEN XM = X(J)
10200 IF Y(J) > YM THEN YM = Y(J)
10210 NEXT EL
10250 XS = 450 / (XM - XZ)
10255 IF YM - YZ = 0 THEN GOTO 10280
10260 YS = 225 / (YM - YZ)
10270 IF XS > YS THEN XS = YS
10280 REM THE TRUSS WILL NOW BE DRAWM
10350 FOR EL = 1 TO ES
10360 I = XI(EL)
```

```
10370 J = XJ(EL)
10375 IF Z9 = 99 THEN GOTO 10430
10380 XXI = (X(I) - XZ) * XS
10390 YI = (Y(I) - YZ) * XS
10400 XXJ = (X(J) - XZ) * XS
10410 YJ = (Y(J) - YZ) * XS
10420 GOTO 10450
10430 XXI = X(I) - XZ * XS: XXJ = X(J) - XZ * XS
10440 YI = Y(I) - YZ * XS: YJ = Y(J) - YZ * XS
10450 REM
10500 LINE (XXI + 20, 300 - YI)-(XXJ + 20, 300 - YJ)
10650 NEXT EL
10670 IF Z9 = 77 THEN GOTO 10800
10680 A$ = INKEY$: IF A$ = "" THEN GOTO 10680
10690 IF A$ = "Y" OR A$ = "y" THEN GOTO 10760
10700 GOTO 10680
10760 SCREEN 0, 0
10800 RETURN
11000 REM PLOT OF DEFLECTED FORM OF FRAME
11010 CLS
11020 PRINT : PRINT "THE DEFLECTED FRAME WILL NOW BE DRAWN"
11030 Z9 = 77
11050 GOSUB 10000
11100 UM = -1E+12
11110 FOR II = 1 TO N2
11160 IF ABS(U(II)) > UM THEN UM = ABS(U(II))
11170 NEXT II
11200 SC = 15
11250 FOR II = 1 TO NN
11260 X(II) = X(II) * XS + SC * U(2 * II - 1) / UM
11270 Y(II) = Y(II) * XS + SC * U(2 * II) / UM
11275 Y(II) = Y(II)
11280 NEXT II
11290 Z9 = 99
11300 GOSUB 10000
11400 RETURN
11999 REM END
```

Appendix 2

Computer Program for Beams "BEAMSF"

1 REM COPYRIGHT OF DR.C.T.F.ROSS
6 CLS : GOSUB 9000
8 PRINT : PRINT "BENDING MOMENTS IN CONTINUOUS BEAMS"
10 PRINT : PRINT "PROGRAM BY DR.C.T.F.ROSS"
11 PRINT : INPUT "TYPE IN THE NAME OF YOUR INPUT FILE"; FILENAME$
PRINT : PRINT "IF YOU ARE SATISFIED WITH THIS FILENAME, TYPE Y; ELSE N"
12 A$ = INKEY$: IF A$ = "" THEN GOTO 12
IF A$ = "Y" OR A$ = "y" THEN GOTO 18
IF A$ = "N" OR A$ = "n" THEN GOTO 11
GOTO 12
18 OPEN FILENAME$ FOR INPUT AS #1
20 PRINT : PRINT "TYPE IN THE NAME OF YOUR OUTPUT FILE"
21 INPUT OUT$
22 PRINT : PRINT "IF YOU ARE SATISFIED WITH THIS FILENAME, TYPE Y; ELSE N"
23 B$ = INKEY$: IF B$ = "" THEN GOTO 23
24 IF B$ = "Y" OR B$ = "y" THEN GOTO 28
25 IF B$ = "N" OR B$ = "n" THEN GOTO 20
GOTO 23
28 OPEN OUT$ FOR OUTPUT AS #2
29 REM "TYPE IN THE NUMBER OF ELEMENTS"
30 INPUT #1, LS
32 IF LS < 1 OR LS > 499 THEN GOSUB 9995: PRINT "NO. OF ELEMENTS="; LS: STOP
40 REM   PRINT"TYPE IN THE NUMBER OF NODES WHICH HAVE ZERO DISPLACEMENTS"
42 INPUT #1, NF
44 IF NF < 1 OR NF > LS + 1 THEN GOSUB 9995: PRINT "NO. OF NODES WITH ZERO DISPLACEMENTS="; NF: STOP
50 NW = 4
55 NN = 2 * LS + 2: NJ = LS + 1
60 NT = NN + NW
62 REM "TYPE IN THE NUMBER OF NODES WITH CONCENTRATED LOADS & COUPLES"
63 INPUT #1, NC
64 IF NC < 0 OR NC > NJ THEN GOSUB 9995: PRINT "NO. OF NODES WITH CONCENTRATED LOADS & COUPLES="; NC: STOP
70 DIM A(NT, NW), Q(NT), C(NT), U(NN), ST(4, 4), UD(LS), UP(LS), SA(LS), XL(LS)
80 DIM DI(4), XAS(LS), B(4), NS(NF * 2), RS(2), RE(NJ), DL(NN)
90 DIM QC(NN), POSW(NC), IX(LS, 2), BM(LS, 2)
130 ND = NJ
180 REM "TYPE IN THE NUMBERS OF THE NODES WITH ZERO DISPLACEMENTS"
185 FOR II = 1 TO NF * 2: NS(II) = 0: NEXT II
190 FOR II = 1 TO NF
195 REM "TYPE IN THE NODE NUMBER OF NODAL POSITION ";II
200 INPUT #1, SU, SUY, SUT
201 IF SU < 1 OR SU > NJ THEN GOSUB 9995: PRINT "NODE NO. WITH ZERO DISPLACEMENT="; SU: GOTO 195
205 REM "IS THE VERTICAL DISPLACEMENT ZERO AT NODE ";SU;" ?
206 IF SUY = 1 THEN SUP$ = "Y": GOTO 208
IF SUY = 0 THEN SUP$ = "N": GOTO 208
PRINT "THE NUMBER DEFINING VERTICAL BOUNDARY CONDITIONS MUST=1 OR 0 AT

332

NODE "; SU: STOP
208 IF SUP$ = "Y" OR SUP$ = "y" THEN NS(II * 2 - 1) = SU * 2 - 1
211 REM "IS THE ROTATIONAL DISPLACEMENT ZERO AT NODE ";SU;" ?
212 IF SUT = 1 THEN SUP$ = "Y": GOTO 214
IF SUT = 0 THEN SUP$ = "N": GOTO 214
214 IF SUP$ = "Y" OR SUP$ = "N" OR SUP$ = "y" OR SUP$ = "n" THEN GOTO 219
215 PRINT "THE NUMBER DEFINING ROTATIONAL BOUNDARY CONDITIONS MUST=1
OR 0 AT NODE"; SU: STOP
219 IF SUP$ = "Y" OR SUP$ = "y" THEN NS(II * 2) = SU * 2
220 NEXT II
230 REM "TYPE IN YOUNG'S MODULUS ";E
232 INPUT #1, E
235 IF E < 1 THEN GOSUB 9995: PRINT "YOUNG'S MODULUS="; E: STOP
240 REM "IF HYDROSTATIC LOAD, TYPE 1; ELSE TYPE 0"
242 INPUT #1, NL
245 IF NL = 0 OR NL = 1 THEN GOTO 250
248 PRINT : PRINT "ERROR IN INPUT OF DATA FOR HYDROSTATIC LOAD. TYPE 1 OR
0 !"
250 REM
260 IF NC = 0 THEN GOTO 298
265 REM "TYPE IN THE NODAL POSITIONS & VALUES OF THE CONCENTRATED LOADS
& COUPLES"
267 FOR II = 1 TO NC: POSW(II) = 0: NEXT II
268 FOR II = 1 TO NN: QC(II) = 0: NEXT II
270 FOR II = 1 TO NC
275 REM "TYPE IN THE NODE NUMBER FOR THE CONCENTRATED 'LOADS' AT NODAL
POSITION ";II
277 INPUT #1, WP, VW, CW
278 IF WP < 1 OR WP > NJ THEN GOSUB 9995: PRINT "NODE NO. WITH
CONCENTRATED 'LOADS'="; WP: STOP
280 REM "TYPE IN THE VERTICAL LOAD (+VE UPWARDS) AT NODE ";WP:INPUT VW
285 REM "TYPE IN THE COUPLE (+VE CLOCKWISE) AT NODE ";WP:INPUT CW
288 QC(WP * 2 - 1) = VW: QC(WP * 2) = CW
289 POSW(II) = WP
290 NEXT II
298 GOSUB 4000
REM START OF ELEMENT CYCLE
300 FOR LE = 1 TO LS
350 REM "TYPE IN 2ND MOMENT OF AREA FOR ELEMENT ";LE
360 INPUT #1, SA(LE), XL(LE)
365 IF SA(LE) < = 0 THEN GOSUB 9995: PRINT "2ND MOA OF ELEMENT NO."; LE; "=";
SA(LE): STOP
370 REM "TYPE IN ELEMENTAL LENGTH FOR ELEMENT ";LE
385 IF XL(LE) < = 0 THEN GOSUB 9995: PRINT "LENGTH OF ELEMENT NO."; LE; "=";
XL(LE): STOP
390 IF NL = 1 THEN GOTO 430
400 REM "TYPE IN DISTRIBUTED LOAD FOR ELEMENT ";LE
410 INPUT #1, UD(LE)
420 GOTO 490
430 REM " TYPE IN HYDROSTATIC LOAD AT 'A' (ON LEFT) FOR ELEMENT ";LE;(UD)
440 INPUT #1, UD(LE), UP(LE), XAS(LE)
450 REM "TYPE IN HYDROSTATIC LOAD (ON RIGHT) FOR ELEMENT ";LE;(UP)
470 REM "TYPE IN THE LENGTH (A) FOR ELEMENT ";LE;(XAS)

```
485 IF XAS(LE) < 0 OR XAS(LE) > XL(LE) THEN GOSUB 9995: PRINT "THE DISTANCE
    THAT LOAD STARTS FROM 'LEFT' END (XA)="; XAS(LE); " FOR ELEMENT"; LE: STOP
490 I = LE
500 J = I + 1
510 L = XL(LE)
517 GOSUB 5310
520 IF NL = 0 THEN SW = UD(LE)
530 ST(1, 1) = 12 / L ^ 3
540 ST(1, 2) = -6 / L ^ 2
550 ST(1, 4) = ST(1, 2)
560 ST(2, 1) = ST(1, 2)
570 ST(4, 1) = ST(1, 2)
580 ST(3, 3) = ST(1, 1)
590 ST(1, 3) = -12 / L ^ 3
600 ST(3, 1) = ST(1, 3)
610 ST(2, 2) = 4 / L
620 ST(4, 4) = 4 / L
630 ST(2, 3) = 6 / L ^ 2
640 ST(3, 2) = ST(2, 3)
650 ST(2, 4) = 2 / L
660 ST(4, 2) = ST(2, 4)
670 ST(3, 4) = ST(2, 3)
680 ST(4, 3) = ST(3, 4)
690 CN = E * SA(LE)
700 FOR II = 1 TO 4
710 FOR JJ = 1 TO 4
720 ST(II, JJ) = ST(II, JJ) * CN
730 NEXT JJ
740 NEXT II
750 I1 = 2 * I - 2
760 J1 = 2 * J - 2
770 IF NL = 1 THEN GOTO 830
780 Q(I1 + 1) = Q(I1 + 1) + SW * L / 2
790 Q(I1 + 2) = Q(I1 + 2) - SW * L ^ 2 / 12
800 Q(J1 + 1) = Q(J1 + 1) + SW * L / 2
810 Q(J1 + 2) = Q(J1 + 2) + SW * L ^ 2 / 12
820 GOTO 980
830 WA = UD(LE)
840 WB = UP(LE)
850 AL = XAS(LE)
860 X3 = (L - AL) ^ 3
870 X2 = (L - AL) ^ 2
880 RA = WA * X3 * (L + AL) / (2 * L ^ 3)
890 RA = RA + (WB - WA) * X3 * (3 * L + 2 * AL) / (20 * L ^ 3)
900 RB = (WB + WA) * (L - AL) / 2 - RA
910 BA = -WA * X3 * (L + 3 * AL) / (12 * L ^ 2)
920 BA = BA - (WB - WA) * X3 * (2 * L + 3 * AL) / (60 * L ^ 2)
930 BB = RA * L + BA - WA * X2 / 2 - (WB - WA) * X2 / 6
940 Q(I1 + 1) = Q(I1 + 1) + RA
950 Q(J1 + 1) = Q(J1 + 1) + RB
960 Q(I1 + 2) = Q(I1 + 2) + BA
970 Q(J1 + 2) = Q(J1 + 2) - BB
980 FOR II = 1 TO 4
```

```
990 FOR JJ = II TO 4
1000 MG = I1 + II
1010 TR = I1 + JJ - MG + 1
1020 A(MG, TR) = A(MG, TR) + ST(II, JJ)
1030 NEXT JJ
1040 NEXT II
1050 NEXT LE
CLOSE #1
1060 IF NC = 0 THEN GOTO 1200
1170 FOR II = 1 TO NN
1180 Q(II) = Q(II) + QC(II)
1190 NEXT II
1200 FOR II = 1 TO NF * 2
1210 N9 = NS(II)
1220 IF N9 = 0 THEN GOTO 1240
1230 A(N9, 1) = A(N9, 1) * 1E+12
1240 NEXT II
1242 FOR II = 1 TO NN
1244 C(II) = Q(II)
1246 NEXT II
1255 N = NN
1260 GOSUB 6000
1270 FOR II = 1 TO NN
1280 U(II) = C(II)
1320 NEXT II
1330 PRINT #2, : PRINT #2, "THE NODAL NUMBERS & NODAL DISPLACEMENTS (V &
THETA ARE AS FOLLOWS:-"
1340 FOR II = 1 TO ND
1350 PRINT #2, "NODE="; II; "    "; U(2 * II - 1); "     "; U(2 * II)
1360 NEXT II
1364 PRINT #2, : PRINT #2, "THE NODAL BENDING MOMENTS ARE GIVEN BELOW :-"
1370 FOR LE = 1 TO LS
1380 I = LE
1390 J = I + 1
1400 L = XL(LE)
1410 IF NL = 0 THEN SW = UD(LE)
1420 JJ = 2 * LE - 2
1430 FOR II = 1 TO 4
1440 DI(II) = U(JJ + II)
1450 NEXT II
1451 ST(1, 1) = 12 / L ^ 3: ST(1, 3) = -ST(1, 1): ST(3, 1) = ST(1, 3): ST(3, 3) = ST(1, 1)
1452 ST(1, 2) = -6 / L ^ 2: ST(1, 4) = ST(1, 2): ST(3, 2) = -ST(1, 2): ST(3, 4) = ST(3, 2)
1453 CN = E * SA(LE)
1454 FOR II = 1 TO 4: ST(1, II) = ST(1, II) * CN: ST(3, II) = ST(3, II) * CN: NEXT II
1455 RS(1) = 0: RS(2) = 0
1456 FOR KK = 1 TO 4
1457 RS(1) = RS(1) + ST(1, KK) * DI(KK)
1458 RS(2) = RS(2) + ST(3, KK) * DI(KK)
1459 NEXT KK
1460 XI = -1
1470 FOR IJ = 1 TO 2
1480 XI = XI + 1
1490 B(1) = -6 + 12 * XI
```

```
1500 B(2) = L * (4 - 6 * XI)
1510 B(3) = 6 - 12 * XI
1520 B(4) = L * (2 - 6 * XI)
1530 CN = -1 / L ^ 2 * E * SA(LE)
1540 FOR KK = 1 TO 4
1550 B(KK) = B(KK) * CN
1560 NEXT KK
1570 TH = 0
1580 FOR KK = 1 TO 4
1590 TH = TH + B(KK) * DI(KK)
1600 NEXT KK
1610 IF NL = 1 THEN GOTO 1640
1620 TH = TH - SW * L ^ 2 / 12
1630 GOTO 1770
1640 WA = UD(LE)
1650 WB = UP(LE)
1660 AL = XAS(LE)
1670 X3 = (L - AL) ^ 3
1680 X2 = (L - AL) ^ 2
1690 RA = WA * X3 * (L + AL) / (2 * L ^ 3)
1700 RA = RA + (WB - WA) * X3 * (3 * L + 2 * AL) / (20 * L ^ 3)
1710 RB = (WB + WA) * (L - AL) / 2 - RA
1720 BA = -WA * X3 * (L + 3 * AL) / (12 * L ^ 2)
1730 BA = BA - (WB - WA) * X3 * (2 * L + 3 * AL) / (60 * L ^ 2)
1740 BB = RA * L + BA - WA * X2 / 2 - (WB - WA) * X2 / 6
1750 IF IJ = 1 THEN TH = TH + BA
1760 IF IJ = 2 THEN TH = TH + BB
1770 IF IJ = 1 THEN II = I
1780 IF IJ = 2 THEN II = J
1790 PRINT #2, : PRINT #2, "BENDING MOMENT AT NODE "; II; "="; TH
1791 TH = -TH
1793 BM(LE, IJ) = TH
1800 NEXT IJ
1801 RE(I) = RE(I) + RS(1): RE(J) = RE(J) + RS(2)
1810 NEXT LE
1822 FOR II = 1 TO NJ
1823 RE(II) = RE(II) - Q(II * 2 - 1): PRINT #2, : PRINT #2, "NODE="; II; "    REACTION=";
RE(II)
1824 NEXT II
1827 GOSUB 10000
1830 STOP
4000 REM OUTPUT 1
4010 PRINT #2, : PRINT #2, "BENDING MOMENTS IN CONTINUOUS BEAMS"
4020 PRINT #2, : PRINT #2, "PROGRAM BY DR.C.T.F.ROSS"
4040 PRINT #2, : PRINT #2, "NUMBER OF ELEMENTS="; LS
4050 PRINT #2, "NUMBER OF NODES WITH ZERO DISPLACEMENTS="; NF
4060 PRINT #2, "NUMBER OF NODES WITH CONCENTRATED LOADS & COUPLES="; NC
4070 PRINT #2, : PRINT #2, "THE NODE NUMBERS OF THE NODES WITH ZERO
DISPLACEMENTS ARE AS FOLLOWS:-"
4090 FOR II = 1 TO NF
4100 SU = INT((NS(II * 2 - 1) + 1) / 2)
4110 IF SU = 0 THEN SU = INT(NS(II * 2) / 2)
4120 PRINT #2, "NODE="; SU
```

4140 IF (NS(II * 2 - 1)) > 0 THEN PRINT #2, "THE VERTICAL DISPLACEMENT IS ZERO AT NODE "; SU
4150 IF (NS(II * 2)) > 0 THEN PRINT #2, "THE ROTATIONAL DISPLACEMENT IS ZERO AT NODE "; SU
4170 NEXT II
4180 PRINT #2, : PRINT #2, "YOUNG'S MODULUS="; E
4190 IF NL = 0 THEN PRINT #2, "UNIFORMLY DISTRIBUTED LOAD CASE"
4200 IF NL = 1 THEN PRINT #2, "HYDROSTATIC LOAD CASE"
4230 IF NC = 0 THEN GOTO 4330
4240 PRINT #2, : PRINT #2, "THE NODAL POSITIONS & VALUES OF THE CONCENTRATED LOADS & COUPLES ARE:-"
4250 FOR II = 1 TO NC
4260 PRINT #2, "NODE="; POSW(II)
4270 PRINT #2, "LOAD="; QC(POSW(II) * 2 - 1); "   COUPLE="; QC(POSW(II) * 2)
4280 NEXT II
4330 RETURN
5310 REM***OUTPUT 2***
5340 PRINT #2, : PRINT #2, "2ND MOMENT OF AREA="; SA(LE)
5350 PRINT #2, "ELEMENTAL LENGTH="; XL(LE)
5360 IF NL = 0 THEN PRINT #2, "UNIFORMLY DISTRIBUTED LOAD="; UD(LE): GOTO 5430
5380 PRINT #2, "HYDROSTATIC LOAD AT 'A' (ON LEFT) OF ELEMENT "; LE; "="; UD(LE)
5390 PRINT #2, "HYDROSTATIC LOAD (ON RIGHT) OF ELEMENT "; LE; "="; UP(LE)
5410 PRINT #2, "THE DISTANCE BETWEEN THE LEFT END OF THE HYDROSTATIC LOAD & THE LEFT END OF THE BEAM OF ELEMENT "; LE; "="; XAS(LE)
5430 RETURN
REM SOLUTION OF SIMULTANEOUS EQUATIONS
6000 FOR II = 1 TO N
6010 IK = II
6020 FOR JJ = 2 TO NW
6030 IK = IK + 1
6040 CN = A(II, JJ) / A(II, 1)
6050 JK = 0
6060 FOR KK = JJ TO NW
6070 JK = JK + 1
6080 A(IK, JK) = A(IK, JK) - CN * A(II, KK)
6090 NEXT KK
6100 A(II, JJ) = CN
6110 C(IK) = C(IK) - CN * C(II)
6120 NEXT JJ
6130 C(II) = C(II) / A(II, 1)
6140 NEXT II
6150 FOR IZ = 2 TO N
6160 II = N - IZ + 1
6170 FOR KK = 2 TO NW
6180 JJ = II + KK - 1
6190 C(II) = C(II) - A(II, KK) * C(JJ)
6200 NEXT KK: NEXT IZ
6220 RETURN
9000 PRINT : PRINT "COPYRIGHT OF DR.C.T.F.ROSS"
9010 PRINT : PRINT "NOT TO BE COPIED"
9020 PRINT : PRINT "HAVE YOU READ"

```
9030 PRINT : PRINT "FINITE ELEMENT PROGRAMS IN STRUCTURAL ENGINEERING &
CONTINUUM MECHANICS"
9040 PRINT : PRINT "BY"
9050 PRINT : PRINT "C.T.F.ROSS."
9060 PRINT : PRINT "TYPE Y/N"
9070 A$ = INKEY$: IF A$ = "" THEN GOTO 9070
9080 IF A$ = "Y" OR A$ = "y" THEN GOTO 9150
9100 PRINT : PRINT "GO BACK & READ IT!": STOP
9150 RETURN
9995 REM ***WARNING***
9996 PRINT : PRINT "INCORRECT DATA": PRINT
9997 RETURN
10000 REM GRAPHICS
10010 PRINT : PRINT "TO CONTINUE, TYPE Y"
10020 A$ = INKEY$: IF A$ = "" THEN GOTO 10020
10030 IF A$ = "Y" OR A$ = "y" THEN GOTO 10040
10035 GOTO 10020
10040 SCREEN 9, 0: WIDTH 80
10050 AA = 0
10060 FOR II = 1 TO LS
10070 IX(II, 1) = AA: IX(II, 2) = AA + XL(II)
10080 AA = AA + XL(II)
10090 NEXT II
10100 SC = 400 / AA
10110 BI = 0: AP = 0: PA = 0: IB = 0
10120 FOR JJ = 1 TO LS
10130 IF ABS(RE(JJ)) > IB THEN IB = ABS(RE(JJ))
10140 XC = BM(JJ, 1): XD = BM(JJ, 2)
10150 SL = (BM(JJ, 2) - BM(JJ, 1)) / (XD - XC)
10170 IF ABS(XC) > BI THEN BI = ABS(XC)
10180 IF ABS(XD) > BI THEN BI = ABS(XD)
10190 NEXT JJ
10200 IF ABS(RE(NJ)) > IB THEN IB = ABS(RE(NJ))
10210 VS = 50 / BI: AA = 0: AB = 0: SV = 50 / IB
10220 FOR II = 1 TO LS
10230 AB = AB + XL(II)
10240 LINE (10 + SC * AA, 70)-(10 + SC * AB, 70)
10250 LINE (10 + SC * AA, 240)-(10 + SC * AB, 240)
10260 AA = XL(II)
10270 NEXT II
10280 AP = 0: PA = 0: SF = 0
10290 FOR GG = 1 TO LS
10300 XC = IX(GG, 1)
10310 XD = IX(GG, 2)
10320 YC = BM(GG, 1)
10330 YD = BM(GG, 2)
10340 SL = (YD - YC) / (XD - XC)
10350 AP = XC
10360 FS = SF + RE(GG) + QC(2 * GG - 1)
10370 FOR WZ = 0 TO 120
10380 WW = XL(GG) * WZ / 120
10390 IF NL = 1 THEN GOTO 10430
10400 UD = (-UD(GG) * XL(GG) * WW / 2) - (-UD(GG) * (WW ^ 2) / 2)
```

```
10410 WD = UD(GG) * WW
10420 GOTO 10480
10430 RA = ((XL(GG) - XAS(GG)) ^ 2) * (2 * UD(GG) + UP(GG)) / (6 * XL(GG))
10440 IF WW < XAS(GG) THEN UD = -RA * WW
10450 IF WW > = XAS(GG) THEN UD = -RA * WW + UD(GG) * (WW - XAS(GG)) ^ 2 / 2
+ (UP(GG) - UD(GG)) * (WW - XAS(GG)) ^ 3 / (6 * (XL(GG) - XAS(GG)))
10460 IF WW < XAS(GG) THEN WD = 0
10470 IF WW > = XAS(GG) THEN WD = UD(GG) + (UP(GG) - UD(GG)) * (WW - XAS(GG))
/ (XL(GG) - XAS(GG))
10475 WD = (WD + UD(GG)) * (WW - XAS(GG)) / 2
10480 EQ = (BM(GG, 1) + (SL * WW) + UD)
10490 SF = FS + WD
10500 LINE (10 + SC * AP, 70)-(10 + SC * AP, 70 - VS * EQ)
10510 AP = XC + WW: PA = EQ
10520 IF GG = 1 AND WW = 0 THEN JG = EQ
10530 LINE (10 + SC * AP, 240)-(10 + SC * AP, 240 - SV * SF)
10540 NEXT WZ
10550 NEXT GG
10560 LOCATE 11, 10: PRINT "BENDING MOMENT DIAGRAM": LOCATE 23, 10: PRINT
"SHEARING FORCE DIAGRAM"
10590 A$ = INKEY$: IF A$ = "" THEN GOTO 10590
10610 IF A$ = "Y" OR A$ = "y" THEN GOTO 10630
10620 GOTO 10590
10630 SCREEN 0, 0
10640 RETURN
25000 REM END
```

Computer Program for Beams on an Elastic Foundation "BEAMERF"

```
1 REM COPYRIGHT OF DR.C.T.F.ROSS
6 CLS : GOSUB 9000
8 PRINT : PRINT "BENDING MOMENTS IN CONTINUOUS BEAMS"
10 PRINT : PRINT "PROGRAM BY DR.C.T.F.ROSS"
11 PRINT : PRINT "TYPE IN THE NAME OF YOUR INPUT FILE"
12 INPUT FILENAME$
13 PRINT "IF YOU ARE SATISFIED WITH YOUR FILENAME, TYPE Y; ELSE TYPE N"
14 A$ = INKEY$: IF A$ = "" THEN GOTO 14
15 IF A$ = "Y" OR A$ = "y" THEN GOTO 18
16 IF A$ = "N" OR A$ = "n" THEN GOTO 11
17 GOTO 14
18 OPEN FILENAME$ FOR INPUT AS #1
20 PRINT : PRINT "TYPE IN THE NAME OF YOUR OUTPUT FILE"
INPUT OUT$
22 PRINT : PRINT "IF YOU ARE SATISFIED WITH THIS FILENAME, TYPE Y; ELSE N"
23 B$ = INKEY$: IF B$ = "" THEN GOTO 23
IF B$ = "Y" OR B$ = "y" THEN GOTO 28
IF B$ = "N" OR B$ = "n" THEN GOTO 20
GOTO 23
28 OPEN OUT$ FOR OUTPUT AS #2
29 REM "TYPE IN THE NUMBER OF ELEMENTS"
30 INPUT #1, LS
32 IF LS < 1 OR LS > 499 THEN GOSUB 9995: PRINT "NO. OF ELEMENTS="; LS: STOP
40 REM TYPE IN THE NUMBER OF NODES WITH ELASTIC & RIGID SUPPORTS
42 INPUT #1, NF
44 IF NF < 1 OR NF > LS + 1 THEN GOSUB 9995: PRINT "NO. OF NODES WITH ZERO
DISPLACEMENTS="; NF: STOP
50 NW = 4
55 NN = 2 * LS + 2: NJ = LS + 1
60 NT = NN + NW
62 REM     "TYPE IN THE NUMBER OF NODES WITH CONCENTRATED LOADS &
COUPLES"
63 INPUT #1, NC
64 IF NC < 0 OR NC > NJ THEN GOSUB 9995: PRINT "NO. OF NODES WITH
CONCENTRATED LOADS & COUPLES="; NC: STOP
70 DIM A(NT, NW), Q(NT), C(NT), U(NN), ST(4, 4), UD(LS), UP(LS), SA(LS), XL(LS)
80 DIM DI(4), XAS(LS), B(4), NS(NF * 2), RS(2), RE(NJ), DL(NN)
90 DIM QC(NN), POSW(NC), IX(LS, 2), BM(LS, 2)
130 ND = NJ
180 REM TYPE IN THE NODAL POSITIONS OF THE ELASTIC SUPPORTS, ETC"
185 FOR II = 1 TO NF * 2: NS(II) = 0: NEXT II
190 FOR II = 1 TO NF
195 REM TYPE IN NODE NUMBER FOR THIS NODAL POSITION
200 INPUT #1, SU, SUV, SUT
201 IF SU < 1 OR SU > NJ THEN GOSUB 9995: PRINT "NO. OF THE NODE WITH ELASTIC
SUPPORTS="; SU: STOP
205 REM TYPE IN THE VERTICAL STIFFNESS OF SUPPORT AT THIS NODE. IF
RIGID,TYPE 1E20. IF FREE, TYPE 0
206 DL(II * 2 - 1) = SUV
```

207 IF DL(II * 2 - 1) < 0 THEN GOSUB 9995: PRINT "VERTICAL STIFFNESS<0 AT NODE
"; SU: STOP
209 NS(II * 2 - 1) = SU * 2 - 1
211 REM TYPE IN THE ROTATIONAL STIFFNESS OF SUPPORT AT THIS NODE. IF RIGID,
TYPE 1E20. IF FREE, TYPE 0
212 DL(II * 2) = SUT
213 IF DL(II * 2) < 0 THEN GOSUB 9995: PRINT "ROTATIONAL STIFFNESS<0 AT NODE
"; SU: STOP
214 NS(II * 2) = SU * 2
220 NEXT II
230 INPUT #1, E
235 IF E < 1 THEN GOSUB 9995: PRINT "YOUNG'S MODULUS="; E: STOP
240 INPUT #1, NL
245 IF NL = 0 OR NL = 1 THEN GOTO 260
248 GOSUB 9995: PRINT "NEITHER HYDROSTATIC NOR UNIFORMLY DISTRIBUTED
LOAD": STOP
260 IF NC = 0 THEN GOTO 298
265 REM "TYPE IN THE NODAL POSITIONS & VALUES OF THE CONCENTRATED LOADS
& COUPLES"
267 FOR II = 1 TO NC: POSW(II) = 0: NEXT II
268 FOR II = 1 TO NN: QC(II) = 0: NEXT II
270 FOR II = 1 TO NC
275 INPUT #1, WP, VW, CW
277 IF WP < 1 OR WP > NJ THEN GOSUB 9995: PRINT "NO. OF THE NODE WITH
CONCENTRATED LOADS & COUPLES="; WP: STOP
288 QC(WP * 2 - 1) = VW: QC(WP * 2) = CW
289 POSW(II) = WP
290 NEXT II
298 GOSUB 4000
300 FOR LE = 1 TO LS
350 REM "TYPE IN 2ND MOMENT OF AREA FOR ELEMENT ";LE
360 INPUT #1, SA(LE), XL(LE)
365 IF SA(LE) < = 0 THEN GOSUB 9995: PRINT "2 ND M.O.A.="; SA(LE); " FOR ELEMENT
"; LE: STOP
370 REM "TYPE IN ELEMENTAL LENGTH FOR ELEMENT ";LE
385 IF XL(LE) < = 0 THEN GOSUB 9995: PRINT "LENGTH FOR ELEMENT "; LE; "=";
XL(LE): STOP
390 IF NL = 1 THEN GOTO 430
400 REM "TYPE IN DISTRIBUTED LOAD FOR ELEMENT ";LE
410 INPUT #1, UD(LE)
420 GOTO 490
430 REM " TYPE IN HYDROSTATIC LOAD AT 'A' (ON LEFT) FOR ELEMENT ";LE ;(UD)
440 INPUT #1, UD(LE), UP(LE), XAS(LE)
450 REM "TYPE IN HYDROSTATIC LOAD (ON RIGHT) FOR ELEMENT ";LE;(UP)
470 REM "TYPE IN THE LENGTH (A) FOR ELEMENT ";LE;(XAS)
485 IF XAS(LE) < 0 OR XAS(LE) > XL(LE) THEN GOSUB 9995: PRINT "THE DISTANCE
OF THE LEFT END OF THE HYDROSTATIC LOAD FROM THE LEFT END OF THE
ELEMENT, NAMELY, 'AS' FOR ELEMENT "; LE; "="; XAS(LE): STOP
490 I = LE
500 J = I + 1
510 L = XL(LE)
517 GOSUB 5310
520 IF NL = 0 THEN SW = UD(LE)

```
530 ST(1, 1) = 12 / L ^ 3
540 ST(1, 2) = -6 / L ^ 2
550 ST(1, 4) = ST(1, 2)
560 ST(2, 1) = ST(1, 2)
570 ST(4, 1) = ST(1, 2)
580 ST(3, 3) = ST(1, 1)
590 ST(1, 3) = -12 / L ^ 3
600 ST(3, 1) = ST(1, 3)
610 ST(2, 2) = 4 / L
620 ST(4, 4) = 4 / L
630 ST(2, 3) = 6 / L ^ 2
640 ST(3, 2) = ST(2, 3)
650 ST(2, 4) = 2 / L
660 ST(4, 2) = ST(2, 4)
670 ST(3, 4) = ST(2, 3)
680 ST(4, 3) = ST(3, 4)
690 CN = E * SA(LE)
700 FOR II = 1 TO 4
710 FOR JJ = 1 TO 4
720 ST(II, JJ) = ST(II, JJ) * CN
730 NEXT JJ
740 NEXT II
750 I1 = 2 * I - 2
760 J1 = 2 * J - 2
770 IF NL = 1 THEN GOTO 830
780 Q(I1 + 1) = Q(I1 + 1) + SW * L / 2
790 Q(I1 + 2) = Q(I1 + 2) - SW * L ^ 2 / 12
800 Q(J1 + 1) = Q(J1 + 1) + SW * L / 2
810 Q(J1 + 2) = Q(J1 + 2) + SW * L ^ 2 / 12
820 GOTO 980
830 WA = UD(LE)
840 WB = UP(LE)
850 AL = XAS(LE)
860 X3 = (L - AL) ^ 3
870 X2 = (L - AL) ^ 2
880 RA = WA * X3 * (L + AL) / (2 * L ^ 3)
890 RA = RA + (WB - WA) * X3 * (3 * L + 2 * AL) / (20 * L ^ 3)
900 RB = (WB + WA) * (L - AL) / 2 - RA
910 BA = -WA * X3 * (L + 3 * AL) / (12 * L ^ 2)
920 BA = BA - (WB - WA) * X3 * (2 * L + 3 * AL) / (60 * L ^ 2)
930 BB = RA * L + BA - WA * X2 / 2 - (WB - WA) * X2 / 6
940 Q(I1 + 1) = Q(I1 + 1) + RA
950 Q(J1 + 1) = Q(J1 + 1) + RB
960 Q(I1 + 2) = Q(I1 + 2) + BA
970 Q(J1 + 2) = Q(J1 + 2) - BB
980 FOR II = 1 TO 4
990 FOR JJ = II TO 4
1000 MG = I1 + II
1010 TR = I1 + JJ - MG + 1
1020 A(MG, TR) = A(MG, TR) + ST(II, JJ)
1030 NEXT JJ
1040 NEXT II
1050 NEXT LE
```

```
1060 IF NC = 0 THEN GOTO 1200
1170 FOR II = 1 TO NN
1180 Q(II) = Q(II) + QC(II)
1190 NEXT II
1200 FOR II = 1 TO NF * 2
1210 N9 = NS(II)
1220 IF N9 = 0 THEN GOTO 1240
1230 A(N9, 1) = A(N9, 1) + DL(II)
1240 NEXT II
1242 FOR II = 1 TO NN
1244 C(II) = Q(II)
1246 NEXT II
1255 N = NN
1260 GOSUB 6000
1270 FOR II = 1 TO NN
1280 U(II) = C(II)
1320 NEXT II
1330 PRINT #2, : PRINT #2, "THE NODAL NUMBERS & NODAL DISPLACEMENTS (V &
THETA ARE AS FOLLOWS:-"
1340 FOR II = 1 TO ND
1350 PRINT #2, "NODE="; II; "    "; U(2 * II - 1); "     "; U(2 * II)
1360 NEXT II
1364 PRINT #2, : PRINT #2, "THE NODAL BENDING MOMENTS ARE GIVEN BELOW :-"
1370 FOR LE = 1 TO LS
1380 I = LE
1390 J = I + 1
1400 L = XL(LE)
1410 IF NL = 0 THEN SW = UD(LE)
1420 JJ = 2 * LE - 2
1430 FOR II = 1 TO 4
1440 DI(II) = U(JJ + II)
1450 NEXT II
1451 ST(1, 1) = 12 / L ^ 3: ST(1, 3) = -ST(1, 1): ST(3, 1) = ST(1, 3): ST(3, 3) = ST(1, 1)
1452 ST(1, 2) = -6 / L ^ 2: ST(1, 4) = ST(1, 2): ST(3, 2) = -ST(1, 2): ST(3, 4) = ST(3, 2)
1453 CN = E * SA(LE)
1454 FOR II = 1 TO 4: ST(1, II) = ST(1, II) * CN: ST(3, II) = ST(3, II) * CN: NEXT II
1455 RS(1) = 0: RS(2) = 0
1456 FOR KK = 1 TO 4
1457 RS(1) = RS(1) + ST(1, KK) * DI(KK)
1458 RS(2) = RS(2) + ST(3, KK) * DI(KK)
1459 NEXT KK
1460 XI = -1
1470 FOR IJ = 1 TO 2
1480 XI = XI + 1
1490 B(1) = -6 + 12 * XI
1500 B(2) = L * (4 - 6 * XI)
1510 B(3) = 6 - 12 * XI
1520 B(4) = L * (2 - 6 * XI)
1530 CN = -1 / L ^ 2 * E * SA(LE)
1540 FOR KK = 1 TO 4
1550 B(KK) = B(KK) * CN
1560 NEXT KK
1570 TH = 0
```

```
1580 FOR KK = 1 TO 4
1590 TH = TH + B(KK) * DI(KK)
1600 NEXT KK
1610 IF NL = 1 THEN GOTO 1640
1620 TH = TH - SW * L ^ 2 / 12
1630 GOTO 1770
1640 WA = UD(LE)
1650 WB = UP(LE)
1660 AL = XAS(LE)
1670 X3 = (L - AL) ^ 3
1680 X2 = (L - AL) ^ 2
1690 RA = WA * X3 * (L + AL) / (2 * L ^ 3)
1700 RA = RA + (WB - WA) * X3 * (3 * L + 2 * AL) / (20 * L ^ 3)
1710 RB = (WB + WA) * (L - AL) / 2 - RA
1720 BA = -WA * X3 * (L + 3 * AL) / (12 * L ^ 2)
1730 BA = BA - (WB - WA) * X3 * (2 * L + 3 * AL) / (60 * L ^ 2)
1740 BB = RA * L + BA - WA * X2 / 2 - (WB - WA) * X2 / 6
1750 IF IJ = 1 THEN TH = TH + BA
1760 IF IJ = 2 THEN TH = TH + BB
1770 IF IJ = 1 THEN II = I
1780 IF IJ = 2 THEN II = J
1790 PRINT #2, : PRINT #2, "BENDING MOMENT AT NODE "; II; "="; TH
1791 TH = -TH
1793 BM(LE, IJ) = TH
1800 NEXT IJ
1801 RE(I) = RE(I) + RS(1): RE(J) = RE(J) + RS(2)
1810 NEXT LE
1822 FOR II = 1 TO NJ
1823 RE(II) = RE(II) - Q(II * 2 - 1): PRINT #2, : PRINT #2, "NODE="; II; "    REACTION=";
RE(II)
1824 NEXT II
1827 GOSUB 10000
1830 STOP
4000 REM OUTPUT 1
4010 PRINT #2, : PRINT #2, "BENDING MOMENTS IN CONTINUOUS BEAMS"
4020 PRINT #2, : PRINT #2, "PROGRAM BY DR.C.T.F.ROSS"
4040 PRINT #2, : PRINT #2, "NUMBER OF ELEMENTS="; LS
4050 PRINT #2, "NUMBER OF NODES WITH ELASTIC & RIGID SUPPORTS="; NF
4060 PRINT #2, "NUMBER OF NODES WITH CONCENTRATED LOADS & COUPLES="; NC
4070 PRINT #2, : PRINT #2, "THE NODAL POSITIONS OF THE ELASTIC SUPPORTS, ETC.
ARE AS FOLLOWS:-"
4090 FOR II = 1 TO NF
4100 SU = INT((NS(II * 2 - 1) + 1) / 2)
4110 IF SU = 0 THEN SU = INT(NS(II * 2) / 2)
4120 PRINT #2, "NODE="; SU
4140 PRINT #2, "THE VERTICAL STIFFNESS AT NODE "; SU; "="; DL(II * 2 - 1)
4150 PRINT #2, "THE ROTATIONAL STIFFNESS AT NODE "; SU; "="; DL(II * 2)
4170 NEXT II
4180 PRINT #2, : PRINT #2, "YOUNG'S MODULUS="; E
4190 IF NL = 0 THEN PRINT #2, "UNIFORMLY DISTRIBUTED LOAD CASE"
4200 IF NL = 1 THEN PRINT #2, "HYDROSTATIC LOAD CASE"
4230 IF NC = 0 THEN GOTO 4330
4240 PRINT #2, : PRINT #2, "THE NODAL POSITIONS & VALUES OF THE CONCENTRATED
```

LOADS & COUPLES ARE:-"
```
4250  FOR II = 1 TO NC
4260  PRINT #2, "NODE="; POSW(II)
4270 PRINT #2, "LOAD="; QC(POSW(II) * 2 - 1); "        COUPLE="; QC(POSW(II) * 2)
4280 NEXT II
4330 RETURN
5310 REM***OUTPUT 2***
5340 PRINT #2, : PRINT #2, "2ND MOMENT OF AREA="; SA(LE)
5350 PRINT #2, "ELEMENTAL LENGTH="; XL(LE)
5360 IF NL = 0 THEN PRINT #2, "UNIFORMLY DISTRIBUTED LOAD="; UD(LE): GOTO
5430
5380 PRINT #2, "HYDROSTATIC LOAD AT 'A' (ON LEFT) OF ELEMENT "; LE; "=";
UD(LE)
5390 PRINT #2, "HYDROSTATIC LOAD (ON RIGHT) OF ELEMENT "; LE; "="; UP(LE)
5410 PRINT #2, "THE DISTANCE BETWEEN THE LEFT END OF THE HYDROSTATIC LOAD
& THE LEFT END OF THE BEAM OF ELEMENT "; LE; "="; XAS(LE)
5430 RETURN
REM SOLUTION OF SIMULTANEOUS EQUATIONS
6000 FOR II = 1 TO N
6010 IK = II
6020 FOR JJ = 2 TO NW
6030 IK = IK + 1
6040 CN = A(II, JJ) / A(II, 1)
6050 JK = 0
6060 FOR KK = JJ TO NW
6070 JK = JK + 1
6080 A(IK, JK) = A(IK, JK) - CN * A(II, KK)
6090 NEXT KK
6100 A(II, JJ) = CN
6110 C(IK) = C(IK) - CN * C(II)
6120 NEXT JJ
6130 C(II) = C(II) / A(II, 1)
6140 NEXT II
6150 FOR IZ = 2 TO N
6160 II = N - IZ + 1
6170 FOR KK = 2 TO NW
6180 JJ = II + KK - 1
6190 C(II) = C(II) - A(II, KK) * C(JJ)
6200 NEXT KK: NEXT IZ
6220 RETURN
9000 PRINT : PRINT "COPYRIGHT OF DR.C.T.F.ROSS"
9010 PRINT : PRINT "NOT TO BE COPIED"
9020 PRINT : PRINT "HAVE YOU READ"
9030 PRINT : PRINT "FINITE ELEMENT PROGRAMS IN STRUCTURAL & CONTINUUM
MECHANICS"
9040 PRINT : PRINT "BY?"
9050 PRINT : PRINT "C.T.F.ROSS."
9060 PRINT : PRINT "TYPE Y/N"
9070 A$ = INKEY$: IF A$ = "" THEN GOTO 9070
9080 IF A$ = "Y" OR A$ = "y" THEN GOTO 9150
9100 PRINT : PRINT "GO BACK & READ IT!": STOP
9150 RETURN
9995 REM ***WARNING***
```

```
9996 PRINT : PRINT "INCORRECT DATA": PRINT
9997 RETURN
10000 REM GRAPHICS
10010 PRINT : PRINT "TO CONTINUE, TYPE Y"
10020 A$ = INKEY$: IF A$ = "" THEN GOTO 10020
10030 IF A$ = "Y" OR A$ = "y" THEN GOTO 10040
10035 GOTO 10020
10040 SCREEN 9, 0: WIDTH 80
10050 AA = 0
10060 FOR II = 1 TO LS
10070 IX(II, 1) = AA: IX(II, 2) = AA + XL(II)
10080 AA = AA + XL(II)
10090 NEXT II
10100 SC = 400 / AA
10110 BI = 0: AP = 0: PA = 0: IB = 0
10120 FOR JJ = 1 TO LS
10130 IF ABS(RE(JJ)) > IB THEN IB = ABS(RE(JJ))
10140 XC = BM(JJ, 1): XD = BM(JJ, 2)
10150 SL = (BM(JJ, 2) - BM(JJ, 1)) / (XD - XC)
10170 IF ABS(XC) > BI THEN BI = ABS(XC)
10180 IF ABS(XD) > BI THEN BI = ABS(XD)
10190 NEXT JJ
10200 IF ABS(RE(NJ)) > IB THEN IB = ABS(RE(NJ))
10210 VS = 50 / BI: AA = 0: AB = 0: SV = 50 / IB
10220 FOR II = 1 TO LS
10230 AB = AB + XL(II)
10240 LINE (10 + SC * AA, 70)-(10 + SC * AB, 70)
10250 LINE (10 + SC * AA, 240)-(10 + SC * AB, 240)
10260 AA = XL(II)
10270 NEXT II
10280 AP = 0: PA = 0: SF = 0
10290 FOR GG = 1 TO LS
10300 XC = IX(GG, 1)
10310 XD = IX(GG, 2)
10320 YC = BM(GG, 1)
10330 YD = BM(GG, 2)
10340 SL = (YD - YC) / (XD - XC)
10350 AP = XC
10360 FS = SF + RE(GG) + QC(2 * GG - 1)
10370 FOR WZ = 0 TO 120
10380 WW = XL(GG) * WZ / 120
10390 IF NL = 1 THEN GOTO 10430
10400 UD = (-UD(GG) * XL(GG) * WW / 2) - (-UD(GG) * (WW ^ 2) / 2)
10410 WD = UD(GG) * WW
10420 GOTO 10480
10430 RA = ((XL(GG) - XAS(GG)) ^ 2) * (2 * UD(GG) + UP(GG)) / (6 * XL(GG))
10440 IF WW < XAS(GG) THEN UD = -RA * WW
10450 IF WW > = XAS(GG) THEN UD = -RA * WW + UD(GG) * (WW - XAS(GG)) ^ 2 / 2
+ (UP(GG) - UD(GG)) * (WW - XAS(GG)) ^ 3 / (6 * (XL(GG) - XAS(GG)))
10460 IF WW < XAS(GG) THEN WD = 0
10470 IF WW > = XAS(GG) THEN WD = UD(GG) + (UP(GG) - UD(GG)) * (WW - XAS(GG))
/ (XL(GG) - XAS(GG))
10475 WD = (WD + UD(GG)) * (WW - XAS(GG)) / 2
```

```
10480 EQ = (BM(GG, 1) + (SL * WW) + UD)
10490 SF = FS + WD
10500 LINE (10 + SC * AP, 70)-(10 + SC * AP, 70 - VS * EQ)
10510 AP = XC + WW: PA = EQ
10520 IF GG = 1 AND WW = 0 THEN JG = EQ
10530 LINE (10 + SC * AP, 240)-(10 + SC * AP, 240 - SV * SF)
10540 NEXT WZ
10550 NEXT GG
10560 LOCATE 11, 10: PRINT "BENDING MOMENT DIAGRAM": LOCATE 23, 10: PRINT
"SHEARING FORCE DIAGRAM"
10590 A$ = INKEY$: IF A$ = "" THEN GOTO 10590
10610 IF A$ = "Y" OR A$ = "y" THEN GOTO 10630
10620 GOTO 10590
10630 SCREEN 0, 0
10640 RETURN
25000 REM END
```

Appendix 4

Computer Program for Rigid-Jointed Frames "FRAME2DF"

```
90 NNODE = 2: NDF = 3
100 CLS
110 REM COPYRIGHT OF DR.C.T.F.ROSS
120 REM 6 HURSTVILLE DRIVE, WATERLOOVILLE, PORTSMOUTH
130 REM PO7 7NB HANTS. ENGLAND
140 GOSUB 4520
150 IF A$ < > "Y" THEN GOTO 3960
160 PRINT "STATIC ANALYSIS OF RIGID JOINTED PLANE FRAMES"
161 PRINT : PRINT "TYPE IN THE NAME OF YOUR INPUT FILE"
162 INPUT FILENAME$
163 PRINT "IF YOU ARE SATISFIED WITH YOUR FILENAME, TYPE Y; ELSE TYPE N"
164 A$ = INKEY$: IF A$ = "" THEN GOTO 164
165 IF A$ = "Y" OR A$ = "y" THEN GOTO 169
166 IF A$ = "N" OR A$ = "n" THEN GOTO 161
167 GOTO 164
169 OPEN FILENAME$ FOR INPUT AS #1
170 PRINT : PRINT "TYPE IN THE NAME OF YOUR OUTPUT FILE"
172 INPUT OUT$
PRINT : PRINT "IF YOU ARE SATISFIED WITH THIS FILENAME, TYPE Y; ELSE N"
174 B$ = INKEY$: IF B$ = "" THEN GOTO 174
IF B$ = "Y" OR B$ = "y" THEN GOTO 178
IF B$ = "N" OR B$ = "n" THEN GOTO 170
GOTO 174
178 OPEN OUT$ FOR OUTPUT AS #2
180 INPUT #1, NJ
190 IF NJ < 2 OR NJ > 499 THEN GOSUB 9000: PRINT "NO. OF NODES="; NJ: STOP
260 N3 = 3 * NJ
270 INPUT #1, MS
290 IF MS < 1 OR MS > 499 THEN GOSUB 9000: PRINT "NO. OF ELEMENTS="; MS: STOP
320 REM  "TYPE IN THE NUMBER OF NODES WITH ZERO 'DISPLACEMENTS'":PRINT
330 INPUT #1, NF
340 IF NF < 1 OR NF > NJ THEN GOSUB 9000: PRINT "NO. OF NODES WITH ZERO
DISPLACEMENTS="; NF: STOP
380 INPUT #1, NC
400 IF NC < 0 OR NC > NJ THEN GOSUB 9000: PRINT "NO.OF NODES WITH
CONCENTRATED LOADS & COUPLES="; NC: STOP
450 DIM U(N3), K(6, 6), NS(NF * 3), I9(MS), J9(MS), DIAG(N3)
460 DIM DC(6, 6), SP(6), SQ(6), SG(6), SL(6), X(NJ), Y(NJ)
470 DIM S9(MS), A9(MS), U9(MS), UP(MS), XAS(MS), BX(2, 6), S2(2)
480 DIM QC(N3), NP(NF), RH(NJ), RV(NJ), POSW(NC), XG(NJ), YG(NJ), BM(2, MS)
500 FOR I = 1 TO NJ
520 INPUT #1, X(I), Y(I)
545 XG(I) = X(I): YG(I) = Y(I)
550 NEXT I
590 FOR I = 1 TO NF * 3: NS(I) = 0: NEXT I
600 FOR I = 1 TO NF
620 INPUT #1, NP(I), SUX, SUY, SUT
630 IF NP(I) < 1 OR NP(I) > NJ THEN GOSUB 9000: PRINT "NO. OF THE NODE WITH
ZERO DISPLACEMENTS="; NP(I): STOP
```

348

680 IF SUX = 1 THEN SU$ = "Y" ELSE SU$ = "N"
690 IF SU$ = "Y" OR SU$ = "N" OR SU$ = "y" OR SU$ = "n" THEN GOTO 720
700 GOSUB 9000: STOP
720 IF SU$ = "Y" OR SU$ = "y" THEN NS(I * 3 - 2) = NP(I) * 3 - 2
740 IF SUY = 1 THEN SU$ = "Y" ELSE SU$ = "N"
750 IF SU$ = "Y" OR SU$ = "N" OR SU$ = "y" OR SU$ = "n" THEN GOTO 770
760 GOSUB 9000: STOP
770 IF SU$ = "Y" OR SU$ = "y" THEN NS(I * 3 - 1) = NP(I) * 3 - 1
790 IF SUT = 1 THEN SU$ = "Y" ELSE SU$ = "N"
800 IF SU$ = "Y" OR SU$ = "N" OR SU$ = "y" OR SU$ = "n" THEN GOTO 820
810 GOSUB 9000: STOP
820 IF SU$ = "Y" OR SU$ = "y" THEN NS(I * 3) = NP(I) * 3
830 NEXT I
860 INPUT #1, E
880 IF E < = 0 THEN GOSUB 9000: PRINT "YOUNG'S MODULUS="; E: STOP
910 REM "IF HYDROSTSTIC LOAD, TYPE 1; ELSE 0"
920 INPUT #1, HY
930 IF HY = 0 OR HY = 1 THEN GOTO 950
940 GOSUB 9000: PRINT "NEITHER DISTRIBUTED OR HYDROSTATIC LOAD": STOP
950 REM ELEMENT DETAILS
960 FOR I = 1 TO MS
970 INPUT #1, I9(I), J9(I), S9(I), A9(I)
980 IF I9(I) < 1 OR I9(I) > NJ THEN GOSUB 9000: PRINT "I NODE FOR ELEMENT "; I;
"="; I9(I): STOP
1020 IF J9(I) < 1 OR J9(I) > NJ THEN GOSUB 9000: PRINT "J NODE FOR ELEMENT "; I;
"="; J9(I): STOP
1050 IN = I9(I): JN = J9(I)
1060 IF IN = JN THEN GOSUB 9000: PRINT "I NODE=J NODE FOR ELEMENT "; I: STOP
1070 REM IF ABS(JN - IN) > MX THEN MX = ABS(JN - IN)
1100 IF S9(I) < = 0 THEN GOSUB 9000: PRINT "2 ND M.O.A FOR ELEMENT "; I; "="; S9(I):
STOP
1130 IF A9(I) < = 0 THEN GOSUB 9000: PRINT "AREA FOR ELEMENT "; I; "="; A9(I):
STOP
1140 IF HY = 0 THEN GOTO 1240
1160 INPUT #1, U9(I), UP(I), XAS(I)
1210 XL = SQR((X(JN) - X(IN)) ^ 2 + (Y(JN) - Y(IN)) ^ 2)
1220 IF XAS(I) < 0 OR XAS(I) > XL THEN GOSUB 9000: PRINT "THE DISTANCE 'AS'
FROM LEFT END OF ELEMENT TO LEFT END OF LOAD="; XAS(I); " FOR ELEMENT";
I: STOP
1230 GOTO 1260
1240 REM UNIFORMLY DISTRIBUTED LOAD
1250 INPUT #1, U9(I)
1260 REM END OF ELEMENT DETAILS
1265 FOR II = 1 TO NNODE: IF II = 1 THEN IA = IN ELSE IA = JN
1270 FOR IJ = 1 TO NDF
1275 II1 = NDF * (IA - 1) + IJ
1280 FOR JJ = 1 TO NNODE: IF JJ = 1 THEN IB = IN ELSE IB = JN
1285 FOR JI = 1 TO NDF
1290 JJ1 = NDF * (IB - 1) + JI
1295 IIJJ = JJ1 - II1 + 1: IF IIJJ > DIAG(JJ1) THEN DIAG(JJ1) = IIJJ
1300 NEXT JI, JJ, IJ, II, I
1302 DIAG(1) = 1: FOR II = 1 TO N3: DIAG(II) = DIAG(II - 1) + DIAG(II): NEXT II
1305 NT = DIAG(N3)

```
1310 DIM A(NT), Q(N3), C(N3)
1330 FOR I = 1 TO NT: A(I) = 0: NEXT I
1340 FOR I = 1 TO N3: Q(I) = 0: NEXT I
1350 FOR I = 1 TO N3: QC(I) = 0: NEXT I
1360 IF NC = 0 THEN GOTO 1490
1370 REM "TYPE IN THE NODAL POSITIONS & VALUES OF THE CONCENTRATED LOADS
& COUPLES"
1380 FOR I = 1 TO NC
1390 INPUT #1, PO, HW, VW, CW
1400 IF PO < 1 OR PO > NJ THEN GOSUB 9000: PRINT "NO OF NODE WITH
CONCENTRATED LOADS & COUPLES="; PO: STOP
1450 QC(PO * 3 - 2) = HW: QC(PO * 3 - 1) = VW: QC(PO * 3) = CW
1455 POSW(I) = PO
1460 NEXT I
1490    GOSUB 4660
1500 Z9 = 0: GOSUB 10000
REM START OF ELEMENT CYCLE
1505 FOR ME = 1 TO MS
1510 PRINT "ELEMENT NO."; ME; " UNDER COMPUTATION": PRINT
1520 I = I9(ME)
1530 J = J9(ME)
1540 SA = S9(ME)
1550 CA = A9(ME)
1560 IF HY = 0 THEN UD = U9(ME)
1570 L = SQR((X(J) - X(I)) ^ 2 + (Y(J) - Y(I)) ^ 2)
1580 CS = (X(J) - X(I)) / L
1590 SN = (Y(J) - Y(I)) / L
1600 I3 = 3 * I
1610 I2 = I3 - 1
1620 I1 = I3 - 2
1630 J3 = 3 * J
1640 J2 = J3 - 1
1650 J1 = J3 - 2
1660 C1 = 12 * E * SA * SN ^ 2 / L ^ 3 + CS ^ 2 * CA * E / L
1670 C2 = 12 * E * SA * CS * SN / L ^ 3 - CS * SN * CA * E / L
1680 C3 = 12 * E * SA * CS ^ 2 / L ^ 3 + SN ^ 2 * CA * E / L
1690 C4 = 6 * E * SA * SN / L ^ 2
1700 C5 = 6 * E * SA * CS / L ^ 2
1710 C6 = 4 * E * SA / L
1720 K(1, 1) = C1
1730 K(2, 1) = -C2
1740 K(1, 2) = -C2
1750 K(3, 1) = C4
1760 K(1, 3) = C4
1770 K(4, 1) = -C1
1780 K(1, 4) = -C1
1790 K(5, 1) = C2
1800 K(1, 5) = C2
1810 K(6, 1) = C4
1820 K(1, 6) = C4
1830 K(2, 2) = C3
1840 K(3, 2) = -C5
1850 K(2, 3) = -C5
```

```
1860 K(4, 2) = C2
1870 K(2, 4) = C2
1880 K(5, 2) = -C3
1890 K(2, 5) = -C3
1900 K(6, 2) = -C5
1910 K(2, 6) = -C5
1920 K(3, 3) = C6
1930 K(4, 3) = -C4
1940 K(3, 4) = -C4
1950 K(5, 3) = C5
1960 K(3, 5) = C5
1970 K(6, 3) = .5 * C6
1980 K(3, 6) = .5 * C6
1990 K(4, 4) = C1
2000 K(5, 4) = -C2
2010 K(4, 5) = -C2
2020 K(6, 4) = -C4
2030 K(4, 6) = -C4
2040 K(5, 5) = C3
2050 K(6, 5) = C5
2060 K(5, 6) = C5
2070 K(6, 6) = C6
2080 SP(1) = 0
2090 SP(4) = 0
2100 IF HY = 1 THEN GOTO 2160
2110 SP(2) = UD * L / 2
2120 SP(5) = UD * L / 2
2130 SP(3) = -UD * L ^ 2 / 12
2140 SP(6) = UD * L ^ 2 / 12
2150 GOTO 2290
2160 WA = U9(ME)
2170 WB = UP(ME)
2180 AL = XAS(ME)
2190 X3 = (L - AL) ^ 3
2200 X2 = (L - AL) ^ 2
2210 SP(2) = WA * X3 * (L + AL) / (2 * L ^ 3)
2220 SP(2) = SP(2) + (WB - WA) * X3 * (3 * L + 2 * AL) / (20 * L ^ 3)
2230 SP(5) = (WA + WB) * (L - AL) / 2 - SP(2)
2240 SP(3) = -WA * X3 * (L + 3 * AL) / (12 * L ^ 2)
2250 SP(3) = SP(3) - (WB - WA) * X3 * (2 * L + 3 * AL) / (60 * L ^ 2)
2260 SP(6) = SP(2) * L + SP(3) - WA * X2 / 2
2270 SP(6) = SP(6) - (WB - WA) * X2 / 6
2280 SP(6) = -SP(6)
2290 FOR II = 1 TO 6
2300 FOR JJ = 1 TO 6
2310 DC(II, JJ) = 0
2320 NEXT JJ
2330 NEXT II
2340 DC(1, 1) = CS
2350 DC(2, 2) = CS
2360 DC(4, 4) = CS
2370 DC(5, 5) = CS
2380 DC(2, 1) = SN
```

```
2390 DC(5, 4) = SN
2400 DC(1, 2) = -SN
2410 DC(4, 5) = -SN
2420 DC(3, 3) = 1
2430 DC(6, 6) = 1
2440 FOR II = 1 TO 6
2450 SQ(II) = 0
2460 FOR JJ = 1 TO 6
2470 SQ(II) = SQ(II) + DC(II, JJ) * SP(JJ)
2480 NEXT JJ
2490 NEXT II
2500 Q(I1) = Q(I1) + SQ(1)
2510 Q(I2) = Q(I2) + SQ(2)
2520 Q(I3) = Q(I3) + SQ(3)
2530 Q(J1) = Q(J1) + SQ(4)
2540 Q(J2) = Q(J2) + SQ(5)
2550 Q(J3) = Q(J3) + SQ(6)
2560 I1 = 3 * I - 3
2570 J1 = 3 * J - 3
2580 FOR II = 1 TO 3
2585 FOR JJ = 1 TO 3
2590 MM = I1 + II: MN = I1 + JJ
2595 NM = J1 + II: N9 = J1 + JJ
2600 IF JJ > II THEN GOTO 2615
2605 A(DIAG(MM) + MN - MM) = A(DIAG(MM) + MN - MM) + K(II, JJ)
2610 A(DIAG(NM) + N9 - NM) = A(DIAG(NM) + N9 - NM) + K(II + 3, JJ + 3)
2615 IF I1 > J1 THEN GOTO 2630
2620 A(DIAG(NM) + MN - NM) = A(DIAG(NM) + MN - NM) + K(II + 3, JJ)
2625 GOTO 2640
2630 A(DIAG(MM) + N9 - MM) = A(DIAG(MM) + N9 - MM) + K(II, JJ + 3)
2640 NEXT JJ, II
2720 NEXT ME
2730 IF NC = 0 THEN GOTO 2770
2740 FOR I = 1 TO N3
2750 Q(I) = Q(I) + QC(I)
2760 NEXT I
2770 SU = 3 * NF
2780 FOR I = 1 TO SU
2790 N9 = NS(I)
2795 DI = DIAG(N9)
2800 IF N9 = 0 THEN GOTO 2830
2810 A(DI) = A(DI) * 1E+12 + 1E+12
2830 NEXT I
2840 FOR II = 1 TO N3
2850 C(II) = Q(II)
2860 NEXT II
2870 N = N3
2880 PRINT : PRINT "THE SIMULTANEOUS EQUATIONS ARE NOW BEING SOLVED":
PRINT
2890 GOSUB 3970
2900 FOR I = 1 TO N3
2910 U(I) = C(I)
2920 NEXT I
```

```
2940 PRINT #2, : PRINT #2, "THE NODAL DISPLACEMENTS (U0,V0 & THETA) ARE:-"
2950 FOR I = 1 TO NJ
2960 PRINT #2, "NODE="; I,
2970 FOR J = 1 TO 3
2980 PRINT #2, U(3 * I - 3 + J), : NEXT J
2990 PRINT #2,
3000 NEXT I
3010 FOR ME = 1 TO MS
3020 I = I9(ME)
3030 J = J9(ME)
3040 SA = S9(ME)
3050 CA = A9(ME)
3060 L = SQR((X(J) - X(I)) ^ 2 + (Y(J) - Y(I)) ^ 2)
3070 CS = (X(J) - X(I)) / L
3080 SN = (Y(J) - Y(I)) / L
3090 IF HY = 1 THEN GOTO 3120
3100 UD = U9(ME)
3110 GOTO 3230
3120 WA = U9(ME)
3130 WB = UP(ME)
3140 AL = XAS(ME)
3150 X3 = (L - AL) ^ 3
3160 X2 = (L - AL) ^ 2
3170 SP(2) = WA * X3 * (L + AL) / (2 * L ^ 3)
3180 SP(2) = SP(2) + (WB - WA) * X3 * (3 * L + 2 * AL) / (20 * L ^ 3)
3190 SP(3) = -WA * X3 * (L + 3 * AL) / (12 * L ^ 2)
3200 SP(3) = SP(3) - (WB - WA) * X3 * (2 * L + 3 * AL) / (60 * L ^ 2)
3220 SP(6) = SP(2) * L + SP(3) - WA * X2 / 2 - (WB - WA) * X2 / 6
3222 SP(2) = WA * X2 / (2 * L) + (WB - WA) * X2 / (6 * L)
3225 SP(5) = (WA + WB) * (L - AL) / 2 - SP(2)
3230 I1 = 3 * I - 3
3240 J1 = 3 * J - 3
3250 FOR II = 1 TO 3
3260 MM = I1 + II
3270 MN = J1 + II
3280 SG(II) = U(MM)
3290 SG(II + 3) = U(MN)
3300 NEXT II
3310 FOR II = 1 TO 6
3320 FOR JJ = 1 TO 6
3330 DC(II, JJ) = 0
3340 NEXT JJ
3350 NEXT II
3360 DC(1, 1) = CS
3370 DC(2, 2) = CS
3380 DC(4, 4) = CS
3390 DC(5, 5) = CS
3400 DC(1, 2) = SN
3410 DC(4, 5) = SN
3420 DC(2, 1) = -SN
3430 DC(5, 4) = -SN
3440 DC(3, 3) = 1
3450 DC(6, 6) = 1
```

```
3460 FOR II = 1 TO 6
3470 SL(II) = 0
3480 FOR JJ = 1 TO 6
3490 SL(II) = SL(II) + DC(II, JJ) * SG(JJ)
3500 NEXT JJ
3510 NEXT II
3520 FOR II = 1 TO 2
3530 IF II = 1 THEN XI = 0: ND = I
3560 IF II = 2 THEN XI = 1: ND = J
3580 FOR JJ = 1 TO 2
3590 FOR KK = 1 TO 6
3600 BX(JJ, KK) = 0
3610 NEXT KK
3620 NEXT JJ
3630 BX(1, 1) = -1 / L
3640 BX(1, 4) = 1 / L
3650 BX(2, 2) = (6 - 12 * XI) / L ^ 2
3660 BX(2, 3) = (-4 + 6 * XI) / L
3670 BX(2, 5) = -(6 - 12 * XI) / L ^ 2
3680 BX(2, 6) = (-2 + 6 * XI) / L
3690 FOR JJ = 1 TO 2
3700 S2(JJ) = 0
3710 FOR KK = 1 TO 6
3720 S2(JJ) = S2(JJ) + BX(JJ, KK) * SL(KK)
3730 NEXT KK
3740 NEXT JJ
3750 S2(1) = E * CA * S2(1)
3760 S2(2) = E * SA * S2(2)
3770 IF HY = 1 THEN GOTO 3820
3780 S2(2) = S2(2) - UD * L ^ 2 / 12
3790 IF II = 1 THEN MA = S2(2)
3800 IF II = 2 THEN MB = S2(2)
3810 GOTO 3840
3820 IF II = 1 THEN S2(2) = S2(2) + SP(3): MA = S2(2)
3830 IF II = 2 THEN S2(2) = S2(2) + SP(6): MB = S2(2)
3840 PRINT #2, "FORCE IN ELEMENT "; I; "-"; J; " AT NODE "; ND; "="; S2(1)
3850 PRINT #2, "BENDING MOMENT IN ELEMENT "; I; "-"; J; " AT NODE "; ND; "=";
S2(2)
3890 NEXT II
3895 BM(1, ME) = -MA: BM(2, ME) = -MB
3900 GOSUB 4190
3910 NEXT ME
3920 GOSUB 4420
3940 GOSUB 11000
3950 GOSUB 11500
3955 GOSUB 12000
3960 GOTO 14000
3970 A(1) = SQR(A(1))
3975 FOR II = 2 TO N
3980 DI = DIAG(II) - II
3985 LL = DIAG(II - 1) - DI + 1
3990 FOR JJ = LL TO II
3995 X = A(DI + JJ)
```

```
4000 DJ = DIAG(JJ) - JJ
4005 IF JJ = 1 THEN GOTO 4035
4010 LB = DIAG(JJ - 1) - DJ + 1
4015 IF LL > LB THEN LB = LL
4020 FOR KK = LB TO JJ - 1
4025 X = X - A(DI + KK) * A(DJ + KK)
4030 NEXT KK
4035 A(DI + JJ) = X / A(DJ + JJ)
4040 NEXT JJ
4045 A(DI + II) = SQR(X)
4050 NEXT II
4055 C(1) = C(1) / A(1)
4060 FOR II = 2 TO N
4065 DI = DIAG(II) - II
4070 LL = DIAG(II - 1) - DI + 1
4075 X = C(II)
4080 FOR JJ = LL TO II - 1
4085 X = X - A(DI + JJ) * C(JJ)
4090 NEXT JJ
4095 C(II) = X / A(DI + II)
4100 NEXT II
4105 FOR II = N TO 2 STEP -1
4110 DI = DIAG(II) - II
4115 X = C(II) / A(DI + II)
4120 C(II) = X
4125 LL = DIAG(II - 1) - DI + 1
4130 FOR KK = LL TO II - 1
4135 C(KK) = C(KK) - X * A(DI + KK)
4140 NEXT KK
4145 NEXT II
4150 C(1) = C(1) / A(1)
4180 RETURN
4190 REM CALCULATIONS OF THE REACTIONS
4200 PRINT #2,
4210 FOR II = 1 TO NF
4220 IF I9(ME) = NP(II) THEN SU = NP(II): GOTO 4250
4230 NEXT II
4240 GOTO 4310
4250 IF HY = 1 THEN YI = -SP(2) + (MA - MB) / L
4260 IF HY = 0 THEN YI = -UD * L / 2 + (MA - MB) / L
4270 XRH = -S2(1) * CS - YI * SN
4280 XRV = YI * CS - S2(1) * SN
4290 RH(SU) = RH(SU) + XRH
4300 RV(SU) = RV(SU) + XRV
4310 FOR II = 1 TO NF
4320 IF J9(ME) = NP(II) THEN SU = NP(II): GOTO 4350
4330 NEXT II
4340 GOTO 4410
4350 IF HY = 1 THEN YI = -SP(5) + (-MA + MB) / L
4360 IF HY = 0 THEN YI = -UD * L / 2 + (-MA + MB) / L
4370 XRH = S2(1) * CS - YI * SN
4380 XRV = YI * CS + S2(1) * SN
4390 RH(SU) = RH(SU) + XRH
```

```
4400 RV(SU) = RV(SU) + XRV
4410 RETURN
4420 REM OUTPUT OF REACTIONS
4430 PRINT #2, : PRINT #2, "THE REACTIONS ARE:-": PRINT #2,
4440 FOR II = 1 TO NJ
4450 IF RH(II) = 0 THEN GOTO 4470
4460 GOTO 4480
4470 IF RV(II) = 0 THEN GOTO 4500
4480 PRINT #2, "NODE="; II
4490 PRINT #2, "HORIZONTAL REACTION="; RH(II); "   VERTICAL REACTION="; RV(II)
4500 NEXT II
4510 RETURN
4520 PRINT "COPYRIGHT OF DR.C.T.F.ROSS"
4530 PRINT : PRINT "NOT TO BE REPRODUCED"
4540 PRINT : PRINT "HAVE YOU READ THE BOOK": PRINT
4550 PRINT : PRINT "FINITE ELEMENT PROGRAMS IN STRUCTURAL ENGINEERING &
CONTINUUM MECHANICS"
4560 PRINT : PRINT "BY"
4570 PRINT : PRINT "C.T.F.ROSS"
4580 PRINT : PRINT "TYPE Y/N"
4590 A$ = INKEY$: IF A$ = "" THEN GOTO 4590
4600 IF A$ = "y" THEN A$ = "Y"
4610 RETURN
4660 REM OUTPUT OF INPUT
4670 PRINT #2,
4680 PRINT #2, : PRINT #2, "STATIC ANALYSIS OF RIGID-JOINTED PLANE FRAMES":
PRINT #2,
4682 IF HY = 0 THEN PRINT #2, "UNIFORMLY DISTRIBUTED LOAD CASE": GOTO 4690
4685 PRINT #2, "HYDROSTATIC LOAD CASE"
4690 PRINT #2, "NUMBER OF NODES="; NJ
4700 PRINT #2, "NUMBER OF ELEMENTS="; MS
4710 PRINT #2, "NUMBER OF NODES WITH ZERO DISPLACEMENTS="; NF
4720 PRINT #2, "NUMBER OF NODES WITH CONCENTRATED LOADS & COUPLES="; NC
4730 PRINT #2, : PRINT #2, "THE NODAL COORDINATES X0 & Y0 ARE:-"
4740 FOR I = 1 TO NJ
4750 PRINT #2, " X("; I; ")="; X(I), "Y("; I; ")="; Y(I)
4760 NEXT I
4780 PRINT #2, : PRINT #2, "THE NODAL POSITIONS OF THE ZERO DISPLACEMENTS":
PRINT #2,
4790 FOR I = 1 TO NF
4800 PRINT #2, "NODE="; NP(I)
4810 SU = NP(I)
4820 N9 = NS(I * 3 - 2)
4830 IF N9 > 0 THEN PRINT #2, "HORIZONTAL DISPLACEMENT AT NODE "; SU; " IS
ZERO"
4850 N9 = NS(I * 3 - 1)
4860 IF N9 > 0 THEN PRINT #2, "VERTICAL DISPLACEMENT AT NODE "; SU; " IS ZERO"
4870 N9 = NS(I * 3)
4880 IF N9 > 0 THEN PRINT #2, "ROTATIONAL DISPLACEMENT AT NODE "; SU; " IS
ZERO"
4890 NEXT I
4910 PRINT #2, "ELASTIC MODULUS="; E
4930 PRINT #2, : PRINT #2, "ELEMENT DETAILS"
```

```
4950 FOR I = 1 TO MS
4960 IN = I9(I): JN = J9(I): PRINT #2, "ELEMENT "; IN; "-"; JN
4980 PRINT #2, "2ND MOMENT OF AREA="; S9(I)
4990 PRINT #2, "CROSS-SECTIONAL AREA="; A9(I)
5010 IF HY = 0 THEN GOTO 5120
5030 PRINT #2, "VALUE OF HYDROSTATIC LOAD ON 'LEFT' OF ELEMENT="; U9(I)
5040 PRINT #2, "VALUE OF HYDROSTATIC LOAD ON 'RIGHT' OF ELEMENT="; UP(I)
5060 PRINT #2, "DISTANCE OF 'LEFT' END OF LOAD FROM 'LEFT' END OF
ELEMENT="; XAS(I)
5090 GOTO 5130
5120 PRINT #2, "VALUE OF UNIFORMLY DISTRIBUTED LOAD="; U9(I)
5130 PRINT #2, : PRINT #2,
5140 NEXT I
5150 PRINT #2, "NUMBER OF NODES WITH CONCENTRATED LOADS & COUPLES"
5160 IF NC = 0 THEN GOTO 5240
5170 FOR I = 1 TO NC
5180 NODEW = POSW(I)
5190 PRINT #2, "NODE="; NODEW
5200 PRINT #2, "CONCENTRATED HORIZONTAL LOAD="; QC(3 * NODEW - 2)
5210 PRINT #2, "CONCENTRATED VERTICAL LOAD="; QC(3 * NODEW - 1)
5220 PRINT #2, "CONCENTRATED COUPLE="; QC(NODEW * 3)
5230 NEXT I
5240 RETURN
9000 REM WARNING
9010 PRINT : PRINT "INCORRECT DATA": PRINT
9020 RETURN
10000 IF Z9 = 77 OR Z9 = 99 THEN GOTO 10280
10010 XZ = YZ = 1E+12
10020 XM = YM = -1E+12
10030 PRINT : PRINT "TO CONTINUE, TYPE Y": PRINT
10040 A$ = INKEY$: IF A$ = "" THEN GOTO 10040
10050 IF A$ = "Y" OR A$ = "y" THEN GOTO 10070
10060 GOTO 10040
10070 FOR EL = 1 TO MS
10080 I = I9(EL)
10090 J = J9(EL)
10100 IF X(I) < XZ THEN XZ = X(I)
10110 IF Y(I) < YZ THEN YZ = Y(I)
10130 IF X(J) < XZ THEN XZ = X(J)
10140 IF Y(J) < YZ THEN YZ = Y(J)
10170 IF X(I) > XM THEN XM = X(I)
10180 IF Y(I) > YM THEN YM = Y(I)
10190 IF X(J) > XM THEN XM = X(J)
10200 IF Y(J) > YM THEN YM = Y(J)
10210 NEXT EL
10230 XS = 500 / (XM - XZ)
10240 IF YM - YZ = 0 THEN GOTO 10280
10250 YS = 250 / (YM - YZ)
10260 IF XS > YS THEN XS = YS
10280 IF Z9 = 0 OR Z9 = 77 THEN SCREEN 9, 0: WIDTH 80
10290 IF Z9 = 99 THEN LOCATE 25, 1: PRINT "DEFLECTED FORM OF FRAME"
10310 FOR EL = 1 TO MS
10320 I = I9(EL)
```

```
10330 J = J9(EL)
10340 IF Z9 = 99 THEN GOTO 10420
10350 XI = (X(I) - XZ) * XS
10360 YI = (Y(I) - YZ) * XS
10370 XJ = (X(J) - XZ) * XS
10380 YJ = (Y(J) - YZ) * XS
10390 GOTO 10500
10420 XI = X(I) - XZ * XS: XJ = X(J) - XZ * XS
10430 YI = Y(I) - YZ * XS: YJ = Y(J) - YZ * XS
10500 LINE (XI + 40, 300 - YI)-(XJ + 40, 300 - YJ)
10550 NEXT EL
10560 IF Z9 = 77 THEN GOTO 10800
10580 PRINT "TO CONTINUE, TYPE Y"
10590 A$ = INKEY$: IF A$ = "" THEN GOTO 10590
10610 IF A$ = "Y" OR A$ = "y" THEN GOTO 10700
10650 GOTO 10590
10700 SCREEN 0, 0
10800 RETURN
11000 REM PLOT OF DEFLECTED FORM OF FRAME
11010 PRINT "THE DEFLECTED FORM OF THE FRAME WILL NOW BE DRAWN"
11020 Z9 = 77
11050 GOSUB 10000
11060 UM = -1E+12
11080 FOR II = 1 TO N3
11090 FP = II / 3
11100 IP = INT(FP + .001)
11110 RM = IP - FP
11120 IF RM = 0 THEN GOTO 11170
11160 IF ABS(U(II)) > UM THEN UM = ABS(U(II))
11170 NEXT II
11180 SC = 20
11200 FOR II = 1 TO NJ
11210 X(II) = X(II) * XS + SC * U(3 * II - 2) / UM
11220 Y(II) = Y(II) * XS + SC * U(3 * II - 1) / UM
11240 NEXT II
11260 Z9 = 99
11280 GOSUB 10000
11300 RETURN
11500 REM PLOT OF BM DIAGRAM
11510 Z9 = 77
11520 FOR II = 1 TO NJ
11530 X(II) = XG(II): Y(II) = YG(II)
11540 NEXT II
11550 GOSUB 10000
11560 BIG = 0
11570 FOR II = 1 TO MS
11580 IF ABS(BM(1, II)) > BIG THEN BIG = ABS(BM(1, II))
11590 IF ABS(BM(2, II)) > BIG THEN BIG = ABS(BM(2, II))
11600 NEXT II
11610 FOR EL = 1 TO MS
11620 I = I9(EL): J = J9(EL)
11630 L = SQR((X(J) - X(I)) ^ 2 + (Y(J) - Y(I)) ^ 2)
11640 IF HY = 0 THEN UD = U9(EL): GOTO 11710
```

```
11650 WA = U9(EL): WB = UP(EL): AL = XAS(EL)
11660 X3 = (L - AL) ^ 3: X2 = (L - AL) ^ 2
11670 RA = WA * X3 * (L + AL) / (2 * L ^ 3) + (WB - WA) * X3 * (3 * L + 2 * AL) / (20
* L ^ 3)
11710 CS = (X(J) - X(I)) / L: SN = (Y(J) - Y(I)) / L
11720 EQN = 0
11730 XI = (X(I) - XZ) * XS
11740 YI = (Y(I) - YZ) * XS
11745 XK = XI: YK = YI
11750 FOR II = 0 TO 10
11760 XXI = II * L / 10
11770 IF HY = 0 THEN EQN = UD * L * XXI / 2 - UD * XXI ^ 2 / 2: GOTO 11850
11780 IF XXI < = AL THEN EQN = RA * XXI
11790 IF XXI > AL THEN EQN = RA * XXI - WA * (XXI - AL) ^ 2 / 2 - (WB - WA) * (XXI
- AL) ^ 3 / (6 * (L - AL))
11850 MOM = BM(1, EL) + (BM(2, EL) - BM(1, EL)) * XXI / L - EQN
11855 XJ = XI + XXI * CS * XS: YJ = YI + XXI * SN * XS
11860 BMX = MOM * SN
11870 BMY = MOM * CS
11880 XJ = XJ - SC * BMX / BIG: YJ = YJ + SC * BMY / BIG
11900 LINE (XK + 40, 300 - YK)-(XJ + 40, 300 - YJ)
11910 XK = XJ
11915 YK = YJ
11920 NEXT II
11930 NEXT EL
11935 LOCATE 25, 1: PRINT "              BENDING MOMENT DIAGRAM"
11940 A$ = INKEY$: IF A$ = "" THEN GOTO 11940
11950 IF A$ = "Y" OR A$ = "y" THEN GOTO 11970
11960 GOTO 11940
11970 SCREEN 0, 0
11980 RETURN
12000 REM PLOT OF SF DIAGRAM
12010 Z9 = 77
12020 FOR II = 1 TO NJ
12030 X(II) = XG(II): Y(II) = YG(II)
12040 NEXT II
12050 GOSUB 10000
12060 BIG = 0
12070 FOR EL = 1 TO MS
12080 I = I9(EL): J = J9(EL)
12090 L = SQR((X(J) - X(I)) ^ 2 + (Y(J) - Y(I)) ^ 2)
12100 RA = (BM(1, EL) - BM(2, EL)) / L
12110 IF ABS(RA) > BIG THEN BIG = ABS(RA)
12120 NEXT EL
12140 FOR EL = 1 TO MS
12150 I = I9(EL): J = J9(EL)
12160 L = SQR((X(J) - X(I)) ^ 2 + (Y(J) - Y(I)) ^ 2)
12170 CS = (X(J) - X(I)) / L: SN = (Y(J) - Y(I)) / L
12180 EQN = 0
12200 IF HY = 0 THEN UD = U9(EL): RA = (BM(2, EL) - BM(1, EL)) / L - UD * L / 2: GOTO
12250
12210 AL = XAS(EL): WA = U9(EL): WB = UP(EL)
12220 RA = (BM(2, EL) - BM(1, EL)) / L - WA * (L - AL) ^ 2 / (2 * L) - (WB - WA) * (L - AL)
```

359

```
^ 2 / (6 * L)
12250 EQN = 0
12260 XI = (X(I) - XZ) * XS
12270 YI = (Y(I) - YZ) * XS
12280 XK = XI - SC * RA * SN / BIG: YK = YI + SC * RA * CS / BIG
12300 FOR II = 0 TO 10
12310 XXI = II * L / 10
12320 IF HY = 0 THEN EQN = UD * XXI: GOTO 12400
12330 IF XXI < AL THEN EQN = 0
12340 IF XXI > AL THEN EQN = WA * (XXI - AL) + (WB - WA) * (XXI - AL) ^ 2 / (2 * (L
- AL))
12400 SF = RA + EQN
12440 XJ = XI + XXI * CS * XS: YJ = YI + XXI * SN * XS
12460 SFX = SF * SN
12470 SFY = SF * CS
12490 XJ = XJ - SC * SFX / BIG: YJ = YJ + SC * SFY / BIG
12510 LINE (XK + 40, 300 - YK)-(XJ + 40, 300 - YJ)
12520 XK = XJ
12530 YK = YJ
12535 NEXT II
12540 NEXT EL
12545 LOCATE 25, 1: PRINT "            SHEARING FORCE DIAGRAM"
12550 A$ = INKEY$: IF A$ = "" THEN GOTO 12550
12560 IF A$ = "Y" OR A$ = "y" THEN GOTO 12580
12570 GOTO 12550
12580 SCREEN 0, 0
12590 RETURN
14000 REM END
```

Computer Program for Rigid/Pin-Jointed Plane Frames "FRAMERP"

```
73 REM CORRECTED 03-02-90
REM IMPROVED 09-03-94
90 NNODE = 2: NDF = 3
100 CLS
110 REM COPYRIGHT OF DR.C.T.F.ROSS
120 REM 6 HURSTVILLE DRIVE, WATERLOOVILLE, PORTSMOUTH
130 REM PO7 7NB HANTS. ENGLAND
140 GOSUB 4520
150 IF A$ < > "Y" THEN GOTO 3960
152 PRINT "STATIC ANALYSIS OF RIGID JOINTED PLANE FRAMES"
154 INPUT "TYPE IN THE NAME OF YOUR INPUT FILE "; FILENAME$
PRINT "IF YOU ARE SATISFIED WITH THIS FILENAME, TYPE Y; ELSE N"
155 A$ = INKEY$: IF A$ = "" THEN GOTO 155
IF A$ = "Y" OR A$ = "y" THEN GOTO 157
IF A$ = "N" OR A$ = "n" THEN GOTO 154
GOTO 155
157 OPEN FILENAME$ FOR INPUT AS #1
159 INPUT "TYPE IN THE NAME OF YOUR OUTPUT FILE "; OUT$
PRINT "IF YOU ARE SATISFIED WITH THIS FILENAME, TYPE Y; ELSE N"
160 B$ = INKEY$: IF B$ = "" THEN GOTO 160
IF B$ = "Y" OR B$ = "y" THEN GOTO 165
IF B$ = "N" OR B$ = "n" THEN GOTO 159
GOTO 160
165 OPEN OUT$ FOR OUTPUT AS #2
170 INPUT #1, NJ
190 IF NJ < 2 OR NJ > 199 THEN GOSUB 9000: STOP
260 N3 = 3 * NJ
270 INPUT #1, MS
290 IF MS < 1 OR MS > 199 THEN GOSUB 9000: STOP
320 REM. "TYPE IN THE NUMBER OF NODES WITH ZERO 'DISPLACEMENTS'":PRINT
330 INPUT #1, NF
340 IF NF < 1 OR NF > NJ THEN GOSUB 9000: STOP
380 INPUT #1, NC
400 IF NC < 0 OR NC > NJ THEN GOSUB 9000: STOP
450 DIM U(N3), K(6, 6), NS(NF * 3), I9(MS), J9(MS), DIAG(N3), I(MS), J(MS)
460 DIM DC(6, 6), SP(6), SQ(6), SG(6), SL(6), X(NJ), Y(NJ)
470 DIM S9(MS), A9(MS), U9(MS), UP(MS), BX(2, 6), S2(2), E(MS)
480 DIM QC(N3), NP(NF), RH(NJ), RV(NJ), POSW(NC), XG(NJ), YG(NJ), BM(2, MS)
482 DIM IPIN(MS), JPIN(MS)
485 DIM NUD(MS), NWLOAD(MS), UDA(MS, 10), UDB(MS, 10), XA(MS, 10), XB(MS, 10)
490 DIM WLOAD(MS, 20), XAW(MS, 20)
495 DIM ITEMP(MS), JTEMP(MS)
500 FOR I = 1 TO NJ
520 INPUT #1, X(I), Y(I)
545 XG(I) = X(I): YG(I) = Y(I)
550 NEXT I
590 FOR I = 1 TO NF * 3: NS(I) = 0: NEXT I
600 FOR I = 1 TO NF
620 INPUT #1, NP(I)
```

```
630 IF NP(I) < 1 OR NP(I) > NJ THEN GOSUB 9000: STOP
680 INPUT #1, SUU, SUV, SUTHETA
SU$ = "FALSE"
IF SUU = 1 THEN SU$ = "Y" ELSE IF SUU = 0 THEN SU$ = "N"
690 IF SU$ = "Y" OR SU$ = "N" OR SU$ = "y" OR SU$ = "n" THEN GOTO 720
700 GOSUB 9000: STOP
720 IF SU$ = "Y" OR SU$ = "y" THEN NS(I * 3 - 2) = NP(I) * 3 - 2
740 SU$ = "FALSE"
IF SUV = 1 THEN SU$ = "Y" ELSE IF SUV = 0 THEN SU$ = "N"
750 IF SU$ = "Y" OR SU$ = "N" OR SU$ = "y" OR SU$ = "n" THEN GOTO 770
760 GOSUB 9000: STOP
770 IF SU$ = "Y" OR SU$ = "y" THEN NS(I * 3 - 1) = NP(I) * 3 - 1
790 SU$ = "FALSE"
IF SUTHETA = 1 THEN SU$ = "Y" ELSE IF SUTHETA = 0 THEN SU$ = "N"
800 IF SU$ = "Y" OR SU$ = "N" OR SU$ = "y" OR SU$ = "n" THEN GOTO 820
810 GOSUB 9000: STOP
820 IF SU$ = "Y" OR SU$ = "y" THEN NS(I * 3) = NP(I) * 3
830 NEXT I
860 REM VARYING YOUNG'S MODULII READ E
880 REM IF E< =0 THEN GOSUB 9000:STOP
950 REM ELEMENT DETAILS
960 FOR I = 1 TO MS
970 INPUT #1, I9(I), J9(I), S9(I), A9(I), E(I): ITEMP(I) = I9(I): I(I) = I9(I)
972 IPIN(I) = 0: JPIN(I) = 0
974 IF I9(I) < 0 THEN I9(I) = -I9(I): IPIN(I) = -1
980 IF I9(I) = 0 OR I9(I) > NJ THEN GOSUB 9000: STOP
1010 JTEMP(I) = J9(I): J(I) = J9(I)
1012 IF J9(I) < 0 THEN J9(I) = -J9(I): JPIN(I) = -1
1020 IF J9(I) = 0 OR J9(I) > NJ THEN GOSUB 9000: STOP
1050 IN = I9(I): JN = J9(I)
1060 IF IN = JN THEN GOSUB 9000: STOP
1100 IF S9(I) < = 0 THEN GOSUB 9000: STOP
1130 IF A9(I) < = 0 THEN GOSUB 9000: STOP
1135 IF E(I) < 0 THEN GOSUB 9000: STOP
1140 INPUT #1, NUD(I), NWLOAD(I)
1150 IF NUD(I) = 0 THEN GOTO 1205
1160 FOR II = 1 TO NUD(I)
1180 INPUT #1, UDA(I, II), UDB(I, II), XA(I, II), XB(I, II)
1200 NEXT II
1205 IF NWLOAD(I) = 0 THEN GOTO 1260
1210 FOR II = 1 TO NWLOAD(I)
1220 INPUT #1, WLOAD(I, II), XAW(I, II)
1230 NEXT II
1260 REM END OF ELEMENT DETAILS
1265 FOR II = 1 TO NNODE: IF II = 1 THEN IA = IN ELSE IA = JN
1270 FOR IJ = 1 TO NDF
1275 II1 = NDF * (IA - 1) + IJ
1280 FOR JJ = 1 TO NNODE: IF JJ = 1 THEN IB = IN ELSE IB = JN
1285 FOR JI = 1 TO NDF
1290 JJ1 = NDF * (IB - 1) + JI
1295 IIJJ = JJ1 - II1 + 1: IF IIJJ > DIAG(JJ1) THEN DIAG(JJ1) = IIJJ
1300 NEXT JI, JJ, IJ, II, I
1302 DIAG(1) = 1: FOR II = 1 TO N3: DIAG(II) = DIAG(II - 1) + DIAG(II): NEXT II
```

```
1305 NT = DIAG(N3)
1310 DIM A(NT), Q(N3), C(N3)
1330 FOR I = 1 TO NT: A(I) = 0: NEXT I
1340 FOR I = 1 TO N3: Q(I) = 0: NEXT I
1350 FOR I = 1 TO N3: QC(I) = 0: NEXT I
1360 IF NC = 0 THEN GOTO 1490
1370 REM "TYPE IN THE NODAL POSITIONS & VALUES OF THE CONCENTRATED LOADS
& COUPLES"
1380 FOR I = 1 TO NC
1390 INPUT #1, PO, HW, VW, CW
1400 IF PO < 1 OR PO > NJ THEN GOSUB 9000: STOP
1450 QC(PO * 3 - 2) = HW: QC(PO * 3 - 1) = VW: QC(PO * 3) = CW
1455 POSW(I) = PO
1460 NEXT I
1490 GOSUB 4660
1500 Z9 = 0: GOSUB 10000
1505 FOR ME = 1 TO MS
1510 PRINT "ELEMENT NO."; ME; " UNDER COMPUTATION": PRINT
1520 I = I9(ME)
1530 J = J9(ME)
1540 SA = S9(ME)
1550 CA = A9(ME)
1555 E = E(ME)
1570 L = SQR((X(J) - X(I)) ^ 2 + (Y(J) - Y(I)) ^ 2)
1580 CS = (X(J) - X(I)) / L
1590 SN = (Y(J) - Y(I)) / L
1600 I3 = 3 * I
1610 I2 = I3 - 1
1620 I1 = I3 - 2
1630 J3 = 3 * J
1640 J2 = J3 - 1
1650 J1 = J3 - 2
1652 IF IPIN(ME) = -1 AND JPIN(ME) = 0 THEN GOSUB 6000: GOTO 1720
1655 IF IPIN(ME) = 0 AND JPIN(ME) = -1 THEN GOSUB 6000: GOTO 1720
1656 IF IPIN(ME) = -1 AND JPIN(ME) = -1 AND NUD(ME) = 0 AND NWLOAD(ME) = 0
THEN SA = 0
1660 C1 = 12 * E * SA * SN ^ 2 / L ^ 3 + CS ^ 2 * CA * E / L
1670 C2 = 12 * E * SA * CS * SN / L ^ 3 - CS * SN * CA * E / L
1680 C3 = 12 * E * SA * CS ^ 2 / L ^ 3 + SN ^ 2 * CA * E / L
1690 C4 = 6 * E * SA * SN / L ^ 2
1700 C5 = 6 * E * SA * CS / L ^ 2
1710 C6 = 4 * E * SA / L
1720 K(1, 1) = C1
1730 K(2, 1) = -C2
1740 K(1, 2) = -C2
1750 K(3, 1) = C4
1760 K(1, 3) = C4
1770 K(4, 1) = -C1
1780 K(1, 4) = -C1
1790 K(5, 1) = C2
1800 K(1, 5) = C2
1810 K(6, 1) = C4
1820 K(1, 6) = C4
```

```
1830 K(2, 2) = C3
1840 K(3, 2) = -C5
1850 K(2, 3) = -C5
1860 K(4, 2) = C2
1870 K(2, 4) = C2
1880 K(5, 2) = -C3
1890 K(2, 5) = -C3
1900 K(6, 2) = -C5
1910 K(2, 6) = -C5
1920 K(3, 3) = C6
1930 K(4, 3) = -C4
1940 K(3, 4) = -C4
1950 K(5, 3) = C5
1960 K(3, 5) = C5
1970 K(6, 3) = .5 * C6
1980 K(3, 6) = .5 * C6
1990 K(4, 4) = C1
2000 K(5, 4) = -C2
2010 K(4, 5) = -C2
2020 K(6, 4) = -C4
2030 K(4, 6) = -C4
2040 K(5, 5) = C3
2050 K(6, 5) = C5
2060 K(5, 6) = C5
2070 K(6, 6) = C6
2080 IF JPIN(ME) = -1 THEN K(6, 1) = 0: K(6, 2) = 0: K(6, 3) = 0: K(6, 4) = 0: K(6, 5) =
0: K(6, 6) = C6 / 1E+20
2090 IF JPIN(ME) = -1 THEN K(1, 6) = 0: K(2, 6) = 0: K(3, 6) = 0: K(4, 6) = 0: K(5, 6) =
0
2100 IF IPIN(ME) = -1 THEN K(3, 1) = 0: K(3, 2) = 0: K(3, 3) = C6 / 1E+20: K(3, 4) = 0:
K(3, 5) = 0: K(3, 6) = 0
2110 IF IPIN(ME) = -1 THEN K(1, 3) = 0: K(2, 3) = 0: K(4, 3) = 0: K(5, 3) = 0: K(6, 3) =
0
2120 FOR II = 1 TO 6: SP(II) = 0: NEXT II
2122 IF NUD(ME) = 0 GOTO 2200
2124 FOR II = 1 TO NUD(ME)
2126 WA = UDA(ME, II): WB = UDB(ME, II)
2128 AL = XA(ME, II): BL = XB(ME, II)
2130 X3 = (L - AL) ^ 3: X4 = (L - AL) ^ 4: X5 = (L - AL) ^ 5
2132 X3B = (L - BL) ^ 3: X4B = (L - BL) ^ 4: X5B = (L - BL) ^ 5
2134 K1 = WA * X3 / 6 + (WB - WA) * X4 / (24 * (BL - AL)) - WB * X3B / 6 - (WB - WA) *
X4B / (24 * (BL - AL))
2136 K2 = WA * X4 / 24 + (WB - WA) * X5 / (120 * (BL - AL)) - WB * X4B / 24 - (WB - WA)
* X5B / (120 * (BL - AL))
2138 IF IPIN(ME) = -1 AND JPIN(ME) = 0 THEN GOSUB 6100: GOTO 2160
2140 IF IPIN(ME) = 0 AND JPIN(ME) = -1 THEN GOSUB 6200: GOTO 2160
2142 Y1 = -6 * K1 / L ^ 2 + 12 * K2 / L ^ 3
2144 M1 = -Y1 * L / 2 - K1 / L
2146 M2 = -M1 - Y1 * L - WA * (BL - AL) * (L - BL + (BL - AL) / 2) - (WB - WA) * (BL - AL)
* (L - BL + (BL - AL) / 3) / 2
2148 Y2 = -Y1 - (WA + WB) * (BL - AL) / 2
2160 SP(2) = SP(2) - Y1
2162 SP(5) = SP(5) - Y2
```

```
2164 SP(3) = SP(3) - M1
2166 SP(6) = SP(6) - M2
2168 NEXT II
2200 IF NWLOAD(ME) = 0 THEN 2290
2202 FOR II = 1 TO NWLOAD(ME)
2204 WI = WLOAD(ME, II): AL = XAW(ME, II): BL = L - AL
2206 IF IPIN(ME) = -1 AND JPIN(ME) = 0 THEN GOSUB 6400: GOTO 2260
2208 IF IPIN(ME) = 0 AND JPIN(ME) = -1 THEN GOSUB 6500: GOTO 2260
2210 Y1 = -12 * WI * BL ^ 2 * (L / 4 - BL / 6) / L ^ 3
2212 M1 = -Y1 * L / 2 - WI * BL ^ 2 / (2 * L)
2214 Y2 = -Y1 - WI
2216 M2 = -M1 - Y1 * L - WI * BL
2260 SP(2) = SP(2) - Y1
2262 SP(5) = SP(5) - Y2
2264 SP(3) = SP(3) - M1
2266 SP(6) = SP(6) - M2
2268 NEXT II
2290 IF IPIN(ME) = -1 AND JPIN(ME) = -1 THEN SP(3) = 0: SP(6) = 0
2295 FOR II = 1 TO 6
2300 FOR JJ = 1 TO 6
2310 DC(II, JJ) = 0
2320 NEXT JJ
2330 NEXT II
2340 DC(1, 1) = CS
2350 DC(2, 2) = CS
2360 DC(4, 4) = CS
2370 DC(5, 5) = CS
2380 DC(2, 1) = SN
2390 DC(5, 4) = SN
2400 DC(1, 2) = -SN
2410 DC(4, 5) = -SN
2420 DC(3, 3) = 1
2430 DC(6, 6) = 1
2440 FOR II = 1 TO 6
2450 SQ(II) = 0
2460 FOR JJ = 1 TO 6
2470 SQ(II) = SQ(II) + DC(II, JJ) * SP(JJ)
2480 NEXT JJ
2490 NEXT II
2500 Q(I1) = Q(I1) + SQ(1)
2510 Q(I2) = Q(I2) + SQ(2)
2520 Q(I3) = Q(I3) + SQ(3)
2530 Q(J1) = Q(J1) + SQ(4)
2540 Q(J2) = Q(J2) + SQ(5)
2550 Q(J3) = Q(J3) + SQ(6)
2560 I1 = 3 * I - 3
2570 J1 = 3 * J - 3
2580 FOR II = 1 TO 3
2585 FOR JJ = 1 TO 3
2590 MM = I1 + II: MN = I1 + JJ
2595 NM = J1 + II: N9 = J1 + JJ
2600 IF JJ > II THEN GOTO 2615
2605 A(DIAG(MM) + MN - MM) = A(DIAG(MM) + MN - MM) + K(II, JJ)
```

```
2610 A(DIAG(NM) + N9 - NM) = A(DIAG(NM) + N9 - NM) + K(II + 3, JJ + 3)
2615 IF I1 > J1 THEN GOTO 2630
2620 A(DIAG(NM) + MN - NM) = A(DIAG(NM) + MN - NM) + K(II + 3, JJ)
2625 GOTO 2640
2630 A(DIAG(MM) + N9 - MM) = A(DIAG(MM) + N9 - MM) + K(II, JJ + 3)
2640 NEXT JJ, II
2720 NEXT ME
2730 IF NC = 0 THEN GOTO 2770
2740 FOR I = 1 TO N3
2750 Q(I) = Q(I) + QC(I)
2760 NEXT I
2770 SU = 3 * NF
2780 FOR I = 1 TO SU
2790 N9 = NS(I)
2795 DI = DIAG(N9)
2800 IF N9 = 0 THEN GOTO 2830
2810 A(DI) = A(DI) * 1E+12 + 1E+12
2830 NEXT I
2840 FOR II = 1 TO N3
2850 C(II) = Q(II)
2860 NEXT II
2870 n = N3
2880 PRINT : PRINT "THE SIMULTANEOUS EQUATIONS ARE NOW BEING SOLVED":
PRINT
2890 GOSUB 3970
2900 FOR I = 1 TO N3
2910 U(I) = C(I)
2920 NEXT I
2940 PRINT #2, : PRINT #2, "THE NODAL DISPLACEMENTS (U0,V0 & THETA) ARE:-"
2950 FOR I = 1 TO NJ
2960 PRINT #2, "NODE="; I,
2970 FOR J = 1 TO 3
2980 PRINT #2, U(3 * I - 3 + J), : NEXT J
2990 PRINT #2,
3000 NEXT I
3010 FOR ME = 1 TO MS
3015 COND = -1
3020 I = I9(ME)
3030 J = J9(ME)
3031 SA = S9(ME)
3032 CA = A9(ME)
3033 E = E(ME)
3034 L = SQR((X(J) - X(I)) ^ 2 + (Y(J) - Y(I)) ^ 2)
3035 CS = (X(J) - X(I)) / L
3036 SN = (Y(J) - Y(I)) / L
3040 FOR II = 1 TO NF
3050 SUU = 3 * II - 2: SUV = 3 * II - 1: SUT = 3 * II
3055 REM IF CS < > 1 AND NS(SUU)=3*I-2 AND NS(SUV)=3*I-1 AND NS(SUT)=0 THEN
IPIN(ME)=-1
3060 REM IF CS < > 1 AND NS(SUU)=3*J-2 AND NS(SUV)=3*J-1 AND NS(SUT)=0 THEN
JPIN(ME)=-1
3062 IF MS = 1 AND NS(SUV) = 3 * I - 1 AND NS(SUT) = 0 THEN IPIN(ME) = -1
3063 IF MS = 1 AND NS(SUV) = 3 * J - 1 AND NS(SUT) = 0 THEN JPIN(ME) = -1
```

```
3065 NEXT II
3070 FOR II = 1 TO 6: SP(II) = 0: NEXT II
3084 IF IPIN(ME) = -1 AND JPIN(ME) = -1 THEN GOSUB 6900: GOTO 3230
3086 IF NUD(ME) = 0 GOTO 3160
3088 FOR II = 1 TO NUD(ME)
3090 WA = UDA(ME, II): WB = UDB(ME, II)
3092 AL = XA(ME, II): BL = XB(ME, II)
3094 X3 = (L - AL) ^ 3: X4 = (L - AL) ^ 4: X5 = (L - AL) ^ 5
3096 X3B = (L - BL) ^ 3: X4B = (L - BL) ^ 4: X5B = (L - BL) ^ 5
3098 K1 = WA * X3 / 6 + (WB - WA) * X4 / (24 * (BL - AL)) - WB * X3B / 6 - (WB - WA) *
X4B / (24 * (BL - AL))
3100 K2 = WA * X4 / 24 + (WB - WA) * X5 / (120 * (BL - AL)) - WB * X4B / 24 - (WB - WA)
* X5B / (120 * (BL - AL))
3102 IF IPIN(ME) = -1 AND JPIN(ME) = 0 THEN GOSUB 6100: GOTO 3140
3104 IF IPIN(ME) = 0 AND JPIN(ME) = -1 THEN GOSUB 6200: GOTO 3140
3106 Y1 = -6 * K1 / L ^ 2 + 12 * K2 / L ^ 3
3108 M1 = -Y1 * L / 2 - K1 / L
3110 M2 = -M1 - Y1 * L - WA * (BL - AL) * (L - BL + (BL - AL) / 2) - (WB - WA) * (BL - AL)
* (L - BL + (BL - AL) / 3) / 2
3112 Y2 = -Y1 - (WA + WB) * (BL - AL) / 2
3140 SP(2) = SP(2) - Y1
3142 SP(5) = SP(5) - Y2
3144 SP(3) = SP(3) - M1
3146 SP(6) = SP(6) - M2
3148 NEXT II
3160 IF NWLOAD(ME) = 0 THEN 3230
3162 FOR II = 1 TO NWLOAD(ME)
3164 WI = WLOAD(ME, II): AL = XAW(ME, II): BL = L - AL
3166 IF IPIN(ME) = -1 AND JPIN(ME) = 0 THEN GOSUB 6400: GOTO 3210
3168 IF IPIN(ME) = 0 AND JPIN(ME) = -1 THEN GOSUB 6500: GOTO 3210
3170 Y1 = -12 * WI * BL ^ 2 * (L / 4 - BL / 6) / L ^ 3
3172 M1 = -Y1 * L / 2 - WI * BL ^ 2 / (2 * L)
3174 Y2 = -Y1 - WI
3176 M2 = -M1 - Y1 * L - WI * BL
3210 SP(2) = SP(2) - Y1
3212 SP(5) = SP(5) - Y2
3214 SP(3) = SP(3) - M1
3216 SP(6) = SP(6) - M2
3218 NEXT II
3230 I1 = 3 * I - 3
3240 J1 = 3 * J - 3
3250 FOR II = 1 TO 3
3260 MM = I1 + II
3270 MN = J1 + II
3280 SG(II) = U(MM)
3290 SG(II + 3) = U(MN)
3300 NEXT II
3310 FOR II = 1 TO 6
3320 FOR JJ = 1 TO 6
3330 DC(II, JJ) = 0
3340 NEXT JJ
3350 NEXT II
3360 DC(1, 1) = CS
```

```
3370 DC(2, 2) = CS
3380 DC(4, 4) = CS
3390 DC(5, 5) = CS
3400 DC(1, 2) = SN
3410 DC(4, 5) = SN
3420 DC(2, 1) = -SN
3430 DC(5, 4) = -SN
3440 DC(3, 3) = 1
3450 DC(6, 6) = 1
3460 FOR II = 1 TO 6
3470 SL(II) = 0
3480 FOR JJ = 1 TO 6
3490 SL(II) = SL(II) + DC(II, JJ) * SG(JJ)
3500 NEXT JJ
3510 NEXT II
3520 FOR II = 1 TO 2
3530 IF II = 1 THEN XI = 0: ND = I
3560 IF II = 2 THEN XI = 1: ND = J
3580 FOR JJ = 1 TO 2
3590 FOR KK = 1 TO 6
3600 BX(JJ, KK) = 0
3610 NEXT KK
3620 NEXT JJ
3630 BX(1, 1) = -1 / L
3640 BX(1, 4) = 1 / L
3650 BX(2, 2) = (6 - 12 * XI) / L ^ 2
3660 BX(2, 3) = (-4 + 6 * XI) / L
3670 BX(2, 5) = -(6 - 12 * XI) / L ^ 2
3680 BX(2, 6) = (-2 + 6 * XI) / L
3682 IF IPIN(ME) = -1 AND JPIN(ME) = 0 THEN GOSUB 6700
3685 IF IPIN(ME) = 0 AND JPIN(ME) = -1 THEN GOSUB 6800
3690 FOR JJ = 1 TO 2
3700 S2(JJ) = 0
3710 FOR KK = 1 TO 6
3720 S2(JJ) = S2(JJ) + BX(JJ, KK) * SL(KK)
3730 NEXT KK
3740 NEXT JJ
3750 S2(1) = E * CA * S2(1)
3760 S2(2) = E * SA * S2(2)
3820 IF II = 1 THEN S2(2) = S2(2) + SP(3): MA = S2(2)
3830 IF II = 2 THEN S2(2) = S2(2) - SP(6): MB = S2(2)
3840 PRINT #2, "FORCE IN ELEMENT "; I; "-"; J; " AT NODE "; ND; "="; S2(1)
3850 PRINT #2, "BENDING MOMENT IN ELEMENT "; I; "-"; J; " AT NODE "; ND; "=";
S2(2)
3890 NEXT II
3892 BM(1, ME) = -MA: BM(2, ME) = -MB
3894 SP(2) = 0: SP(5) = 0
3895 IF NUD(ME) = 0 THEN GOTO 3904
3896 FOR II = 1 TO NUD(ME)
3897 WA = UDA(ME, II): WB = UDB(ME, II)
3898 AL = XA(ME, II): BL = XB(ME, II)
3899 Y1 = -(WA * (BL - AL) * (L - BL + (BL - AL) / 2) + (WB - WA) * (L - BL + (BL - AL)
/ 3) / 2) / L
```

```
3900 Y2 = -Y1 - .5 * (WA + WB) * (BL - AL)
3901 SP(2) = SP(2) - Y1
3902 SP(5) = SP(5) - Y2
3903 NEXT II
3904 IF NWLOAD(ME) = 0 THEN GOTO 3912
3905 FOR II = 1 TO NWLOAD(ME)
3906 WI = WLOAD(ME, II): AL = XAW(ME, II): BL = L - AL
3907 Y1 = -WI * BL / L
3908 Y2 = -Y1 - WI
3909 SP(2) = SP(2) - Y1
3910 SP(5) = SP(5) - Y2
3911 NEXT II
3912 SP(2) = SP(2) + (-MA + MB) / L
3913 SP(5) = SP(5) + (MA - MB) / L
3916 GOSUB 4190
3918 NEXT ME
3920 GOSUB 4420
3940 GOSUB 11000
3950 GOSUB 11500
3955 GOSUB 12000
3960 GOTO 14000
3970 A(1) = SQR(A(1))
3975 FOR II = 2 TO n
3980 DI = DIAG(II) - II
3985 LL = DIAG(II - 1) - DI + 1
3990 FOR JJ = LL TO II
3995 X = A(DI + JJ)
4000 DJ = DIAG(JJ) - JJ
4005 IF JJ = 1 THEN GOTO 4035
4010 LB = DIAG(JJ - 1) - DJ + 1
4015 IF LL > LB THEN LB = LL
4020 FOR KK = LB TO JJ - 1
4025 X = X - A(DI + KK) * A(DJ + KK)
4030 NEXT KK
4035 A(DI + JJ) = X / A(DJ + JJ)
4040 NEXT JJ
4045 A(DI + II) = SQR(X)
4050 NEXT II
4055 C(1) = C(1) / A(1)
4060 FOR II = 2 TO n
4065 DI = DIAG(II) - II
4070 LL = DIAG(II - 1) - DI + 1
4075 X = C(II)
4080 FOR JJ = LL TO II - 1
4085 X = X - A(DI + JJ) * C(JJ)
4090 NEXT JJ
4095 C(II) = X / A(DI + II)
4100 NEXT II
4105 FOR II = n TO 2 STEP -1
4110 DI = DIAG(II) - II
4115 X = C(II) / A(DI + II)
4120 C(II) = X
4125 LL = DIAG(II - 1) - DI + 1
```

```
4130 FOR KK = LL TO II - 1
4135 C(KK) = C(KK) - X * A(DI + KK)
4140 NEXT KK
4145 NEXT II
4150 C(1) = C(1) / A(1)
4180 RETURN
4190 REM CALCULATIONS OF THE REACTIONS
4200 PRINT #2,
4210 FOR II = 1 TO NF
4220 IF I9(ME) = NP(II) THEN SU = NP(II): GOTO 4250
4230 NEXT II
4240 GOTO 4310
4250 YI = -SP(2)
4270 XRH = -S2(1) * CS - YI * SN
4280 XRV = YI * CS - S2(1) * SN
4290 RH(SU) = RH(SU) + XRH
4300 RV(SU) = RV(SU) + XRV
4310 FOR II = 1 TO NF
4320 IF J9(ME) = NP(II) THEN SU = NP(II): GOTO 4350
4330 NEXT II
4340 GOTO 4410
4350 YI = -SP(5)
4370 XRH = S2(1) * CS - YI * SN
4380 XRV = YI * CS + S2(1) * SN
4390 RH(SU) = RH(SU) + XRH
4400 RV(SU) = RV(SU) + XRV
4410 RETURN
4420 REM OUTPUT OF REACTIONS
4430 PRINT #2, : PRINT #2, "THE REACTIONS ARE:-": PRINT #2,
4440 FOR II = 1 TO NJ
4450 IF RH(II) = 0 THEN GOTO 4470
4460 GOTO 4480
4470 IF RV(II) = 0 THEN GOTO 4500
4480 PRINT #2, "NODE="; II
4490 PRINT #2, "HORIZONTAL REACTION="; RH(II); "   VERTICAL REACTION="; RV(II)
4500 NEXT II
4510 RETURN
4520 PRINT "COPYRIGHT OF DR.C.T.F.ROSS"
4530 PRINT : PRINT "NOT TO BE REPRODUCED"
4540 PRINT : PRINT "HAVE YOU READ THE BOOK": PRINT
4550 PRINT : PRINT "FINITE ELEMENT PROGRAMS IN STRUCTURAL ENGINEERING &
CONTINUUM MECHANICS"
4560 PRINT : PRINT "BY"
4570 PRINT : PRINT "C.T.F.ROSS"
4580 PRINT : PRINT "TYPE Y/N"
4590 A$ = INKEY$: IF A$ = "" THEN GOTO 4590
4600 IF A$ = "y" THEN A$ = "Y"
4610 RETURN
4660 REM OUTPUT OF INPUT
4670 PRINT #2,
4680 PRINT #2, : PRINT #2, "STATIC ANALYSIS OF RIGID-JOINTED PLANE FRAMES":
PRINT #2,
4685 PRINT #2, "VARYING YOUNG'S MODULII": PRINT #2,
```

```
4690 PRINT #2, "NUMBER OF NODES="; NJ
4700 PRINT #2, "NUMBER OF ELEMENTS="; MS
4710 PRINT #2, "NUMBER OF NODES WITH ZERO DISPLACEMENTS="; NF
4720 PRINT #2, "NUMBER OF NODES WITH CONCENTRATED LOADS & COUPLES="; NC
4730 PRINT #2, : PRINT #2, "THE NODAL COORDINATES X0 & Y0 ARE:-"
4740 FOR I = 1 TO NJ
4750 PRINT #2, " X("; I; ")="; X(I), "Y("; I; ")="; Y(I)
4760 NEXT I
4780 PRINT #2, : PRINT #2, "THE NODAL POSITIONS OF THE ZERO DISPLACEMENTS":
PRINT #2,
4790 FOR I = 1 TO NF
4800 PRINT #2, "NODE="; NP(I)
4810 SU = NP(I)
4820 N9 = NS(I * 3 - 2)
4830 IF N9 > 0 THEN PRINT #2, "HORIZONTAL DISPLACEMENT AT NODE "; SU; " IS
ZERO"
4850 N9 = NS(I * 3 - 1)
4860 IF N9 > 0 THEN PRINT #2, "VERTICAL DISPLACEMENT AT NODE "; SU; " IS ZERO"
4870 N9 = NS(I * 3)
4880 IF N9 > 0 THEN PRINT #2, "ROTATIONAL DISPLACEMENT AT NODE "; SU; " IS
ZERO"
4890 NEXT I
4910 REM PRINT #2,"ELASTIC MODULUS=";E
4930 PRINT #2, : PRINT #2, "ELEMENT DETAILS"
4950 FOR I = 1 TO MS
4960 IN = I(I): JN = J(I): PRINT #2, "ELEMENT("; IN; ")-("; JN; ")"
4980 PRINT #2, "2ND MOMENT OF AREA="; S9(I)
4990 PRINT #2, "CROSS-SECTIONAL AREA="; A9(I)
5000 PRINT #2, "ELASTIC MODULUS="; E(I)
5010 PRINT #2, "NO. OF HYDROSTATIC LOADS ON ELEMENT="; NUD(I)
5015 PRINT #2, "NO. OF CONCENTRATED LOADS ON ELEMENT="; NWLOAD(I)
5020 IF NUD(I) = 0 THEN 5100
5025 FOR II = 1 TO NUD(I)
5030 PRINT #2, "HYDROSTATIC LOAD AT 'LEFT'="; UDA(I, II)
5035 PRINT #2, "HYDROSTATIC LOAD AT 'RIGHT'="; UDB(I, II)
5040 PRINT #2, "DISTANC FROM THE 'LEFT' END OF THE ELEMENT TO LEFT OF
HYDRO. LOAD="; XA(I, II)
5045 PRINT #2, "DISTANCE FROM THE 'LEFT' END OF THE ELEMENT TO RIGHT OF
HYDRO. LOAD="; XB(I, II)
5050 NEXT II
5100 IF NWLOAD(I) = 0 THEN 5130
5105 FOR II = 1 TO NWLOAD(I)
5110 PRINT #2, "VALUE OF CONCENTRATED LOAD "; II; "="; WLOAD(I, II)
5115 PRINT #2, "DISTANCE FROM THE 'LEFT' END OF THE ELEMENT OF CONC.
LOAD="; XAW(I, II)
5120 NEXT II
5130 PRINT #2, : PRINT #2,
5140 NEXT I
5150 PRINT #2, "NUMBER OF NODES WITH CONCENTRATED LOADS & COUPLES"
5160 IF NC = 0 THEN GOTO 5240
5170 FOR I = 1 TO NC
5180 NODEW = POSW(I)
5190 PRINT #2, "NODE="; NODEW
```

```
5200 PRINT #2, "CONCENTRATED HORIZONTAL LOAD="; QC(3 * NODEW - 2)
5210 PRINT #2, "CONCENTRATED VERTICAL LOAD="; QC(3 * NODEW - 1)
5220 PRINT #2, "CONCENTRATED COUPLE="; QC(NODEW * 3)
5230 NEXT I
5240 RETURN
6000 C1 = 3 * E * SA * SN ^ 2 / L ^ 3 + CS ^ 2 * CA * E / L
6002 C2 = 3 * E * SA * CS * SN / L ^ 3 - CS * SN * CA * E / L
6004 C3 = 3 * E * SA * CS ^ 2 / L ^ 3 + SN ^ 2 * CA * E / L
6006 C4 = 3 * E * SA * SN / L ^ 2
6008 C5 = 3 * E * SA * CS / L ^ 2
6010 C6 = 3 * E * SA / L
6012 RETURN
6100 Y1 = -3 * K1 / L ^ 2 + 3 * K2 / L ^ 3
6102 M1 = 0
6104 M2 = -Y1 * L - WA * (BL - AL) * (L - BL + (BL - AL) / 2) - (WB - WA) * (BL - AL) *
(L - BL + (BL - AL) / 3) / 2
6106 IF COND = -1 THEN M2 = -M2
6108 IF COND = -1 THEN Y1 = -WA * (BL - AL) * (L - BL + (BL - AL) / 2) / L
6110 IF COND = -1 THEN Y1 = Y1 - (WB - WA) * (BL - AL) * (L - BL + (BL - AL) / 3) / (2
* L)
6112 Y2 = -Y1 - (WA + WB) * (BL - AL) / 2
6114 RETURN
6200 Y1 = -3 * WA * (BL - AL) * (L - BL + (BL - AL) / 2) / (2 * L)
6202 Y1 = Y1 - 3 * (WB - WA) * (BL - AL) * (L - BL + (BL - AL) / 3) / (4 * L)
6204 Y1 = Y1 + 3 * K2 / L ^ 3
6206 M1 = -Y1 * L - WA * (BL - AL) * (L - BL + (BL - AL) / 2) - (WB - WA) * (BL - AL) *
(L - BL + (BL - AL) / 3) / 2
6208 M2 = 0
6210 IF COND = -1 THEN Y1 = -WA * (BL - AL) * (L - BL + (BL - AL) / 2) / L
6212 IF COND = -1 THEN Y1 = Y1 - (WB - WA) * (BL - AL) * (L - BL + (BL - AL) / 3) / (2
* L)
6214 Y2 = -Y1 - (WA + WB) * (BL - AL) / 2
6216 RETURN
6400 Y1 = -3 * WI * BL ^ 2 * (L / 2 - BL / 6) / L ^ 3
6402 Y2 = -Y1 - WI
6404 M1 = 0
6406 M2 = -Y1 * L - WI * BL
6408 REM IF COND=-1 THEN M2=-M2
6410 RETURN
6500 Y1 = -3 * WI * BL * (1 - BL ^ 2 / (3 * L ^ 2)) / (2 * L)
6502 Y2 = -Y1 - WI
6504 M1 = -Y1 * L - WI * BL
6506 M2 = 0
6508 RETURN
6700 SL(3) = 0
6702 BX(2, 2) = (3 * XI / L ^ 2)
6704 BX(2, 3) = 0
6706 BX(2, 5) = (-3 * XI / L ^ 2)
6708 BX(2, 6) = (-3 * XI / L)
6709 FOR III = 1 TO 6: BX(2, III) = -BX(2, III): NEXT III
6710 RETURN
6800 SL(6) = 0
6802 BX(2, 2) = (3 - 3 * XI) / L ^ 2
```

```
6804 BX(2, 3) = (-3 + 3 * XI) / L
6806 BX(2, 5) = (-3 + 3 * XI) / L ^ 2
6808 BX(2, 6) = 0
6810 RETURN
6900 SA = SA / 1E+20: SP(3) = 0: SP(6) = 0
6910 IF NUD(ME) = 0 THEN GOTO 7000
6920 FOR II = 1 TO NUD(ME)
6930 WA = UDA(ME, II): WB = UDB(ME, II)
6940 AL = XA(ME, II): BL = XB(ME, II)
6950 BA = BL - AL
6960 Y1 = -(WA * BA * (L - BL + BA / 2) + (WB - WA) * BA * (L - BL + BA / 3) / 2) / L
6970 Y2 = -(WA + WB) * BA / 2 - Y1
6975 SP(2) = SP(2) - Y1: SP(5) = SP(5) - Y2
6980 NEXT II
7000 IF NWLOAD(ME) = 0 THEN GOTO 7060
7010 FOR II = 1 TO NWLOAD(ME)
7020 WI = WLOAD(ME, II): AL = XAW(ME, II): BL = L - AL
7030 Y1 = -WI * BL / L
7040 Y2 = -Y1 - WI
7045 SP(2) = SP(2) - Y1: SP(5) = SP(5) - Y2
7050 NEXT II
7060 RETURN
9000 REM WARNING
9010 PRINT : PRINT "INCORRECT DATA": PRINT
9020 RETURN
10000 IF Z9 = 77 OR Z9 = 99 THEN GOTO 10280
10010 XZ = YZ = 1E+12
10020 XM = YM = -1E+12
10030 PRINT : PRINT "TO CONTINUE, TYPE Y": PRINT
10040 A$ = INKEY$: IF A$ = "" THEN GOTO 10040
10050 IF A$ = "Y" OR A$ = "y" THEN GOTO 10070
10060 GOTO 10040
10070 FOR EL = 1 TO MS
10080 I = I9(EL)
10090 J = J9(EL)
10100 IF X(I) < XZ THEN XZ = X(I)
10110 IF Y(I) < YZ THEN YZ = Y(I)
10130 IF X(J) < XZ THEN XZ = X(J)
10140 IF Y(J) < YZ THEN YZ = Y(J)
10170 IF X(I) > XM THEN XM = X(I)
10180 IF Y(I) > YM THEN YM = Y(I)
10190 IF X(J) > XM THEN XM = X(J)
10200 IF Y(J) > YM THEN YM = Y(J)
10210 NEXT EL
10230 XS = 400 / (XM - XZ)
10240 IF YM - YZ = 0 THEN GOTO 10280
10250 YS = 200 / (YM - YZ)
10260 IF XS > YS THEN XS = YS
10280 IF Z9 = 0 OR Z9 = 77 THEN SCREEN 9
10290 IF Z9 = 99 THEN PRINT "DEFLECTED FORM OF FRAME"
10310 FOR EL = 1 TO MS
10320 I = I9(EL)
10330 J = J9(EL)
```

10340 IF Z9 = 99 THEN GOTO 10400
10350 XI = (X(I) - XZ) * XS
10360 YI = (Y(I) - YZ) * XS
10370 XJ = (X(J) - XZ) * XS
10380 YJ = (Y(J) - YZ) * XS
10390 GOTO 10500
10400 L = SQR((XG(J) - XG(I)) ^ 2 + (YG(J) - YG(I)) ^ 2)
10402 SMA = S9(EL): AREA = A9(EL)
10404 CS = (XG(J) - XG(I)) / L: SN = (YG(J) - YG(I)) / L
10406 XI = X(I) - XZ * XS: YI = Y(I) - YZ * XS
10408 XK = XI: YK = YI
10410 VI = -SN * U(I * 3 - 2) + CS * U(I * 3 - 1): VJ = -SN * U(J * 3 - 2) + CS * U(J * 3 - 1)
10412 UI = CS * U(I * 3 - 2) + SN * U(I * 3 - 1): UJ = CS * U(J * 3 - 2) + SN * U(J * 3 - 1)
10414 THETAI = U(I * 3): THETAJ = U(J * 3)
10416 XCONST = E * SMA
10418 FOR II = 0 TO 50
10420 XXI = II * L / 50
10422 EQN = 0
10424 IF NUD(EL) = 0 THEN 10452
10426 FOR III = 1 TO NUD(EL)
10428 WA = UDA(EL, III): WB = UDB(EL, III)
10430 AL = XA(EL, III): BL = XB(EL, III)
10432 X3 = (L - AL) ^ 3: X4 = (L - AL) ^ 4: X5 = (L - AL) ^ 5
10434 X3B = (L - BL) ^ 3: X4B = (L - BL) ^ 4: X5B = (L - BL) ^ 5
10436 K1 = WA * X3 / 6 + (WB - WA) * X4 / (24 * (BL - AL)) - WB * X3B / 6 - (WB - WA) * X4B / (24 * (BL - AL))
10438 K2 = WA * X4 / 24 + (WB - WA) * X5 / (120 * (BL - AL)) - WB * X4B / 24 - (WB - WA) * X5B / (120 * (BL - AL))
10440 Y1 = -6 * XCONST * (THETAI + THETAJ) / L ^ 2 + 12 * XCONST * (VI - VJ) / L ^ 3 - 6 * K1 / L ^ 2 + 12 * K2 / L ^ 3
10442 M1 = XCONST * (THETAI - THETAJ) / L - Y1 * L / 2 - K1 / L
10444 EQN = EQN + Y1 * XXI ^ 3 / 6 + M1 * XXI ^ 2 / 2 - XCONST * THETAI * XXI
10445 IF CS = 1 THEN EQN = EQN + XCONST * VI
10446 IF XXI > AL THEN EQN = EQN + WA * (XXI - AL) ^ 4 / 24 + (WB - WA) * (XXI - AL) ^ 5 / (120 * (BL - AL))
10448 IF XXI > BL THEN EQN = EQN - WB * (XXI - BL) ^ 4 / 24 - (WB - WA) * (XXI - BL) ^ 5 / (120 * (BL - AL))
10450 NEXT III
10452 IF NWLOAD(EL) = 0 THEN 10468
10454 FOR III = 1 TO NWLOAD(EL)
10456 WI = WLOAD(EL, III): AL = XAW(EL, III): BL = L - AL
10458 Y1 = -6 * XCONST * (THETAI + THETAJ) / L ^ 2 + 12 * XCONST * (VI - VJ) / L ^ 3 - 12 * WI * BL ^ 2 * (L / 4 - BL / 6) / L ^ 3
10460 M1 = XCONST * (THETAI - THETAJ) / L - WI * BL ^ 2 / (2 * L) - Y1 * L / 2
10462 EQN = EQN + Y1 * XXI ^ 3 / 6 + M1 * XXI ^ 2 / 2 - XCONST * THETAI * XXI
10463 IF NUD(EL) = 0 AND CS = 1 THEN EQN = EQN + XCONST * VI
10464 IF XXI > AL THEN EQN = EQN + WI * (XXI - AL) ^ 3 / 6
10466 NEXT III
10468 EQN = EQN / XCONST
10470 IF NUD(EL) = 0 AND NWLOAD(EL) = 0 THEN EQN = EQN + (VJ - VI) * XXI / L
10472 XJ = XI + XXI * CS * XS: YJ = YI + XXI * SN * XS
10474 XJ = XJ - SC * EQN * SN / UM: YJ = YJ + SC * EQN * CS / UM

```
10476 LINE (XK + 40, 250 - YK)-(XJ + 40, 250 - YJ)
10478 XK = XJ: YK = YJ
10480 NEXT II
10482 GOTO 10550
10500 LINE (XI + 40, 250 - YI)-(XJ + 40, 250 - YJ)
10550 NEXT EL
10560 IF Z9 = 77 THEN GOTO 10800
10580 PRINT "TO CONTINUE, TYPE Y"
10590 A$ = INKEY$: IF A$ = "" THEN GOTO 10590
10610 IF A$ = "Y" OR A$ = "y" THEN GOTO 10700
10650 GOTO 10590
10700 SCREEN 0, 0: WIDTH 80
10800 RETURN
11000 REM PLOT OF DEFLECTED FORM OF FRAME
11010 PRINT "THE DEFLECTED FORM OF THE FRAME WILL NOW BE DRAWN"
11020 Z9 = 77
11050 GOSUB 10000
11060 UM = -1E+12
11080 FOR II = 1 TO N3
11090 FP = II / 3
11100 IP = INT(FP + .001)
11110 RM = IP - FP
11120 IF RM = 0 THEN GOTO 11170
11160 IF ABS(U(II)) > UM THEN UM = ABS(U(II))
11170 NEXT II
11175 INPUT "TYPE IN MAXIMUM POSSIBLE DEFLECTION"; UM: UM = ABS(UM)
11180 SC = 20
11200 FOR II = 1 TO NJ
11210 X(II) = X(II) * XS + SC * U(3 * II - 2) / UM
11220 Y(II) = Y(II) * XS + SC * U(3 * II - 1) / UM
11240 NEXT II
11260 Z9 = 99
11280 GOSUB 10000
11300 RETURN
11500 REM PLOT OF BM DIAGRAM
11510 Z9 = 77
11520 FOR II = 1 TO NJ
11530 X(II) = XG(II): Y(II) = YG(II)
11540 NEXT II
11550 GOSUB 10000
11560 BIG = 0
11561 PRINT "DO YOU REQUIRE YOUR OWN BENDING MOMENT SCALE ? TYPE Y/N"
11562 INPUT BM$
11563 IF BM$ = "y" THEN BM$ = "Y"
11564 IF BM$ = "n" THEN BM$ = "N"
11565 IF BM$ = "Y" OR BM$ = "N" THEN GOTO 11567
11566 GOTO 11561
11567 IF BM$ = "Y" THEN INPUT "TYPE IN THE MAXIMUM VALUE OF THE BENDING
MOMENT"; BIG
11568 BIG = ABS(BIG)
11569 IF BM$ = "Y" THEN GOTO 11610
11570 FOR II = 1 TO MS
11580 IF ABS(BM(1, II)) > BIG THEN BIG = ABS(BM(1, II))
```

```
11590 IF ABS(BM(2, II)) > BIG THEN BIG = ABS(BM(2, II))
11600 NEXT II
11610 FOR EL = 1 TO MS
11620 I = I9(EL): J = J9(EL)
11630 L = SQR((X(J) - X(I)) ^ 2 + (Y(J) - Y(I)) ^ 2)
11640 CS = (X(J) - X(I)) / L: SN = (Y(J) - Y(I)) / L
11730 XI = (X(I) - XZ) * XS
11740 YI = (Y(I) - YZ) * XS
11745 XK = XI: YK = YI
11750 FOR II = 0 TO 10
11760 XXI = II * L / 10
11761 EQN = 0: RA = 0
11762 IF NUD(EL) = 0 THEN GOTO 11790
11763 FOR III = 1 TO NUD(EL)
11764 WA = UDA(EL, III): WB = UDB(EL, III)
11766 AL = XA(EL, III): BL = XB(EL, III)
11767 X3 = (L - AL) ^ 3: X4 = (L - AL) ^ 4: X5 = (L - AL) ^ 5
11769 X3B = (L - BL) ^ 3: X4B = (L - BL) ^ 4: X5B = (L - BL) ^ 5
11770 BA = BL - AL
11772 RA = RA - WA * BA * (L - BL + BA / 2) / L
11775 RA = RA - (WB - WA) * BA * (L - BL + BA / 3) / (2 * L)
11780 IF XXI > AL THEN EQN = EQN - WA * (XXI - AL) ^ 2 / 2 - (WB - WA) * (XXI - AL)
^ 3 / (6 * (BL - AL))
11781 IF XXI > BL THEN EQN = EQN + WB * (XXI - BL) ^ 2 / 2 + (WB - WA) * (XXI - BL)
^ 3 / (6 * (BL - AL))
11782 NEXT III
11790 IF NWLOAD(EL) = 0 THEN GOTO 11850
11792 FOR III = 1 TO NWLOAD(EL)
11794 WI = WLOAD(EL, III): AL = XAW(EL, III): BL = L - AL
11800 RA = RA - WI * BL / L
11802 IF XXI > AL THEN EQN = EQN - WI * (XXI - AL)
11810 NEXT III
11850 RA = RA + (BM(2, EL) - BM(1, EL)) / L
11852 MOM = BM(1, EL) - EQN + RA * XXI
11855 XJ = XI + XXI * CS * XS: YJ = YI + XXI * SN * XS
11860 BMX = MOM * SN
11870 BMY = MOM * CS
11880 XJ = XJ - SC * BMX / BIG: YJ = YJ + SC * BMY / BIG
11900 LINE (XK + 40, 250 - YK)-(XJ + 40, 250 - YJ)
11910 XK = XJ
11915 YK = YJ
11920 NEXT II
11930 NEXT EL
11935 PRINT : PRINT : PRINT "          BENDING MOMENT DIAGRAM"
11940 A$ = INKEY$: IF A$ = "" THEN GOTO 11940
11950 IF A$ = "Y" OR A$ = "y" THEN GOTO 11970
11960 GOTO 11940
11970 SCREEN 0: WIDTH 80
11980 RETURN
12000 REM PLOT OF SF DIAGRAM
12010 Z9 = 77
12015 SC = 10
12020 FOR II = 1 TO NJ
```

```
12030 X(II) = XG(II): Y(II) = YG(II)
12040 NEXT II
12050 GOSUB 10000
12060 BIG = 0
12070 FOR EL = 1 TO MS
12080 I = I9(EL): J = J9(EL)
12090 L = SQR((X(J) - X(I)) ^ 2 + (Y(J) - Y(I)) ^ 2)
12100 RA = (BM(1, EL) - BM(2, EL)) / L
12110 IF ABS(RA) > BIG THEN BIG = ABS(RA)
12120 NEXT EL
12122 PRINT "DO YOU REQUIRE YOUR OWN SCALE FOR SHEARING FORCE ? TYPE Y/N"
12124 INPUT F$
12125 IF F$ = "y" THEN F$ = "Y"
12126 IF F$ = "n" THEN F$ = "N"
12128 IF F$ = "Y" OR F$ = "N" THEN GOTO 12130
12129 GOTO 12122
12130 IF F$ = "Y" THEN INPUT "TYPE IN THE MAXIMUM VALUE OF SHEARING
FORCE"; BIG
12132 BIG = ABS(BIG)
12140 FOR EL = 1 TO MS
12150 I = I9(EL): J = J9(EL)
12160 L = SQR((X(J) - X(I)) ^ 2 + (Y(J) - Y(I)) ^ 2)
12170 CS = (X(J) - X(I)) / L: SN = (Y(J) - Y(I)) / L
12172 XI = (X(I) - XZ) * XS
12174 YI = (Y(I) - YZ) * XS
12176 XK = XI: YK = YI
12178 FOR II = 0 TO 50
12180 XXI = II * L / 50
12182 EQN = 0: RA = 0
12184 IF NUD(EL) = 0 THEN GOTO 12220
12186 FOR III = 1 TO NUD(EL)
12188 WA = UDA(EL, III): WB = UDB(EL, III)
12190 AL = XA(EL, III): BL = XB(EL, III)
12192 RA = RA - WA * (BL - AL) * (L - BL + (BL - AL) / 2) / L
12194 RA = RA - (WB - WA) * (BL - AL) * (L - BL + (BL - AL) / 3) / (2 * L)
12196 IF XXI > AL THEN EQN = EQN + WA * (XXI - AL) + (WB - WA) * (XXI - AL) ^ 2
/ (2 * (BL - AL))
12198 IF XXI > BL THEN EQN = EQN - WB * (XXI - BL) + (WB - WA) * (XXI - BL) ^ 2 /
(2 * (BL - AL))
12200 NEXT III
12220 IF NWLOAD(EL) = 0 THEN GOTO 12260
12222 FOR III = 1 TO NWLOAD(EL)
12224 WI = WLOAD(EL, III): AL = XAW(EL, III): BL = L - AL
12226 RA = RA - WI * BL / L
12228 IF XXI > AL THEN EQN = EQN + WI
12230 NEXT III
12260 RA = RA + (BM(2, EL) - BM(1, EL)) / L
12270 SFF = RA + EQN
12440 XJ = XI + XXI * CS * XS: YJ = YI + XXI * SN * XS
12460 SFX = SFF * SN
12470 SFY = SFF * CS
12490 XJ = XJ - SC * SFX / BIG: YJ = YJ + SC * SFY / BIG
12510 LINE (XK + 40, 250 - YK)-(XJ + 40, 250 - YJ)
```

```
12520 XK = XJ
12530 YK = YJ
12535 NEXT II
12540 NEXT EL
12545 PRINT : PRINT : PRINT "          SHEARING FORCE DIAGRAM"
12550 A$ = INKEY$: IF A$ = "" THEN GOTO 12550
12560 IF A$ = "Y" OR A$ = "y" THEN GOTO 12580
12570 GOTO 12550
12580 SCREEN 0: WIDTH 80
12590 RETURN
14000 END
```

Appendix 6

Computer Program for Pin-Jointed Space Trusses "STRUSSF"

```
10 GOSUB 9000
20 IF A$ = "Y" OR A$ = "y" THEN GOTO 50
30 PRINT : PRINT "GO BACK & READ IT!": PRINT : STOP
50 REM COPYRIGHT OF DR.C.T.F.ROSS
60 REM 6 HURSTVILLE DRIVE,
70 REM WATERLOOVILLE, PORTSMOUTH
80 REM HANTS. PO7 7NB ENGLAND
90 NNODE = 2: NDF = 3
100 CLS
110 PRINT : PRINT "FORCES IN PIN-JOINTED SPACE TRUSSES": PRINT
120 PRINT : PRINT "TYPE IN THE NAME OF YOUR INPUT FILE"
130 INPUT FILENAME$
135 PRINT "IF YOU ARE SATISFIED WITH YOUR FILENAME, TYPE Y; ELSE TYPE N"
140 A$ = INKEY$: IF A$ = "" THEN GOTO 140
150 IF A$ = "Y" OR A$ = "y" THEN GOTO 180
160 IF A$ = "N" OR A$ = "n" THEN GOTO 120
170 GOTO 140
180 OPEN FILENAME$ FOR INPUT AS #1
182 PRINT : PRINT "TYPE IN THE NAME OF YOUR OUTPUT FILE"
184 INPUT OUT$
PRINT : PRINT "IF YOU ARE SATISFIED WITH THIS FILENAME, TYPE Y; ELSE N"
186 B$ = INKEY$: IF B$ = "" THEN GOTO 186
IF B$ = "Y" OR B$ = "y" THEN GOTO 190
IF B$ = "N" OR B$ = "n" THEN GOTO 182
GOTO 186
190 OPEN OUT$ FOR OUTPUT AS #2
199 INPUT #1, NJ
200 IF NJ < 4 OR NJ > 499 THEN GOSUB 9995: PRINT "NO. OF PIN-JOINTS="; NJ;
":STOP"
210 INPUT #1, MS
220 IF MS < 3 OR MS > 499 THEN GOSUB 9995: PRINT "NO. OF MEMBERS="; MS: STOP
230 INPUT #1, NF
240 IF NF < 3 OR NF > NJ THEN GOSUB 9995: PRINT "NO. OF NODES WITH ZERO
DISPLACEMENTS="; NF: STOP
250 N3 = NJ * 3
260 DIM U(N3), AA(MS), EA(MS), IA(MS), JA(MS)
270 DIM ST(6, 6), X(NJ), Y(NJ), Z(NJ), NS(NF * 3), SG(3), SH(3), DC(1, 3)
280 DIM RX(NJ), RY(NJ), RZ(NJ), POSS(NF), DIAG(N3)
300 REM PRINT:PRINT"TYPE IN THE NODAL COORDINATES":PRINT
310 FOR I = 1 TO NJ
320 REM PRINT"TYPE IN X COORD FOR NODE ";I
330 INPUT #1, X(I), Y(I), Z(I)
340 REM PRINT"TYPE IN Y COORD FOR NODE ";I
360 REM PRINT"TYPE IN Z COORD FOR NODE ";I
380 NEXT I
390 REM PRINT:PRINT"TYPE IN THE NODAL POSITIONS OF THE NODES WITH ZERO
DISPLACEMENTS":PRINT
395 FOR I = 1 TO NF * 3: NS(I) = 0: NEXT I
397 FOR I = 1 TO NF: POSS(I) = 0: NEXT I
```

```
400 FOR I = 1 TO NF
410 REM PRINT:PRINT"TYPE IN THE NODE NUMBER OF 'SUPPRESSED' NODE ";I
INPUT #1, SU, SUX, SUY, SUZ
413 POSS(I) = SU
415 IF SU < 1 OR SU > NJ THEN GOSUB 9995: PRINT "NO. OF THE NODE WITH THE
ZERO DISPLACEMENTS="; SU: STOP
416 REM PRINT:PRINT"IS U0 ZERO AT NODE ";SU;"? TYPE Y/N"
417 IF SUX = 1 THEN Y$ = "Y" ELSE Y$ = "N"
419 IF Y$ = "Y" OR Y$ = "N" OR Y$ = "y" OR Y$ = "n" THEN GOTO 422
420 GOSUB 9995: STOP
422 IF Y$ = "Y" OR Y$ = "y" THEN NS(I * 3 - 2) = SU * 3 - 2
423 REM PRINT:PRINT"IS V0 ZERO AT NODE ";SU;"? TYPE Y/N"
424 IF SUY = 1 THEN Y$ = "Y" ELSE Y$ = "N"
425 IF Y$ = "Y" OR Y$ = "N" OR Y$ = "y" OR Y$ = "n" THEN GOTO 427
426 GOSUB 9995: STOP
427 IF Y$ = "Y" OR Y$ = "y" THEN NS(I * 3 - 1) = SU * 3 - 1
428 REM PRINT:PRINT"IS W0 ZERO AT NODE ";SU;"? TYPE Y/N"
429 IF SUZ = 1 THEN Y$ = "Y" ELSE Y$ = "N"
430 IF Y$ = "Y" OR Y$ = "N" OR Y$ = "y" OR Y$ = "n" THEN GOTO 432
431 GOSUB 9995: STOP
432 IF Y$ = "Y" OR Y$ = "y" THEN NS(I * 3) = SU * 3
433 NEXT I
440 REM PRINT:PRINT"TYPE IN MEMBER DETAILS":PRINT
450 FOR I = 1 TO MS
460 REM PRINT"TYPE IN THE I NODE FOR MEMBER ";I
470 INPUT #1, IA(I), JA(I), AA(I), EA(I)
480 IF IA(I) < 1 OR IA(I) > NJ THEN GOSUB 9995: PRINT "I NODE FOR MEMBER "; I;
"="; IA(I): STOP
485 REM TYPE IN J NODE
492 IF JA(I) < 1 OR JA(I) > NJ THEN GOSUB 9995: PRINT "J NODE FOR MEMBER "; I;
"="; JA(I): STOP
495 IF IA(I) = JA(I) THEN GOSUB 9995: STOP
497 REM CROSS-SECTIONAL AREA
510 IF AA(I) < = 0 THEN GOSUB 9995: PRINT "AREA FOR MEMBER "; I; "="; AA(I): STOP
525 REM ELASTIC MODULUS
530 IF EA(I) < 1 THEN GOSUB 9995: PRINT "YOUNG'S MODULUS FOR MEMBER "; I; "=";
EA(I): STOP
537 I9 = IA(I): J9 = JA(I)
538 FOR II = 1 TO NNODE: IF II = 1 THEN IN = I9 ELSE IN = J9
539 FOR IJ = 1 TO NDF
540 II1 = NDF * (IN - 1) + IJ
541 FOR JJ = 1 TO NNODE: IF JJ = 1 THEN JN = I9 ELSE JN = J9
542 FOR JI = 1 TO NDF
543 JJ1 = NDF * (JN - 1) + JI
544 IIJJ = JJ1 - II1 + 1: IF IIJJ > DIAG(JJ1) THEN DIAG(JJ1) = IIJJ
545 NEXT JI, JJ, IJ, II
546 NEXT I
547 DIAG(1) = 1: FOR II = 1 TO N3: DIAG(II) = DIAG(II - 1) + DIAG(II): NEXT II
548 GOTO 600
550 REM OUTPUT OF MEMBER DETAILS
555 FOR I = 1 TO MS
560 PRINT #2, "DETAILS OF MEMBER "; IA(I); "-"; JA(I)
565 PRINT #2, "CROSS-SECTIONAL AREA="; AA(I)
```

```
570 PRINT #2, "ELASTIC MODULUS="; EA(I)
575 NEXT I
580 RETURN
600 NT = DIAG(N3)
610 DIM A(NT), Q(N3), C(N3)
630 REM NO. OF CONCENTRATED LOADS
635 INPUT #1, NC
636 IF NC < 1 OR NC > NJ THEN GOSUB 9995: PRINT "NO. OF NODES WITH
CONCENTRATED LOADS="; NC: STOP
638 DIM POSW(NC)
639 FOR I = 1 TO N3: Q(I) = 0: NEXT I
640 FOR I = 1 TO NC
641 REM NODAL POSITION
642 INPUT #1, POSW(I), XW, YW, ZW
643 IF POSW(I) < 1 OR POSW(I) > NJ THEN GOSUB 9995: PRINT "NO. OF THE NODE
WITH CONCENTRATED LOADS="; POSW(I): STOP
646 REM LOADS IN X0,Y0 & Z0 DIRECTIONS
649 Q(POSW(I) * 3 - 2) = XW: Q(POSW(I) * 3 - 1) = YW: Q(POSW(I) * 3) = ZW
650 NEXT I
657 GOSUB 4000
658 Z9 = 0: GOSUB 10000
660 FOR ME = 1 TO MS
670 PRINT : PRINT "MEMBER NUMBER "; ME; " UNDER COMPUTATION": PRINT
680 I = IA(ME)
690 J = JA(ME)
700 A = AA(ME)
710 E = EA(ME)
720 L = SQR((X(J) - X(I)) ^ 2 + (Y(J) - Y(I)) ^ 2 + (Z(J) - Z(I)) ^ 2)
730 XC = (X(J) - X(I)) / L
740 YC = (Y(J) - Y(I)) / L
750 ZC = (Z(J) - Z(I)) / L
760 ST(1, 1) = XC ^ 2
770 ST(1, 2) = XC * YC
780 ST(2, 1) = XC * YC
790 ST(2, 2) = YC ^ 2
800 ST(1, 3) = XC * ZC
810 ST(3, 1) = XC * ZC
820 ST(2, 3) = YC * ZC
830 ST(3, 2) = YC * ZC
840 ST(3, 3) = ZC ^ 2
850 ST(4, 1) = -XC * XC
860 ST(4, 2) = -XC * YC
870 ST(4, 3) = -XC * ZC
880 ST(5, 1) = -XC * YC
890 ST(5, 2) = -YC * YC
900 ST(5, 3) = -YC * ZC
910 ST(6, 1) = -XC * ZC
920 ST(6, 2) = -YC * ZC
930 ST(6, 3) = -ZC * ZC
940 FOR II = 1 TO 3
950 FOR JJ = 1 TO 3
960 ST(II + 3, JJ + 3) = ST(II, JJ)
970 ST(II, JJ + 3) = ST(JJ + 3, II)
```

```
980 NEXT JJ
990 NEXT II
1000 CN = A * E / L
1010 FOR II = 1 TO 6
1020 FOR JJ = 1 TO 6
1030 ST(II, JJ) = ST(II, JJ) * CN
1040 NEXT JJ
1050 NEXT II
1060 I1 = 3 * I - 3
1070 J1 = 3 * J - 3
1075 FOR II = 1 TO 3
1080 FOR JJ = 1 TO 3
1085 MM = I1 + II: MN = I1 + JJ
1090 NM = J1 + II: N9 = J1 + JJ
1095 IF JJ > II THEN GOTO 1110
1100 A(DIAG(MM) + MN - MM) = A(DIAG(MM) + MN - MM) + ST(II, JJ)
1105 A(DIAG(NM) + N9 - NM) = A(DIAG(NM) + N9 - NM) + ST(II + 3, JJ + 3)
1110 IF I1 > J1 THEN GOTO 1125
1115 A(DIAG(NM) + MN - NM) = A(DIAG(NM) + MN - NM) + ST(II + 3, JJ)
1120 GOTO 1130
1125 A(DIAG(MM) + N9 - MM) = A(DIAG(MM) + N9 - MM) + ST(II, JJ + 3)
1130 NEXT JJ, II
1160 NEXT ME
1210 FOR I = 1 TO NF * 3
1215 N9 = NS(I)
1217 IF N9 = 0 THEN GOTO 1230
1218 DI = DIAG(N9)
1220 A(DI) = A(DI) * 1E+12 + 1E+12
1225 Q(NS(I)) = 0
1230 NEXT I
1240 N = N3
1250 FOR II = 1 TO N3: C(II) = Q(II): NEXT II
1255 PRINT : PRINT "SOLUTION OF THE SIMULTANEOUS EQUATIONS IS NOW BEING
CARRIED OUT": PRINT
1260 GOSUB 2940
1270 FOR I = 1 TO N3
1280 U(I) = C(I)
1290 NEXT I
1330 PRINT #2, : PRINT #2, "THE NODAL DISPLACEMENTS (U0,V0 & W0) ARE AS
FOLLOWS:": PRINT #2,
1340 FOR I = 1 TO NJ: PRINT #2, "NODE="; I; "   ";
1350 FOR J = 1 TO 3
1360 PRINT #2, U(I * 3 - 3 + J); "   ";
1370 NEXT J
1380 PRINT #2, : NEXT I
1390 PRINT #2, : PRINT #2, "THE FORCES IN THE MEMBERS ARE AS FOLLOWS:-": PRINT
#2,
1400 FOR ME = 1 TO MS
1410 I = IA(ME)
1420 J = JA(ME)
1430 A = AA(ME)
1440 E = EA(ME)
1450 L = SQR((X(J) - X(I)) ^ 2 + (Y(J) - Y(I)) ^ 2 + (Z(J) - Z(I)) ^ 2)
```

```
1460 XC = (X(J) - X(I)) / L
1470 YC = (Y(J) - Y(I)) / L
1480 ZC = (Z(J) - Z(I)) / L
1490 DC(1, 1) = XC
1500 DC(1, 2) = YC
1510 DC(1, 3) = ZC
1520 I1 = 3 * I - 3
1530 J1 = 3 * J - 3
1540 FOR I3 = 1 TO 3
1550 J3 = I1 + I3
1560 J2 = J1 + I3
1570 SG(I3) = U(J3)
1580 SH(I3) = U(J2)
1590 NEXT I3
1600 A1 = 0
1610 A2 = 0
1620 FOR II = 1 TO 3
1630 A1 = A1 + DC(1, II) * SG(II)
1640 A2 = A2 + DC(1, II) * SH(II)
1650 NEXT II
1660 FC = A * E * (A2 - A1) / L
1670 PRINT #2, "FORCE IN MEMBER "; I; "-"; J; "="; FC
1680 GOSUB 4500
1690 NEXT ME
1700 GOSUB 4700
1750 GOSUB 11000
1790 STOP
2940 N = N3
2945 A(1) = SQR(A(1))
2950 FOR II = 2 TO N
2955 DI = DIAG(II) - II
2960 LL = DIAG(II - 1) - DI + 1
2965 FOR JJ = LL TO II
2970 X = A(DI + JJ)
2975 DJ = DIAG(JJ) - JJ
2980 IF JJ = 1 THEN GOTO 3010
2985 LB = DIAG(JJ - 1) - DJ + 1
2990 IF LL > LB THEN LB = LL
2995 FOR KK = LB TO JJ - 1
3000 X = X - A(DI + KK) * A(DJ + KK)
3005 NEXT KK
3010 A(DI + JJ) = X / A(DJ + JJ)
3015 NEXT JJ
3020 A(DI + II) = SQR(X)
3025 NEXT II
3030 C(1) = C(1) / A(1)
3035 FOR II = 2 TO N
3040 DI = DIAG(II) - II
3045 LL = DIAG(II - 1) - DI + 1
3050 X = C(II)
3055 FOR JJ = LL TO II - 1
3060 X = X - A(DI + JJ) * C(JJ)
3065 NEXT JJ
```

```
3070 C(II) = X / A(DI + II)
3075 NEXT II
3080 FOR II = N TO 2 STEP -1
3085 DI = DIAG(II) - II
3090 X = C(II) / A(DI + II)
3095 C(II) = X
3100 LL = DIAG(II - 1) - DI + 1
3105 FOR KK = LL TO II - 1
3110 C(KK) = C(KK) - X * A(DI + KK)
3115 NEXT KK
3120 NEXT II
3125 C(1) = C(1) / A(1)
3180 RETURN
4000 REM ***OUTPUT***
4040 PRINT #2, : PRINT #2, "STATIC ANALYSIS OF PIN-JOINTED SPACE TRUSSES": PRINT
#2,
4050 PRINT #2, : PRINT #2, "NUMBER OF PIN-JOINTS ="; NJ
4060 PRINT #2, "NUMBER OF MEMBERS ="; MS
4070 PRINT #2, "NUMBER OF NODES WITH ZERO DISPLACEMENTS ="; NF
4080 PRINT #2, : PRINT #2, "THEN NODAL COORDINATES (X0,Y0 & Z0) ARE AS
FOLLOWS:": PRINT #2,
4090 FOR I = 1 TO NJ
4100 PRINT #2, "X("; I; ")="; X(I); "    Y("; I; ")="; Y(I); "    Z("; I; ")="; Z(I)
4110 NEXT I
4120 PRINT #2, : PRINT #2, "THE NODE NUMBERS OF THE NODES WITH ZERO
DISPLACEMENTS ARE:": PRINT #2,
4130 FOR I = 1 TO NF
4140 SU = INT((NS(I * 3 - 2) + 2) / 3)
4150 IF SU = 0 THEN SU = INT((NS(I * 3 - 1) + 1) / 3)
4160 IF SU = 0 THEN SU = INT(NS(I * 3) / 3)
4170 PRINT #2, "NODE ="; SU
4180 N9 = NS(I * 3 - 2)
4190 IF N9 > 0 THEN PRINT #2, "U0 DISPLACEMENT IS ZERO AT NODE "; SU
4200 N9 = NS(I * 3 - 1)
4210 IF N9 > 0 THEN PRINT #2, "V0 DISPLACEMENT IS ZERO AT NODE "; SU
4220 N9 = NS(I * 3)
4230 IF N9 > 0 THEN PRINT #2, "W0 DISPLACEMENT IS ZERO AT NODE "; SU
4240 NEXT I
4245 GOSUB 550
4250 PRINT #2, : PRINT #2, "THE NODAL POSITIONS & VALUES OF THE CONCENTRATED
LOADS ARE :": PRINT #2,
4260 FOR I = 1 TO NC
4270 PRINT #2, "AT NODE "; POSW(I); " LOADS ARE:    ";
4280 FOR J = 1 TO 3
4290 II = POSW(I) * 3 - 3 + J
4300 PRINT #2, Q(II); "   ";
4310 NEXT J
4320 PRINT #2, : NEXT I
4360 RETURN
4500 REM CALCULATIONS OF THE REACTIONS
4510 REX = 0: REY = 0: REZ = 0
4520 FOR II = 1 TO NF
4530 IF IA(ME) = POSS(II) THEN SUP = POSS(II): GOTO 4560
```

```
4540 NEXT II
4550 GOTO 4600
4560 REX = -FC * XC: REY = -FC * YC: REZ = -FC * ZC
4570 RX(SUP) = RX(SUP) + REX
4580 RY(SUP) = RY(SUP) + REY
4590 RZ(SUP) = RZ(SUP) + REZ
4600 FOR II = 1 TO NF
4610 IF JA(ME) = POSS(II) THEN SUP = POSS(II): GOTO 4640
4620 NEXT II
4630 GOTO 4680
4640 REX = FC * XC: REY = FC * YC: REZ = FC * ZC
4650 RX(SUP) = RX(SUP) + REX
4660 RY(SUP) = RY(SUP) + REY
4670 RZ(SUP) = RZ(SUP) + REZ
4680 RETURN
4700 REM OUTPUT OF REACTIONS
4710 PRINT #2, : PRINT #2, "THE REACTIONS ARE:-"
4720 FOR II = 1 TO NJ
4730 IF RX(II) = 0 THEN GOTO 4750
4740 GOTO 4780
4750 IF RY(II) = 0 THEN GOTO 4770
4760 GOTO 4780
4770 IF RZ(II) = 0 THEN GOTO 4820
4780 PRINT #2, "AT NODE "; II
4790 PRINT #2, "REACTION IN X0 DIRECTION="; RX(II)
4800 PRINT #2, "REACTION IN Y0 DIRECTION="; RY(II)
4810 PRINT #2, "REACTION IN Z0 DIRECTION="; RZ(II)
4820 NEXT II
4830 RETURN
9000 CLS : PRINT : PRINT "COPYRIGHT OF DR.C.T.F.ROSS": PRINT
9010 PRINT "NOT TO BE REPRODUCED": PRINT
9020 PRINT "HAVE YOU READ THE BOOK:-": PRINT
9030 PRINT "FINITE ELEMENT PROGRAMS IN STRUCTURAL ENGINEERING &
CONTINUUM MECHANICS": PRINT
9040 PRINT "BY": PRINT
9050 PRINT "C.T.F.ROSS.": PRINT
9060 PRINT "TYPE Y/N"
9070 A$ = INKEY$: IF A$ = "" THEN GOTO 9070
9080 IF A$ = "y" THEN A$ = "Y"
9090 RETURN
9500 PRINT : PRINT "ARE YOU SATISFIED WITH THE INPUT OF YOUR LAST BATCH OF
DATA ? TYPE Y/N": PRINT
9510 X$ = INKEY$: IF X$ = "" THEN GOTO 9510
9520 IF X$ = "y" THEN X$ = "Y"
9530 RETURN
9995 REM ***WARNING***
9996 PRINT : PRINT "INCORRECT DATA": PRINT
9997 RETURN
10000 IF Z9 = 99 THEN GOTO 10350
10010 XZ = 1E+12: YZ = 1E+12: ZZ = 1E+12
10020 XM = -1E+12: YM = -1E+12: ZM = -1E+12
10040 CLS
10050 PRINT "THE FRAME WILL NOW BE DRAWN. TO CONTINUE, TYPE Y"
```

```
10060 A$ = INKEY$: IF A$ = "" THEN GOTO 10060
10070 IF A$ = "Y" OR A$ = "y" THEN GOTO 10090
10080 GOTO 10050
10090 PRINT : PRINT "TYPE IN THE VIEWING ANGLES IN DEGREES": INPUT "ALPHA=";
ALPHA: INPUT "BETA="; BETA
10100 PI = 3.1415926536#: ALPHA = ALPHA * PI / 180: BETA = -BETA * PI / 180
10105 SCREEN 9, 0: WIDTH 80
10110 FOR EL = 1 TO MS
10120 I = IA(EL): J = JA(EL)
10130 IF X(I) < XZ THEN XZ = X(I)
10140 IF Y(I) < YZ THEN YZ = Y(I)
10150 IF Z(I) < ZZ THEN ZZ = Z(I)
10160 IF X(J) < XZ THEN XZ = X(J)
10170 IF Y(J) < YZ THEN YZ = Y(J)
10180 IF Z(J) < ZZ THEN ZZ = Z(J)
10190 IF X(I) > XM THEN XM = X(I)
10200 IF Y(I) > YM THEN YM = Y(I)
10210 IF Z(I) > ZM THEN ZM = Z(I)
10220 IF X(J) > XM THEN XM = X(J)
10230 IF Y(J) > YM THEN YM = Y(J)
10240 IF Z(J) > ZM THEN ZM = Z(J)
10245 NEXT EL
10250 XS = 500 / (XM - XZ)
10260 IF YM - YZ = 0 OR ZM - ZZ = 0 THEN GOTO 10280
10265 YS = 250 / (YM - YZ): ZS = 300 / (ZM - ZZ)
10270 IF YS > ZS THEN YS = ZS
10275 IF XS > YS THEN XS = YS
10280 CONX = 1E+12
10290 CONZ = 1E+12
10300 JUMP = 0
10350 FOR EL = 1 TO MS
10360 I = IA(EL)
10370 J = JA(EL)
10375 IF Z9 = 99 THEN GOTO 10417
10380 XI = (X(I) - XZ) * XS
10390 YI = (Y(I) - YZ) * XS
10400 XJ = (X(J) - XZ) * XS
10410 YJ = (Y(J) - YZ) * XS
10412 ZI = (Z(I) - ZZ) * XS
10415 ZJ = (Z(J) - ZZ) * XS
10416 GOTO 10420
10417 XI = X(I) - XZ * XS: XJ = X(J) - XZ * XS
10418 YI = Y(I) - YZ * XS: YJ = Y(J) - YZ * XS
10419 ZI = Z(I) - ZZ * XS: ZJ = Z(J) - ZZ * XS
10420 XII = -XI * SIN(BETA) + YI * COS(BETA)
10421 XIJ = -XJ * SIN(BETA) + YJ * COS(BETA)
10422 ZII = -XI * SIN(ALPHA) * COS(BETA) - YI * SIN(ALPHA) * SIN(BETA) + ZI *
COS(ALPHA)
10423 ZIJ = -XJ * SIN(ALPHA) * COS(BETA) - YJ * SIN(ALPHA) * SIN(BETA) + ZJ *
COS(ALPHA)
10424 IF Z9 = 99 THEN GOTO 10500
10425 IF XII < CONX THEN CONX = XII
10426 IF XIJ < CONX THEN CONX = XIJ
```

```
10427 IF ZII < CONZ THEN CONZ = ZII
10428 IF ZIJ < CONZ THEN CONZ = ZIJ
10460 IF JUMP = 0 THEN GOTO 10550
10500 LINE (XII - CONX + 40, 300 + CONZ - ZII)-(XIJ - CONX + 40, 300 + CONZ - ZIJ)
10550 NEXT EL
10560 JUMP = JUMP + 1
10570 IF JUMP = 1 THEN GOTO 10350
10670 IF Z9 = 77 THEN GOTO 10800
10690 PRINT "TO CONTINUE, TYPE Y"
10700 A$ = INKEY$: IF A$ = "" THEN GOTO 10700
10720 IF A$ = "Y" OR A$ = "y" THEN GOTO 10770
10750 GOTO 10700
10770 SCREEN 0, 0
10800 RETURN
11000 REM PLOT OF DEFLECTED FORM OF FRAME
11010 CLS
11030 Z9 = 77
11040 GOSUB 10000
11050 UM = -1E+12
11060 FOR II = 1 TO N3
11070 IF ABS(U(II)) > UM THEN UM = ABS(U(II))
11080 NEXT II
11090 SC = 15
11100 FOR II = 1 TO NJ
11110 X(II) = X(II) * XS + SC * U(3 * II - 2) / UM
11120 Y(II) = Y(II) * XS + SC * U(3 * II - 1) / UM
11130 Z(II) = Z(II) * XS + SC * U(3 * II) / UM
11140 NEXT II
11150 Z9 = 99
11160 GOSUB 10000
11170 RETURN
```

Computer Program for Three-Dimensional Rigid-Jointed Frames "3DFRAMEF"

```
10 GOSUB 9000
20 IF A$ = "Y" OR A$ = "y" THEN GOTO 40
30 PRINT : PRINT "GO BACK & READ IT!": PRINT : STOP
40 INPUT "TYPE IN THE NAME OF YOUR INPUT FILE"; FILENAME$
41 PRINT "THE NAME OF YOUR INPUT FILE IS "; FILENAME$
42 PRINT "IF THIS FILENAME IS CORRECT TYPE Y; ELSE TYPE N"
43 A$ = INKEY$: IF A$ = "" THEN GOTO 43
44 IF A$ = "Y" OR A$ = "y" THEN GOTO 49
45 IF A$ = "N" OR A$ = "n" THEN GOTO 40
46 GOTO 43
49 OPEN FILENAME$ FOR INPUT AS #1
50 INPUT "TYPE IN THE NAME OF YOUR OUTPUT FILE "; OUT$
PRINT "IF YOU ARE SATISFIED WITH THIS FILENAME, TYPE Y; ELSE N"
51 B$ = INKEY$: IF B$ = "" THEN GOTO 51
IF B$ = "Y" OR B$ = "y" THEN GOTO 57
IF B$ = "N" OR B$ = "n" THEN GOTO 50
GOTO 51
57 OPEN OUT$ FOR OUTPUT AS #2
59 REM COPYRIGHT OF DR.C.T.F.ROSS
60 REM 6 HURSTVILLE DRIVE,
70 REM WATERLOOVILLE, PORTSMOUTH
80 REM HANTS. PO7 7NB ENGLAND
90 NNODE = 2: NDF = 6
100 CLS
160 INPUT #1, NJ
165 NJ6 = 6 * NJ
170 INPUT #1, NF
180 INPUT #1, NIMK
200 NJI = NJ + NIMK
210 INPUT #1, NMATL
220 INPUT #1, MEMS
230 M6 = 3 * MEMS
240 IF NMATL = 1 THEN MZ = 1
250 IF NMATL > 1 THEN MZ = MEMS
260 DIM UL(12), UG(12), BM(4, 12), SR(4)
270 DIM DC(12, 12), DT(12, 12), SK(12, 12), SKX(12, 12), X(NJI), Y(NJI), Z(NJI)
280 DIM XA(3), YA(3), ZA(3)
290 DIM ELAST(NMATL), RIGD(NMATL), AREA(NMATL), XIN(NMATL), YIN(NMATL),
ZIN(NMATL)
300 DIM UDY(NMATL), UDZ(NMATL)
310 DIM U(NJ6), NS(NF * 6), POSS(NF)
320 DIM NMA(MZ), IJK(M6), DIAG(NJ6)
325 DIM ZETA1(3), ZETA2(3), ZETA3(3), SY21(3), SY31(3), SY34(3), ZETAP(3, 3), ZETAN(3,
3), UNIT(3, 3)
330 REM NODAL COORDINATES
340 FOR II = 1 TO NJI
350 REM X0,Y0 & Z0
360 INPUT #1, X(II), Y(II), Z(II)
PRINT X(II), Y(II), Z(II)
```

370 NEXT II
380 REM MEMBER TYPES
390 FOR II = 1 TO NMATL
400 REM E,G,A,IX,IY,IZ, UDY,UDZ FOR EACH MEMBER TYPE
410 INPUT #1, ELAST(II), RIGD(II), AREA(II), XIN(II), YIN(II), ZIN(II), UDY(II), UDZ(II)
420 NEXT II
430 REM ELEMENT TOPOLOGY
450 FOR MEM = 1 TO MEMS
460 REM I,J & K NODES
470 INPUT #1, I, J, K
PRINT I, J, K
480 M8 = 3 * MEM
490 IJK(M8 - 2) = I
500 IJK(M8 - 1) = J
510 IJK(M8) = K
511 FOR II = 1 TO NNODE: IF II = 1 THEN IN = I ELSE IN = J
512 FOR IJ = 1 TO NDF
513 II1 = NDF * (IN - 1) + IJ
514 FOR JJ = 1 TO NNODE: IF JJ = 1 THEN JN = I ELSE JN = J
515 FOR JI = 1 TO NDF
516 JJ1 = NDF * (JN - 1) + JI
517 IIJJ = JJ1 - II1 + 1: IF IIJJ > DIAG(JJ1) THEN DIAG(JJ1) = IIJJ
520 NEXT JI, JJ, IJ, II
530 IF NMATL = 1 THEN NMA(1) = 1: GOTO 560
540 REM MEMBER TYPE
550 INPUT #1, NMA(MEM)
560 NEXT MEM
565 DIAG(1) = 1: FOR II = 1 TO NJ6: DIAG(II) = DIAG(II - 1) + DIAG(II): NEXT II
570 NT = DIAG(NJ6)
590 DIM A(NT), Q(NJ6), CV(NJ6)
600 FOR II = 1 TO NT: A(II) = 0: NEXT II
610 FOR II = 1 TO N6: Q(II) = 0: CV(II) = 0: NEXT II
660 FOR II = 1 TO NF * 3
670 NS(II) = 0
680 NEXT II
690 REM NO. OF CONCENTRATED LOADS
700 INPUT #1, NCONC
710 IF NCONC = 0 THEN GOTO 900
720 DIM WPOS(NCONC)
730 FOR II = 1 TO NCONC
740 REM NODAL POSITION OF CONCENTRATED LOAD & THE 3 LOADS & 3 COUPLES
750 INPUT #1, PW, XLW, YLW, ZLW, XTHLW, YTHLW, ZTHLW
760 WPOS(II) = PW
780 Q(PW * 6 - 5) = XLW
800 Q(PW * 6 - 4) = YLW
820 Q(PW * 6 - 3) = ZLW
840 Q(PW * 6 - 2) = XTHLW
860 Q(PW * 6 - 1) = YTHLW
880 Q(PW * 6) = ZTHLW
890 NEXT II
900 REM NODAL POSITIONS, ETC. OF THE ZERO DISPLACEMENTS
910 FOR II = 1 TO NF
930 INPUT #1, POSS(II), SUU, SUV, SUW, SUTX, SUTY, SUTZ

```
940 IF SUU = 1 THEN NS(II * 6 - 5) = POSS(II) * 6 - 5
960 IF SUV = 1 THEN NS(II * 6 - 4) = POSS(II) * 6 - 4
980 IF SUW = 1 THEN NS(II * 6 - 3) = POSS(II) * 6 - 3
1000 IF SUTX = 1 THEN NS(II * 6 - 2) = POSS(II) * 6 - 2
1020 IF SUTY = 1 THEN NS(II * 6 - 1) = POSS(II) * 6 - 1
1040 IF SUTZ = 1 THEN NS(II * 6) = POSS(II) * 6
1050 NEXT II
1060 GOSUB 4220
1065 Z9 = 0: GOSUB 10000
1070 FOR MEM = 1 TO MEMS
1080 PRINT : PRINT "ELEMENT NO."; MEM; "UNDER COMPUTATION": PRINT
1090 M8 = 3 * MEM
1100 I = IJK(M8 - 2)
1110 J = IJK(M8 - 1)
1120 K = IJK(M8)
1130 MATL = 1
1140 IF NMATL > 1 THEN MATL = NMA(MEM)
1150 E = ELAST(MATL)
1160 G = RIGD(MATL)
1170 AR = AREA(MATL)
1180 TC = XIN(MATL)
1190 SMAY = YIN(MATL)
1200 SMAZ = ZIN(MATL)
1210 WY = UDY(MATL): WZ = UDZ(MATL)
1220 FOR II = 1 TO 12
1230 FOR JJ = 1 TO 12
1240 DC(II, JJ) = 0
1250 DT(II, JJ) = 0
1260 SK(II, JJ) = 0
1270 SKX(II, JJ) = 0
1280 NEXT JJ
1290 UL(II) = 0
1300 UG(II) = 0
1310 NEXT II
1330 GOSUB 1350
1340 GOTO 2150
1350 XA(1) = X(I)
1360 YA(1) = Y(I)
1370 ZA(1) = Z(I)
1380 XA(2) = X(J)
1390 YA(2) = Y(J)
1400 ZA(2) = Z(J)
1410 XA(3) = X(K)
1420 YA(3) = Y(K)
1430 ZA(3) = Z(K)
1440 REM DIRECTION COSINES
1450 SY21(1) = XA(2) - XA(1)
1460 SY21(2) = YA(2) - YA(1)
1470 SY21(3) = ZA(2) - ZA(1)
1480 SY31(1) = XA(3) - XA(1)
1490 SY31(2) = YA(3) - YA(1)
1500 SY31(3) = ZA(3) - ZA(1)
1510 L21 = SQR(SY21(1) ^ 2 + SY21(2) ^ 2 + SY21(3) ^ 2)
```

```
1520 FOR II = 1 TO 3
1530 ZETA1(II) = SY21(II) / L21
1540 NEXT II
1550 FOR II = 1 TO 3
1560 FOR JJ = 1 TO 3
1570 UNIT(II, JJ) = 0
1580 NEXT JJ
1590 UNIT(II, II) = 1
1600 NEXT II
1610 FOR II = 1 TO 3
1620 FOR JJ = 1 TO 3
1630 ZETAP(II, JJ) = 0
1635 NEXT JJ
1640 FOR JJ = 1 TO 3
1650 ZETAP(II, JJ) = ZETAP(II, JJ) + ZETA1(II) * ZETA1(JJ)
1660 ZETAN(II, JJ) = -ZETAP(II, JJ)
1670 NEXT JJ
1680 NEXT II
1690 FOR II = 1 TO 3
1700 FOR JJ = 1 TO 3
1710 ZETAP(II, JJ) = UNIT(II, JJ) + ZETAN(II, JJ)
1720 NEXT JJ
1730 NEXT II
1740 FOR II = 1 TO 3
1750 SY34(II) = 0
1760 NEXT II
1780 FOR II = 1 TO 3
1790 FOR JJ = 1 TO 3
1800 SY34(II) = SY34(II) + ZETAP(II, JJ) * SY31(JJ)
1810 NEXT JJ
1820 NEXT II
1830 L21 = SQR(SY34(1) ^ 2 + SY34(2) ^ 2 + SY34(3) ^ 2)
1840 FOR II = 1 TO 3
1850 ZETA2(II) = SY34(II) / L21
1860 NEXT II
1870 AXY = (XA(1) * (YA(2) - YA(3)) - YA(1) * (XA(2) - XA(3)) + (XA(2) * YA(3) - YA(2) *
XA(3))) / 2
1880 AYZ = (YA(1) * (ZA(2) - ZA(3)) - ZA(1) * (YA(2) - YA(3)) + (YA(2) * ZA(3) - ZA(2) *
YA(3))) / 2
1890 AZX = (ZA(1) * (XA(2) - XA(3)) - XA(1) * (ZA(2) - ZA(3)) + (ZA(2) * XA(3) - XA(2) *
ZA(3))) / 2
1900 TRIANG = SQR(AYZ ^ 2 + AZX ^ 2 + AXY ^ 2)
1910 ZETA3(1) = AYZ / TRIANG
1920 ZETA3(2) = AZX / TRIANG
1930 ZETA3(3) = AXY / TRIANG
1940 FOR II = 1 TO 3
1950 DC(1, II) = ZETA1(II)
1960 DC(2, II) = ZETA2(II)
1970 DC(3, II) = ZETA3(II)
1980 NEXT II
1990 FOR II = 1 TO 3
2000 MM = 1 + 3 * II
2010 MN = MM + 1
```

```
2020 NN = MN + 1
2030 DC(MM, MM) = DC(1, 1)
2040 DC(MM, MN) = DC(1, 2)
2050 DC(MM, NN) = DC(1, 3)
2060 DC(MN, MM) = DC(2, 1)
2070 DC(MN, MN) = DC(2, 2)
2080 DC(MN, NN) = DC(2, 3)
2090 DC(NN, MM) = DC(3, 1)
2100 DC(NN, MN) = DC(3, 2)
2110 DC(NN, NN) = DC(3, 3)
2120 NEXT II
2130 REM DIRCOS
2140 RETURN
2150 LE = SQR((XA(2) - XA(1)) ^ 2 + (YA(2) - YA(1)) ^ 2 + (ZA(2) - ZA(1)) ^ 2)
2160 L2 = LE ^ 2
2170 L3 = LE ^ 3
2180 UL(2) = WY * LE / 2
2190 UL(8) = UL(2)
2200 UL(3) = WZ * LE / 2
2210 UL(9) = UL(3)
2220 UL(6) = WY * L2 / 12
2230 UL(12) = -UL(6)
2240 UL(5) = -WZ * L2 / 12
2250 UL(11) = -UL(5)
2260 FOR II = 1 TO 12
2270 UG(II) = 0
2280 FOR JJ = 1 TO 12
2290 UG(II) = UG(II) + DC(JJ, II) * UL(JJ)
2300 NEXT JJ, II
2310 REM ELEMENTAL STIFFNESS MATRIX
2320 SK(1, 1) = AR * E / LE
2330 SK(7, 7) = SK(1, 1)
2340 SK(7, 1) = -AR * E / LE
2350 SK(2, 2) = 12 * E * SMAZ / L3
2360 SK(8, 8) = SK(2, 2)
2370 SK(8, 2) = -SK(2, 2)
2380 SK(3, 3) = 12 * E * SMAY / L3
2390 SK(9, 9) = SK(3, 3)
2400 SK(9, 3) = -SK(3, 3)
2410 SK(4, 4) = G * TC / LE
2420 SK(10, 10) = SK(4, 4)
2430 SK(10, 4) = -SK(4, 4)
2440 SK(5, 5) = 4 * E * SMAY / LE
2450 SK(11, 11) = SK(5, 5)
2460 SK(11, 5) = SK(5, 5) / 2
2470 SK(6, 6) = 4 * E * SMAZ / LE
2480 SK(12, 12) = 4 * E * SMAZ / LE
2490 SK(12, 6) = SK(6, 6) / 2
2495 SK(6, 2) = 6 * E * SMAZ / L2
2500 SK(12, 2) = SK(6, 2)
2520 SK(8, 6) = -SK(6, 2)
2530 SK(12, 8) = SK(8, 6)
2540 SK(9, 5) = 6 * E * SMAY / L2
```

```
2550 SK(11, 9) = SK(9, 5)
2560 SK(5, 3) = -SK(9, 5)
2570 SK(11, 3) = -SK(9, 5)
2580 FOR II = 1 TO 12
2590 FOR JJ = II TO 12
2600 SK(II, JJ) = SK(JJ, II)
2610 NEXT JJ
2620 NEXT II
2630 FOR II = 1 TO 12
2640 FOR JJ = 1 TO 12
2650 DT(JJ, II) = DC(II, JJ)
2660 NEXT JJ
2670 NEXT II
2680 FOR II = 1 TO 12
2690 FOR JJ = 1 TO 12
2700 SKX(II, JJ) = 0
2710 NEXT JJ
2720 FOR JJ = 1 TO 12
2730 FOR KK = 1 TO 12
2740 SKX(II, JJ) = SKX(II, JJ) + SK(II, KK) * DC(KK, JJ)
2750 NEXT KK
2760 NEXT JJ
2770 NEXT II
2780 FOR II = 1 TO 12
2790 FOR JJ = 1 TO 12
2800 SK(II, JJ) = 0
2810 NEXT JJ
2820 FOR JJ = 1 TO 12
2830 FOR KK = 1 TO 12
2840 SK(II, JJ) = SK(II, JJ) + DT(II, KK) * SKX(KK, JJ)
2850 NEXT KK
2860 NEXT JJ
2870 NEXT II
2880 REM (K) IS NOW IN GLOBAL COORDINATES
2890 REM (K) WILL NOW BE ASSEMBLED IN 'SKYLINE' FORM
2900 I3 = 6 * I - 6
2910 J3 = 6 * J - 6
2920 FOR II = 1 TO 6
2925 FOR JJ = 1 TO 6
2930 MM = I3 + II: MN = I3 + JJ
2935 NM = J3 + II: N9 = J3 + JJ
2940 IF JJ > II THEN GOTO 2955
2945 A(DIAG(MM) + MN - MM) = A(DIAG(MM) + MN - MM) + SK(II, JJ)
2950 A(DIAG(NM) + N9 - NM) = A(DIAG(NM) + N9 - NM) + SK(II + 6, JJ + 6)
2955 IF I3 > J3 THEN GOTO 2970
2960 A(DIAG(NM) + MN - NM) = A(DIAG(NM) + MN - NM) + SK(II + 6, JJ)
2965 GOTO 2975
2970 A(DIAG(MM) + N9 - MM) = A(DIAG(MM) + N9 - MM) + SK(II, JJ + 6)
2975 NEXT JJ
3080 Q(MM) = Q(MM) + UG(II): Q(NM) = Q(NM) + UG(II + 6)
3090 NEXT II
3110 NEXT MEM
3120 FOR II = 1 TO NF * 6
```

```
3130 IF NS(II) = 0 THEN GOTO 3160
3135 DI = DIAG(NS(II))
3140 A(DI) = A(DI) * 1E+12
3150 Q(NS(II)) = 0
3160 NEXT II
3170 FOR II = 1 TO NJ6
3180 CV(II) = Q(II)
3190 NEXT II
3195 CLOSE #1
3200 PRINT : PRINT "THE SIMULTANEOUS EQUATIONS ARE NOW BEING SOLVED":
PRINT
3210 A(1) = SQR(A(1))
3215 FOR II = 2 TO NJ6
3220 DI = DIAG(II) - II
3225 LL = DIAG(II - 1) - DI + 1
3230 FOR JJ = LL TO II
3235 X = A(DI + JJ)
3240 DJ = DIAG(JJ) - JJ
3245 IF JJ = 1 THEN GOTO 3275
3250 LB = DIAG(JJ - 1) - DJ + 1
3255 IF LL > LB THEN LB = LL
3260 FOR KK = LB TO JJ - 1
3265 X = X - A(DI + KK) * A(DJ + KK)
3270 NEXT KK
3275 A(DI + JJ) = X / A(DJ + JJ)
3280 NEXT JJ
3285 A(DI + II) = SQR(X)
3290 NEXT II
3295 CV(1) = CV(1) / A(1)
3300 FOR II = 2 TO NJ6
3305 DI = DIAG(II) - II
3310 LL = DIAG(II - 1) - DI + 1
3315 X = CV(II)
3320 FOR JJ = LL TO II - 1
3325 X = X - A(DI + JJ) * CV(JJ)
3330 NEXT JJ
3335 CV(II) = X / A(DI + II)
3340 NEXT II
3345 FOR II = NJ6 TO 2 STEP -1
3350 DI = DIAG(II) - II
3355 X = CV(II) / A(DI + II)
3360 CV(II) = X
3365 LL = DIAG(II - 1) - DI + 1
3370 FOR KK = LL TO II - 1
3375 CV(KK) = CV(KK) - X * A(DI + KK)
3380 NEXT KK
3385 NEXT II
3390 CV(1) = CV(1) / A(1)
3430 PRINT #2,
3440 PRINT #2, "THE NODAL DISPLACEMENTS U0,V0,W0,THETAX0,THETAY0 &
THETAZ0 (IN GLOBAL COORDINATES) ARE:-": PRINT #2,
3450 FOR I = 1 TO NJ: PRINT #2, "NODE="; I; "   ";
3460 FOR J = 1 TO 6
```

```
3470 II = 6 * I - 6 + J
3480 U(II) = CV(II)
3490 PRINT #2, U(II),
3500 NEXT J
3510 PRINT #2,
3520 NEXT I
3530 PRINT : PRINT "THE CALCULATION FOR STRESSES HAS NOW COMMENCED":
PRINT
3540 FOR MEM = 1 TO MEMS
3550 M8 = 3 * MEM
3560 I = IJK(M8 - 2)
3570 J = IJK(M8 - 1)
3580 K = IJK(M8)
3590 MATL = 1
3600 IF NMATL > 1 THEN MATL = NMA(MEM)
3610 E = ELAST(MATL)
3620 G = RIGD(MATL)
3630 AR = AREA(MATL)
3640 TC = XIN(MATL)
3650 SMAY = YIN(MATL)
3660 SMAZ = ZIN(MATL)
3670 WY = UDY(MATL)
3680 WZ = UDZ(MATL)
3690 GOSUB 1350
3700 LE = SQR((X(J) - X(I)) ^ 2 + (Y(J) - Y(I)) ^ 2 + (Z(J) - Z(I)) ^ 2)
3710 I3 = 6 * I - 6
3720 J3 = 6 * J - 6
3730 FOR II = 1 TO 6
3740 UG(II) = U(I3 + II)
3750 UG(II + 6) = U(J3 + II)
3760 NEXT II
3770 FOR II = 1 TO 12
3780 UL(II) = 0
3790 FOR JJ = 1 TO 12
3800 UL(II) = UL(II) + DC(II, JJ) * UG(JJ)
3810 NEXT JJ
3820 NEXT II
3830 FOR CC = 1 TO 2
3840 XI = 1
3850 IF CC = 1 THEN XI = 0
3860 BM(1, 1) = -1
3870 BM(1, 7) = 1
3880 BM(2, 4) = -1
3890 BM(2, 10) = 1
3900 BM(3, 3) = (6 - 12 * XI) / LE ^ 2
3910 BM(3, 5) = (-4 + 6 * XI) / LE
3920 BM(3, 9) = -(6 - 12 * XI) / LE ^ 2
3930 BM(3, 11) = (-2 + 6 * XI) / LE
3940 BM(4, 2) = -BM(3, 3)
3950 BM(4, 6) = BM(3, 5)
3960 BM(4, 8) = -BM(3, 9)
3970 BM(4, 12) = BM(3, 11)
3980 FOR JJ = 1 TO 12
```

```
3990 BM(1, JJ) = BM(1, JJ) * AR * E / LE
4000 BM(2, JJ) = BM(2, JJ) * G * TC / LE
4010 BM(3, JJ) = BM(3, JJ) * E * SMAY
4020 BM(4, JJ) = BM(4, JJ) * E * SMAZ
4030 NEXT JJ
4040 PRINT #2,
4050 PRINT #2, "THE FOLLOWING FOUR VALUES ARE 1)AXIAL FORCE 2)TORQUE
3)MOMENT ABOUT THE XY PLANE 4)MOMENT ABOUT THE XZ PLANE FOR MEMBER
"; I; "-"; J; "-"; K
4060 IF CC = 1 THEN PRINT #2, "THIS SET IS AT NODE "; I
4070 IF CC = 2 THEN PRINT #2, "THIS SET IS AT NODE "; J
4080 FOR II = 1 TO 4
4090 SR(II) = 0
4100 FOR JJ = 1 TO 12
4110 SR(II) = SR(II) + BM(II, JJ) * UL(JJ)
4120 NEXT JJ
4130 NEXT II
4140 SR(3) = SR(3) - WZ * LE ^ 2 / 12
4150 SR(4) = SR(4) + WY * LE ^ 2 / 12
4160 FOR II = 1 TO 4
4170 PRINT #2, SR(II),
4180 NEXT II
4190 NEXT CC
4200 NEXT MEM
4210 PRINT #2, : GOSUB 11000
4215 END
4220 REM OUTPUT OF INPUT 1
4250 PRINT #2, : PRINT #2, "FORCES IN RIGID-JOINTED SPACE FRAMES"
4260 PRINT #2, : PRINT #2, "PROGRAM BY DR.C.T.F.ROSS": PRINT #2,
4270 PRINT #2, "NO. OF REAL NODES="; NJ
4280 PRINT #2, "NO. OF NODES WITH ZERO DISPLACEMENTS="; NF
4290 PRINT #2, "NO. OF IMAGINARY NODES="; NIMK
4300 PRINT #2, "NO. OF MEMBER TYPES="; NMATL
4310 PRINT #2, "NO. OF MEMBERS="; MEMS
4320 PRINT #2,
4330 PRINT #2, "NODAL COORDINATES"
4340 FOR II = 1 TO NJI
4350 PRINT #2, "NODE="; II
4360 PRINT #2, "X,Y & Z ARE   "; X(II), Y(II), Z(II)
4370 NEXT II
4380 PRINT #2,
4390 PRINT #2, "MEMBER PROPERTIES"
4400 FOR II = 1 TO NMATL
4410 PRINT #2, "MEMBER TYPE="; II
4420 PRINT #2, "ELASTIC MODULUS="; ELAST(II)
4430 PRINT #2, "RIGIDITY MODULUS="; RIGD(II)
4440 PRINT #2, "SECTIONAL AREA="; AREA(II)
4450 PRINT #2, "TORSIONAL CONSTANT="; XIN(II)
4460 PRINT #2, "2ND MOA ABOUT LOCAL X-Y PLANE="; YIN(II)
4470 PRINT #2, "2ND MOA ABOUT LOCAL X-Z PLANE="; ZIN(II)
4480 PRINT #2, "UDL IN LOCAL Y DIRECTION="; UDY(II)
4490 PRINT #2, "UDL IN LOCAL Z DIRECTION="; UDZ(II)
4500 NEXT II
```

```
4510 PRINT #2,
4520 PRINT #2, "ELEMENT TOPOLOGY, ETC."
4530 FOR II = 1 TO MEMS
4540 M8 = 3 * II
4550 I = IJK(M8 - 2)
4560 J = IJK(M8 - 1)
4570 K = IJK(M8)
4580 PRINT #2, "I,J & K NODES ARE   "; I; "-"; J; "-"; K
4590 IF NMATL = 1 THEN GOTO 4610
4600 PRINT #2, "MEMBER TYPE="; NMA(II)
4610 NEXT II
4620 PRINT #2, "NO. OF NODES WITH CONCENTRATED LOADS="; NCONC
4630 IF NCONC = 0 THEN GOTO 4740
4640 PRINT #2, "THE NODAL POSITIONS & VALUES OF THE CONCENTRATED LOADS &
COUPLES ARE:-"
4650 FOR II = 1 TO NCONC
4660 PRINT #2, "NODE="; WPOS(II)
4670 PRINT #2, "LOAD IN X0 DIRECTION="; Q(WPOS(II) * 6 - 5)
4680 PRINT #2, "LOAD IN Y0 DIRECTION="; Q(WPOS(II) * 6 - 4)
4690 PRINT #2, "LOAD IN Z0 DIRECTION="; Q(WPOS(II) * 6 - 3)
4700 PRINT #2, "COUPLE IN X0 'DIRECTION'="; Q(WPOS(II) * 6 - 2)
4710 PRINT #2, "COUPLE IN Y0 'DIRECTION'="; Q(WPOS(II) * 6 - 1)
4720 PRINT #2, "COUPLE IN Z0 'DIRECTION'="; Q(WPOS(II) * 6)
4730 NEXT II
4740 PRINT #2, "THE NODAL POSITIONS, ETC. OF THE ZERO DISPLACEMENTS ARE"
4750 FOR II = 1 TO NF
4760 PRINT #2, "NODE="; POSS(II)
4770 IF NS(II * 6 - 5) < > 0 THEN PRINT #2, "U0 IS ZERO"
4780 IF NS(II * 6 - 4) < > 0 THEN PRINT #2, "V0 IS ZERO"
4790 IF NS(II * 6 - 3) < > 0 THEN PRINT #2, "W0 IS ZERO"
4800 IF NS(II * 6 - 2) < > 0 THEN PRINT #2, "THETAX0 IS ZERO"
4810 IF NS(II * 6 - 1) < > 0 THEN PRINT #2, "THETAY0 IS ZERO"
4820 IF NS(II * 6) < > 0 THEN PRINT #2, "THETAZ0 IS ZERO"
4830 NEXT II
4840 PRINT #2, : PRINT #2,
4860 RETURN
9000 CLS : PRINT : PRINT "PROGRAM BY DR.C.T.F.ROSS": PRINT
9010 PRINT "NOT TO BE REPRODUCED": PRINT
9020 PRINT "HAVE YOU READ THE BOOK:-": PRINT
9030 PRINT "FINITE ELEMENT PROGRAMS  IN STRUCTURAL ENGINEERING &
CONTINUUM MECHANICS": PRINT
9040 PRINT "BY": PRINT
9050 PRINT "C.T.F.ROSS": PRINT
9060 PRINT "TYPE Y/N"
9070 A$ = INKEY$: IF A$ = "" THEN GOTO 9070
9080 IF A$ = "y" THEN A$ = "Y"
9090 RETURN
9500 PRINT : PRINT "ARE YOU SATISFIED WITH THE INPUT OF YOUR LAST BATCH OF
DATA ? TYPE Y/N": PRINT
9510 X$ = INKEY$: IF X$ = "" THEN GOTO 9510
9520 IF X$ = "y" THEN X$ = "Y"
9530 RETURN
9995 REM ***WARNING***
```

```
9996 PRINT : PRINT "INCORRECT DATA": PRINT
9997 RETURN
10000 IF Z9 = 99 THEN GOTO 10350
10010 XZ = 1E+12: YZ = 1E+12: ZZ = 1E+12
10020 XM = -1E+12: YM = -1E+12: ZM = -1E+12
10040 CLS
10050 PRINT "THE FRAME WILL NOW BE DRAWN. TO CONTINUE, TYPE Y"
10060 A$ = INKEY$: IF A$ = "" THEN GOTO 10060
10070 IF A$ = "Y" OR A$ = "y" THEN GOTO 10090
10080 GOTO 10050
10090 PRINT : PRINT "TYPE IN THE VIEWING ANGLES IN DEGREES": INPUT "ALPHA=";
ALPHA: INPUT "BETA="; BETA
10100 PI = 3.1415926536#: ALPHA = ALPHA * PI / 180: BETA = -BETA * PI / 180
10105 SCREEN 9
10110 FOR EL = 1 TO MEMS
10120 I = IJK(EL * 3 - 2): J = IJK(EL * 3 - 1)
10130 IF X(I) < XZ THEN XZ = X(I)
10140 IF Y(I) < YZ THEN YZ = Y(I)
10150 IF Z(I) < ZZ THEN ZZ = Z(I)
10160 IF X(J) < XZ THEN XZ = X(J)
10170 IF Y(J) < YZ THEN YZ = Y(J)
10180 IF Z(J) < ZZ THEN ZZ = Z(J)
10190 IF X(I) > XM THEN XM = X(I)
10200 IF Y(I) > YM THEN YM = Y(I)
10210 IF Z(I) > ZM THEN ZM = Z(I)
10220 IF X(J) > XM THEN XM = X(J)
10230 IF Y(J) > YM THEN YM = Y(J)
10240 IF Z(J) > ZM THEN ZM = Z(J)
10245 NEXT EL
10250 XS = 400 / (XM - XZ)
10260 IF YM - YZ = 0 OR ZM - ZZ = 0 THEN GOTO 10280
10265 YS = 200 / (YM - YZ): ZS = 200 / (ZM - ZZ)
10270 IF YS > ZS THEN YS = ZS
10275 IF XS > YS THEN XS = YS
10280 CONX = 1E+12
10290 CONZ = 1E+12
10300 JUMP = 0
10350 FOR EL = 1 TO MEMS
10360 I = IJK(EL * 3 - 2)
10370 J = IJK(EL * 3 - 1)
10375 IF Z9 = 99 THEN GOTO 10417
10380 XI = (X(I) - XZ) * XS
10390 YI = (Y(I) - YZ) * XS
10400 XJ = (X(J) - XZ) * XS
10410 YJ = (Y(J) - YZ) * XS
10412 ZI = (Z(I) - ZZ) * XS
10415 ZJ = (Z(J) - ZZ) * XS
10416 GOTO 10420
10417 XI = X(I) - XZ * XS: XJ = X(J) - XZ * XS
10418 YI = Y(I) - YZ * XS: YJ = Y(J) - YZ * XS
10419 ZI = Z(I) - ZZ * XS: ZJ = Z(J) - ZZ * XS
10420 XII = -XI * SIN(BETA) + YI * COS(BETA)
10421 XIJ = -XJ * SIN(BETA) + YJ * COS(BETA)
```

```
10422 ZII = -XI * SIN(ALPHA) * COS(BETA) - YI * SIN(ALPHA) * SIN(BETA) + ZI *
COS(ALPHA)
10423 ZIJ = -XJ * SIN(ALPHA) * COS(BETA) - YJ * SIN(ALPHA) * SIN(BETA) + ZJ *
COS(ALPHA)
10424 IF Z9 = 99 THEN GOTO 10500
10425 IF XII < CONX THEN CONX = XII
10426 IF XIJ < CONX THEN CONX = XIJ
10427 IF ZII < CONZ THEN CONZ = ZII
10428 IF ZIJ < CONZ THEN CONZ = ZIJ
10460 IF JUMP = 0 THEN GOTO 10550
10500 LINE (XII - CONX + 20, 275 + CONZ - ZII)-(XIJ - CONX + 20, 275 + CONZ - ZIJ)
10550 NEXT EL
10560 JUMP = JUMP + 1
10570 IF JUMP = 1 THEN GOTO 10350
10670 IF Z9 = 77 THEN GOTO 10800
10690 PRINT "TO CONTINUE, TYPE Y"
10700 A$ = INKEY$: IF A$ = "" THEN GOTO 10700
10720 IF A$ = "Y" OR A$ = "y" THEN GOTO 10770
10750 GOTO 10700
10770 SCREEN 0, 0: WIDTH 80
10800 RETURN
11000 REM PLOT OF DEFLECTED FORM OF FRAME
11010 CLS
11030 Z9 = 77
11040 GOSUB 10000
11050 UM = -1E+12
11060 FOR II = 1 TO NJ
11062 IX = II * 6 - 5
11063 IF ABS(U(IX)) > UM THEN UM = ABS(U(IX))
11064 IF ABS(U(IX + 1)) > UM THEN UM = ABS(U(IX + 1))
11070 IF ABS(U(IX + 2)) > UM THEN UM = ABS(U(IX + 2))
11080 NEXT II
11090 SC = 15
11100 FOR II = 1 TO NJ
11110 X(II) = X(II) * XS + SC * U(6 * II - 5) / UM
11120 Y(II) = Y(II) * XS + SC * U(6 * II - 4) / UM
11130 Z(II) = Z(II) * XS + SC * U(6 * II - 3) / UM
11140 NEXT II
11150 Z9 = 99
11160 GOSUB 10000
11170 RETURN
```

Appendix 8

Computer Program for Vibration of Rigid-Jointed Space Frames "VIB3DSFN"

```
10 CLS
15 PRINT "FREE VIBRATION OF RIGID-JOINTED SPACE FRAMES": PRINT : PRINT
"PROGRAM BY DR.C.T.F.ROSS"
20 REM VIBRATIONS OF RIGID-JOINTED SPACE FRAMES
30 REM COPYRIGHT OF DR.C.T.F.ROSS
40 REM 6 HURSTVILLE DRIVE, WATERLOOVILLE,
50 REM PORTSMOUTH. HANTS. PO7 7NB
60 REM ENGLAND
61 INPUT "TYPE IN THE NAME OF YOUR INPUT FILE "; FILENAME$
62 PRINT "IF YOU ARE SATISFIED WITH THIS NAME, TYPE Y; ELSE N"
63 A$ = INKEY$: IF A$ = "" THEN GOTO 63
64 IF A$ = "Y" OR A$ = "y" THEN GOTO 67
65 IF A$ = "N" OR A$ = "n" THEN GOTO 61
66 GOTO 63
67 OPEN FILENAME$ FOR INPUT AS #1
68 INPUT "TYPE IN THE NAME OF YOUR OUTPUT FILE "; OUT$
69 PRINT "IF YOU ARE SATISFIED WITH THIS NAME, TYPE Y; ELSE N"
70 B$ = INKEY$: IF B$ = "" THEN GOTO 70
71 IF B$ = "Y" OR B$ = "y" THEN GOTO 76
72 IF B$ = "N" OR B$ = "n" THEN GOTO 68
73 GOTO 70
76 OPEN OUT$ FOR OUTPUT AS #2
79 INPUT #1, NJ
80 NJ6 = 130
90 NP = NJ6 + 1
100 INPUT #1, NF
110 INPUT #1, NSUP, NIMK, N, M1, NMATL, MEMS
130 M6 = 3 * MEMS
140 IF NMATL = 1 THEN MZ = 1
150 IF NMATL > 1 THEN MZ = MEMS
160 DIM A(NJ6, NJ6), XX(NJ6, NP), AM(M1), VC(N, M1), X(200), Z(200), IG(N), VE(N)
170 DIM DC(12, 12), DT(12, 12), SK(12, 12), SKX(12, 12), Y(NJ + NIMK)
180 DIM XA(3), YA(3), ZA(3), VECTOR(N)
190 DIM DENS(NMATL), ELAST(NMATL), RIGD(NMATL), AREA(NMATL), XIN(NMATL),
YIN(NMATL), ZIN(NMATL), JP(NMATL)
200 DIM SM(12, 12), SMX(12, 12), NS(200)
210 DIM NMA(MZ), IJK(M6)
220 DIM ZETA1(3), ZETA2(3), ZETA3(3), SY21(3), SY31(3), SY34(3), ZETAP(3, 3), ZETAN(3,
3), UNIT(3, 3)
250 GOTO 4000
380 PRINT "THE EIGENVALUES ARE BEING DETERMINED": PRINT
390 GOSUB 510
410 FOR I = 1 TO M1
420 PRINT #2, "EIGENVALUE="; AM(I)
430 PRINT #2, "FREQUENCY="; SQR(1 / AM(I)) / (2 * 3.141593)
440 PRINT #2, "EIGENVECTOR IS"
450 PRINT #2,
460 FOR J = 1 TO N
470 PRINT #2, VC(J, I); "    ";
```

```
480 PRINT #2, : NEXT J
490 PRINT #2, : NEXT I
500 END
510 MN = N
520 NN = N
530 GOSUB 1390
540 M = 1
550 FOR I = 1 TO NN
560 VC(I, M) = X(I)
570 XX(I, M) = VC(I, M)
580 NEXT I
590 AM(M) = XM
600 IF M1 < 2 THEN RETURN
610 FOR M = 2 TO M1
620 FOR I = 1 TO NN
630 K4 = ABS(XX(I, M - 1) - 1)
640 IF K4 < .00001 THEN IR = I
650 NEXT I
660 IG(M - 1) = IR
670 FOR I = 1 TO NN
680 XX(MN - I + 1, MN - M + 3) = A(IR, I)
690 NEXT I
700 FOR I = 1 TO NN
710 FOR J = 1 TO NN
720 Z1 = MN - J + 1
730 Z2 = MN - M + 3
740 A(I, J) = A(I, J) - XX(I, M - 1) * XX(Z1, Z2)
750 NEXT J
760 NEXT I
770 FOR I = 1 TO NN
780 IF I = IR THEN GOTO 870
790 IF I > IR THEN K1 = I - 1
800 IF I < IR THEN K1 = I
810 FOR J = 1 TO NN
820 IF J = IR THEN GOTO 860
830 IF J > IR THEN K2 = J - 1
840 IF J < IR THEN K2 = J
850 A(K1, K2) = A(I, J)
860 NEXT J
870 NEXT I
880 NN = NN - 1
890 M3 = NN
900 IF M < > MN THEN GOTO 940
910 XM = A(1, 1)
920 X(1) = 1
930 GOTO 950
940 GOSUB 1390
950 FOR I = 1 TO NN
960 XX(I, M) = X(I)
970 NEXT I
980 AM(M) = XM
990 M4 = M - 1
1000 M5 = 1000 - M4
```

```
1010 FOR M8 = M5 TO 999
1020 M6 = M3 + 1
1030 M2 = 1000 - M8
1040 M7 = IG(M2) + 1
1050 IF M6 < M7 THEN GOTO 1120
1060 N9 = 1000 - M7
1070 N8 = 1000 - M6
1080 FOR I3 = N8 TO N9
1090 I = 1000 - I3
1100 X(I) = X(I - 1)
1110 NEXT I3
1120 J = IG(M2)
1130 X(J) = 0
1140 SUM = 0
1150 FOR I = 1 TO M6
1160 Z3 = MN - I + 1
1170 Z4 = MN - M2 + 2
1180 SUM = SUM + XX(Z3, Z4) * X(I)
1190 NEXT I
1200 XK = (AM(M2) - XM) / SUM
1210 FOR I = 1 TO M6
1220 X(I) = XX(I, M2) - XK * X(I)
1230 NEXT I
1240 SUM = 0
1250 FOR I = 1 TO M6
1260 IF ABS(SUM) < ABS(X(I)) THEN SUM = X(I)
1270 NEXT I
1280 FOR I = 1 TO M6
1290 X(I) = X(I) / SUM
1300 NEXT I
1310 M3 = M3 + 1
1320 IF M2 < > 1 THEN GOTO 1360
1330 FOR I = 1 TO M3
1340 VC(I, M) = X(I)
1350 NEXT I
1360 NEXT M8
1370 NEXT M
1380 RETURN
1390 Y1 = 100000!
1400 FOR I = 1 TO NN
1410 X(I) = 1
1420 NEXT I
1430 XM = -100000!
1440 FOR I = 1 TO NN
1450 SG = 0
1460 FOR J = 1 TO NN
1470 SG = SG + A(I, J) * X(J)
1480 NEXT J
1490 Z(I) = SG
1500 NEXT I
1510 XM = 0
1520 FOR I = 1 TO NN
1530 IF ABS(XM) < ABS(Z(I)) THEN XM = Z(I)
```

```
1540 NEXT I
1550 FOR I = 1 TO NN
1560 X(I) = Z(I) / XM
1570 NEXT I
1580 IF ABS((Y1 - XM) / XM) > D THEN GOTO 1600
1590 GOTO 1620
1600 Y1 = XM
1610 GOTO 1440
1620 X3 = 0
1630 FOR I = 1 TO NN
1640 IF ABS(X3) < ABS(X(I)) THEN X3 = X(I)
1650 NEXT I
1660 FOR I = 1 TO NN
1670 X(I) = X(I) / X3
1680 NEXT I
1690 RETURN
1700 N9 = N - 1
1710 FOR NX = 1 TO N
1720 DI = A(NX, 1)
1730 IF DI = 0 THEN PRINT "MATRIX IS SINGULAR"
1740 FOR NY = 1 TO N9
1750 Y9 = NY + 1
1760 A(NX, NY) = A(NX, Y9) / DI
1770 NEXT NY
1780 A(NX, N) = 1 / DI
1790 FOR NZ = 1 TO N
1800 IF NZ = NX THEN GOTO 1870
1810 O = A(NZ, 1)
1820 FOR NY = 1 TO N9
1830 Y9 = NY + 1
1840 A(NZ, NY) = A(NZ, Y9) - A(NX, NY) * O
1850 NEXT NY
1860 A(NZ, N) = -A(NX, N) * O
1870 NEXT NZ
1880 NEXT NX
1890 RETURN
4000 REM START OF MAIN PART OF PROGRAM
4010 REM NODAL COORDS OF NON-ZERO NODES
4015 PRINT "    X        Y        Z"
4020 NJI = NJ + NIMK
4030 FOR II = 1 TO NJI
4040 INPUT #1, X(II), Y(II), Z(II)
4045 PRINT X(II), Y(II), Z(II)
4050 NEXT II
4060 REM MATERIAL PROPERTIES
4070 FOR II = 1 TO NMATL
4080 REM RHO,E,G,AREA,IX,IY,IZ,JP
4090 INPUT #1, DENS(II), ELAST(II), RIGD(II), AREA(II), XIN(II), YIN(II), ZIN(II), JP(II)
4100 NEXT II
4110 REM ELEMENT TOPOLOGY
4115 PRINT "   I        J        K"
4120 FOR MEM = 1 TO MEMS
4130 REM I,J & K NODES
```

```
4140 INPUT #1, I, J, K
4145 PRINT I, J, K
4150 M8 = 3 * MEM
4160 IJK(M8 - 2) = I
4170 IJK(M8 - 1) = J
4180 IJK(M8) = K
4190 IF NMATL = 1 THEN NMA(1) = 1: GOTO 4220
4200 REM MEMBER TYPE
4210 INPUT #1, NMA(MEM)
4220 NEXT MEM
4230 IF NSUP = 0 THEN 4270
4240 FOR II = 1 TO NSUP
4250 READ NS(II)
4260 NEXT II
4270 INPUT #1, NCONC
4280 IF NCONC = 0 THEN 4350
4290 DIM POSN(NCONC), MASS(NCONC)
4300 FOR II = 1 TO NCONC
4310 REM TYPE IN THE FINISHED 'U' DISPLACEMENT POSITION OF THE
CONCENTRATED MASS
4320 INPUT #1, POSN(II), MASS(II)
4340 NEXT II
4350 GOSUB 19000
4360 IDS = 0
4370 IMAX = 0
4375 PRINT "    I         J        K"
4380 FOR MEM = 1 TO MEMS
4390 PRINT "ELEMENT NO. "; MEM; " UNDER COMPUTATION"
5200 ISUP = 0
5210 JSUP = 0
5220 M8 = 3 * MEM
5230 I = IJK(M8 - 2)
5240 J = IJK(M8 - 1)
5250 K = IJK(M8)
5255 PRINT I, J, K
5260 MATL = 1
5270 IF NMATL > 1 THEN MATL = NMA(MEM)
5280 RHO = DENS(MATL): E = ELAST(MATL): G = RIGD(MATL)
5290 AR = AREA(MATL)
5300 TC = XIN(MATL)
5310 SMAY = YIN(MATL)
5320 SMAZ = ZIN(MATL)
5330 POLAR = JP(MATL)
5340 GOSUB 20000
5350 FOR II = 1 TO 12
5360 FOR JJ = 1 TO 12
5370 DC(II, JJ) = 0
5380 DT(II, JJ) = 0
5390 SK(II, JJ) = 0
5400 SKX(II, JJ) = 0
5410 SM(II, JJ) = 0
5420 SMX(II, JJ) = 0
5430 NEXT JJ
```

```
5440 NEXT II
5450 IF I < > 0 THEN 5840
5810 REM COORDS OF FIXED I NODE
5820 INPUT #1, XA(1), YA(1), ZA(1): GOTO 5870
5840 XA(1) = X(I)
5850 YA(1) = Y(I)
5860 ZA(1) = Z(I)
5870 IF J < > 0 THEN 5900
5880 REM COORDS OF FIXED J NODE
5890 INPUT #1, XA(2), YA(2), ZA(2): GOTO 5930
5900 XA(2) = X(J)
5910 YA(2) = Y(J)
5920 ZA(2) = Z(J)
5930 IF K < > 0 THEN 5960
5940 REM COORDS OF FIXED K NODE
5950 INPUT #1, XA(3), YA(3), ZA(3): GOTO 5990
5960 XA(3) = X(K)
5970 YA(3) = Y(K)
5980 ZA(3) = Z(K)
5990 IF I = 0 OR J = 0 OR K = 0 THEN GOSUB 19500
6000 REM DIRECTION COSINES
6020 SY21(1) = XA(2) - XA(1)
6030 SY21(2) = YA(2) - YA(1)
6040 SY21(3) = ZA(2) - ZA(1)
6050 SY31(1) = XA(3) - XA(1)
6060 SY31(2) = YA(3) - YA(1)
6070 SY31(3) = ZA(3) - ZA(1)
6080 L21 = SQR(SY21(1) ^ 2 + SY21(2) ^ 2 + SY21(3) ^ 2)
6090 FOR II = 1 TO 3
6100 ZETA1(II) = SY21(II) / L21
6110 NEXT II
6120 FOR II = 1 TO 3
6130 FOR JJ = 1 TO 3
6140 UNIT(II, JJ) = 0
6150 NEXT JJ
6160 UNIT(II, II) = 1
6170 NEXT II
6180 FOR II = 1 TO 3
6190 FOR JJ = 1 TO 3
6200 ZETAP(II, JJ) = 0
6210 NEXT JJ
6220 FOR JJ = 1 TO 3
6230 ZETAP(II, JJ) = ZETAP(II, JJ) + ZETA1(II) * ZETA1(JJ)
6240 ZETAN(II, JJ) = -ZETAP(II, JJ)
6250 NEXT JJ, II
6260 FOR II = 1 TO 3
6270 FOR JJ = 1 TO 3
6280 ZETAP(II, JJ) = UNIT(II, JJ) + ZETAN(II, JJ)
6290 NEXT JJ, II
6300 FOR II = 1 TO 3
6310 SY34(II) = 0
6320 NEXT II
6330 FOR II = 1 TO 3
```

```
6340 FOR JJ = 1 TO 3
6350 SY34(II) = SY34(II) + ZETAP(II, JJ) * SY31(JJ)
6360 NEXT JJ, II
6370 L21 = SQR(SY34(1) ^ 2 + SY34(2) ^ 2 + SY34(3) ^ 2)
6380 FOR II = 1 TO 3
6390 ZETA2(II) = SY34(II) / L21
6400 NEXT II
6410 AXY = (XA(1) * (YA(2) - YA(3)) - YA(1) * (XA(2) - XA(3)) + (XA(2) * YA(3) - YA(2) *
XA(3))) / 2
6420 AYZ = (YA(1) * (ZA(2) - ZA(3)) - ZA(1) * (YA(2) - YA(3)) + (YA(2) * ZA(3) - ZA(2) *
YA(3))) / 2
6430 AZX = (ZA(1) * (XA(2) - XA(3)) - XA(1) * (ZA(2) - ZA(3)) + (ZA(2) * XA(3) - XA(2) *
ZA(3))) / 2
6440 TRIANG = SQR(AYZ ^ 2 + AZX ^ 2 + AXY ^ 2)
6450 ZETA3(1) = AYZ / TRIANG
6460 ZETA3(2) = AZX / TRIANG
6470 ZETA3(3) = AXY / TRIANG
6480 FOR II = 1 TO 3
6490 DC(1, II) = ZETA1(II)
6500 DC(2, II) = ZETA2(II)
6510 DC(3, II) = ZETA3(II)
6520 NEXT II
6530 FOR II = 1 TO 3
6540 MM = 1 + 3 * II
6550 MN = MM + 1: NN = MN + 1
6560 DC(MM, MM) = DC(1, 1)
6565 DC(MM, MN) = DC(1, 2)
6570 DC(MM, NN) = DC(1, 3)
6580 DC(MN, MM) = DC(2, 1)
6590 DC(MN, MN) = DC(2, 2)
6600 DC(MN, NN) = DC(2, 3)
6610 DC(NN, MM) = DC(3, 1)
6620 DC(NN, MN) = DC(3, 2)
6630 DC(NN, NN) = DC(3, 3)
6640 NEXT II
6650 REM END OF DIRCOC
6660 LE = SQR((XA(2) - XA(1)) ^ 2 + (YA(2) - YA(1)) ^ 2 + (ZA(2) - ZA(1)) ^ 2)
6670 L2 = LE ^ 2
6680 L3 = LE ^ 3
6690 REM ELEMENTAL STIFFNESS & MASS MATRICES
6700 SK(1, 1) = AR * E / LE
6710 SK(7, 7) = SK(1, 1)
6720 SK(7, 1) = -AR * E / LE
6730 SK(2, 2) = 12 * E * SMAZ / L3
6740 SK(8, 8) = SK(2, 2)
6745 SK(8, 2) = -SK(2, 2)
6750 SK(3, 3) = 12 * E * SMAY / L3
6760 SK(9, 9) = SK(3, 3)
6770 SK(9, 3) = -SK(3, 3)
6780 SK(4, 4) = G * TC / LE
6790 SK(10, 10) = SK(4, 4)
6800 SK(10, 4) = -SK(4, 4)
6810 SK(5, 5) = 4 * E * SMAY / LE
```

```
6820 SK(11, 11) = SK(5, 5)
6830 SK(11, 5) = SK(5, 5) / 2
6840 SK(6, 6) = 4 * E * SMAZ / LE
6850 SK(12, 12) = 4 * E * SMAZ / LE
6860 SK(12, 6) = SK(6, 6) / 2
6870 SK(6, 2) = 6 * E * SMAZ / L2
6880 SK(12, 2) = SK(6, 2)
6890 SK(8, 6) = -SK(6, 2)
6900 SK(12, 8) = SK(8, 6)
6910 SK(9, 5) = 6 * E * SMAY / L2
6920 SK(11, 9) = SK(9, 5)
6930 SK(5, 3) = -SK(9, 5)
6940 SK(11, 3) = -SK(9, 5)
7000 SM(1, 1) = 1 / 3
7010 SM(7, 7) = 1 / 3
7020 SM(7, 1) = 1 / 6
7030 SM(2, 2) = 13 / 35 + 6 * SMAZ / (5 * AR * L2)
7040 SM(8, 8) = SM(2, 2)
7050 SM(6, 2) = 11 * LE / 210 + SMAZ / (10 * AR * LE)
7060 SM(8, 2) = 9 / 70 - 6 * SMAZ / (5 * AR * L2)
7070 SM(12, 2) = -13 * LE / 420 + SMAZ / (10 * AR * LE)
7080 SM(3, 3) = 13 / 35 + 6 * SMAY / (5 * AR * L2)
7090 SM(9, 9) = SM(3, 3)
7100 SM(5, 3) = -11 * LE / 210 - SMAY / (10 * AR * LE)
7110 SM(9, 3) = 9 / 70 - 6 * SMAY / (5 * AR * L2)
7120 SM(11, 3) = 13 * LE / 420 - SMAY / (10 * AR * LE)
7130 SM(4, 4) = POLAR / (3 * AR)
7140 SM(10, 10) = SM(4, 4)
7150 SM(10, 4) = POLAR / (6 * AR)
7160 SM(5, 5) = L2 / 105 + 2 * SMAY / (15 * AR)
7170 SM(11, 11) = SM(5, 5)
7180 SM(9, 5) = -13 * LE / 420 + SMAY / (10 * AR * LE)
7190 SM(11, 5) = -L2 / 140 - SMAY / (30 * AR)
7200 SM(6, 6) = L2 / 105 + 2 * SMAZ / (15 * AR)
7210 SM(12, 12) = SM(6, 6)
7220 SM(8, 6) = 13 * LE / 420 - SMAZ / (10 * AR * LE)
7230 SM(12, 6) = -L2 / 140 - SMAZ / (30 * AR)
7240 SM(12, 8) = -SM(6, 2)
7250 SM(11, 9) = -SM(5, 3)
7260 FOR II = 1 TO 12
7270 FOR JJ = II TO 12
7280 SK(II, JJ) = SK(JJ, II)
7290 SM(II, JJ) = SM(JJ, II)
7300 NEXT JJ
7310 NEXT II
7320 FOR II = 1 TO 12
7330 FOR JJ = 1 TO 12
7340 SM(II, JJ) = SM(II, JJ) * RHO * AR * LE
7350 DT(JJ, II) = DC(II, JJ)
7360 NEXT JJ
7370 NEXT II
7380 FOR II = 1 TO 12
7390 FOR JJ = 1 TO 12
```

```
7400 SKX(II, JJ) = 0
7410 SMX(II, JJ) = 0
7420 NEXT JJ
7430 FOR JJ = 1 TO 12
7440 FOR KK = 1 TO 12
7450 SKX(II, JJ) = SKX(II, JJ) + SK(II, KK) * DC(KK, JJ)
7460 SMX(II, JJ) = SMX(II, JJ) + SM(II, KK) * DC(KK, JJ)
7470 NEXT KK
7480 NEXT JJ
7490 NEXT II
7500 FOR II = 1 TO 12
7510 FOR JJ = 1 TO 12
7520 SK(II, JJ) = 0
7530 SM(II, JJ) = 0
7540 NEXT JJ
7550 FOR JJ = 1 TO 12
7560 FOR KK = 1 TO 12
7570 SK(II, JJ) = SK(II, JJ) + DT(II, KK) * SKX(KK, JJ)
7580 SM(II, JJ) = SM(II, JJ) + DT(II, KK) * SMX(KK, JJ)
7590 NEXT KK
7600 NEXT JJ
7610 NEXT II
7620 REM [K] & [M] ARE NOW IN GLOBAL COORDINATES
7630 REM [K] & [M] WILL NOW BE ASSEMBLED
7640 FOR II = 1 TO 2
7650 IF II = 1 THEN MM = 6 * I - 6
7660 IF II = 2 THEN MM = 6 * J - 6
7670 IF MM < 0 THEN 7930
7740 FOR JJ = 1 TO 2
7750 IF JJ = 1 THEN MN = 6 * I - 6
7760 IF JJ = 2 THEN MN = 6 * J - 6
7770 IF MN < 0 THEN 7920
7780 NM = 6 * II - 6
7790 K1 = MM - IDS
7800 FOR IP1 = 1 TO 6
7810 K1 = K1 + 1
7820 L1 = MN - IDS
7830 NM = NM + 1
7840 N2 = 6 * JJ - 6
7850 FOR JP1 = 1 TO 6
7860 L1 = L1 + 1
7870 N2 = N2 + 1
7880 A(K1, L1) = A(K1, L1) + SK(NM, N2)
7890 XX(K1, L1) = XX(K1, L1) + SM(NM, N2)
7900 NEXT JP1
7910 NEXT IP1
7920 NEXT JJ
7930 NEXT II
7940 REM BRANCHING TO EIGENVALUE ECONOMISER
7950 IF I > IMAX THEN IMAX = I
7960 IF J > IMAX THEN IMAX = J
7970 NINST = 6 * IMAX - IDS
7980 IF ISUP = 0 THEN 8170
```

```
7990 IF IX = 0 THEN 8020
8000 S = 6 * I - IDS - 5
8010 GOSUB 21000
8020 IF IY = 0 THEN 8050
8030 S = 6 * I - IDS - 4
8040 GOSUB 21000
8050 IF IZ = 0 THEN 8080
8060 S = 6 * I - IDS - 3
8070 GOSUB 21000
8080 IF IAX = 0 THEN 8110
8090 S = 6 * I - IDS - 2
8100 GOSUB 21000
8110 IF IAY = 0 THEN 8140
8120 S = 6 * I - IDS - 1
8130 GOSUB 21000
8140 IF IAZ = 0 THEN 8170
8150 S = 6 * I - IDS
8160 GOSUB 21000
8170 IF JSUP = 0 THEN 8360
8180 IF JX = 0 THEN 8210
8190 S = 6 * J - IDS - 5
8200 GOSUB 21000
8210 IF JY = 0 THEN 8240
8220 S = 6 * J - IDS - 4
8230 GOSUB 21000
8240 IF JZ = 0 THEN 8270
8250 S = 6 * J - IDS - 3
8260 GOSUB 21000
8270 IF JAX = 0 THEN 8300
8280 S = 6 * J - IDS - 2
8290 GOSUB 21000
8300 IF JAY = 0 THEN 8330
8310 S = 6 * J - IDS - 1
8320 GOSUB 21000
8330 IF JAZ = 0 THEN 8360
8340 S = 6 * J - IDS
8350 GOSUB 21000
8360 REM END OF ECONOMISING TECHNIQUE
8380 NEXT MEM
8390 IF NSUP = 0 THEN 8490
8400 FOR II = 1 TO NSUP
8410 N9 = NS(II)
8420 A(N9, N9) = A(N9, N9) * 1E+12
8430 NEXT II
8490 PRINT "THE STIFFNESS MATRIX IS BEING INVERTED"
8500 GOSUB 1700
8510 IF NCONC = 0 THEN 8580
8520 FOR II = 1 TO NCONC
8530 N9 = POSN(II)
8540 XX(N9, N9) = XX(N9, N9) + MASS(II)
8550 XX(N9 + 1, N9 + 1) = XX(N9 + 1, N9 + 1) + MASS(II)
8560 XX(N9 + 2, N9 + 2) = XX(N9 + 2, N9 + 2) + MASS(II)
8570 NEXT II
```

```
8580 FOR I = 1 TO N
8590 FOR J = 1 TO N
8600 VECTOR(J) = 0
8610 FOR I1 = 1 TO N
8620 VECTOR(J) = VECTOR(J) + A(I, I1) * XX(I1, J)
8630 NEXT I1
8640 NEXT J
8650 FOR J = 1 TO N
8660 A(I, J) = VECTOR(J)
8670 NEXT J
8680 NEXT I
8690 D = .001
8700 GOTO 380
8710 REM THE EIGENVALUES ARE NOW BEING CALCULATED
19000 REM OUTPUT OF INPUT 1
19010 PRINT #2, "VIBRATION OF RIGID-JOINTED SPACE FRAMES"
19020 PRINT #2, "PROGRAM BY DR.C.T.F.ROSS"
19030 PRINT #2, "NO. OF FREE NODES="; NJ
19040 PRINT #2, "NUMBER OF FIXED NODES="; NF
19050 PRINT #2, "NO. OF SUPPRESSED DISPLACEMENTS AT PIN-JOINTED SUPPORTS,
ETC.="; NSUP
19060 PRINT #2, "NO. OF IMAGINARY NODES="; NIMK
19080 PRINT #2, "NO. OF FREE DISPLACEMENTS="; N
19090 PRINT #2, "NO. OF FREQUENCIES="; M1
19110 PRINT #2, "NO. OF MEMBER TYPES="; NMATL
19120 PRINT #2, "NO. OF MEMBERS="; MEMS
19130 PRINT #2, : PRINT #2, "NODAL COORDINATES X0,Y0 & Z0"
19140 FOR II = 1 TO NJI
19150 PRINT #2, "NODE="; II
19160 PRINT #2, "X0,Y0 & Z0 ARE:  "; X(II), Y(II), Z(II)
19170 NEXT II
19180 PRINT #2, : PRINT #2, "MATERIAL PROPERTIES"
19190 FOR II = 1 TO NMATL
19200 PRINT #2, "MATERIAL TYPE="; II
19210 PRINT #2, "DENSITY="; DENS(II)
19220 PRINT #2, "ELASTIC MODULUS="; ELAST(II)
19230 PRINT #2, "RIGIDITY MODULUS="; RIGD(II)
19240 PRINT #2, "SECTIONAL AREA="; AREA(II)
19250 PRINT #2, "TORSIONAL CONSTANT="; XIN(II)
19260 PRINT #2, "2ND MOMENT OF AREA ABOUT XY PLANE="; YIN(II)
19270 PRINT #2, "2ND MOMENT OF AREA ABOUT XZ PLANE="; ZIN(II)
19280 PRINT #2, "POLAR 2ND MOMENT OF AREA="; JP(II)
19290 NEXT II
19300 PRINT #2,
19310 PRINT #2, "ELEMENT TOPOLOGY"
19320 FOR II = 1 TO MEMS
19330 M8 = 3 * II
19340 I = IJK(M8 - 2)
19350 J = IJK(M8 - 1)
19360 K = IJK(M8)
19370 PRINT #2, "I,J & K NODES ARE:  "; I, J, K
19380 IF NMATL = 1 THEN 19400
19390 PRINT #2, "MEMBER TYPE="; NMA(II)
```

```
19400 NEXT II
19410 IF NSUP = 0 THEN 19416
19411 IF NSUP = 0 THEN 19416
19412 PRINT #2, "THE DISPLACEMENT POSITIONS OF THE ZERO DISPLACEMENTS AT
PINNED SUPPORTS, ETC ARE:"
19413 FOR II = 1 TO NSUP
19414 PRINT #2, NS(II)
19415 NEXT II
19416 PRINT #2,
19418 PRINT #2, "NO. OF CONCENTRATED MASSES="; NCONC
19419 IF NCONC = 0 THEN 19430
19420 FOR II = 1 TO NCONC
19421 PRINT #2, "CONCENTRATED MASS AT THE 'FINISHED' DISPLACEMENT
POSITIONS "; POSN(II); ","; POSN(II) + 1; ","; POSN(II) + 2; "="; MASS(II)
19422 NEXT II
19430 RETURN
19500 REM OUTPUT OF COORDINATES OF THE FIXED NODES
19510 PRINT #2, : PRINT #2, "COORDS OF FIXED NODE/S FOR ELEMENT "; MEM
19520 IF I < > 0 THEN 19540
19530 PRINT #2, "X0,Y0 & Z0 FOR I NODE ARE   "; XA(1), YA(1), ZA(1)
19540 IF J < > 0 THEN 19560
19550 PRINT #2, "X0,Y0 & Z0 FOR J NODE ARE   "; XA(2), YA(2), ZA(2)
19560 IF K < > 0 THEN 19580
19570 PRINT #2, "X0,Y0 & Z0 FOR K NODE ARE   ", XA(3), YA(3), ZA(3)
19580 RETURN
20000 REM EIGENVALUE ECONOMISER (PART1)
20010 IF I > = 0 THEN 20110
20020 I = -I
20030 ISUP = 1
20040 IX = 0
20050 IY = 0
20060 IZ = 0
20070 IAX = 1
20080 IAY = 1
20090 IAZ = 1
20100 GOTO 20210
20110 IF I < 999 THEN 20210
20120 I = I / 1000
20130 ISUP = 1
20140 IX = 1
20150 IY = 1
20160 IZ = 1
20170 IAX = 1
20180 IAY = 1
20190 IAZ = 1
20200 GOTO 20310
20210 IF J > = 0 THEN 20310
20220 J = -J
20230 JSUP = 1
20240 JX = 0
20250 JY = 0
20260 JZ = 0
20270 JAX = 1
```

```
20280 JAY = 1
20290 JAZ = 1
20300 GOTO 20400
20310 IF J < 999 THEN 20400
20320 J = J / 1000
20330 JSUP = 1
20340 JX = 1
20350 JY = 1
20360 JZ = 1
20370 JAX = 1
20380 JAY = 1
20390 JAZ = 1
20400 RETURN
21000 REM EIGENVALUE ECONOMISER (PART2)
21010 FOR II = 1 TO NINST
21020 IF II = S THEN 21080
21030 FOR JJ = 1 TO NINST
21040 IF JJ = S THEN 21070
21050 A(II, JJ) = A(II, JJ) - A(II, S) * A(S, JJ) / A(S, S)
21060 XX(II, JJ) = XX(II, JJ) - XX(II, S) * A(S, JJ) / A(S, S) - XX(S, JJ) * A(II, S) / A(S, S) +
XX(S, S) * A(S, JJ) * A(II, S) / (A(S, S) ^ 2)
21070 NEXT JJ
21080 NEXT II
21090 NINST = NINST - 1
21100 IDS = IDS + 1
21110 IF S > NINST THEN RETURN
21120 FOR II = 1 TO S - 1
21130 FOR JJ = S TO NINST
21140 A(II, JJ) = A(II, JJ + 1)
21150 A(JJ, II) = A(JJ + 1, II)
21160 XX(II, JJ) = XX(II, JJ + 1)
21170 XX(JJ, II) = XX(JJ + 1, II)
21180 NEXT JJ
21190 NEXT II
21200 FOR II = S TO NINST
21210 FOR JJ = S TO NINST
21220 A(II, JJ) = A(II + 1, JJ + 1): XX(II, JJ) = XX(II + 1, JJ + 1)
21230 NEXT JJ
21240 NEXT II
21250 FOR II = 1 TO NINST + 1
21260 A(NINST + 1, II) = 0
21270 A(II, NINST + 1) = 0
21280 XX(NINST + 1, II) = 0
21290 XX(II, NINST + 1) = 0
21300 NEXT II
21310 RETURN
```

Computer Program for Bending Moments in Grillages "GRIDSG"

```
50 REM COPYRIGHT OF DR.C.T.F.ROSS
60 REM 6 HURSTVILLE DRIVE,
70 REM WATERLOOVILLE, PORTSMOUTH
80 REM HANTS. PO7 7NB
90 REM ENGLAND
95 NNODE = 2: NDF = 3
100 CLS
110 PRINT "STATIC ANALYSIS OF GRILLAGES"
120 PRINT "PROGRAM BY DR.C.T.F.ROSS"
PRINT : PRINT "PLEASE ENSURE THAT YOU HAVE READ THE BOOK:-": PRINT
PRINT "FINITE ELEMENT PROGRAMS IN STRUCTURAL ENGINEERING & CONTINUUM
MECHANICS": PRINT
PRINT "BY": PRINT
PRINT "C.T.F.ROSS": PRINT
122 INPUT "TYPE IN THE NAME OF YOUR INPUT FILE "; FILENAME$
PRINT "IF YOU ARE SATISFIED WITH THIS FILE NAME, TYPE Y; ELSE N"
123 A$ = INKEY$: IF A$ = "" THEN GOTO 123
IF A$ = "Y" OR A$ = "y" THEN GOTO 125
IF A$ = "N" OR A$ = "n" THEN GOTO 122
GOTO 123
125 OPEN FILENAME$ FOR INPUT AS #1
126 INPUT "TYPE IN THE NAME OF YOUR OUTPUT FILE "; OUT$
PRINT "IF YOU ARE SATISFIED WITH THIS FILENAME, TYPE Y; ELSE N"
128 B$ = INKEY$: IF B$ = "" THEN GOTO 128
IF B$ = "Y" OR B$ = "y" THEN GOTO 129
IF B$ = "N" OR B$ = "n" THEN GOTO 126
GOTO 128
129 OPEN OUT$ FOR OUTPUT AS #2
130 INPUT #1, NJ: REM NJ=NO. OF NODES
140 INPUT #1, NF: REM NF=NO. OF NODES WITH ELASTIC OR RIGID SUPPORTS
150 N3 = NJ * 3
160 INPUT #1, NC: REM NC=NO. OF NODES WITH CONCENTRATED LOADS
170 DIM DC(6, 6), SK(6, 6), SX(6, 6), SP(6), SQ(6), SG(6), SL(6), DIAG(N3)
180 DIM BX(2, 6), ST(2), X(NJ), Y(NJ), Z(NJ), NS(NF), IJ(300), SUW(NF), SUTX(NF),
SUTY(NF)
PRINT "X & Y COORDINATES"
190 FOR I = 1 TO NJ
200 INPUT #1, X(I), Y(I): REM X & Y COORDINATES
PRINT X(I), Y(I)
220 NEXT I
230 INPUT #1, MS: REM MS=NO OF ELEMENTS
240 INPUT #1, NTYPE: REM NTYPE=NO. OF DIFFERENT 'TYPES' OF ELEMENT
250 IF NTYPE = 1 THEN MZ = 1: GOTO 270
260 MZ = MS
270 DIM NMA(MZ), ELAST(NTYPE), RIGD(NTYPE), YIN(NTYPE), XIN(NTYPE),
UDL(NTYPE)
275 DIM MI(MS), MJ(MS)
PRINT "MEMBER DETAILS"
280 FOR II = 1 TO NTYPE
```

290 REM MEMBER DETAILS - ELAST=YOUNG'S MOD.; RIGD=RIGIDITY MOD.;
YIN=2ND MOA ABOUT X-Y PLANE; XIN=TORSIONAL CONSTANT; UDL=VALUE OF
UNIFORMLY DISTRIBUTED LOAD
300 INPUT #1, ELAST(II), RIGD(II), YIN(II), XIN(II), UDL(II)
PRINT ELAST(II), RIGD(II), YIN(II), XIN(II), UDL(II)
310 NEXT II
PRINT "ELEMENT TOPOLOGY"
330 FOR ME = 1 TO MS
340 REM INPUT OF I & J NODES
350 INPUT #1, I, J
PRINT I, J
360 IF NTYPE = 1 THEN GOTO 367
365 INPUT #1, NMA(ME)
PRINT "ELEMENT TYPE="; NMA(ME)
367 REM ONLY ONE TYPE OF ELEMENT
370 M8 = ME * 2 - 1
380 IJ(M8) = I
382 IJ(M8 + 1) = J
383 FOR II = 1 TO NNODE: IF II = 1 THEN IN = I ELSE IN = J
384 FOR IJ = 1 TO NDF
385 II1 = NDF * (IN - 1) + IJ
386 FOR JJ = 1 TO NNODE: IF JJ = 1 THEN JN = I ELSE JN = J
387 FOR JI = 1 TO NDF
388 JJ1 = NDF * (JN - 1) + JI
389 IIJJ = JJ1 - II1 + 1: IF IIJJ > DIAG(JJ1) THEN DIAG(JJ1) = IIJJ
390 NEXT JI, JJ, IJ, II
395 NEXT ME
400 DIAG(1) = 1: FOR II = 1 TO N3: DIAG(II) = DIAG(II - 1) + DIAG(II): NEXT II
410 NT = DIAG(N3)
450 DIM A(NT), Q(N3), C(N3), U(N3)
460 IF NC = 0 THEN GOTO 570
470 DIM CONC(NC), POSW(NC)
480 REM DETAILS OF NODAL POSITIONS & VALUES OF CONCENTRATED LOADS
490 FOR I = 1 TO NC
500 INPUT #1, POW, QC
510 POSW(I) = POW
520 POW = POW * 3 - 2
540 CONC(I) = QC
550 Q(POW) = QC
560 NEXT I
570 REM DETAILS OF ELASTIC & RIGID SUPPORTS
580 FOR II = 1 TO NF
590 NS(II) = 0
600 NEXT II
610 FOR II = 1 TO NF
620 INPUT #1, SUN, SUW, SUTX, SUTY
630 REM SUN=NODE NUMBER OF ELASTIC OR RIGID SUPPORTS
640 REM SUW=MAGNITUDE OF VERTICAL STIFFNESS - IF RIGID,TYPE 1E30 - IF FREE,
TYPE 0
650 REM SUTX=ROTATIONAL STIFFNESS ALONG X AXIS - SUTY=ROTATIONAL
STIFFNESS ALONG Y AXIS - IF RIGID, TYPE 1E30 - IF FREE, TYPE 0
655 IF SUW < 0 OR SUTX < 0 OR SUTY < 0 THEN PRINT : PRINT "INCORRECT
STIFFNESS AT NODE "; II: END

660 NS(II) = SUN: SUW(II) = SUW: SUTX(II) = SUTX: SUTY(II) = SUTY
670 NEXT II
680 GOSUB 2740
685 GOSUB 10000
690 FOR ME = 1 TO MS
700 PRINT "MEMBER NO. "; ME; " UNDER COMPUTATION"
710 M8 = ME * 2 - 1
720 I = IJ(M8)
730 J = IJ(M8 + 1)
740 MATL = 1
750 IF NTYPE = 1 THEN GOTO 770
760 MATL = NMA(ME)
770 E = ELAST(MATL)
780 G = RIGD(MATL)
790 SA = YIN(MATL)
800 TC = XIN(MATL)
810 UD = UDL(MATL)
820 FOR II = 1 TO 6
830 FOR JJ = 1 TO 6
840 DC(II, JJ) = 0
850 SK(II, JJ) = 0
860 SX(II, JJ) = 0
870 NEXT JJ
880 NEXT II
890 LE = SQR((X(J) - X(I)) ^ 2 + (Y(J) - Y(I)) ^ 2)
900 DC(1, 1) = 1
910 DC(4, 4) = 1
920 DC(2, 3) = (Y(J) - Y(I)) / LE
930 DC(5, 6) = DC(2, 3)
940 DC(3, 2) = -DC(2, 3)
950 DC(6, 5) = DC(3, 2)
960 DC(2, 2) = (X(J) - X(I)) / LE
970 DC(3, 3) = DC(2, 2)
980 DC(5, 5) = DC(2, 2)
990 DC(6, 6) = DC(2, 2)
1000 SK(1, 1) = 12 * E * SA / LE ^ 3
1010 SK(4, 4) = SK(1, 1)
1020 SK(4, 1) = -SK(1, 1)
1030 SK(1, 4) = SK(4, 1)
1040 SK(2, 2) = G * TC / LE
1050 SK(5, 5) = SK(2, 2)
1060 SK(5, 2) = -SK(2, 2)
1070 SK(2, 5) = SK(5, 2)
1080 SK(3, 3) = 4 * E * SA / LE
1090 SK(6, 6) = SK(3, 3)
1100 SK(6, 3) = SK(3, 3) / 2
1110 SK(3, 6) = SK(6, 3)
1120 SK(4, 3) = 6 * E * SA / LE ^ 2
1130 SK(3, 4) = SK(4, 3)
1140 SK(6, 4) = SK(4, 3)
1150 SK(4, 6) = SK(6, 4)
1160 SK(3, 1) = -SK(4, 3)
1170 SK(1, 3) = SK(3, 1)

```
1180 SK(6, 1) = SK(3, 1)
1190 SK(1, 6) = SK(6, 1)
1200 FOR II = 1 TO 6
1210 FOR JJ = 1 TO 6
1220 SX(II, JJ) = 0
1230 NEXT JJ
1240 FOR JJ = 1 TO 6
1250 FOR KK = 1 TO 6
1260 SX(II, JJ) = SX(II, JJ) + SK(II, KK) * DC(KK, JJ)
1270 NEXT KK, JJ, II
1280 FOR II = 1 TO 6
1290 FOR JJ = 1 TO 6
1300 SK(II, JJ) = 0
1310 NEXT JJ
1320 FOR JJ = 1 TO 6
1330 FOR KK = 1 TO 6
1340 SK(II, JJ) = SK(II, JJ) + DC(KK, II) * SX(KK, JJ)
1350 NEXT KK, JJ, II
1360 SP(1) = UD * LE / 2
1370 SP(4) = SP(1)
1380 SP(2) = 0: SP(5) = 0
1390 SP(3) = -UD * LE ^ 2 / 12
1400 SP(6) = -SP(3)
1410 FOR II = 1 TO 6
1420 SQ(II) = 0
1430 FOR JJ = 1 TO 6
1440 SQ(II) = SQ(II) + DC(JJ, II) * SP(JJ)
1450 NEXT JJ, II
1460 I1 = 3 * I - 3
1470 J1 = 3 * J - 3
1480 Q(I1 + 1) = Q(I1 + 1) + SQ(1)
1490 Q(I1 + 2) = Q(I1 + 2) + SQ(2)
1500 Q(I1 + 3) = Q(I1 + 3) + SQ(3)
1510 Q(J1 + 1) = Q(J1 + 1) + SQ(4)
1520 Q(J1 + 2) = Q(J1 + 2) + SQ(5)
1530 Q(J1 + 3) = Q(J1 + 3) + SQ(6)
1540 FOR II = 1 TO 3
1550 FOR JJ = 1 TO 3
1560 MM = I1 + II: MN = I1 + JJ: NM = J1 + II: N9 = J1 + JJ
1570 IF JJ > II THEN 1600
1580 A(DIAG(MM) + MN - MM) = A(DIAG(MM) + MN - MM) + SK(II, JJ)
1590 A(DIAG(NM) + N9 - NM) = A(DIAG(NM) + N9 - NM) + SK(II + 3, JJ + 3)
1600 IF I1 > J1 THEN 1630
1610 A(DIAG(NM) + MN - NM) = A(DIAG(NM) + MN - NM) + SK(II + 3, JJ)
1620 GOTO 1640
1630 A(DIAG(MM) + N9 - MM) = A(DIAG(MM) + N9 - MM) + SK(II, JJ + 3)
1640 NEXT JJ, II
1680 NEXT ME
1690 FOR I = 1 TO NF
1695 II = NS(I)
1700 FOR J = 1 TO 3
1705 N9 = 3 * II - 3 + J: DI = DIAG(N9)
1710 STIFF = SUTY(I)
```

1715 IF J = 1 THEN STIFF = SUW(I)
1720 IF J = 2 THEN STIFF = SUTX(I)
1725 A(DI) = A(DI) + STIFF
1730 NEXT J
1735 NEXT I
1740 FOR II = 1 TO N3
1750 C(II) = Q(II)
1760 NEXT II
1770 N = N3
1780 PRINT : PRINT "THE SIMULTANEOUS EQUATIONS ARE NOW BEING SOLVED":
PRINT
1790 GOSUB 2510
1800 FOR I = 1 TO N3
1810 U(I) = C(I)
1820 NEXT I
1830 PRINT #2, "THE NODAL DISPLACEMENTS W, THETAX0 & THETAY0 ARE AS
FOLLOWS:"
1840 FOR I = 1 TO NJ
1850 II = 3 * I - 3: PRINT #2, "NODE="; I
1860 PRINT #2, "W="; U(II + 1), "THETAX0="; U(II + 2), "THETAY0="; U(II + 3)
1910 NEXT I
1920 FOR ME = 1 TO MS
1930 M8 = ME * 2 - 1
1940 I = IJ(M8)
1950 J = IJ(M8 + 1)
1960 MATL = 1
1970 IF NTYPE = 1 THEN GOTO 1990
1980 MATL = NMA(ME)
1990 E = ELAST(MATL)
2000 G = RIGD(MATL)
2010 SA = YIN(MATL)
2020 TC = XIN(MATL)
2030 UD = UDL(MATL)
2040 LE = SQR((X(J) - X(I)) ^ 2 + (Y(J) - Y(I)) ^ 2)
2050 I1 = 3 * I - 3
2060 J1 = 3 * J - 3
2070 FOR II = 1 TO 3
2080 MM = I1 + II
2090 NM = J1 + II
2100 SG(II) = U(MM)
2110 SG(II + 3) = U(NM)
2120 NEXT II
2130 FOR II = 1 TO 6
2140 FOR JJ = 1 TO 6
2150 DC(II, JJ) = 0
2160 NEXT JJ, II
2170 DC(1, 1) = 1: DC(4, 4) = 1
2180 DC(2, 3) = (Y(J) - Y(I)) / LE
2190 DC(5, 6) = DC(2, 3)
2200 DC(3, 2) = -DC(2, 3)
2210 DC(6, 5) = DC(3, 2)
2220 DC(2, 2) = (X(J) - X(I)) / LE
2230 DC(3, 3) = DC(2, 2)

```
2240 DC(5, 5) = DC(2, 2)
2250 DC(6, 6) = DC(2, 2)
2260 FOR II = 1 TO 6
2270 SL(II) = 0
2280 FOR JJ = 1 TO 6
2290 SL(II) = SL(II) + DC(II, JJ) * SG(JJ)
2300 NEXT JJ, II
2310 FOR II = 1 TO 2
2320 XI = 1
2330 IF II = 1 THEN XI = 0
2340 BX(1, 2) = -G * TC / LE
2350 BX(1, 5) = -BX(1, 2)
2360 BX(2, 1) = E * SA * (6 - 12 * XI) / LE ^ 2
2370 BX(2, 3) = E * SA * (-4 + 6 * XI) / LE
2380 BX(2, 6) = E * SA * (-2 + 6 * XI) / LE
2390 BX(2, 4) = -BX(2, 1)
2400 FOR JJ = 1 TO 2
2410 ST(JJ) = 0
2420 FOR KK = 1 TO 6
2430 ST(JJ) = ST(JJ) + BX(JJ, KK) * SL(KK)
2440 NEXT KK, JJ
2450 ST(2) = ST(2) - UD * LE ^ 2 / 12
2460 PRINT #2, "MEMBER="; I; "-"; J, "TORQUE="; ST(1), "MOMENT="; ST(2)
2465 IF II = 1 THEN MI(ME) = ST(2)
2467 IF II = 2 THEN MJ(ME) = ST(2)
2470 NEXT II
2480 NEXT ME
2485 GOSUB 14000: GOSUB 15000
2490 END
2510 N = N3
2515 A(1) = SQR(A(1))
2520 FOR II = 2 TO N
2525 DI = DIAG(II) - II
2530 LL = DIAG(II - 1) - DI + 1
2535 FOR JJ = LL TO II
2540 X = A(DI + JJ)
2545 DJ = DIAG(JJ) - JJ
2550 IF JJ = 1 THEN 2580
2555 LB = DIAG(JJ - 1) - DJ + 1
2560 IF LL > LB THEN LB = LL
2565 FOR KK = LB TO JJ - 1
2570 X = X - A(DI + KK) * A(DJ + KK)
2575 NEXT KK
2580 A(DI + JJ) = X / A(DJ + JJ)
2585 NEXT JJ
2590 A(DI + II) = SQR(X)
2595 NEXT II
2600 C(1) = C(1) / A(1)
2605 FOR II = 2 TO N
2610 DI = DIAG(II) - II
2615 LL = DIAG(II - 1) - DI + 1
2620 X = C(II)
2625 FOR JJ = LL TO II - 1
```

```
2630 X = X - A(DI + JJ) * C(JJ)
2635 NEXT JJ
2640 C(II) = X / A(DI + II)
2645 NEXT II
2650 FOR II = N TO 2 STEP -1
2655 DI = DIAG(II) - II
2660 X = C(II) / A(DI + II)
2665 C(II) = X
2670 LL = DIAG(II - 1) - DI + 1
2675 FOR KK = LL TO II - 1
2680 C(KK) = C(KK) - X * A(DI + KK)
2685 NEXT KK
2690 NEXT II
2695 C(1) = C(1) / A(1)
2720 RETURN
2740 RETURN: REM OUTPUT OF INPUT
2750 PRINT #2, : PRINT #2, "STATIC ANALYSIS OF GRILLAGES": PRINT #2,
2760 PRINT #2, "NO. OF NODES="; NJ
2770 PRINT #2, "NO. OF NODES WITH ELASTIC OR RIGID SUPPORTS="; NF
2780 PRINT #2, "NO. OF OUT-OF-PLANE CONCENTRATED LOADS="; NC
2800 PRINT #2, "NODAL COORDINATES"
2810 FOR II = 1 TO NJ
2820 PRINT #2, "X("; II; ")="; X(II), "Y("; II; ")="; Y(II)
2830 NEXT II
2850 PRINT #2, "NO. OF ELEMENTS="; MS
2860 PRINT #2, "NO. OF DIFFERENT ELEMENT 'TYPES' ="; NTYPE
2870 PRINT #2, "DETAILS OF THE ELEMENT TYPES"
2880 FOR II = 1 TO NTYPE
2890 PRINT #2, "ELEMENT 'TYPE'="; II
2900 PRINT #2, "YOUNG'S MODULUS="; ELAST(II)
2910 PRINT #2, "RIGIDITY MODULUS="; RIGD(II)
2920 PRINT #2, "2ND MOMENT OF AREA="; YIN(II)
2930 PRINT #2, "TORSIONAL CONSTANT="; XIN(II)
2940 PRINT #2, "DISTRIBUTED LOAD="; UDL(II)
2950 NEXT II
2960 PRINT #2, "ELEMENT TOPOLOGY"
2970 FOR ME = 1 TO MS
2980 M8 = 2 * ME - 1
2990 I = IJ(M8)
3000 J = IJ(M8 + 1)
3010 PRINT #2, "ELEMENT "; I; "-"; J
3020 IF NTYPE = 1 THEN GOTO 3040
3030 PRINT #2, "ELEMENT TYPE="; NMA(ME)
3040 NEXT ME
3050 IF NC = 0 THEN GOTO 3110
3060 FOR II = 1 TO NC
3070 PRINT #2, "NODAL POSITION OF LOAD="; POSW(II)
3080 PRINT #2, "VALUE OF OUT-OF-PLANE CONCENTRATED LOAD="; CONC(II)
3090 NEXT II
3100 PRINT #2, "DETAILS OF ELASTIC & RIGID SUPPORTS ARE:"
3110 FOR II = 1 TO NF
3120 SUN = NS(II)
3150 PRINT #2, "NODE="; SUN
```

```
3160 PRINT #2, "VERTICAL STIFFNESS="; SUW(II)
3170 PRINT #2, "ROTATIONAL STIFFNESS ALONG X AXIS="; SUTX(II)
3180 PRINT #2, "ROTATIONAL STIFFNESS ALONG Y AXIS="; SUTY(II)
3190 NEXT II
3200 RETURN
10000 REM PLOT OF GRID
10010 XZ = 1E+12
10020 YZ = 1E+12
10030 XM = -1E+12
10040 YM = -1E+12
10050 CLS : PRINT "THE GRID WILL NOW BE DRAWN"
10060 SCREEN 9
10065 PRINT : PRINT "PLAN VIEW OF GRID": PRINT
10070 FOR EL = 1 TO MS
10080 M8 = 2 * EL
10090 I = IJ(M8 - 1)
10100 J = IJ(M8)
10110 IF X(I) < XZ THEN XZ = X(I)
10120 IF Y(I) < YZ THEN YZ = Y(I)
10130 IF X(J) < XZ THEN XZ = X(J)
10140 IF Y(J) < YZ THEN YZ = Y(J)
10150 IF X(I) > XM THEN XM = X(I)
10160 IF Y(I) > YM THEN YM = Y(I)
10170 IF X(J) > XM THEN XM = X(J)
10180 IF Y(J) > YM THEN YM = Y(J)
10200 NEXT EL
10240 XS = 350 / (XM - XZ)
10250 IF YM - YZ = 0 THEN GOTO 10280
10260 YS = 175 / (YM - YZ)
10270 IF XS > YS THEN XS = YS
10280 REM
10350 FOR EL = 1 TO MS
10360 M8 = 2 * EL
10370 I = IJ(M8 - 1): J = IJ(M8)
10380 XI = (X(I) - XZ) * XS
10390 YI = (Y(I) - YZ) * XS
10400 XJ = (X(J) - XZ) * XS
10410 YJ = (Y(J) - YZ) * XS
10500 LINE (XI + 40, 280 - YI)-(XJ + 40, 280 - YJ)
10650 NEXT EL
10660 PRINT "TO CONTINUE, TYPE Y"
10670 A$ = INKEY$: IF A$ = "" THEN GOTO 10670
10680 IF A$ = "Y" OR A$ = "y" THEN GOTO 10750
10690 GOTO 10670
10750 SCREEN 0: WIDTH 80
10800 RETURN
14000 REM DEFLECTED FORM OF GRID
14007 FOR I = 1 TO NJ
14010 Z(I) = 0
14020 NEXT I
14030 PRINT : PRINT "TO CONTINUE TYPE Y"
14035 A$ = INKEY$: IF A$ = "" THEN GOTO 14035
14037 IF A$ = "y" OR A$ = "Y" THEN GOTO 14039
```

```
14038 GOTO 14030
14039 SCREEN 9
14040 FOR I = 1 TO NJ
14050 Z(I) = U(I * 3 - 2)
14060 NEXT I
14062 DMAX = -1E+12: DMIN = 1E+12
14063 FOR I = 1 TO NJ
14064 IF Z(I) > DMAX THEN DMAX = Z(I)
14065 IF Z(I) < DMIN THEN DMIN = Z(I)
14066 NEXT
14067 DSCALE = 20 / (DMAX - DMIN)
14070 CLS
14075 PRINT "DEFLECTED FORM OF GRILLAGE"
XS = .7 * XS
14080 FOR EL = 1 TO MS
14090 M8 = 2 * EL
14100 I = IJ(M8 - 1)
14110 J = IJ(M8)
14120 XI = X(I) * XS: XJ = X(J) * XS
14130 YI = Y(I) * XS: YJ = Y(J) * XS
14140 ZI = Z(I) * DSCALE: ZJ = Z(J) * DSCALE
14150 LINE (40 + (XI + YI) * .866, 160 - (YI - XI))-(40 + (XJ + YJ) * .866, 160 - (YJ - XJ))
14160 LINE (40 + (XI + YI) * .866, 160 - (YI - XI) - ZI)-(40 + (XJ + YJ) * .866, 160 - (YJ - XJ) - ZJ)
14170 NEXT EL
14180 RETURN
15000 REM BENDING MOMENT DIAGRAM
15005 MMAX = -1E+12: MMIN = 1E+12
15010 FOR EL = 1 TO MS
15020 M8 = 2 * EL
15030 I = IJ(M8 - 1)
15040 J = IJ(M8)
15070 IF MI(EL) > MMAX THEN MMAX = MI(EL)
15080 IF MJ(EL) > MMAX THEN MMAX = MJ(EL)
15090 IF MI(EL) < MMIN THEN MMIN = MI(EL)
15100 IF MJ(EL) < MMIN THEN MMIN = MJ(EL)
15110 NEXT EL
15120 MSCALE = 30 / (MMAX - MMIN)
15200 REM BM DIAGRAM
15210 A$ = INKEY$: IF A$ = "" THEN GOTO 15210
15220 IF A$ = "y" OR A$ = "Y" THEN GOTO 15240
15230 GOTO 15200
15240 CLS
PRINT "BENDING MOMENT DIAGRAM"
15250 FOR EL = 1 TO MS
15260 M8 = 2 * EL
15270 I = IJ(M8 - 1)
15280 J = IJ(M8)
15290 XI = X(I) * XS: XJ = X(J) * XS
15300 YI = Y(I) * XS: YJ = Y(J) * XS
15310 MI = MI(EL) * MSCALE: MJ = MJ(EL) * MSCALE
15330 LINE (40 + (XI + YI) * .866, 160 - (YI - XI))-(40 + (XJ + YJ) * .866, 160 - (YJ - XJ))
15340 LINE (40 + (XI + YI) * .866, 160 - (YI - XI) - MI)-(40 + (XJ + YJ) * .866, 160 - (YJ -
```

XJ) - MJ)
15345 LINE (40 + (XI + YI) * .866, 160 - (YI - XI))-(40 + (XI + YI) * .866, 160 - (YI - XI) - MI)
15347 LINE (40 + (XJ + YJ) * .866, 160 - (YJ - XJ))-(40 + (XJ + YJ) * .866, 160 - (YJ - XJ) - MJ)
15350 NEXT EL
15355 A$ = INKEY$: IF A$ = "" THEN GOTO 15355
15357 IF A$ = "y" OR A$ = "Y" THEN GOTO 15360
15358 GOTO 15355
15360 SCREEN 0: WIDTH 80
15370 RETURN

Computer Program for the Vibration of Grillages "VIBGRIDG"

```
50 REM VIBRATIONS OF GRILLAGES
60 REM COPYRIGHT OF DR.C.T.F.ROSS
70 REM 6 HURSTVILLE DRIVE, WATERLOOVILLE,
80 REM PORTSMOUTH. PO7 7NB
90 REM ENGLAND
92 CLS : PRINT : PRINT "FREE VIBRATION OF GRILLAGES": PRINT
PRINT "PLEASE ENSURE THAT YOU HAVE READ THE BOOK:-": PRINT
PRINT "FINITE ELEMENT PROGRAMS IN STRUCTURAL ENGINEERING & CONTINUUM
MECHANICS": PRINT
PRINT "BY"
PRINT : PRINT "C.T.F.ROSS"
PRINT : PRINT "BEFORE YOU USE THIS PROGRAM !": PRINT
94 INPUT "TYPE IN THE NAME OF YOUR INPUT FILE "; FILENAME$
PRINT "IF YOU ARE SATISFIED WITH THIS FILENAME, TYPE Y; ELSE N"
95 A$ = INKEY$: IF A$ = "" THEN GOTO 95
IF A$ = "Y" OR A$ = "y" THEN GOTO 97
IF A$ = "N" OR A$ = "n" THEN GOTO 94
GOTO 95
97 OPEN FILENAME$ FOR INPUT AS #1
98 INPUT "TYPE IN THE NAME OF YOUR OUTPUT FILE "; OUT$
PRINT "IF YOU ARE SATISFIED WITH THIS FILE NAME, TYPE Y; ELSE N"
100 B$ = INKEY$: IF B$ = "" THEN GOTO 100
IF B$ = "Y" OR B$ = "y" THEN GOTO 105
IF B$ = "N" OR B$ = "n" THEN GOTO 98
GOTO 100
105 OPEN OUT$ FOR OUTPUT AS #2
106 INPUT #1, NJ
110 NJ6 = NJ * 3
120 NP = NJ6 + 1
130 REM NUMBER OF NODES WITH ZERO DISPLACEMENTS
140 INPUT #1, NSUP
145 IF NSUP < = 0 THEN PRINT "INCORRECT DATA": PRINT : STOP
150 N = NJ6
160 INPUT #1, M1
170 INPUT #1, MEMX, MEMY
190 MEMS = MEMX + MEMY
200 M6 = 2 * MEMS
210 DIM A(NJ6, NJ6), XX(NJ6, NP), AM(M1), VC(N, M1), X(100), Z(100), IG(N), VE(N)
220 DIM DC(6, 6), DT(6, 6), SK(6, 6), SKX(6, 6), XG(NJ), Y(NJ), W(NJ)
230 DIM VECTOR(N), XA(2), YA(2)
240 DIM SM(6, 6), SMX(6, 6)
250 DIM IJ(M6), NFIX(NSUP), WFIX(NSUP), THETAX(NSUP), THETAY(NSUP)
370 GOTO 1900
380 PRINT "THE EIGENVALUES ARE BEING DETERMINED": PRINT
390 GOSUB 510
410 FOR III = 1 TO M1
420 PRINT #2, "EIGENVALUE="; AM(III)
430 PRINT #2, "FREQUENCY="; SQR(1 / AM(III)) / (2 * 3.141593)
440 PRINT #2, "EIGENVECTOR IS"
```

```
450 PRINT #2,
CLS
PRINT "FREQUENCY="; SQR(1 / AM(III)) / (2 * 3.141593); "Hz.(MODE="; III; ")"
460 FOR KK = 1 TO NJ
FOR JJ = 1 TO 3
J = KK * 3 - 3 + JJ
470 PRINT #2, VC(J, III); "   ";
IF JJ = 1 THEN W(KK) = VC(J, III)
NEXT JJ
480 PRINT #2,
NEXT KK
Z9 = 77: GOSUB 10000
490 PRINT #2,
NEXT III
500
SCREEN 0: WIDTH 80
END
510 MN = N
520 NN = N
530 GOSUB 1390
540 M = 1
550 FOR I = 1 TO NN
560 VC(I, M) = X(I)
570 XX(I, M) = VC(I, M)
580 NEXT I
590 AM(M) = XM
600 IF M1 < 2 THEN RETURN
610 FOR M = 2 TO M1
620 FOR I = 1 TO NN
630 K4 = ABS(XX(I, M - 1) - 1)
640 IF K4 < .00001 THEN IR = I
650 NEXT I
660 IG(M - 1) = IR
670 FOR I = 1 TO NN
680 XX(MN - I + 1, MN - M + 3) = A(IR, I)
690 NEXT I
700 FOR I = 1 TO NN
710 FOR J = 1 TO NN
720 Z1 = MN - J + 1
730 Z2 = MN - M + 3
740 A(I, J) = A(I, J) - XX(I, M - 1) * XX(Z1, Z2)
750 NEXT J
760 NEXT I
770 FOR I = 1 TO NN
780 IF I = IR THEN GOTO 870
790 IF I > IR THEN K1 = I - 1
800 IF I < IR THEN K1 = I
810 FOR J = 1 TO NN
820 IF J = IR THEN GOTO 860
830 IF J > IR THEN K2 = J - 1
840 IF J < IR THEN K2 = J
850 A(K1, K2) = A(I, J)
860 NEXT J
```

```
870 NEXT I
880 NN = NN - 1
890 M3 = NN
900 IF M < > MN THEN GOTO 940
910 XM = A(1, 1)
920 X(1) = 1
930 GOTO 950
940 GOSUB 1390
950 FOR I = 1 TO NN
960 XX(I, M) = X(I)
970 NEXT I
980 AM(M) = XM
990 M4 = M - 1
1000 M5 = 1000 - M4
1010 FOR M8 = M5 TO 999
1020 M6 = M3 + 1
1030 M2 = 1000 - M8
1040 M7 = IG(M2) + 1
1050 IF M6 < M7 THEN GOTO 1120
1060 N9 = 1000 - M7
1070 N8 = 1000 - M6
1080 FOR I3 = N8 TO N9
1090 I = 1000 - I3
1100 X(I) = X(I - 1)
1110 NEXT I3
1120 J = IG(M2)
1130 X(J) = 0
1140 SUM = 0
1150 FOR I = 1 TO M6
1160 Z3 = MN - I + 1
1170 Z4 = MN - M2 + 2
1180 SUM = SUM + XX(Z3, Z4) * X(I)
1190 NEXT I
1200 XK = (AM(M2) - XM) / SUM
1210 FOR I = 1 TO M6
1220 X(I) = XX(I, M2) - XK * X(I)
1230 NEXT I
1240 SUM = 0
1250 FOR I = 1 TO M6
1260 IF ABS(SUM) < ABS(X(I)) THEN SUM = X(I)
1270 NEXT I
1280 FOR I = 1 TO M6
1290 X(I) = X(I) / SUM
1300 NEXT I
1310 M3 = M3 + 1
1320 IF M2 < > 1 THEN GOTO 1360
1330 FOR I = 1 TO M3
1340 VC(I, M) = X(I)
1350 NEXT I
1360 NEXT M8
1370 NEXT M
1380 RETURN
1390 Y1 = 100000!
```

```
1400 FOR I = 1 TO NN
1410 X(I) = 1
1420 NEXT I
1430 XM = -100000!
1440 FOR I = 1 TO NN
1450 SG = 0
1460 FOR J = 1 TO NN
1470 SG = SG + A(I, J) * X(J)
1480 NEXT J
1490 Z(I) = SG
1500 NEXT I
1510 XM = 0
1520 FOR I = 1 TO NN
1530 IF ABS(XM) < ABS(Z(I)) THEN XM = Z(I)
1540 NEXT I
1550 FOR I = 1 TO NN
1560 X(I) = Z(I) / XM
1570 NEXT I
1580 IF ABS((Y1 - XM) / XM) > D THEN GOTO 1600
1590 GOTO 1620
1600 Y1 = XM
1610 GOTO 1440
1620 X3 = 0
1630 FOR I = 1 TO NN
1640 IF ABS(X3) < ABS(X(I)) THEN X3 = X(I)
1650 NEXT I
1660 FOR I = 1 TO NN
1670 X(I) = X(I) / X3
1680 NEXT I
1690 RETURN
1700 N9 = N - 1
1710 FOR NX = 1 TO N
1720 DI = A(NX, 1)
1730 IF DI = 0 THEN PRINT "MATRIX IS SINGULAR"
1740 FOR NY = 1 TO N9
1750 Y9 = NY + 1
1760 A(NX, NY) = A(NX, Y9) / DI
1770 NEXT NY
1780 A(NX, N) = 1 / DI
1790 FOR NZ = 1 TO N
1800 IF NZ = NX THEN GOTO 1870
1810 O = A(NZ, 1)
1820 FOR NY = 1 TO N9
1830 Y9 = NY + 1
1840 A(NZ, NY) = A(NZ, Y9) - A(NX, NY) * O
1850 NEXT NY
1860 A(NZ, N) = -A(NX, N) * O
1870 NEXT NZ
1880 NEXT NX
1890 RETURN
1900 REM START OF MAIN PART OF PROGRAM
1910 REM NODAL COORDINATES FOR NON-ZERO NODES
PRINT "NODAL COORDINATES"
```

```
1920 FOR II = 1 TO NJ
1930 INPUT #1, X(II), Y(II)
PRINT X(II), Y(II)
XG(II) = X(II)
1940 NEXT II
1950 REM MATERIAL PROPERTIES
PRINT "MATERIAL PROPERTIES"
1960 INPUT #1, E, G, RHO
INPUT #1, CSAX, CSAY, SMAX, SMAY, TCX, TCY, POLX, POLY
PRINT CSAX, CSAY, SMAX, SMAY, TCX, TCY, POLX, POLY
2060 REM ELEMENT TOPOLOGY
PRINT "ELEMENT TOPOLOGY"
2070 FOR MEM = 1 TO MEMS
2080 INPUT #1, I, J
PRINT I, J
2090 M8 = 2 * MEM
2100 IJ(M8 - 1) = I
2110 IJ(M8) = J
2120 NEXT MEM
PRINT "DETAILS OF ZERO DISPLACEMENTS, ETC.,"
2140 FOR II = 1 TO NSUP
2145 INPUT #1, NFIX(II), WFIX(II), THETAX(II), THETAY(II)
PRINT NFIX(II), WFIX(II), THETAX(II), THETAY(II)
2160 NEXT II
2170 INPUT #1, NCONC
2180 IF NCONC = 0 THEN 2230
2190 DIM POSN(NCONC), MASS(NCONC)
2200 FOR II = 1 TO NCONC
2210 INPUT #1, POSN(II), MASS(II)
2220 NEXT II
2230 GOSUB 4150
Z9 = 0: GOSUB 10000
2260 FOR MEM = 1 TO MEMS
2270 PRINT "ELEMENT NO. "; MEM; " UNDER COMPUTATION"
2300 M8 = 2 * MEM
2310 I = IJ(M8 - 1)
2320 J = IJ(M8)
2330 IF MEM > MEMX THEN 2390
2340 AR = CSAX
2350 SMA = SMAX
2360 TC = TCX
2370 POLAR = POLX
2380 GOTO 2430
2390 AR = CSAY
2400 SMA = SMAY
2410 TC = TCY
2420 POLAR = POLY
2430 REM
2440 FOR II = 1 TO 6
2450 FOR JJ = 1 TO 6
2460 DC(II, JJ) = 0
2470 DT(II, JJ) = 0
2480 SK(II, JJ) = 0
```

```
2490 SKX(II, JJ) = 0
2500 SM(II, JJ) = 0
2510 SMX(II, JJ) = 0
2520 NEXT JJ
2530 NEXT II
2600 XA(1) = X(I): XA(2) = X(J)
2610 YA(1) = Y(I): YA(2) = Y(J)
2660 REM DIRECTION COSINES
2670 LE = SQR((XA(2) - XA(1)) ^ 2 + (YA(2) - YA(1)) ^ 2)
2680 DC(1, 1) = 1
2690 DC(4, 4) = 1
2700 DC(2, 3) = (YA(2) - YA(1)) / LE
2710 DC(5, 6) = DC(2, 3)
2720 DC(3, 2) = -DC(2, 3)
2730 DC(6, 5) = DC(3, 2)
2740 DC(2, 2) = (XA(2) - XA(1)) / LE
2750 DC(3, 3) = DC(2, 2)
2760 DC(5, 5) = DC(2, 2)
2770 DC(6, 6) = DC(2, 2)
2780 REM END DIRCOS
2790 L2 = LE ^ 2
2800 L3 = LE ^ 3
2810 REM ELEMENTAL STIFFNESS & MASS MATRICES
2820 SK(1, 1) = 12 * E * SMA / L3
2830 SK(4, 4) = SK(1, 1)
2840 SK(4, 1) = -SK(1, 1)
2850 SK(2, 2) = G * TC / LE
2860 SK(5, 5) = SK(2, 2)
2870 SK(5, 2) = -SK(2, 2)
2880 SK(3, 3) = 4 * E * SMA / LE
2890 SK(6, 6) = SK(3, 3)
2900 SK(6, 3) = SK(3, 3) / 2
2910 SK(4, 3) = 6 * E * SMA / L2
2920 SK(6, 4) = SK(4, 3)
2930 SK(3, 1) = -SK(4, 3)
2940 SK(6, 1) = -SK(4, 3)
2950 SM(1, 1) = 13 / 35 + 6 * SMA / (5 * AR * L2)
2960 SM(4, 4) = SM(1, 1)
2970 SM(2, 2) = POLAR / (3 * AR)
2980 SM(5, 5) = SM(2, 2)
2990 SM(3, 1) = -11 * LE / 210 - SMA / (10 * AR * LE)
3000 SM(3, 3) = L2 / 105 + 2 * SMA / (15 * AR)
3010 SM(6, 6) = SM(3, 3)
3020 SM(4, 1) = 9 / 70 - 6 * SMA / (5 * AR * L2)
3030 SM(4, 3) = -13 * LE / 420 + SMA / (10 * AR * LE)
3040 SM(5, 2) = POLAR / (6 * AR)
3050 SM(6, 1) = 13 * LE / 420 - SMA / (10 * AR * LE)
3060 SM(6, 3) = L2 / 140 - SMA / (30 * AR)
3070 SM(6, 4) = -11 * LE / 210 + SMA / (10 * AR * LE)
3080 FOR II = 1 TO 6
3090 FOR JJ = II TO 6
3100 SK(II, JJ) = SK(JJ, II)
3110 SM(II, JJ) = SM(JJ, II)
```

```
3120 NEXT JJ
3130 NEXT II
3140 FOR II = 1 TO 6
3150 FOR JJ = 1 TO 6
3160 SM(II, JJ) = SM(II, JJ) * RHO * AR * LE
3170 DT(JJ, II) = DC(II, JJ)
3180 NEXT JJ
3190 NEXT II
3200 FOR II = 1 TO 6
3210 FOR JJ = 1 TO 6
3220 SKX(II, JJ) = 0
3230 SMX(II, JJ) = 0
3240 NEXT JJ
3250 FOR JJ = 1 TO 6
3260 FOR KK = 1 TO 6
3270 SKX(II, JJ) = SKX(II, JJ) + SK(II, KK) * DC(KK, JJ)
3280 SMX(II, JJ) = SMX(II, JJ) + SM(II, KK) * DC(KK, JJ)
3290 NEXT KK
3300 NEXT JJ
3310 NEXT II
3320 FOR II = 1 TO 6
3330 FOR JJ = 1 TO 6
3340 SK(II, JJ) = 0
3350 SM(II, JJ) = 0
3360 NEXT JJ
3370 FOR JJ = 1 TO 6
3380 FOR KK = 1 TO 6
3390 SK(II, JJ) = SK(II, JJ) + DT(II, KK) * SKX(KK, JJ)
3400 SM(II, JJ) = SM(II, JJ) + DT(II, KK) * SMX(KK, JJ)
3410 NEXT KK
3420 NEXT JJ
3430 NEXT II
3440 REM (K) * (M) ARE NOW IN GLOBAL COORDINATES
3460 REM (K) & (M) WILL NOW BE ASSEMBLED
3470 FOR II = 1 TO 2
3480 IF II = 1 THEN MM = 3 * I - 3
3490 IF II = 2 THEN MM = 3 * J - 3
3500 IF MM < 0 THEN 3690
3510 FOR JJ = 1 TO 2
3520 IF JJ = 1 THEN MN = 3 * I - 3
3530 IF JJ = 2 THEN MN = 3 * J - 3
3540 IF MN < 0 THEN 3680
3550 NM = 3 * II - 3
3560 K1 = MM
3570 FOR IP1 = 1 TO 3
3580 K1 = K1 + 1
3590 L1 = MN
3600 NM = NM + 1
3610 N2 = 3 * JJ - 3
3620 FOR JP1 = 1 TO 3
3630 L1 = L1 + 1
3640 N2 = N2 + 1
3645 A(K1, L1) = A(K1, L1) + SK(NM, N2)
```

3650 XX(K1, L1) = XX(K1, L1) + SM(NM, N2)
3660 NEXT JP1
3670 NEXT IP1
3680 NEXT JJ
3690 NEXT II
3890 NEXT MEM
3910 FOR II = 1 TO NSUP
3920 N9 = NFIX(II) * 3 - 2
3922 IF WFIX(II) = 1 THEN A(N9, N9) = A(N9, N9) * 1E+12
3924 IF THETAX(II) = 1 THEN A(N9 + 1, N9 + 1) = A(N9 + 1, N9 + 1) * 1E+12
3926 IF THETAY(II) = 1 THEN A(N9 + 2, N9 + 2) = A(N9 + 2, N9 + 2) * 1E+12
3940 NEXT II
3950 PRINT "THE STIFFNESS MATRIX IS BEING INVERTED"
3960 GOSUB 1700
3970 IF NCONC = 0 THEN 4020
3980 FOR II = 1 TO NCONC
3990 N9 = POSN(II) * 3 - 2
4000 XX(N9, N9) = XX(N9, N9) + MASS(II)
4010 NEXT II
4020 FOR I = 1 TO N
4030 FOR J = 1 TO N
4040 VECTOR(J) = 0
4045 FOR I1 = 1 TO N
4050 VECTOR(J) = VECTOR(J) + A(I, I1) * XX(I1, J)
4060 NEXT I1
4070 NEXT J
4080 FOR J = 1 TO N
4090 A(I, J) = VECTOR(J)
4100 NEXT J
4110 NEXT I
4120 D = .001
4130 GOTO 380
4140 REM THE EIGENVALUES ARE NOW BEING DETERMINED
4150 REM OUTPUT OF INPUT 1
4190 PRINT #2, "VIBRATION OF GRILLAGES"
4200 PRINT #2,
4210 PRINT #2, "NO. OF NODES="; NJ
4230 PRINT #2, "NO. OF NODES WITH ZERO DISPLACEMENTS=", NSUP
4250 PRINT #2, "NO. OF FREQUENCIES="; M1
4260 PRINT #2, "NO. OF MEMBERS IN X DIRECTION="; MEMX
4270 PRINT #2, "NO. OF MEMBERS IN Y DIRECTION="; MEMY
4280 PRINT #2,
4290 PRINT #2, "NODAL COORDINATES"
4300 FOR II = 1 TO NJ
4310 PRINT #2, "NODE="; II
4320 PRINT #2, "X & Y ARE "; X(II), Y(II)
4330 NEXT II
4340 PRINT #2,
4350 PRINT #2, "MATERIAL PROPERTIES"
4370 PRINT #2, "ELASTIC MODULUS="; E
4380 PRINT #2, "RIGIDITY MODULUS="; G
PRINT #2, "DENSITY="; RHO
4390 PRINT #2, "CROSS SECTIONAL AREA FOR X MEMBERS="; CSAX

```
4400 PRINT #2, "CROSS SECTIONAL AREAS FOR Y MEMBERS="; CSAY
4410 PRINT #2, "2ND MOMENT OF AREA FOR X MEMBERS="; SMAX
4420 PRINT #2, "2ND MOMENT OF AREA FOR Y MEMBERS="; SMAY
4430 PRINT #2, "TORSIONAL CONSTANT FOR X MEMBERS="; TCX
4440 PRINT #2, "TORSIONAL CONSTANT FOR Y MEMBERS="; TCY
4450 PRINT #2, "POLAR 2ND MOMENT OF AREA FOR X MEMBERS="; POLX
4460 PRINT #2, "POLAR 2ND MOMENT OF AREA FOR Y MEMBERS="; POLY
4470 PRINT #2,
4480 PRINT #2, "ELEMENT TOPOLOGY, ETC"
4490 FOR II = 1 TO MEMS
4500 M8 = 2 * II
4510 I = IJ(M8 - 1)
4520 J = IJ(M8)
4530 PRINT #2, "I & J NODES ARE "; I, J
4540 NEXT II
4550 PRINT #2,
4570 PRINT #2, "THE DETAILS OF THE ZERO DISPLACEMENTS ARE"
4575 FOR II = 1 TO NSUP
4580 PRINT #2, "NODAL POSITION OF THE ZERO DISPLACEMENTS="; NFIX(II)
4582 IF WFIX(II) = 1 THEN PRINT #2, "W=0" ELSE PRINT #2, "W IS FREE"
4584 IF THETAX(II) = 1 THEN PRINT #2, "THETAX=0" ELSE PRINT #2, "THETAX IS
FREE"
4586 IF THETAY(II) = 1 THEN PRINT #2, "THETAY=0" ELSE PRINT #2, "THETAY IS
FREE"
4590 NEXT II
4600 PRINT #2,
4610 PRINT #2, "NO. OF CONCENTRATED MASSES="; NCONC
4620 IF NCONC = 0 THEN 4680
4630 PRINT #2,
4640 FOR II = 1 TO NCONC
4650 PRINT #2, "NODAL POSITION OF MASS="; POSN(II)
4660 PRINT #2, "VALUE OF MASS="; MASS(II)
4670 NEXT II
4680 RETURN
5240 REM END
10000 REM EIGENMODES OF GRID
MAX = -1E+12
MIN = 1E+12
FOR EL = 1 TO MEMS
M8 = 2 * EL
I = IJ(M8 - 1)
J = IJ(M8)
IF Z9 = 77 THEN GOTO 10001
IF XG(I) > MAX THEN MAX = XG(I)
IF XG(J) > MAX THEN MAX = XG(J)
IF Y(I) > MAX THEN MAX = Y(I)
IF Y(J) > MAX THEN MAX = Y(J)
IF XG(I) < MIN THEN MIN = XG(I)
IF XG(J) < MIN THEN MIN = XG(J)
IF Y(I) < MIN THEN MIN = Y(I)
IF Y(J) < MIN THEN MIN = Y(J)
NEXT EL
XS = 200 / (MAX - MIN)
```

```
IF Z9 = 0 THEN GOTO 10499
10001 DMAX = -1E+12
DMIN = 1E+12
FOR II = 1 TO NJ
IF W(II) > DMAX THEN DMAX = W(II)
IF W(II) < DMIN THEN DMIN = W(II)
NEXT II
DSCALE = 20 / (DMAX - DMIN)
10499 PRINT "TO CONTINUE TYPE Y"
10500 A$ = INKEY$: IF A$ = "" THEN GOTO 10500
IF A$ = "y" OR A$ = "Y" THEN GOTO 10600
IF Z9 = 77 THEN GOTO 10602
GOTO 10500
10600 SCREEN 9
IF Z9 = 0 THEN PRINT "THE GRID"
10601 FOR EL = 1 TO MEMS
M8 = 2 * EL
I = IJ(M8 - 1)
J = IJ(M8)
XI = XG(I) * XS: XJ = XG(J) * XS
YI = Y(I) * XS: YJ = Y(J) * XS
IF Z9 = 0 THEN ZI = 0 AND ZJ = 0: GOTO 10602
ZI = W(I) * DSCALE: ZJ = W(J) * DSCALE
10602 LINE (40 + (XI + YI) * .866, 190 - (YI - XI))-(40 + (XJ + YJ) * .866, 190 - (YJ - XJ))
LINE (40 + (XI + YI) * .866, 190 - (YI - XI) - ZI)-(40 + (XJ + YJ) * .866, 190 - (YJ - XJ) - ZJ)
NEXT EL
10700 A$ = INKEY$: IF A$ = "" THEN GOTO 10700
IF A$ = "y" OR A$ = "Y" THEN GOTO 10800
GOTO 10700
10800 IF Z9 = 77 THEN RETURN
REM SCREEN 0: WIDTH 80
RETURN
```

Computer Program for a Slab on an Elastic Foundation "SLABELAG"

```
100 CLS
110 PRINT "STRESSES IN THICK PLATES ON ELASTIC SUPPORTS (USING 8 NODES) BY
PROF. C.T.F.ROSS"
PRINT : PRINT "PLEASE ENSURE THAT YOU HAVE READ THE BOOK:-"
PRINT : PRINT "FINITE ELEMENT PROGRAMS IN STRUCTURAL ENGINEERING &
CONTINUUM MECHANICS"
PRINT : PRINT "BY"
PRINT : PRINT "C.T.F.ROSS."
112 INPUT "TYPE IN THE NAME OF YOUR INPUT FILE "; FILENAME$
PRINT "IF YOU ARE SATISFIED WITH THIS FILENAME, TYPE Y, ELSE N"
115 A$ = INKEY$: IF A$ = "" THEN GOTO 115
IF A$ = "Y" OR A$ = "y" THEN GOTO 120
IF A$ = "N" OR A$ = "n" THEN GOTO 112
GOTO 115
120 OPEN FILENAME$ FOR INPUT AS #1
122 INPUT "TYPE IN THE NAME OF YOUR OUTPUT FILE "; OUT$
PRINT "IF YOU ARE SATISFIED WITH THIS FILENAME, TYPE Y; ELSE N"
125 B$ = INKEY$: IF B$ = "" THEN GOTO 125
IF B$ = "Y" OR B$ = "y" THEN GOTO 130
IF B$ = "N" OR B$ = "n" THEN GOTO 122
GOTO 125
130 OPEN OUT$ FOR OUTPUT AS #2
135 NODES = 8: NDF = 3
140 REM NNODE=NO. OF NODES
150 INPUT #1, NNODE: NN3 = NNODE * 3
NJ = NNODE
160 REM ELEMS=NO. OF ELEMENTS
170 INPUT #1, ELEMS
175 INPUT #1, ALL$
177 IF ALL$ = "Y" OR ALL$ = "y" THEN NFIX = NNODE: GOTO 192
180 REM NFIX=N0. OF NODES WITH ELASTIC SUPPORTS (I.E. 1 TRANSLATIONAL & 2
ROTATIONAL)
185 IF ALL$ = "Y" OR ALL$ = "y" THEN NFIX = NNODE: GOTO 192
190 INPUT #1, NFIX
192 REM IF PRESSURE LOAD, TYPE "Y", ELSE TYPE "N"
193 INPUT #1, PRESS$
200 DIM IN(ELEMS), JN(ELEMS), KN(ELEMS), L4(ELEMS), M8(ELEMS), N8(ELEMS),
O8(ELEMS), P8(ELEMS), PRESSURE(ELEMS)
210 DIM BCURL(5, 24), BCURLT(24, 5), DCURL(5, 5), STRESS(5)
220 DIM XC(NNODE), YC(NNODE), COORD(8, 2), TH(NNODE)
DIM X(NNODE), Y(NNODE)
DIM VG(NNODE), VEC(NNODE), SHEAR(NNODE), PRINCIPAL(NNODE),
MINIMUM(NNODE)
230 DIM JMX(2, 2), NM(8), DN(2, 8), AM(8), BM(8)
240 DIM POSN(NFIX), NSUP(NN3), SK(24, 24), SL(24), SG(24), BB(5, 24), BDB(24, 24)
245 DIM DIAG(NN3)
250 REM NCONC=NO. OF CONCENTRATED LOADS
260 INPUT #1, NCONC
270 DIM CONC(NCONC), QW(NN3)
```

```
280 REM ELAST=ELASTIC MODULUS
290 INPUT #1, ELAST
300 REM NU=POISSON'S RATIO
310 INPUT #1, NU
320 REM TYPE IN NODAL THICKNESS
330 REM IF THICKNESS IS CONSTANT, TYPE "Y", ELSE TYPE "N"
340 INPUT #1, THICKNESS$
345 IF THICKNESS$ = "N" OR THICKNESS$ = "n" THEN 350
347 IF THICKNESS$ = "Y" OR THICKNESS$ = "y" THEN 349
348 PRINT : PRINT "THE DATA FOR WHETHER OR NOT THE THICKNESS IS CONSTANT,
IS INCORRECT": END
349 INPUT #1, THICK
350 FOR I = 1 TO NNODE
355 IF THICKNESS$ = "Y" OR THICKNESS$ = "y" THEN 365
360 INPUT #1, TH(I)
365 TH(I) = THICK
370 NEXT I
380 GMOD = ELAST / (2 * (1 + NU))
390 REM TYPE IN COORDINATES
400 FOR I = 1 TO NNODE
410 INPUT #1, XC(I), YC(I)
X(I) = XC(I): Y(I) = YC(I)
420 NEXT I
430 REM TYPE IN THE NODAL POSITIONS OF THE ELASTIC SUPPORTS & STIFFNESSES
440 IF ALL$ = "Y" OR ALL$ = "y" THEN INPUT #1, SUPPSTIFF, ROTX, ROTY
450 FOR I = 1 TO NFIX
460 IF ALL$ = "Y" OR ALL$ = "y" THEN POSN(I) = I: GOTO 500
470 INPUT #1, POSN(I), SUPPSTIFF, ROTX, ROTY
500 NSUP(3 * POSN(I) - 2) = SUPPSTIFF
510 NSUP(3 * POSN(I) - 1) = ROTX
520 NSUP(3 * POSN(I)) = ROTY
530 NEXT I
540 IF NCONC = 0 THEN 690
550 REM TYPE IN NODE NUMBERS & VALUES OF CONCENTRATED LOADS
560 FOR I = 1 TO NCONC
565 INPUT #1, CONC(I), QW(CONC(I) * 3 - 2), QW(CONC(I) * 3 - 1), QW(CONC(I) * 3)
590 NEXT I
690 REM NO CONCENTRATED LOADS
700 REM TYPE IN THE NODAL NUMBERS IN AN ANTI-CLOCKWISE DIRECTION, &
PRESSURE LOADS IF APPLICABLE
710 FOR EL = 1 TO ELEMS
720 INPUT #1, IN(EL), JN(EL), KN(EL), L4(EL), M8(EL), N8(EL), O8(EL), P8(EL)
730 INODE = IN(EL)
740 JNODE = JN(EL)
750 KNODE = KN(EL)
760 LNODE = L4(EL)
761 MNODE = M8(EL)
762 NNOD = N8(EL)
763 ONODE = O8(EL)
764 PNODE = P8(EL)
765 IF PRESS$ = "Y" OR PRESS$ = "y" THEN 767
766 GOTO 770
767 INPUT #1, PRESSURE(EL)
```

```
770 FOR II = 1 TO NODES: IA = PNODE
772 IF II = 1 THEN IA = INODE
774 IF II = 2 THEN IA = JNODE
776 IF II = 3 THEN IA = KNODE
777 IF II = 4 THEN IA = LNODE
778 IF II = 5 THEN IA = MNODE
780 IF II = 6 THEN IA = NNOD
782 IF II = 7 THEN IA = ONODE
786 FOR IJ = 1 TO NDF
788 II1 = NDF * (IA - 1) + IJ
790 FOR JJ = 1 TO NODES: IB = PNODE
792 IF JJ = 1 THEN IB = INODE
794 IF JJ = 2 THEN IB = JNODE
796 IF JJ = 3 THEN IB = KNODE
797 IF JJ = 4 THEN IB = LNODE
798 IF JJ = 5 THEN IB = MNODE
800 IF JJ = 6 THEN IB = NNOD
802 IF JJ = 7 THEN IB = ONODE
810 FOR JI = 1 TO NDF
812 JJ1 = NDF * (IB - 1) + JI
814 IIJJ = JJ1 - II1 + 1
816 IF IIJJ > DIAG(JJ1) THEN DIAG(JJ1) = IIJJ
818 NEXT JI, JJ, IJ, II
840 NEXT EL
850 DIAG(1) = 1: FOR II = 1 TO NN3: DIAG(II) = DIAG(II - 1) + DIAG(II): NEXT II
860 NT = DIAG(NN3)
Z9 = 0: GOSUB 10000
870 DIM A(NT)
940 GOSUB 5000
950 FOR II = 1 TO 5: FOR JJ = 1 TO 5
960 DCURL(II, JJ) = 0
970 NEXT JJ
980 NEXT II
990 REM ELEMENT LOOP
1000 FOR EL = 1 TO ELEMS
1010 PRINT : PRINT "ELEMENT NO. "; EL; " UNDER COMPUTATION": PRINT
1020 INODE = IN(EL)
1030 JNODE = JN(EL)
1040 KNODE = KN(EL)
1050 LNODE = L4(EL)
1060 MNODE = M8(EL)
1070 NNOD = N8(EL)
1080 ONODE = O8(EL)
1090 PNODE = P8(EL)
I = INODE: J = JNODE: K = KNODE: L = LNODE
M = MNODE: N = NNOD: O = ONODE: P = PNODE
VEC(I) = VEC(I) + 1: VEC(K) = VEC(K) + 1: VEC(M) = VEC(M) + 1: VEC(O) = VEC(O)
+ 1
VEC(J) = 1: VEC(L) = 1: VEC(N) = 1: VEC(P) = 1
1100 FOR II = 1 TO 24
1110 SL(II) = 0: SG(II) = 0
1120 FOR JJ = 1 TO 24
1130 SK(II, JJ) = 0
```

435

```
1140 NEXT JJ
1150 NEXT II
1160 COORD(1, 1) = XC(INODE): COORD(1, 2) = YC(INODE)
1165 COORD(2, 1) = XC(JNODE): COORD(2, 2) = YC(JNODE)
1170 COORD(3, 1) = XC(KNODE): COORD(3, 2) = YC(KNODE)
1175 COORD(4, 1) = XC(LNODE): COORD(4, 2) = YC(LNODE)
1180 COORD(5, 1) = XC(MNODE): COORD(5, 2) = YC(MNODE)
1185 COORD(6, 1) = XC(NNOD): COORD(6, 2) = YC(NNOD)
1190 COORD(7, 1) = XC(ONODE): COORD(7, 2) = YC(ONODE)
1195 COORD(8, 1) = XC(PNODE): COORD(8, 2) = YC(PNODE)
1200 FOR IJ = 1 TO 3
1210 FOR JI = 1 TO 3
1215 XSTRESS = 0
1220 GOSUB 6500
1240 GOSUB 7510
1300 GOSUB 8000
1350 SL(1) = PRESSURE(EL) * NM(1)
1360 SL(4) = PRESSURE(EL) * NM(2)
1370 SL(7) = PRESSURE(EL) * NM(3)
1380 SL(10) = PRESSURE(EL) * NM(4)
1390 SL(13) = PRESSURE(EL) * NM(5)
1400 SL(16) = PRESSURE(EL) * NM(6)
1410 SL(19) = PRESSURE(EL) * NM(7)
1420 SL(22) = PRESSURE(EL) * NM(8)
1980 FOR II = 1 TO 24
1990 FOR JJ = 1 TO 24
2000 BDB(II, JJ) = 0
2010 FOR KK = 1 TO 5
2020 BDB(II, JJ) = BDB(II, JJ) + BCURLT(II, KK) * BB(KK, JJ)
2030 NEXT KK
2040 NEXT JJ
2050 NEXT II
2060 FOR II = 1 TO 24
2070 FOR JJ = 1 TO 24
2080 SK(II, JJ) = SK(II, JJ) + BDB(II, JJ) * AI * AJ * CON
2090 NEXT JJ
2100 SG(II) = SG(II) + SL(II) * AI * AJ * CON
2110 NEXT II
2120 NEXT JI
2130 NEXT IJ
2150 I1 = 3 * INODE - 3: J1 = 3 * JNODE - 3: K1 = 3 * KNODE - 3: L1 = 3 * LNODE - 3
2160 M1 = 3 * MNODE - 3: N1 = 3 * NNOD - 3: O1 = 3 * ONODE - 3: P1 = 3 * PNODE -
3
2170 FOR II = 1 TO 3
2172 FOR JJ = 1 TO 3
2174 MM = I1 + II: MN = I1 + JJ
2176 NM = J1 + II: N9 = J1 + JJ
2178 IF JJ > II THEN 2184
2180 A(DIAG(MM) + MN - MM) = A(DIAG(MM) + MN - MM) + SK(II, JJ)
2182 A(DIAG(NM) + N9 - NM) = A(DIAG(NM) + N9 - NM) + SK(II + 3, JJ + 3)
2184 IF I1 > J1 THEN 2190
2186 A(DIAG(NM) + MN - NM) = A(DIAG(NM) + MN - NM) + SK(II + 3, JJ)
2188 GOTO 2192
```

```
2190 A(DIAG(MM) + N9 - MM) = A(DIAG(MM) + N9 - MM) + SK(II, JJ + 3)
2192 NM = K1 + II: N9 = K1 + JJ
2194 IF JJ > II THEN 2198
2196 A(DIAG(NM) + N9 - NM) = A(DIAG(NM) + N9 - NM) + SK(II + 6, JJ + 6)
2198 IF I1 > K1 THEN 2204
2200 A(DIAG(NM) + MN - NM) = A(DIAG(NM) + MN - NM) + SK(II + 6, JJ)
2202 GOTO 2206
2204 A(DIAG(MM) + N9 - MM) = A(DIAG(MM) + N9 - MM) + SK(II, JJ + 6)
2206 NM = L1 + II: N9 = L1 + JJ
2208 IF JJ > II THEN 2212
2210 A(DIAG(NM) + N9 - NM) = A(DIAG(NM) + N9 - NM) + SK(II + 9, JJ + 9)
2212 IF I1 > L1 THEN 2222
2214 A(DIAG(NM) + MN - NM) = A(DIAG(NM) + MN - NM) + SK(II + 9, JJ)
2216 GOTO 2224
2222 A(DIAG(MM) + N9 - MM) = A(DIAG(MM) + N9 - MM) + SK(II, JJ + 9)
2224 NM = M1 + II: N9 = M1 + JJ
2226 IF JJ > II THEN 2230
2228 A(DIAG(NM) + N9 - NM) = A(DIAG(NM) + N9 - NM) + SK(II + 12, JJ + 12)
2230 IF I1 > M1 THEN 2236
2232 A(DIAG(NM) + MN - NM) = A(DIAG(NM) + MN - NM) + SK(II + 12, JJ)
2234 GOTO 2238
2236 A(DIAG(MM) + N9 - MM) = A(DIAG(MM) + N9 - MM) + SK(II, JJ + 12)
2238 NM = N1 + II: N9 = N1 + JJ
2240 IF JJ > II THEN 2244
2242 A(DIAG(NM) + N9 - NM) = A(DIAG(NM) + N9 - NM) + SK(II + 15, JJ + 15)
2244 IF I1 > N1 THEN 2252
2248 A(DIAG(NM) + MN - NM) = A(DIAG(NM) + MN - NM) + SK(II + 15, JJ)
2250 GOTO 2254
2252 A(DIAG(MM) + N9 - MM) = A(DIAG(MM) + N9 - MM) + SK(II, JJ + 15)
2254 NM = O1 + II: N9 = O1 + JJ
2256 IF JJ > II THEN 2260
2258 A(DIAG(NM) + N9 - NM) = A(DIAG(NM) + N9 - NM) + SK(II + 18, JJ + 18)
2260 IF I1 > O1 THEN 2266
2262 A(DIAG(NM) + MN - NM) = A(DIAG(NM) + MN - NM) + SK(II + 18, JJ)
2264 GOTO 2268
2266 A(DIAG(MM) + N9 - MM) = A(DIAG(MM) + N9 - MM) + SK(II, JJ + 18)
2268 NM = P1 + II: N9 = P1 + JJ
2270 IF JJ > II THEN 2274
2272 A(DIAG(NM) + N9 - NM) = A(DIAG(NM) + N9 - NM) + SK(II + 21, JJ + 21)
2274 IF I1 > P1 THEN 2280
2276 A(DIAG(NM) + MN - NM) = A(DIAG(NM) + MN - NM) + SK(II + 21, JJ)
2278 GOTO 2282
2280 A(DIAG(MM) + N9 - MM) = A(DIAG(MM) + N9 - MM) + SK(II, JJ + 21)
2282 MM = J1 + II: MN = J1 + JJ
2284 NM = K1 + II: N9 = K1 + JJ
2356 IF J1 > K1 THEN 2362
2358 A(DIAG(NM) + MN - NM) = A(DIAG(NM) + MN - NM) + SK(II + 6, JJ + 3)
2360 GOTO 2364
2362 A(DIAG(MM) + N9 - MM) = A(DIAG(MM) + N9 - MM) + SK(II + 3, JJ + 6)
2364 NM = L1 + II: N9 = L1 + JJ
2366 IF J1 > L1 THEN 2372
2368 A(DIAG(NM) + MN - NM) = A(DIAG(NM) + MN - NM) + SK(II + 9, JJ + 3)
2370 GOTO 2374
```

2372 A(DIAG(MM) + N9 - MM) = A(DIAG(MM) + N9 - MM) + SK(II + 3, JJ + 9)
2374 NM = M1 + II: N9 = M1 + JJ
2376 IF J1 > M1 THEN 2382
2378 A(DIAG(NM) + MN - NM) = A(DIAG(NM) + MN - NM) + SK(II + 12, JJ + 3)
2380 GOTO 2384
2382 A(DIAG(MM) + N9 - MM) = A(DIAG(MM) + N9 - MM) + SK(II + 3, JJ + 12)
2384 NM = N1 + II: N9 = N1 + JJ
2386 IF J1 > N1 THEN 2392
2388 A(DIAG(NM) + MN - NM) = A(DIAG(NM) + MN - NM) + SK(II + 15, JJ + 3)
2390 GOTO 2394
2392 A(DIAG(MM) + N9 - MM) = A(DIAG(MM) + N9 - MM) + SK(II + 3, JJ + 15)
2394 NM = O1 + II: N9 = O1 + JJ
2396 IF J1 > O1 THEN 2402
2398 A(DIAG(NM) + MN - NM) = A(DIAG(NM) + MN - NM) + SK(II + 18, JJ + 3)
2400 GOTO 2404
2402 A(DIAG(MM) + N9 - MM) = A(DIAG(MM) + N9 - MM) + SK(II + 3, JJ + 18)
2404 NM = P1 + II: N9 = P1 + JJ
2406 IF J1 > P1 THEN 2412
2408 A(DIAG(NM) + MN - NM) = A(DIAG(NM) + MN - NM) + SK(II + 21, JJ + 3)
2410 GOTO 2416
2412 A(DIAG(MM) + N9 - MM) = A(DIAG(MM) + N9 - MM) + SK(II + 3, JJ + 21)
2416 MM = K1 + II: MN = K1 + JJ
2418 NM = L1 + II: N9 = L1 + JJ
2420 IF K1 > L1 THEN 2426
2422 A(DIAG(NM) + MN - NM) = A(DIAG(NM) + MN - NM) + SK(II + 9, JJ + 6)
2424 GOTO 2428
2426 A(DIAG(MM) + N9 - MM) = A(DIAG(MM) + N9 - MM) + SK(II + 6, JJ + 9)
2428 NM = M1 + II: N9 = M1 + JJ
2430 IF K1 > M1 THEN 2436
2432 A(DIAG(NM) + MN - NM) = A(DIAG(NM) + MN - NM) + SK(II + 12, JJ + 6)
2434 GOTO 2438
2436 A(DIAG(MM) + N9 - MM) = A(DIAG(MM) + N9 - MM) + SK(II + 6, JJ + 12)
2438 NM = N1 + II: N9 = N1 + JJ
2440 IF K1 > N1 THEN 2448
2442 A(DIAG(NM) + MN - NM) = A(DIAG(NM) + MN - NM) + SK(II + 15, JJ + 6)
2446 GOTO 2450
2448 A(DIAG(MM) + N9 - MM) = A(DIAG(MM) + N9 - MM) + SK(II + 6, JJ + 15)
2450 NM = O1 + II: N9 = O1 + JJ
2452 IF K1 > O1 THEN 2458
2454 A(DIAG(NM) + MN - NM) = A(DIAG(NM) + MN - NM) + SK(II + 18, JJ + 6)
2456 GOTO 2460
2458 A(DIAG(MM) + N9 - MM) = A(DIAG(MM) + N9 - MM) + SK(II + 6, JJ + 18)
2460 NM = P1 + II: N9 = P1 + JJ
2462 IF K1 > P1 THEN 2468
2464 A(DIAG(NM) + MN - NM) = A(DIAG(NM) + MN - NM) + SK(II + 21, JJ + 6)
2466 GOTO 2470
2468 A(DIAG(MM) + N9 - MM) = A(DIAG(MM) + N9 - MM) + SK(II + 6, JJ + 21)
2470 MM = L1 + II: MN = L1 + JJ
2474 NM = M1 + II: N9 = M1 + JJ
2476 IF L1 > M1 THEN 2482
2478 A(DIAG(NM) + MN - NM) = A(DIAG(NM) + MN - NM) + SK(II + 12, JJ + 9)
2480 GOTO 2484
2482 A(DIAG(MM) + N9 - MM) = A(DIAG(MM) + N9 - MM) + SK(II + 9, JJ + 12)

```
2484 NM = N1 + II: N9 = N1 + JJ
2486 IF L1 > N1 THEN 2492
2488 A(DIAG(NM) + MN - NM) = A(DIAG(NM) + MN - NM) + SK(II + 15, JJ + 9)
2490 GOTO 2494
2492 A(DIAG(MM) + N9 - MM) = A(DIAG(MM) + N9 - MM) + SK(II + 9, JJ + 15)
2494 NM = O1 + II: N9 = O1 + JJ
2496 IF L1 > O1 THEN 2502
2498 A(DIAG(NM) + MN - NM) = A(DIAG(NM) + MN - NM) + SK(II + 18, JJ + 9)
2500 GOTO 2504
2502 A(DIAG(MM) + N9 - MM) = A(DIAG(MM) + N9 - MM) + SK(II + 9, JJ + 18)
2504 NM = P1 + II: N9 = P1 + JJ
2506 IF L1 > P1 THEN 2512
2508 A(DIAG(NM) + MN - NM) = A(DIAG(NM) + MN - NM) + SK(II + 21, JJ + 9)
2510 GOTO 2516
2512 A(DIAG(MM) + N9 - MM) = A(DIAG(MM) + N9 - MM) + SK(II + 9, JJ + 21)
2516 MM = M1 + II: MN = M1 + JJ
2518 NM = N1 + II: N9 = N1 + JJ
2520 IF M1 > N1 THEN 2526
2522 A(DIAG(NM) + MN - NM) = A(DIAG(NM) + MN - NM) + SK(II + 15, JJ + 12)
2524 GOTO 2528
2526 A(DIAG(MM) + N9 - MM) = A(DIAG(MM) + N9 - MM) + SK(II + 12, JJ + 15)
2528 NM = O1 + II: N9 = O1 + JJ
2530 IF M1 > O1 THEN 2536
2532 A(DIAG(NM) + MN - NM) = A(DIAG(NM) + MN - NM) + SK(II + 18, JJ + 12)
2534 GOTO 2538
2536 A(DIAG(MM) + N9 - MM) = A(DIAG(MM) + N9 - MM) + SK(II + 12, JJ + 18)
2538 NM = P1 + II: N9 = P1 + JJ
2540 IF M1 > P1 THEN 2546
2542 A(DIAG(NM) + MN - NM) = A(DIAG(NM) + MN - NM) + SK(II + 21, JJ + 12)
2544 GOTO 2550
2546 A(DIAG(MM) + N9 - MM) = A(DIAG(MM) + N9 - MM) + SK(II + 12, JJ + 21)
2550 MM = N1 + II: MN = N1 + JJ
2552 NM = O1 + II: N9 = O1 + JJ
2554 IF N1 > O1 THEN 2560
2556 A(DIAG(NM) + MN - NM) = A(DIAG(NM) + MN - NM) + SK(II + 18, JJ + 15)
2558 GOTO 2562
2560 A(DIAG(MM) + N9 - MM) = A(DIAG(MM) + N9 - MM) + SK(II + 15, JJ + 18)
2562 NM = P1 + II: N9 = P1 + JJ
2564 IF N1 > P1 THEN 2572
2568 A(DIAG(NM) + MN - NM) = A(DIAG(NM) + MN - NM) + SK(II + 21, JJ + 15)
2570 GOTO 2576
2572 A(DIAG(MM) + N9 - MM) = A(DIAG(MM) + N9 - MM) + SK(II + 15, JJ + 21)
2576 MM = O1 + II: MN = O1 + JJ
2578 NM = P1 + II: N9 = P1 + JJ
2580 IF O1 > P1 THEN 2586
2582 A(DIAG(NM) + MN - NM) = A(DIAG(NM) + MN - NM) + SK(II + 21, JJ + 18)
2584 GOTO 2590
2586 A(DIAG(MM) + N9 - MM) = A(DIAG(MM) + N9 - MM) + SK(II + 18, JJ + 21)
2590 NEXT JJ, II
2592 FOR ROW = 1 TO 8
2593 IF ROW = 1 THEN NR = INODE
2594 IF ROW = 2 THEN NR = JNODE
2595 IF ROW = 3 THEN NR = KNODE
```

```
2596 IF ROW = 4 THEN NR = LNODE
2597 IF ROW = 5 THEN NR = MNODE
2598 IF ROW = 6 THEN NR = NNOD
2599 IF ROW = 7 THEN NR = ONODE
2600 IF ROW = 8 THEN NR = PNODE
2601 NR = 3 * NR - 3
2602 FOR ROWP = 1 TO 3
2603 NR = NR + 1: INR = (ROW - 1) * 3 + ROWP
2604 QW(NR) = QW(NR) + SG(INR)
2606 NEXT ROWP, ROW
2612 NEXT EL
2614 FOR II = 1 TO NFIX
2615 N9 = 3 * POSN(II) - 3
2616 FOR JJ = 1 TO 3
2617 N9 = N9 + 1
2618 DI = DIAG(N9)
2619 A(DI) = A(DI) + NSUP(N9)
2620 NEXT JJ
2621 NEXT II
2622 PRINT : PRINT "THE SIMULTANEOUS EQUATIONS ARE NOW BEING SOLVED":
PRINT
2624 N = NN3
2626 A(1) = SQR(A(1))
2628 FOR II = 2 TO N
2630 DI = DIAG(II) - II
2632 LL = DIAG(II - 1) - DI + 1
2634 FOR JJ = LL TO II
2636 X = A(DI + JJ)
2638 DJ = DIAG(JJ) - JJ
2640 IF JJ = 1 THEN 2652
2642 LB = DIAG(JJ - 1) - DJ + 1
2644 IF LL > LB THEN LB = LL
2646 FOR KK = LB TO JJ - 1
2648 X = X - A(DI + KK) * A(DJ + KK)
2650 NEXT KK
2652 A(DI + JJ) = X / A(DJ + JJ)
2654 NEXT JJ
2656 A(DI + II) = SQR(X)
2658 NEXT II
2660 QW(1) = QW(1) / A(1)
2662 FOR II = 2 TO N
2664 DI = DIAG(II) - II
2666 LL = DIAG(II - 1) - DI + 1
2668 X = QW(II)
2670 FOR JJ = LL TO II - 1
2672 X = X - A(DI + JJ) * QW(JJ)
2674 NEXT JJ
2676 QW(II) = X / A(DI + II)
2678 NEXT II
2680 FOR II = N TO 2 STEP -1
2682 DI = DIAG(II) - II
2684 X = QW(II) / A(DI + II): QW(II) = X
2686 LL = DIAG(II - 1) - DI + 1
```

```
2688 FOR KK = LL TO II - 1
2690 QW(KK) = QW(KK) - X * A(DI + KK)
2692 NEXT KK
2694 NEXT II
2696 QW(1) = QW(1) / A(1)
2750 PRINT #2, : PRINT #2, "THE NODAL NUMBERS & DISPLACEMENTS (W,THETAX &
THETAY) ARE:="
2760 FOR II = 1 TO NNODE
2770 JJ = 3 * II - 3
2780 PRINT #2, ; II, QW(JJ + 1), QW(JJ + 2), QW(JJ + 3)
2790 NEXT II
3000 REM CALCULATION OF THE 'STRESSES', ETC.
3030 XSTRESS = 1
3060 FOR NEL = 1 TO ELEMS
3070 INODE = IN(NEL)
3080 JNODE = JN(NEL)
3090 KNODE = KN(NEL)
3100 LNODE = L4(NEL)
3101 MNODE = M8(NEL)
3102 NNOD = N8(NEL)
3103 ONODE = O8(NEL)
3104 PNODE = P8(NEL)
3110 FOR II = 1 TO 24
3120 SL(II) = 0
3130 NEXT II
3160 COORD(1, 1) = XC(INODE): COORD(1, 2) = YC(INODE)
3170 COORD(2, 1) = XC(JNODE): COORD(2, 2) = YC(JNODE)
3180 COORD(3, 1) = XC(KNODE): COORD(3, 2) = YC(KNODE)
3190 COORD(4, 1) = XC(LNODE): COORD(4, 2) = YC(LNODE)
3192 COORD(5, 1) = XC(MNODE): COORD(5, 2) = YC(MNODE)
3193 COORD(6, 1) = XC(NNOD): COORD(6, 2) = YC(NNOD)
3195 COORD(7, 1) = XC(ONODE): COORD(7, 2) = YC(ONODE)
3200 COORD(8, 1) = XC(PNODE): COORD(8, 2) = YC(PNODE)
3210 SL(1) = QW(3 * INODE - 2)
3220 SL(2) = QW(3 * INODE - 1)
3230 SL(3) = QW(3 * INODE)
3310 SL(4) = QW(3 * JNODE - 2)
3320 SL(5) = QW(3 * JNODE - 1)
3330 SL(6) = QW(3 * JNODE)
3410 SL(7) = QW(3 * KNODE - 2)
3420 SL(8) = QW(3 * KNODE - 1)
3430 SL(9) = QW(3 * KNODE)
3510 SL(10) = QW(3 * LNODE - 2)
3520 SL(11) = QW(3 * LNODE - 1)
3530 SL(12) = QW(3 * LNODE)
3535 SL(13) = QW(3 * MNODE - 2)
3540 SL(14) = QW(3 * MNODE - 1)
3545 SL(15) = QW(3 * MNODE)
3550 SL(16) = QW(3 * NNOD - 2)
3555 SL(17) = QW(3 * NNOD - 1)
3560 SL(18) = QW(3 * NNOD)
3565 SL(19) = QW(3 * ONODE - 2)
3570 SL(20) = QW(3 * ONODE - 1)
```

```
3575 SL(21) = QW(3 * ONODE)
3580 SL(22) = QW(3 * PNODE - 2)
3585 SL(23) = QW(3 * PNODE - 1)
3590 SL(24) = QW(3 * PNODE)
3600 FOR IJ = 1 TO 3
3610 IF IJ = 1 THEN XI = -1
3620 IF IJ = 2 THEN XI = 0
3625 IF IJ = 3 THEN XI = 1
3630 FOR JI = 1 TO 3
3640 IF JI = 1 THEN ETA = -1
3645 IF JI = 2 THEN ETA = 0
3650 IF JI = 3 THEN ETA = 1
3670 GOSUB 7510
3680 GOSUB 8000
3700 FOR II = 1 TO 5
3710 STRESS(II) = 0
3720 FOR JJ = 1 TO 24
3730 STRESS(II) = STRESS(II) + BB(II, JJ) * SL(JJ)
3740 NEXT JJ
3750 NEXT II
3770 PRINT #2, : PRINT #2, "THE 'STRESSES' AT ELEMENT "; INODE; "-"; JNODE; "-";
PRINT #2, KNODE; "-"; LNODE; "-"; MNODE; "-"; NNOD; "-"; ONODE; "-"; PNODE
PRINT #2, " FOR XI="; XI; " & ETA="; ETA; " ARE:"
3780 PRINT #2, ; "SIGMA X="; STRESS(1)
3790 PRINT #2, ; "SIGMA Y="; STRESS(2)
3800 PRINT #2, ; "TAU X-Y="; STRESS(3)
3810 PRINT #2, ; "TAU Y-Z="; STRESS(4)
3820 PRINT #2, ; "TAU X-Z="; STRESS(5)
ET = ETA: GOSUB 13220
3900 END
5000 REM OUTPUT OF INPUT
5020 PRINT #2, : PRINT #2, "FLAT PLATES OF IRREGULAR SHAPE UNDER LATERAL
LOADING & SUPPORTED ON ELASTIC SUPPORTS": PRINT #2,
5030 PRINT #2, "NUMBER OF NODES="; NNODE
5040 PRINT #2, "NUMBER OF ELEMENTS="; ELEMS
5045 IF ALL$ = "Y" OR ALL$ = "y" THEN PRINT #2, "ALL SUPPORTS HAVE EQUAL
MAGNITUDES OF STIFFNESS": PRINT #2, : GOTO 5055
5050 PRINT #2, "NUMBER OF NODES WITH ZERO DISPLACEMENTS="; NFIX
5055 IF PRESS$ = "Y" OR PRESS$ = "y" THEN PRINT #2, "THIS SLAB IS PRESSURE
LOADED"
5060 PRINT #2, "NUMBER OF NODES WITH CONCENTRATED LOADS="; NCONC
5070 PRINT #2, "ELASTIC MODULUS="; ELAST
5080 PRINT #2, "POISSON'S RATIO="; NU
5085 IF THICKNESS$ = "Y" OR THICKNESS$ = "y" THEN PRINT #2, "THICKNESS=";
THICK: GOTO 5130
5090 PRINT #2, "THE NODAL THICKNESSES ARE:-"
5100 FOR II = 1 TO NNODE
5110 PRINT #2, "THICKNESS("; II; ")="; TH(II)
5120 NEXT II
5130 PRINT #2, "THE GLOBAL COORDINATES ARE:-"
5140 FOR II = 1 TO NNODE
5150 PRINT #2, "X("; II; ")="; XC(II), "Y("; II; ")="; YC(II)
5160 NEXT II
```

5165 IF ALL$ = "Y" OR ALL$ = "y" THEN PRINT #2, "AT THE SUPPORTS, VERTICAL STIFFNESSES ="; SUPPSTIFF: PRINT #2, "ROTATIONAL STIFFNESSES IN THE 'THETAX' DIRECTIONS ="; ROTX: PRINT #2, "ROTATIONAL STIFFNESSES IN THE 'THETAY' DIRECTIONS ="; ROTY: GOTO _
5235
5170 PRINT #2, "THE NODAL POSITIONS & VALUES OF THE VERTICAL STIFFNESSES AT THE ELASTIC SUPPORTS ARE:"
5180 FOR II = 1 TO NFIX
5190 PRINT #2, "NODE ="; POSN(II)
5200 PRINT #2, "VERTICAL STIFFNESS ="; NSUP(3 * POSN(II) - 2)
5210 PRINT #2, "ROTATIONAL STIFFNESS IN 'THETAX' DIRECTION ="; NSUP(3 * POSN(II) - 1)
5220 PRINT #2, "ROTATIONAL STIFFNESS IN 'THETAY' DIRECTION ="; NSUP(3 * POSN(II))
5230 NEXT II
5235 IF NCONC = 0 THEN GOTO 5280
5240 PRINT #2, "THE VALUE OF THE ADDITIONAL CONCENTRATED LOADS ARE:-"
5250 FOR II = 1 TO NCONC
5260 PRINT #2, "NODE ="; CONC(II), "OUT-OF-PLANE LOAD ="; QW(CONC(II) * 3 - 2)
5265 PRINT #2, "ROTATIONAL COUPLE IN THE 'THETAX' DIRECTION ="; QW(CONC(II) * 3 - 1)
5267 PRINT #2, "ROTATIONAL COUPLE IN THE 'THETAY' DIRECTION ="; QW(CONC(II) * 3)
5270 NEXT II
5280 PRINT #2, "ELEMENT TOPOLOGY"
5290 FOR EL = 1 TO ELEMS
5300 PRINT #2, "ELEMENT "; EL, "NODES- "; IN(EL); "-"; JN(EL); "-"; KN(EL); "-"; L4(EL); "-"; M8(EL); "-"; N8(EL); "-"; O8(EL); "-"; P8(EL)
5302 IF PRESS$ = "Y" OR PRESS$ = "y" THEN 5305
5303 GOTO 5310
5305 PRINT #2, "PRESSURE ON ELEMENT "; EL; "="; PRESSURE(EL)
5310 NEXT EL
5390 RETURN
6500 REM GAUSS POINTS
6510 IF IJ = 1 THEN XI = -.7745967: AI = .5555556
6515 IF IJ = 2 THEN XI = 0: AI = .8888889
6520 IF IJ = 3 THEN XI = .7745967: AI = .5555556
6530 IF JI = 1 THEN ETA = -.7745967: AJ = .5555556
6535 IF JI = 2 THEN ETA = 0: AJ = .8888889
6540 IF JI = 3 THEN ETA = .7745967: AJ = .5555556
6590 RETURN
7000 REM BCURL
7010 FOR II = 1 TO 5
7020 FOR JJ = 1 TO 24
7030 BCURL(II, JJ) = 0
7040 NEXT JJ
7050 NEXT II
7060 FOR II = 1 TO 8
7070 IF XSTRESS = 1 THEN 7200
7080 JJ = 3 * II - 3
7090 BCURL(1, JJ + 2) = -AM(II)
7100 BCURL(2, JJ + 3) = -BM(II)
7110 BCURL(3, JJ + 2) = -BM(II)

```
7120 BCURL(3, JJ + 3) = -AM(II)
7130 BCURL(4, JJ + 1) = BM(II)
7140 BCURL(4, JJ + 3) = -NM(II)
7150 BCURL(5, JJ + 1) = AM(II)
7160 BCURL(5, JJ + 2) = -NM(II)
7190 GOTO 7300
7200 JJ = 3 * II - 3
7210 BCURL(1, JJ + 2) = -AM(II) * THICK / 2
7220 BCURL(2, JJ + 3) = -BM(II) * THICK / 2
7230 BCURL(3, JJ + 2) = -BM(II) * THICK / 2
7240 BCURL(3, JJ + 3) = -AM(II) * THICK / 2
7250 BCURL(4, JJ + 1) = BM(II)
7260 BCURL(4, JJ + 3) = -NM(II)
7270 BCURL(5, JJ + 1) = AM(II)
7280 BCURL(5, JJ + 2) = -NM(II)
7300 NEXT II
7390 RETURN
7510 NM(1) = .25 * (1 - XI) * (1 - ETA) - .25 * (1 - XI ^ 2) * (1 - ETA) - .25 * (1 - XI) * (1 -
ETA ^ 2)
7515 NM(2) = .5 * (1 - XI ^ 2) * (1 - ETA)
7520 NM(3) = .25 * (1 + XI) * (1 - ETA) - .25 * (1 - XI ^ 2) * (1 - ETA) - .25 * (1 + XI) * (1
- ETA ^ 2)
7525 NM(4) = .5 * (1 + XI) * (1 - ETA ^ 2)
7530 NM(5) = .25 * (1 + XI) * (1 + ETA) - .25 * (1 - XI ^ 2) * (1 + ETA) - .25 * (1 + XI) *
(1 - ETA ^ 2)
7535 NM(6) = .5 * (1 - XI ^ 2) * (1 + ETA)
7540 NM(7) = .25 * (1 - XI) * (1 + ETA) - .25 * (1 - XI ^ 2) * (1 + ETA) - .25 * (1 - XI) * (1
- ETA ^ 2)
7545 NM(8) = .5 * (1 - XI) * (1 - ETA ^ 2)
7590 RETURN
8000 REM [D]*[B]
8020 DN(1, 1) = .25 * (1 - ETA) * (2 * XI + ETA)
8025 DN(1, 2) = -XI * (1 - ETA)
8030 DN(1, 3) = .25 * (1 - ETA) * (2 * XI - ETA)
8035 DN(1, 4) = .5 * (1 - ETA ^ 2)
8040 DN(1, 5) = .25 * (1 + ETA) * (2 * XI + ETA)
8045 DN(1, 6) = -XI * (1 + ETA)
8047 DN(1, 7) = .25 * (1 + ETA) * (2 * XI - ETA)
8050 DN(1, 8) = -.5 * (1 - ETA ^ 2)
8060 DN(2, 1) = .25 * (1 - XI) * (XI + 2 * ETA)
8065 DN(2, 2) = -.5 * (1 - XI ^ 2)
8070 DN(2, 3) = .25 * (1 + XI) * (-XI + 2 * ETA)
8075 DN(2, 4) = -ETA * (1 + XI)
8080 DN(2, 5) = .25 * (1 + XI) * (XI + 2 * ETA)
8085 DN(2, 6) = .5 * (1 - XI ^ 2)
8087 DN(2, 7) = .25 * (1 - XI) * (-XI + 2 * ETA)
8090 DN(2, 8) = -ETA * (1 - XI)
8100 FOR II = 1 TO 2
8110 FOR JJ = 1 TO 2
8120 JMX(II, JJ) = 0
8130 FOR KK = 1 TO 8
8140 JMX(II, JJ) = JMX(II, JJ) + DN(II, KK) * COORD(KK, JJ)
8150 NEXT KK
```

```
8160 NEXT JJ
8170 NEXT II
8180 CON = JMX(1, 1) * JMX(2, 2) - JMX(1, 2) * JMX(2, 1)
8190 KON = JMX(1, 1)
8200 JMX(1, 1) = JMX(2, 2)
8210 JMX(2, 2) = KON
8220 JMX(1, 2) = -JMX(1, 2)
8230 JMX(2, 1) = -JMX(2, 1)
8240 FOR II = 1 TO 2
8250 FOR JJ = 1 TO 2
8260 JMX(II, JJ) = JMX(II, JJ) / CON
8270 NEXT JJ
8280 NEXT II
8290 FOR II = 1 TO 8
8300 AM(II) = 0
8310 BM(II) = 0
8320 AM(II) = JMX(1, 1) * DN(1, II) + JMX(1, 2) * DN(2, II)
8330 BM(II) = JMX(2, 1) * DN(1, II) + JMX(2, 2) * DN(2, II)
8340 NEXT II
8350 THICK = TH(INODE) * NM(1) + TH(JNODE) * NM(2) + TH(KNODE) * NM(3) +
TH(LNODE) * NM(4) + TH(MNODE) * NM(5) + TH(NNOD) * NM(6) + TH(ONODE) * NM(7)
+ TH(PNODE) * NM(8)
8400 IF XSTRESS = 1 THEN 8470
8410 DCURL(1, 1) = ELAST * THICK ^ 3 / (12 * (1 - NU ^ 2)): DCURL(2, 2) = DCURL(1, 1)
8420 DCURL(1, 2) = NU * DCURL(1, 1): DCURL(2, 1) = DCURL(1, 2)
8430 DCURL(3, 3) = GMOD * THICK ^ 3 / 12
8440 DCURL(4, 4) = GMOD * THICK / 1.2
8450 DCURL(5, 5) = DCURL(4, 4)
8460 GOTO 8790
8470 DCURL(1, 1) = ELAST / (1 - NU ^ 2)
8480 DCURL(1, 2) = NU * DCURL(1, 1): DCURL(2, 1) = DCURL(1, 2)
8510 DCURL(2, 2) = DCURL(1, 1)
8520 DCURL(3, 3) = GMOD
8530 DCURL(4, 4) = GMOD
8540 DCURL(5, 5) = DCURL(4, 4)
8790 GOSUB 7000
8810 FOR II = 1 TO 5
8820 FOR JJ = 1 TO 24
8830 BCURLT(JJ, II) = BCURL(II, JJ)
8840 NEXT JJ
8850 NEXT II
8900 FOR II = 1 TO 5
8910 FOR JJ = 1 TO 24
8920 BB(II, JJ) = 0
8930 FOR KK = 1 TO 5
8940 BB(II, JJ) = BB(II, JJ) + DCURL(II, KK) * BCURL(KK, JJ)
8950 NEXT KK
8960 NEXT JJ
8970 NEXT II
8980 RETURN
13220 SR(1) = STRESS(1): SR(2) = STRESS(2): SR(3) = STRESS(3)
13241 SHEAR = .5 * SQR((SR(1) - SR(2)) ^ 2 + 4 * SR(3) ^ 2)
13242 PRINCIPAL = .5 * (SR(1) + SR(2)) + SHEAR
```

```
13243 MINIMUM = .5 * (SR(1) + SR(2)) - SHEAR
13244 PRINT #2, "SIGMA1="; PRINCIPAL, "SIGMA2="; MINIMUM
13245 IF XI = 0 OR ET = 0 THEN SHEAR = 0: GOTO 13270
13246 IF SR(1) < SR(2) THEN SHEAR = -SHEAR
I = INODE: J = JNODE: K = KNODE: L = LNODE
M = MNODE: N = NNOD: O = ONODE: P = PNODE
13248 IF XI = -1 AND ET = -1 THEN SHEAR(I) = SHEAR(I) + SHEAR: PRINCIPAL(I) =
PRINCIPAL(I) + PRINCIPAL: MINIMUM(I) = MINIMUM(I) + MINIMUM
13250 IF XI = 1 AND ET = -1 THEN SHEAR(K) = SHEAR(K) + SHEAR: PRINCIPAL(K) =
PRINCIPAL(K) + PRINCIPAL: MINIMUM(K) = MINIMUM(K) + MINIMUM
13252 IF XI = 1 AND ET = 1 THEN SHEAR(M) = SHEAR(M) + SHEAR: PRINCIPAL(M) =
PRINCIPAL(M) + PRINCIPAL: MINIMUM(M) = MINIMUM(M) + MINIMUM
13254 IF XI = -1 AND ET = 1 THEN SHEAR(O) = SHEAR(O) + SHEAR: PRINCIPAL(O) =
PRINCIPAL(O) + PRINCIPAL: MINIMUM(O) = MINIMUM(O) + MINIMUM
13270 NEXT JI
13280 NEXT IJ
13290 NEXT NEL
13292 REM GOSUB 11000
13295 FOR II = 1 TO NJ STEP 2
13297 VG(II) = SHEAR(II) / VEC(II)
13300 NEXT II
13310 GOSUB 16000
RETURN
13320 END
16000 LF = 10: Z9 = 99
DIM YY(4), XX(4), FF(LF + 1)
16001 GOSUB 16006
16002 FOR II = 1 TO NJ STEP 2: VG(II) = PRINCIPAL(II) / VEC(II): NEXT II: Z9 = 999:
GOSUB 16006
16003 FOR II = 1 TO NJ STEP 2: VG(II) = MINIMUM(II) / VEC(II): NEXT II: Z9 = 9999:
GOSUB 16006
16004 SCREEN 0, 0
16005 END
16006 MAX = MAXX = -1E+12: MAXU = -1E+12: MINU = 1E+12
16007 SCREEN 9
16010 CLS
16011 IF Z9 = 99 THEN PRINT "LINES OF CONSTANT MAXIMUM SHEAR STRESS"
16012 IF Z9 = 999 THEN PRINT "LINES OF CONSTANT MAXIMUM PRINCIPAL STRESS"
16013 IF Z9 = 9999 THEN PRINT "LINES OF CONSTANT MINIMUM PRINCIPAL STRESS"
16020 FOR I = 1 TO NJ STEP 2
16030 IF VG(I) > MAXU THEN MAXU = VG(I)
16040 IF VG(I) < MINU THEN MINU = VG(I)
16045 IF ABS(X(I)) > MAXX THEN MAXX = ABS(X(I))
16050 IF ABS(Y(I)) > MAXY THEN MAXY = ABS(Y(I))
16055 IF MAXX > MAX THEN MAX = MAXX
16060 IF MAXY > MAX THEN MAX = MAXY
16065 NEXT I
16070 LF = 10: XS = 200 / MAX
16080 DIF = MAXU - MINU: STP = DIF / LF: AG = 1
16090 FOR I = MINU TO MAXU STEP STP
16100 FF(AG) = I: IF ABS(FF(AG)) < .000001 THEN FF(AG) = 0
16110 AG = AG + 1
NEXT I
```

```
16120 FOR N = 1 TO LF
16130 FOR EL = 1 TO ELEMS
16135 AA = 0: BB = 0: CC = 0: DD = 0
16140 I = IN(EL): J = KN(EL): K = M8(EL): L = O8(EL)
16145 IF FF(N) = (INT(VG(I) * 1000)) / 1000 THEN GOTO 16155
16147 IF FF(N) = (INT(VG(J) * 1000)) / 1000 THEN GOTO 16155
16149 IF FF(N) = (INT(VG(K) * 1000)) / 1000 THEN GOTO 16155
16151 IF FF(N) = (INT(VG(L) * 1000)) / 1000 THEN GOTO 16155
16153 GOTO 16160
16155 FF(N) = FF(N) + .000001
16160 IF FF(N) > VG(I) AND FF(N) < VG(J) THEN GOSUB 17000
16165 IF FF(N) < VG(I) AND FF(N) > VG(J) THEN GOSUB 17000
16170 IF FF(N) > VG(I) AND FF(N) < VG(L) THEN GOSUB 17100
16175 IF FF(N) < VG(I) AND FF(N) > VG(L) THEN GOSUB 17100
16180 IF FF(N) > VG(J) AND FF(N) < VG(K) THEN GOSUB 17200
16185 IF FF(N) < VG(J) AND FF(N) > VG(K) THEN GOSUB 17200
16190 IF FF(N) > VG(K) AND FF(N) < VG(L) THEN GOSUB 17300
16195 IF FF(N) < VG(K) AND FF(N) > VG(L) THEN GOSUB 17300
16220 IF AA = 1 AND BB = 1 THEN LINE (20 + XS * (XX(1)), 280 - XS * (YY(1)))-(20 + XS
* (XX(2)), 280 - XS * (YY(2)))
16230 IF AA = 1 AND CC = 1 THEN LINE (20 + XS * (XX(1)), 280 - XS * (YY(1)))-(20 + XS
* (XX(3)), 280 - XS * (YY(3)))
16240 IF AA = 1 AND DD = 1 THEN LINE (20 + XS * (XX(1)), 280 - XS * (YY(1)))-(20 +
XS * (XX(4)), 280 - XS * (YY(4)))
16250 IF BB = 1 AND CC = 1 THEN LINE (20 + XS * (XX(2)), 280 - XS * (YY(2)))-(20 + XS
* (XX(3)), 280 - XS * (YY(3)))
16260 IF BB = 1 AND DD = 1 THEN LINE (20 + XS * (XX(2)), 280 - XS * (YY(2)))-(20 + XS
* (XX(4)), 280 - XS * (YY(4)))
16270 IF CC = 1 AND DD = 1 THEN LINE (20 + XS * (XX(3)), 280 - XS * (YY(3)))-(20 + XS
* (XX(4)), 280 - XS * (YY(4)))
16290 NEXT EL
16300 NEXT N
16310 A$ = INKEY$: IF A$ = "" THEN GOTO 16310
16320 IF A$ = "Y" OR A$ = "y" THEN RETURN
16330 GOTO 16310
16400 END
17000 DF = FF(N) - VG(I): DE = VG(J) - VG(I)
17010 DX = X(J) - X(I): DY = Y(J) - Y(I)
17020 IF DF = 0 AND DE = 0 THEN GOTO 17050
17030 XX(1) = (DF * DX / DE) + X(I): YY(1) = (DF * DY / DE) + Y(I)
17040 AA = 1: RETURN
17050 XX(1) = X(I): YY(1) = Y(I): AA = 1: RETURN
17100 DF = FF(N) - VG(I): DE = VG(L) - VG(I)
17110 DX = X(L) - X(I): DY = Y(L) - Y(I)
17120 IF DF = 0 AND DE = 0 THEN GOTO 17150
17130 XX(2) = (DF * DX / DE) + X(I): YY(2) = (DF * DY / DE) + Y(I)
17140 BB = 1: RETURN
17150 XX(2) = X(I): YY(2) = Y(I): BB = 1: RETURN
17200 DF = FF(N) - VG(J): DE = VG(K) - VG(J)
17210 DX = X(K) - X(J): DY = Y(K) - Y(J)
17220 IF DF = 0 AND DE = 0 THEN GOTO 17250
17230 XX(3) = (DF * DX / DE) + X(J): YY(3) = (DF * DY / DE) + Y(J)
17240 CC = 1: RETURN
```

```
17250 XX(3) = X(J): YY(3) = Y(J): CC = 1: RETURN
17300 DF = FF(N) - VG(K): DE = VG(L) - VG(K)
17310 DX = X(L) - X(K): DY = Y(L) - Y(K)
17320 IF DF = 0 AND DE = 0 THEN GOTO 17350
17330 XX(4) = (DF * DX / DE) + X(K): YY(4) = (DF * DY / DE) + Y(K)
17340 DD = 1: RETURN
17350 XX(4) = X(K): YY(3) = Y(K): DD = 1: RETURN
19995 REM WARNING
19996 PRINT : PRINT "INCORRECT DATA": PRINT
19997 RETURN
10000 IF Z9 = 77 OR Z9 = 99 THEN 10295
10010 CLS : XZ = 1E+12: YZ = 1E+12: XM = -1E+12: YM = -1E+12
10020 PRINT "TO CONTINUE, TYPE Y"
10060 A$ = INKEY$: IF A$ = "" THEN 10060
10070 IF A$ = "Y" OR A$ = "y" THEN 10090
10080 GOTO 10060
10090 FOR EL = 1 TO ELEMS
10100 I = IN(EL): J = JN(EL): K = KN(EL): L = L4(EL)
 M = M8(EL): N = N8(EL): O = O8(EL): P = P8(EL)
10104 IF X(I) < XZ THEN XZ = X(I)
10106 IF X(J) < XZ THEN XZ = X(J)
10108 IF X(K) < XZ THEN XZ = X(K)
10110 IF X(L) < XZ THEN XZ = X(L)
10112 IF X(M) < XZ THEN XZ = X(M)
10114 IF X(N) < XZ THEN XZ = X(N)
10116 IF X(O) < XZ THEN XZ = X(O)
10120 IF X(P) < XZ THEN XZ = X(P)
10122 IF Y(I) < YZ THEN YZ = Y(I)
10124 IF Y(J) < YZ THEN YZ = Y(J)
10126 IF Y(K) < YZ THEN YZ = Y(K)
10128 IF Y(L) < YZ THEN YZ = Y(L)
10130 IF Y(M) < YZ THEN YZ = Y(M)
10132 IF Y(N) < YZ THEN YZ = Y(N)
10134 IF Y(O) < YZ THEN YZ = Y(O)
10140 IF Y(P) < YZ THEN YZ = Y(P)
10150 IF X(I) > XM THEN XM = X(I)
10152 IF X(J) > XM THEN XM = X(J)
10154 IF X(K) > XM THEN XM = X(K)
10156 IF X(L) > XM THEN XM = X(L)
10158 IF X(M) > XM THEN XM = X(M)
10160 IF X(N) > XM THEN XM = X(N)
10162 IF X(O) > XM THEN XM = X(O)
10170 IF X(P) > XM THEN XM = X(P)
10172 IF Y(I) > YM THEN YM = Y(I)
10174 IF Y(J) > YM THEN YM = Y(J)
10176 IF Y(K) > YM THEN YM = Y(K)
10178 IF Y(L) > YM THEN YM = Y(L)
10180 IF Y(M) > YM THEN YM = Y(M)
10182 IF Y(N) > YM THEN YM = Y(N)
10184 IF Y(O) > YM THEN YM = Y(O)
10186 IF Y(P) > YM THEN YM = Y(P)
10200 NEXT EL
10210 XS = 400 / (XM - XZ): YS = 200 / (YM - YZ)
```

```
10220 IF XS > YS THEN XS = YS
10295 IF Z9 = 0 OR Z9 = 77 THEN SCREEN 9
10296 IF Z9 = 0 THEN PRINT "THE MESH WILL NOW BE DRAWN"
10298 IF Z9 = 77 THEN PRINT "DEFLECTED FORM OF MESH"
10300 FOR EL = 1 TO ELEMS
10310 I = IN(EL): J = JN(EL): K = KN(EL): L = L4(EL)
M = M8(EL): N = N8(EL): O = O8(EL): P = P8(EL)
10315 FOR IV = 1 TO 8
10318 IF IV = 1 THEN II = I: JJ = J
10320 IF IV = 2 THEN II = J: JJ = K
10322 IF IV = 3 THEN II = K: JJ = L
10324 IF IV = 4 THEN II = L: JJ = M
10326 IF IV = 5 THEN II = M: JJ = N
10330 IF IV = 6 THEN II = N: JJ = O
10332 IF IV = 7 THEN II = O: JJ = P
10340 IF IV = 8 THEN II = P: JJ = I
10342 IF Z9 = 99 THEN 10450
10345 XI = (X(II) - XZ) * XS
10350 XJ = (X(JJ) - XZ) * XS
10355 YI = (Y(II) - YZ) * XS
10360 YJ = (Y(JJ) - YZ) * XS
10365 GOTO 10480
10450 XI = X(II) - XZ * XS: XJ = X(JJ) - XZ * XS
10455 YI = Y(II) - YZ * XS: YJ = Y(JJ) - YZ * XS
10480 LINE (XI + 20, 280 - YI)-(XJ + 20, 280 - YJ)
10490 NEXT IV
10495 NEXT EL
10510 IF Z9 = 77 THEN 10600
10515 PRINT "TO CONTINUE, TYPE Y"
10520 A$ = INKEY$: IF A$ = "" THEN 10520
10530 IF A$ = "Y" OR A$ = "y" THEN 10560
10540 GOTO 10520
10560 SCREEN 0: WIDTH 80
10600 RETURN
11000 REM DEFLECTED FORM OF MESH
11010 CLS : PRINT "TO CONTINUE, TYPE Y"
11020 A$ = INKEY$: IF A$ = "" THEN 11020
11030 IF A$ = "Y" OR A$ = "y" THEN 11050
11040 GOTO 11020
11050 REM
11070 Z9 = 77
11080 GOSUB 10000
11090 UM = -1E+12
11100 FOR II = 1 TO NN
11110 IF ABS(UG(II)) > UM THEN UM = ABS(UG(II))
11120 NEXT II
11130 SC = 20
11140 FOR II = 1 TO NJ
11150 X(II) = X(II) * XS + SC * UG(II * 2 - 1) / UM
11160 Y(II) = Y(II) * XS + SC * UG(II * 2) / UM
11170 NEXT II
11180 Z9 = 99: GOSUB 10000
11190 FOR II = 1 TO NJ
```

```
11200 X(II) = XG(II)
11210 Y(II) = YG(II)
11220 NEXT II
11230 RETURN
```

Computer Program for In-Plane Stresses in Plates "STRESS3N"

```
100 REM PROGRAM BY DR.C.T.F.ROSS
110 REM 6 HURSTVILLE DRIVE,
120 REM WATERLOOVILLE, PORTSMOUTH.
130 REM PO7 7NB ENGLAND
140 CLS
145 NNODE = 3: NDF = 2
150 PRINT : PRINT "PLANE STRESS & PLANE STRAIN": PRINT
151 PRINT "PROGRAM BY DR.C.T.F.ROSS.": PRINT : PRINT "NOT TO BE COPIED"
152 INPUT "TYPE IN YOUR INPUT FILE NAME"; FILENAME$
153 PRINT "YOUR INPUT FILENAME IS "; FILENAME$
154 PRINT "IF YOU ARE SATISFIED WITH THIS NAME, TYPE Y; ELSE N"
155 A$ = INKEY$: IF A$ = "" THEN GOTO 155
156 IF A$ = "Y" OR A$ = "y" THEN GOTO 159
157 IF A$ = "N" OR A$ = "n" THEN GOTO 152
158 GOTO 155
159 OPEN FILENAME$ FOR INPUT AS #1
160 INPUT "TYPE IN THE NAME OF YOUR OUTPUT FILE "; OUT$
161 PRINT "YOUR OUTPUT FILENAME IS "; OUT$
162 PRINT "IF YOU ARE SATISFIED WITH THIS NAME, TYPE Y; ELSE N"
163 B$ = INKEY$: IF B$ = "" THEN GOTO 163
164 IF B$ = "Y" OR B$ = "y" THEN GOTO 168
165 IF B$ = "N" OR B$ = "n" THEN GOTO 160
166 GOTO 163
168 OPEN OUT$ FOR OUTPUT AS #2
169 REM NO. OF ELEMENTS
170 INPUT #1, MS
175 IF MS < 1 THEN PRINT "NO. OF ELEMENTS="; MS: GOSUB 9000: STOP
180 REM NO. OF NODES
190 INPUT #1, NN
195 IF NN < 3 THEN PRINT "NO. OF NODES="; NN: GOSUB 9000: STOP
200 M3 = 3 * MS
210 N2 = 2 * NN
220 REM NO. OF NODES WITH ZERO DISPLACEMENTS
230 INPUT #1, NF
235 IF NF < 1 THEN PRINT "NO. OF NODES WITH ZERO DISPLACEMENTS="; NF:
GOSUB 9000: STOP
240 DIM SK(6, 6), IJ(M3), NS(2 * NF), UG(N2), VG(NN)
250 DIM B(3), C(3), U(3), V(3), VEC(NN), SHEAR(NN)
260 DIM BM(3, 6), BD(3, 6), D(3, 3), SR(3), DIAG(N2)
265 DIM PRINCIPAL(NN), MINIMUM(NN)
270 REM ELASTIC MODULUS
280 INPUT #1, E
285 IF E < 1 THEN PRINT "YOUNG'S MODULUS="; E: GOSUB 9000: STOP
290 REM POISSON'S RATIO
300 INPUT #1, NU
305 IF NU < 0 OR NU > .5 THEN PRINT "POISSON'S RATIO="; NU: GOSUB 9000: STOP
310 REM PLATE THICKNESS
320 INPUT #1, T
325 IF T < = 0 THEN PRINT "PLATE THICKNESS="; T: GOSUB 9000: STOP
```

```
330 REM IF PLANE STRESS, TYPE 1, BUT IF PLANE STRAIN, TYPE 0
340 INPUT #1, SS
342 IF SS = 0 OR SS = 1 THEN GOTO 350
345 PRINT "NEITHER PLANE STRESS NOR PLANE STRAIN. SS="; SS: STOP
350 IF SS = 0 THEN GOTO 400
360 CN = E * T / (1 - NU ^ 2)
370 ZU = NU
380 MU = (1 - NU) / 2
390 GOTO 430
400 CN = E * T * (1 - NU) / ((1 + NU) * (1 - 2 * NU))
410 ZU = NU / (1 - NU)
420 MU = (1 - 2 * NU) / (2 * (1 - NU))
430 REM ELEMENT TOPOLOGY
440 PRINT " I          J          K"
450 FOR ME = 1 TO MS
460 REM I,J & K NODES
470 INPUT #1, IN, JN, KN
471 PRINT IN, JN, KN
472 IF IN < 1 OR IN > NN THEN PRINT "INODE="; IN: GOSUB 9000: STOP
475 IF JN < 1 OR JN > NN THEN PRINT "JNODE="; JN: GOSUB 9000: STOP
477 IF KN < 1 OR KN > NN THEN PRINT "KNODE="; KN: GOSUB 9000: STOP
480 I8 = 3 * ME
490 IJ(I8 - 2) = IN
500 IJ(I8 - 1) = JN
510 IJ(I8) = KN
515 FOR II = 1 TO NNODE: IA = KN
517 IF II = 1 THEN IA = IN
518 IF II = 2 THEN IA = JN
520 FOR IJ = 1 TO NDF
521 II1 = NDF * (IA - 1) + IJ
522 FOR JJ = 1 TO NNODE: IB = KN
523 IF JJ = 1 THEN IB = IN
525 IF JJ = 2 THEN IB = JN
527 FOR JI = 1 TO NDF
529 JJ1 = NDF * (IB - 1) + JI
531 IIJJ = JJ1 - II1 + 1
533 IF IIJJ > DIAG(JJ1) THEN DIAG(JJ1) = IIJJ
535 NEXT JI, JJ, IJ, II
540 NEXT ME
545 PRINT "TO CONTINUE, PRESS 'ANY' KEY"
547 C$ = INKEY$: IF C$ = "" THEN GOTO 547
550 DIAG(1) = 1: FOR II = 1 TO N2: DIAG(II) = DIAG(II - 1) + DIAG(II): NEXT II
570 NT = DIAG(N2)
580 DIM A(NT), Q(N2), CV(N2), X(NN), Y(NN), XG(NN), YG(NN)
590 REM NODAL COORDINATES
595 PRINT " X          Y"
600 FOR II = 1 TO NN
610 REM XCOORD & YCOORD
620 INPUT #1, X(II), Y(II): XG(II) = X(II)
622 PRINT X(II), Y(II)
625 IF X(II) < -1000000! OR X(II) > 1000000! THEN PRINT "X("; II; ")="; X(II): GOSUB
9000: STOP
630 IF Y(II) < -1000000! OR Y(II) > 1000000! THEN PRINT "Y("; II; ")="; Y(II): GOSUB
```

```
9000: STOP
640 YG(II) = Y(II)
650 NEXT II
655 PRINT "TO CONTINUE, PRESS 'ANY' KEY"
657 C$ = INKEY$: IF C$ = "" THEN GOTO 657
660 REM NO. OF NODES WITH EXTERNAL LOADS
670 INPUT #1, NC
675 IF NC < 0 OR NC > NN THEN PRINT "NO. OF NODES WITH CONCENTRATED
LOADS="; NC: GOSUB 9000: STOP
680 DIM NP(NC), POSS(NF)
690 REM NODAL POSITIONS & VALUES OF EXTERNAL LOADS
700 FOR II = 1 TO NC
710 REM NODAL POSITION, XLOAD & YLOAD
720 INPUT #1, NP(II), XWC, YWC
730 IF NP(II) < 1 OR NP(II) > NN THEN PRINT " NODE NO. OF CONCENTRATED
LOAD="; NC: GOSUB 900: STOP
750 Q(2 * NP(II) - 1) = XWC
780 Q(2 * NP(II)) = YWC
790 NEXT II
800 REM POSITIONS OF ZERO DISPLACEMENTS
810 FOR II = 1 TO NF * 2: NS(II) = 0: NEXT II
820 FOR II = 1 TO NF
830 REM NODAL POSITION OF ZERO DISPLACEMENTS, & IF XDISP & YDISP ARE ZERO
840 INPUT #1, POSS(II), UX, VY
842 IF POSS(II) < 1 OR POSS(II) > NN THEN PRINT "NODE NO. OF ZERO
DISPLACEMENT="; POSS(II): GOSUB 9000: STOP
844 IF UX = 0 OR UX = 1 THEN GOTO 847
845 PRINT "UX="; UX: GOSUB 9000: STOP
847 IF UY = 0 OR UY = 1 THEN GOTO 850
848 PRINT "UY="; UY: GOSUB 9000: STOP
850 REM IF U DISPLACEMENT IS ZERO,TYPE 1, ELSE TYPE 0; SIMILARLY FOR V
DISPLACEMENT
870 IF UX < > 0 THEN NS(II * 2 - 1) = POSS(II) * 2 - 1
880 IF VY < > 0 THEN NS((II) * 2) = POSS(II) * 2
890 NEXT II
895 GOSUB 2360
900 Z9 = 0: GOSUB 10000
910 FOR ME = 1 TO MS
920 PRINT : PRINT "ELEMENT NO."; ME; " UNDER COMPUTATION"
930 I8 = 3 * ME
940 IN = IJ(I8 - 2)
950 JN = IJ(I8 - 1)
960 KN = IJ(I8)
962 VEC(IN) = VEC(IN) + 1
965 VEC(JN) = VEC(JN) + 1
967 VEC(KN) = VEC(KN) + 1
970 B(1) = Y(JN) - Y(KN)
980 B(2) = Y(KN) - Y(IN)
1000 B(3) = Y(IN) - Y(JN)
1010 C(1) = X(KN) - X(JN)
1020 C(2) = X(IN) - X(KN)
1030 C(3) = X(JN) - X(IN)
1040 DL = X(IN) * B(1) + X(JN) * B(2) + X(KN) * B(3)
```

```
1050 DL = ABS(DL)
1060 FOR I = 1 TO 3
1070 I2 = 2 * I - 2
1080 FOR J = 1 TO 3
1090 J2 = 2 * J - 2
1100 SK(I2 + 1, J2 + 1) = .5 * CN * (B(I) * B(J) + MU * C(I) * C(J)) / DL
1110 SK(I2 + 2, J2 + 2) = .5 * CN * (C(I) * C(J) + MU * B(I) * B(J)) / DL
1120 SK(I2 + 1, J2 + 2) = .5 * CN * (ZU * B(I) * C(J) + MU * C(I) * B(J)) / DL
1130 SK(I2 + 2, J2 + 1) = .5 * CN * (ZU * B(J) * C(I) + MU * C(J) * B(I)) / DL
1140 NEXT J: NEXT I
1150 I1 = 2 * IN - 2
1160 J1 = 2 * JN - 2
1170 K1 = 2 * KN - 2
1180 FOR II = 1 TO 2
1182 FOR JJ = 1 TO 2
1186 MM = I1 + II: MN = I1 + JJ
1188 NM = J1 + II: N9 = J1 + JJ
1190 IF JJ > II THEN GOTO 1196
1192 A(DIAG(MM) + MN - MM) = A(DIAG(MM) + MN - MM) + SK(II, JJ)
1194 A(DIAG(NM) + N9 - NM) = A(DIAG(NM) + N9 - NM) + SK(II + 2, JJ + 2)
1196 IF I1 > J1 THEN GOTO 1202
1198 A(DIAG(NM) + MN - NM) = A(DIAG(NM) + MN - NM) + SK(II + 2, JJ)
1200 GOTO 1204
1202 A(DIAG(MM) + N9 - MM) = A(DIAG(MM) + N9 - MM) + SK(II, JJ + 2)
1204 MM = I1 + II: MN = I1 + JJ
1206 NM = K1 + II: N9 = K1 + JJ
1208 IF JJ > II THEN GOTO 1214
1212 A(DIAG(NM) + N9 - NM) = A(DIAG(NM) + N9 - NM) + SK(II + 4, JJ + 4)
1214 IF I1 > K1 THEN GOTO 1220
1216 A(DIAG(NM) + MN - NM) = A(DIAG(NM) + MN - NM) + SK(II + 4, JJ)
1218 GOTO 1222
1220 A(DIAG(MM) + N9 - MM) = A(DIAG(MM) + N9 - MM) + SK(II, JJ + 4)
1222 MM = J1 + II: MN = J1 + JJ
1224 NM = K1 + II: N9 = K1 + JJ
1232 IF J1 > K1 THEN GOTO 1238
1234 A(DIAG(NM) + MN - NM) = A(DIAG(NM) + MN - NM) + SK(II + 4, JJ + 2)
1236 GOTO 1240
1238 A(DIAG(MM) + N9 - MM) = A(DIAG(MM) + N9 - MM) + SK(II + 2, JJ + 4)
1240 NEXT JJ, II
1370 NEXT ME
1380 FOR II = 1 TO NF * 2
1390 N9 = NS(II)
1400 IF N9 = 0 THEN GOTO 1430
1405 DI = DIAG(N9)
1410 A(DI) = A(DI) * 1E+12 + 1E+12
1420 Q(N9) = 0
1430 NEXT II
1440 FOR II = 1 TO N2
1450 CV(II) = Q(II)
1460 NEXT II
1465 CLOSE #1
1470 PRINT : PRINT "THE SIMULTANEOUS EQUATIONS ARE NOW BEING SOLVED":
PRINT
```

```
1480 GOSUB 2140
1485 PRINT #2, "THE NODAL DISPLACEMENTS U & V ARE:"
1490 FOR I = 1 TO NN
1500 FOR J = 1 TO 2
1510 II = 2 * I - 2 + J
1520 UG(II) = CV(II)
1525 IF J = 1 THEN PRINT #2, "U("; I; ")="; UG(II),
1527 IF J = 2 THEN PRINT #2, "V("; I; ")="; UG(II)
1530 NEXT J
1540 NEXT I
1550 CN = CN / T
1590 FOR ME = 1 TO MS
1600 I8 = 3 * ME
1610 IN = IJ(I8 - 2)
1620 JN = IJ(I8 - 1)
1630 KN = IJ(I8)
1635 PRINT #2, "ELEMENT NO."; ME; "    "; IN; "-"; JN; "-"; KN
1640 U(1) = UG(2 * IN - 1)
1650 U(2) = UG(2 * JN - 1)
1660 U(3) = UG(2 * KN - 1)
1670 V(1) = UG(2 * IN)
1680 V(2) = UG(2 * JN)
1690 V(3) = UG(2 * KN)
1700 B(1) = Y(JN) - Y(KN)
1710 B(2) = Y(KN) - Y(IN)
1720 B(3) = Y(IN) - Y(JN)
1730 C(1) = X(KN) - X(JN)
1740 C(2) = X(IN) - X(KN)
1750 C(3) = X(JN) - X(IN)
1760 DL = X(IN) * B(1) + X(JN) * B(2) + X(KN) * B(3)
1770 DL = ABS(DL)
1780 D(1, 1) = CN
1790 D(2, 2) = CN
1800 D(1, 2) = ZU * CN
1810 D(2, 1) = ZU * CN
1820 D(3, 3) = MU * CN
1830 FOR II = 1 TO 3
1840 BM(1, II) = B(II) / DL
1850 BM(2, II + 3) = C(II) / DL
1860 BM(3, II) = C(II) / DL
1870 BM(3, II + 3) = B(II) / DL
1880 NEXT II
1890 FOR II = 1 TO 3
1900 FOR JJ = 1 TO 6
1910 BD(II, JJ) = 0
1920 FOR KK = 1 TO 3
1930 BD(II, JJ) = BD(II, JJ) + D(II, KK) * BM(KK, JJ)
1940 NEXT KK
1950 NEXT JJ: NEXT II
1960 FOR II = 1 TO 3
1970 SR(II) = 0
1980 FOR JJ = 1 TO 6
1990 IF JJ > 3 THEN GOTO 2020
```

```
2000 SR(II) = SR(II) + BD(II, JJ) * U(JJ)
2010 GOTO 2030
2020 SR(II) = SR(II) + BD(II, JJ) * V(JJ - 3)
2030 NEXT JJ
2035 IF II = 1 THEN PRINT #2, "SIGMA X="; SR(II), : SX = SR(II)
2040 IF II = 2 THEN PRINT #2, "SIGMA Y="; SR(II), : SY = SR(II)
2045 IF II = 3 THEN PRINT #2, "TAU X-Y="; SR(II), : SXY = SR(II)
2047 NEXT II
2050 SHEAR = .5 * SQR((SX - SY) ^ 2 + 4 * SXY ^ 2)
2052 PRINCIPAL = .5 * (SX + SY) + SHEAR
2053 MINIMUM = .5 * (SX + SY) - SHEAR
2054 PRINT #2, "SIGMA1="; PRINCIPAL, "SIGMA2="; MINIMUM
2055 IF SX < SY THEN SHEAR = -SHEAR
2065 SHEAR(IN) = SHEAR(IN) + SHEAR / 3
2067 PRINCIPAL(IN) = PRINCIPAL(IN) + PRINCIPAL / 3
2068 MINIMUM(IN) = MINIMUM(IN) + MINIMUM / 3
2070 SHEAR(JN) = SHEAR(JN) + SHEAR / 3
2072 PRINCIPAL(JN) = PRINCIPAL(JN) + PRINCIPAL / 3
2073 MINIMUM(JN) = MINIMUM(JN) + MINIMUM / 3
2075 SHEAR(KN) = SHEAR(KN) + SHEAR / 3
2077 PRINCIPAL(KN) = PRINCIPAL(KN) + PRINCIPAL / 3
2078 MINIMUM(KN) = MINIMUM(KN) + MINIMUM / 3
2080 NEXT ME
2085 FOR II = 1 TO NN
2090 VG(II) = SHEAR(II) / VEC(II)
2100 NEXT II
2110 GOSUB 11000
2130 GOTO 3000
2140 N = N2
2145 A(1) = SQR(A(1))
2150 FOR II = 2 TO N
2155 DI = DIAG(II) - II
2160 LL = DIAG(II - 1) - DI + 1
2165 FOR JJ = LL TO II
2170 X = A(DI + JJ)
2175 DJ = DIAG(JJ) - JJ
2180 IF JJ = 1 THEN GOTO 2210
2185 LB = DIAG(JJ - 1) - DJ + 1
2190 IF LL > LB THEN LB = LL
2195 FOR KK = LB TO JJ - 1
2200 X = X - A(DI + KK) * A(DJ + KK)
2205 NEXT KK
2210 A(DI + JJ) = X / A(DJ + JJ)
2215 NEXT JJ
2220 A(DI + II) = SQR(X)
2225 NEXT II
2230 CV(1) = CV(1) / A(1)
2235 FOR II = 2 TO N
2240 DI = DIAG(II) - II
2245 LL = DIAG(II - 1) - DI + 1
2250 X = CV(II)
2255 FOR JJ = LL TO II - 1
2260 X = X - A(DI + JJ) * CV(JJ)
```

```
2265 NEXT JJ
2270 CV(II) = X / A(DI + II)
2275 NEXT II
2280 FOR II = N TO 2 STEP -1
2285 DI = DIAG(II) - II
2290 X = CV(II) / A(DI + II)
2295 CV(II) = X
2300 LL = DIAG(II - 1) - DI + 1
2305 FOR KK = LL TO II - 1
2310 CV(KK) = CV(KK) - X * A(DI + KK)
2315 NEXT KK
2320 NEXT II
2325 CV(1) = CV(1) / A(1)
2350 RETURN
2360 REM OUTPUT OF INPUT
2380 PRINT #2, : PRINT #2,
2390 PRINT #2, "PLANE STRESS & PLANE STRAIN USING A 3 NODE TRIANGLE"
2400 PRINT #2,
2410 PRINT #2, "NO. OF ELEMENTS="; MS
2420 PRINT #2, "NO. OF NODES="; NN
2430 PRINT #2, "NO. OF NODES WITH ZERO DISPLACEMENTS="; NF
2440 PRINT #2, "ELASTIC MODULUS="; E
2450 PRINT #2, "POISSON'S RATIO="; NU
2460 PRINT #2, "PLATE THICKNESS="; T
2470 IF SS = 1 THEN PRINT #2, "THIS IS A PLANE STRESS PROBLEM"
2480 IF SS = 0 THEN PRINT #2, "THIS IS A PLANE STRAIN PROBLEM"
2490 PRINT #2, "ELEMENT TOPOLOGY"
2500 FOR II = 1 TO MS
2510 I8 = II * 3
2520 IN = IJ(I8 - 2)
2530 JN = IJ(I8 - 1)
2540 KN = IJ(I8)
2550 PRINT #2, "I,J & K ARE: "; IN; "-"; JN; "-"; KN
2560 NEXT II
2570 PRINT #2, "NODAL COORDINATES"
2580 FOR II = 1 TO NN
2590 PRINT #2, "X("; II; ")="; X(II); "    Y("; II; ")="; Y(II)
2600 NEXT II
2610 PRINT #2, "EXTERNAL LOADS"
2615 PRINT #2, "NO. OF NODES WITH CONCENTRATED LOADS="; NC
2620 FOR II = 1 TO NC
2630 PRINT #2, "NODE="; NP(II)
2640 PRINT #2, "LOAD IN X DIRECTION="; Q(NP(II) * 2 - 1)
2650 PRINT #2, "LOAD IN Y DIRECTION="; Q(NP(II) * 2)
2660 NEXT II
2670 PRINT #2, "DETAILS OF ZERO DISPLACEMENTS"
2680 FOR II = 1 TO NF
2690 PRINT #2, "NODE="; POSS(II)
2700 N9 = NS(II * 2 - 1)
2710 IF N9 < > 0 THEN PRINT #2, "U IS ZERO AT NODE "; POSS(II)
2720 N9 = NS(II * 2)
2730 IF N9 < > 0 THEN PRINT #2, "V IS ZERO AT NODE "; POSS(II)
2740 NEXT II
```

```
2750 RETURN
3000 LF = 10: Z9 = 99
3001 DIM YY(3), XX(3), FF(LF + 1)
3002 GOSUB 3006
3003 FOR II = 1 TO NN: VG(II) = PRINCIPAL(II) / VEC(II): NEXT II: Z9 = 999: GOSUB 3006
3004 FOR II = 1 TO NN: VG(II) = MINIMUM(II) / VEC(II): NEXT II: Z9 = 9999: GOSUB 3006
3005 SCREEN 0, 0: END
3006 CLS : MAX = -1E+12: MAXX = -1E+12: MAXY = -1E+12: MAXU = -1E+12: MINU
= 1E+12
3007 SCREEN 9
3008 IF Z9 = 99 THEN PRINT "LINES OF CONSTANT MAXIMUM SHEAR STRESS"
3009 IF Z9 = 999 THEN PRINT "LINES OF CONSTANT MAXIMUM PRINCIPAL STRESS"
3010 IF Z9 = 9999 THEN PRINT "LINES OF CONSTANT MINIMUM PRINCIPAL STRESS"
3011 FOR I = 1 TO NN
3020 IF VG(I) > MAXU THEN MAXU = VG(I)
3030 IF VG(I) < MINU THEN MINU = VG(I)
3040 IF ABS(X(I)) > MAXX THEN MAXX = ABS(X(I))
3050 IF ABS(Y(I)) > MAXY THEN MAXY = ABS(Y(I))
3060 IF MAXX > MAX THEN MAX = MAXX
3070 IF MAXY > MAX THEN MAX = MAXY
3080 NEXT I
3090 LF = 10
3100 DIF = MAXU - MINU: STP = DIF / LF: AG = 1
3110 FOR I = MINU TO MAXU STEP STP
3120 FF(AG) = I: IF ABS(FF(AG)) < .0000001 THEN FF(AG) = 0
3130 AG = AG + 1: NEXT I
4120 FOR N = 1 TO LF
4130 FOR ME = 1 TO MS
4140 I8 = 3 * ME: AA = 0: BB = 0: CC = 0
4150 I = IJ(I8 - 2): J = IJ(I8 - 1): K = IJ(I8)
4151 IF FF(N) = (INT(VG(I) * 1000)) / 1000 THEN GOTO 4155
4152 IF FF(N) = (INT(VG(J) * 1000)) / 1000 THEN GOTO 4155
4153 IF FF(N) = (INT(VG(K) * 1000)) / 1000 THEN GOTO 4155
4154 GOTO 4160
4155 FF(N) = FF(N) + .0000001
4160 IF FF(N) > VG(I) AND FF(N) < VG(J) THEN GOSUB 5000
4170 IF FF(N) < VG(I) AND FF(N) > VG(J) THEN GOSUB 5000
4180 IF FF(N) > VG(I) AND FF(N) < VG(K) THEN GOSUB 5100
4190 IF FF(N) < VG(I) AND FF(N) > VG(K) THEN GOSUB 5100
4200 IF FF(N) > VG(J) AND FF(N) < VG(K) THEN GOSUB 5200
4210 IF FF(N) < VG(J) AND FF(N) > VG(K) THEN GOSUB 5200
4220 IF AA = 1 AND BB = 1 THEN LINE (40 + XS * (XX(1)), 250 - XS * (YY(1)))-(40 + XS
* (XX(2)), 250 - XS * (YY(2)))
4230 IF AA = 1 AND CC = 1 THEN LINE (40 + XS * (XX(1)), 250 - XS * (YY(1)))-(40 + XS
* (XX(3)), 250 - XS * (YY(3)))
4240 IF BB = 1 AND CC = 1 THEN LINE (40 + XS * (XX(2)), 250 - XS * (YY(2)))-(40 + XS
* (XX(3)), 250 - XS * (YY(3)))
4250 NEXT ME
4260 NEXT N
4270 A$ = INKEY$: IF A$ = "" THEN GOTO 4270
4280 IF A$ = "Y" OR A$ = "y" THEN SCREEN 0: RETURN
4290 GOTO 4270
4300 END
```

```
5000 DF = FF(N) - VG(I): DE = VG(J) - VG(I)
5010 DX = X(J) - X(I): DY = Y(J) - Y(I)
5015 IF DF = 0 AND DE = 0 THEN GOTO 5040
5020 XX(1) = (DF * DX / DE) + X(I): YY(1) = (DF * DY / DE) + Y(I)
5030 AA = 1: RETURN
5040 XX(1) = X(I): YY(1) = Y(I): AA = 1: RETURN
5100 DF = FF(N) - VG(I): DE = VG(K) - VG(I)
5110 DX = X(K) - X(I): DY = Y(K) - Y(I)
5115 IF DF = 0 AND DE = 0 THEN GOTO 5140
5120 XX(2) = (DF * DX / DE) + X(I): YY(2) = (DF * DY / DE) + Y(I)
5130 BB = 1: RETURN
5140 XX(2) = X(I): YY(2) = Y(I): BB = 1: RETURN
5200 DF = FF(N) - VG(J): DE = VG(K) - VG(J)
5210 DX = X(K) - X(J): DY = Y(K) - Y(J)
5215 IF DF = 0 AND DE = 0 THEN GOTO 5240
5220 XX(3) = (DF * DX / DE) + X(J): YY(3) = (DF * DY / DE) + Y(J)
5230 CC = 1: RETURN
5240 XX(3) = X(J): YY(3) = Y(J): CC = 1: RETURN
9000 PRINT "DATA ERROR"
9010 RETURN
10000 IF Z9 = 77 OR Z9 = 99 THEN GOTO 10295
10010 XZ = 1E+12: YZ = 1E+12
10030 XM = -1E+12: YM = -1E+12
10050 PRINT : PRINT "TO CONTINUE, TYPE Y"
10060 A$ = INKEY$: IF A$ = "" THEN GOTO 10060
10070 IF A$ = "Y" OR A$ = "y" THEN GOTO 10090
10080 GOTO 10060
10090 FOR ME = 1 TO MS
10100 I8 = 3 * ME
10110 I = IJ(I8 - 2)
10120 J = IJ(I8 - 1)
10130 K = IJ(I8)
10140 IF X(I) < XZ THEN XZ = X(I)
10150 IF X(J) < XZ THEN XZ = X(J)
10160 IF X(K) < XZ THEN XZ = X(K)
10170 IF Y(I) < YZ THEN YZ = Y(I)
10180 IF Y(J) < YZ THEN YZ = Y(J)
10190 IF Y(K) < YZ THEN YZ = Y(K)
10200 IF X(I) > XM THEN XM = X(I)
10210 IF X(J) > XM THEN XM = X(J)
10220 IF X(K) > XM THEN XM = X(K)
10230 IF Y(I) > YM THEN YM = Y(I)
10240 IF Y(J) > YM THEN YM = Y(J)
10250 IF Y(K) > YM THEN YM = Y(K)
10260 NEXT ME
10270 XS = 350 / (XM - XZ)
10280 YS = 175 / (YM - YZ)
10290 IF XS > YS THEN XS = YS
10295 IF Z9 = 0 OR Z9 = 77 THEN SCREEN 9
10297 IF Z9 = 77 THEN PRINT "DEFLECTED FORM OF MESH"
10300 FOR ME = 1 TO MS
10310 I8 = 3 * ME
10320 II = IJ(I8 - 2)
```

```
10330 JJ = IJ(I8 - 1)
10340 KK = IJ(I8)
10350 FOR IV = 1 TO 3
10360 IF IV = 1 THEN I = II: J = JJ
10370 IF IV = 2 THEN I = JJ: J = KK
10380 IF IV = 3 THEN I = KK: J = II
10390 IF Z9 = 99 THEN GOTO 10450
10400 XI = (X(I) - XZ) * XS
10410 XJ = (X(J) - XZ) * XS
10420 YI = (Y(I) - YZ) * XS
10430 YJ = (Y(J) - YZ) * XS
10440 GOTO 10480
10450 XI = X(I) - XZ * XS: XJ = X(J) - XZ * XS
10460 YI = Y(I) - YZ * XS: YJ = Y(J) - YZ * XS
10470 IF Z9 = 77 THEN GOTO 10490
10480 LINE (XI + 40, 250 - YI)-(XJ + 40, 250 - YJ)
10490 NEXT IV
10500 NEXT ME
10510 IF Z9 = 77 THEN GOTO 10600
10515 PRINT "TO CONTINUE, TYPE Y"
10520 A$ = INKEY$: IF A$ = "" THEN GOTO 10520
10530 IF A$ = "Y" OR A$ = "y" THEN GOTO 10560
10540 GOTO 10520
10560 SCREEN 0: WIDTH 80
10600 RETURN
11000 REM DEFLECTED FORM OF MESH
11020 Z9 = 77
11030 GOSUB 10000
11040 UM = -1E+12
11050 FOR II = 1 TO N2
11060 IF ABS(UG(II)) > UM THEN UM = ABS(UG(II))
11070 NEXT II
11080 SC = 20
11090 FOR II = 1 TO NN
11100 X(II) = X(II) * XS + SC * UG(II * 2 - 1) / UM
11110 Y(II) = Y(II) * XS + SC * UG(II * 2) / UM
11120 NEXT II
11130 Z9 = 99
11140 GOSUB 10000
11150 FOR II = 1 TO NN
11160 X(II) = XG(II)
11170 Y(II) = YG(II)
11180 NEXT II
11190 RETURN
```

Computer Program for In-Plane Stresses in Plates "STRESS4N"

```
50 CLS
60 REM COPYRIGHT OF DR.C.T.F.ROSS
70 REM 6 HURSTVILLE DRIVE, WATERLOOVILLE
80 REM PORTSMOUTH. ENGLAND
90 PRINT : PRINT "PLANE STRESS & PLANE STRAIN (4 NODE QUADRILATERAL)": PRINT
100 PRINT : PRINT "PROGRAM BY DR.C.T.F.ROSS": PRINT
110 PRINT : PRINT "NOT TO BE REPRODUCED": PRINT
111 INPUT "TYPE IN THE NAME OF YOUR INPUT FILE"; FILENAME$
112 PRINT "YOUR INPUT FILE NAME IS "; FILENAME$
113 PRINT "IF THIS NAME IS CORRECT, TYPE Y; ELSE N"
114 A$ = INKEY$: IF A$ = "" THEN GOTO 114
115 IF A$ = "Y" OR A$ = "y" THEN GOTO 119
116 IF A$ = "N" OR A$ = "n" THEN GOTO 111
117 END
119 OPEN FILENAME$ FOR INPUT AS #1
120 INPUT "TYPE IN THE NAME OF YOUR OUTPUT FILE "; OUT$
121 PRINT "YOUR OUTPUT FILE NAME IS "; OUT$
122 PRINT "IF THIS NAME IS CORRECT, TYPE Y, ELSE N"
123 B$ = INKEY$: IF B$ = "" THEN GOTO 123
124 IF B$ = "Y" OR B$ = "y" THEN GOTO 128
125 IF B$ = "N" OR B$ = "n" THEN GOTO 120
126 GOTO 123
128 OPEN OUT$ FOR OUTPUT AS #2
129 REM TYPE IN NO. OF NODES
130 INPUT #1, NJ: NN = 2 * NJ
140 IF NJ < 4 OR NJ > 299 THEN PRINT "NO. OF NODES="; NJ: GOSUB 9995: STOP
145 NNODE = 4: NDF = 2
150 REM TYPE IN THE NO. OF ELEMENTS
160 INPUT #1, ES
170 IF ES < 1 OR ES > 199 THEN PRINT "NO. OF ELEMENTS="; ES: GOSUB 9995: STOP
180 REM TYPE IN THE NO. OF NODES WITH ZERO DISPLACEMENTS
190 INPUT #1, NF
195 IF NF < 2 OR NF > NJ THEN PRINT "NO. OF NODES WITH ZERO
DISPLACEMENTS="; NF: GOSUB 9995: STOP
200 DIM I(ES), J(ES), K(ES), L(ES)
210 DIM SK(8, 8), KS(8, 8), D(3, 3), DB(3, 8), UG(NN), VG(NJ), VEC(NJ), SHEAR(NJ)
220 DIM X(NJ), Y(NJ), XG(NJ), YG(NJ), NS(NF * 2)
230 DIM B(2, 4), BX(3, 8), BT(8, 3), JA(2, 2)
240 DIM CO(4, 2), SU(8), SR(3), DIAG(NN)
245 DIM PRINCIPAL(NJ), MINIMUM(NJ)
250 REM TYPE IN ELASTIC MODULUS & POISSON'S RATIO
260 INPUT #1, E, NU
270 IF E < 1 OR NU < 0 OR NU > .5 THEN PRINT "YOUNG'S MODULUS="; E; "
POISSON'S RATIO="; NU: GOSUB 9995: STOP
280 REM TYPE IN PLATE THICKNESS
282 INPUT #1, T
283 IF T < = 0 THEN PRINT "PLATE THICKNESS="; T: GOSUB 9995: STOP
290 REM IF PLANE STRESS, TYPE 1; BUT IF PLANE STRAIN, TYPE 0
300 INPUT #1, PS
```

```
310 IF PS = 0 OR PS = 1 THEN GOTO 330
320 PRINT "NEITHER PLANE STRESS NOR PLANE STRAIN": GOSUB 9995: STOP
330 IF PS = 0 THEN GOTO 380
340 CN = E * T / (1 - NU ^ 2)
350 ZU = NU
360 MU = (1 - NU) / 2
370 GOTO 410
380 CN = E * T * (1 - NU) / ((1 + NU) * (1 - 2 * NU))
390 ZU = NU / (1 - NU)
400 MU = (1 - 2 * NU) * (2 * (1 - NU))
410 D(1, 1) = CN
420 D(2, 2) = CN
430 D(1, 2) = ZU * CN
440 D(2, 1) = D(1, 2)
450 D(3, 3) = MU * CN
460 REM TYPE IN THE NODAL COORDINATES
465 PRINT " X        Y"
470 FOR II = 1 TO NJ
480 REM TYPE IN X(II) & Y(II)
490 INPUT #1, X(II), Y(II)
495 PRINT X(II), Y(II)
500 IF X(II) < -1000000! OR X(II) > 1000000! THEN PRINT "X("; II; ")="; X(II): GOSUB
9995: STOP
510 IF Y(II) < -1000000! OR Y(II) > 1000000! THEN PRINT "Y("; II; ")="; Y(II): GOSUB
9995: STOP
520 XG(II) = X(II): YG(II) = Y(II)
540 NEXT II
545 PRINT "TO CONTINUE, PRESS 'ANY' KEY"
547 C$ = INKEY$: IF C$ = "" THEN GOTO 547
550 REM TYPE IN THE NODAL POSITIONS OF THE ZERO DISPLACEMENTS, ETC.
560 FOR II = 1 TO NF * 2: NS(II) = 0: NEXT II
570 FOR II = 1 TO NF
580 REM TYPE NODAL POSITION 'SU', & IF XDISP & YDISP ARE ZERO
590 INPUT #1, SU, SUX, SUY
600 IF SU < 1 OR SU > NJ THEN PRINT "NODE="; SU: GOSUB 9995: STOP
602 REM IF THE HORIZONTAL DEFLECTION IS ZERO. TYPE "Y"; ELSE TYPE "N"
603 IF SUX = 1 THEN Y$ = "Y" ELSE Y$ = "N"
604 IF Y$ = "Y" OR Y$ = "y" THEN NS(II * 2 - 1) = SU * 2 - 1: GOTO 607
605 IF Y$ = "N" OR Y$ = "n" THEN GOTO 607
606 PRINT "TYPE 1 OR ZERO FOR SUX": GOSUB 9995: STOP
607 REM IF THE VERTICAL DEFLECTION IS ZERO. TYPE "Y"; ELSE TYPE "N"
608 IF SUY = 1 THEN Y$ = "Y" ELSE Y$ = "N"
609 IF Y$ = "Y" OR Y$ = "y" THEN NS(II * 2) = SU * 2: GOTO 615
610 IF Y$ = "N" OR Y$ = "n" THEN GOTO 615
612 PRINT "TYPE 1 OR ZERO FOR SUY": GOSUB 9995: STOP
615 NEXT II
620 REM CALCULATION OF DIAGONAL POINTER VECTOR
625 PRINT " I   J   K   L"
630 FOR EL = 1 TO ES
640 REM TYPE IN THE NODAL NUMBERS IN AN ANTI-CLOCKWISE DIRECTION
650 INPUT #1, I(EL), J(EL), K(EL), L(EL)
660 I = I(EL): J = J(EL): K = K(EL): L = L(EL)
670 PRINT I; J; K; L
```

```
680 IF I < 1 OR I > NJ THEN PRINT "INODE="; I: GOSUB 9995: STOP
690 IF J < 1 OR J > NJ THEN PRINT "JNODE="; J: GOSUB 9995: STOP
700 IF K < 1 OR K > NJ THEN PRINT "KNODE="; K: GOSUB 9995: STOP
710 IF L < 1 OR L > NJ THEN PRINT "LNODE="; L: GOSUB 9995: STOP
755 IF I = J OR I = K OR I = L THEN PRINT "INODE="; I; "JNODE="; J; "KNODE="; K;
"LNODE="; L: GOSUB 9995: STOP
760 IF J = K OR J = L THEN PRINT "JNODE="; J; "KNODE="; K; "LNODE="; L: GOSUB
9995: STOP
765 IF K = L THEN PRINT "KNODE="; K; "LNODE="; L: GOSUB 9995: STOP
790 FOR II = 1 TO NNODE: IA = L
795 IF II = 1 THEN IA = I
800 IF II = 2 THEN IA = J
805 IF II = 3 THEN IA = K
810 FOR IJ = 1 TO NDF
815 II1 = NDF * (IA - 1) + IJ
820 FOR JJ = 1 TO NNODE: IB = L
825 IF JJ = 1 THEN IB = I
830 IF JJ = 2 THEN IB = J
835 IF JJ = 3 THEN IB = K
840 FOR JI = 1 TO NDF
845 JJ1 = NDF * (IB - 1) + JI
850 IIJJ = JJ1 - II1 + 1
855 IF IIJJ > DIAG(JJ1) THEN DIAG(JJ1) = IIJJ
860 NEXT JI, JJ, IJ, II
940 NEXT EL
945 PRINT "TO CONTINUE, PRESS 'ANY' KEY"
947 C$ = INKEY$: IF C$ = "" THEN GOTO 947
950 DIAG(1) = 1: FOR II = 1 TO NN: DIAG(II) = DIAG(II - 1) + DIAG(II): NEXT II
960 NT = DIAG(NN)
970 DIM A(NT), Q(NN), POW(NJ)
980 REM TYPE IN THE NO. OF NODES WITH CONCENTRATED LOADS
990 INPUT #1, NC
1000 IF NC < 1 OR NC > NJ THEN GOSUB 9995: STOP
1010 FOR II = 1 TO NN: Q(II) = 0: NEXT II
1020 FOR II = 1 TO NJ: POW(II) = 0: NEXT II
1030 FOR II = 1 TO NC
1040 REM TYPE IN THE NODAL POSITION OF THE LOAD, & XLOAD & YLOAD
1050 INPUT #1, POW, XLOAD, YLOAD: POW(II) = POW
1070 Q(2 * POW - 1) = XLOAD
1090 Q(2 * POW) = YLOAD
1160 NEXT II
1180 GOSUB 5000
1185 Z9 = 0: GOSUB 10000
1190 FOR EL = 1 TO ES
1200 PRINT : PRINT "ELEMENT NO."; EL; " UNDER COMPUTATION"
1210 I = I(EL)
1220 J = J(EL)
1230 K = K(EL)
1240 L = L(EL)
1245 VEC(I) = VEC(I) + 1
1250 VEC(J) = VEC(J) + 1
1255 VEC(K) = VEC(K) + 1
1260 VEC(L) = VEC(L) + 1
```

```
1290 CO(1, 1) = X(I)
1300 CO(1, 2) = Y(I)
1310 CO(2, 1) = X(J)
1320 CO(2, 2) = Y(J)
1330 CO(3, 1) = X(K)
1340 CO(3, 2) = Y(K)
1350 CO(4, 1) = X(L)
1360 CO(4, 2) = Y(L)
1450 FOR II = 1 TO 8
1460 FOR JJ = 1 TO 8
1470 SK(II, JJ) = 0
1480 NEXT JJ
1490 NEXT II
1500 FOR LL = 1 TO 2
1510 IF LL = 1 THEN XI = -.5773502700000001#
1520 IF LL = 2 THEN XI = .5773502700000001#
1620 FOR MM = 1 TO 2
1630 IF MM = 1 THEN ET = -.5773502700000001#
1640 IF MM = 2 THEN ET = .5773502700000001#
1740 GOSUB 3330
1750 FOR II = 1 TO 3
1760 FOR JJ = 1 TO 8
1770 BT(JJ, II) = BX(II, JJ)
1780 NEXT JJ, II
1790 FOR II = 1 TO 8
1800 FOR JJ = 1 TO 8
1810 KS(II, JJ) = 0
1820 NEXT JJ
1830 NEXT II
1840 FOR II = 1 TO 8
1850 FOR JJ = 1 TO 8
1860 FOR KK = 1 TO 3
1870 KS(II, JJ) = KS(II, JJ) + BT(II, KK) * DB(KK, JJ) * DT
1880 NEXT KK
1890 NEXT JJ
1900 NEXT II
1910 FOR IJ = 1 TO 8
1920 FOR JI = 1 TO 8
1930 SK(IJ, JI) = SK(IJ, JI) + KS(IJ, JI)
1940 NEXT JI
1950 NEXT IJ
1960 NEXT MM
1970 NEXT LL
1980 I1 = 2 * I - 2: J1 = 2 * J - 2
1985 K1 = 2 * K - 2: L1 = 2 * L - 2
1990 FOR II = 1 TO 2
1995 FOR JJ = 1 TO 2
2000 MM = I1 + II: MN = I1 + JJ
2005 NM = J1 + II: N9 = J1 + JJ
2010 IF JJ > II THEN 2025
2015 A(DIAG(MM) + MN - MM) = A(DIAG(MM) + MN - MM) + SK(II, JJ)
2020 A(DIAG(NM) + N9 - NM) = A(DIAG(NM) + N9 - NM) + SK(II + 2, JJ + 2)
2025 IF I1 > J1 THEN 2040
```

464

2030 A(DIAG(NM) + MN - NM) = A(DIAG(NM) + MN - NM) + SK(II + 2, JJ)
2035 GOTO 2045
2040 A(DIAG(MM) + N9 - MM) = A(DIAG(MM) + N9 - MM) + SK(II, JJ + 2)
2045 MM = I1 + II: MN = I1 + JJ
2050 NM = K1 + II: N9 = K1 + JJ
2055 IF JJ > II THEN 2065
2060 A(DIAG(NM) + N9 - NM) = A(DIAG(NM) + N9 - NM) + SK(II + 4, JJ + 4)
2065 IF I1 > K1 THEN 2085
2070 A(DIAG(NM) + MN - NM) = A(DIAG(NM) + MN - NM) + SK(II + 4, JJ)
2080 GOTO 2090
2085 A(DIAG(MM) + N9 - MM) = A(DIAG(MM) + N9 - MM) + SK(II, JJ + 4)
2090 MM = I1 + II: MN = I1 + JJ
2095 NM = L1 + II: N9 = L1 + JJ
2100 IF JJ > II THEN 2110
2105 A(DIAG(NM) + N9 - NM) = A(DIAG(NM) + N9 - NM) + SK(II + 6, JJ + 6)
2110 IF I1 > L1 THEN 2125
2115 A(DIAG(NM) + MN - NM) = A(DIAG(NM) + MN - NM) + SK(II + 6, JJ)
2120 GOTO 2130
2125 A(DIAG(MM) + N9 - MM) = A(DIAG(MM) + N9 - MM) + SK(II, JJ + 6)
2130 MM = J1 + II: MN = J1 + JJ
2135 NM = K1 + II: N9 = K1 + JJ
2140 IF J1 > K1 THEN 2155
2145 A(DIAG(NM) + MN - NM) = A(DIAG(NM) + MN - NM) + SK(II + 4, JJ + 2)
2150 GOTO 2160
2155 A(DIAG(MM) + N9 - MM) = A(DIAG(MM) + N9 - MM) + SK(II + 2, JJ + 4)
2160 NM = L1 + II: N9 = L1 + JJ
2165 IF J1 > L1 THEN 2180
2170 A(DIAG(NM) + MN - NM) = A(DIAG(NM) + MN - NM) + SK(II + 6, JJ + 2)
2175 GOTO 2185
2180 A(DIAG(MM) + N9 - MM) = A(DIAG(MM) + N9 - MM) + SK(II + 2, JJ + 6)
2185 MM = K1 + II: MN = K1 + JJ
2190 NM = L1 + II: N9 = L1 + JJ
2195 IF K1 > L1 THEN 2210
2200 A(DIAG(NM) + MN - NM) = A(DIAG(NM) + MN - NM) + SK(II + 6, JJ + 4)
2205 GOTO 2220
2210 A(DIAG(MM) + N9 - MM) = A(DIAG(MM) + N9 - MM) + SK(II + 4, JJ + 6)
2220 NEXT JJ, II
2290 NEXT EL
2300 FOR II = 1 TO NF * 2
2310 N9 = NS(II)
2320 IF N9 = 0 THEN GOTO 2350
2330 Q(N9) = 0: DI = DIAG(N9)
2340 A(DI) = A(DI) * 1E+12 + 1E+12
2350 NEXT II
2355 CLOSE #1
2360 PRINT : PRINT "THE SIMULTANEOUS EQUATIONS ARE NOW BEING SOLVED":
PRINT
2370 N = NN
2375 A(1) = SQR(A(1))
2380 FOR II = 2 TO N
2385 DI = DIAG(II) - II
2390 LL = DIAG(II - 1) - DI + 1
2395 FOR JJ = LL TO II

```
2400 X = A(DI + JJ)
2405 DJ = DIAG(JJ) - JJ
2410 IF JJ = 1 THEN 2440
2415 LB = DIAG(JJ - 1) - DJ + 1
2420 IF LL > LB THEN LB = LL
2425 FOR KK = LB TO JJ - 1
2430 X = X - A(DI + KK) * A(DJ + KK)
2435 NEXT KK
2440 A(DI + JJ) = X / A(DJ + JJ)
2445 NEXT JJ
2450 A(DI + II) = SQR(X)
2455 NEXT II
2460 Q(1) = Q(1) / A(1)
2465 FOR II = 2 TO N
2470 DI = DIAG(II) - II
2475 LL = DIAG(II - 1) - DI + 1
2480 X = Q(II)
2485 FOR JJ = LL TO II - 1
2490 X = X - A(DI + JJ) * Q(JJ)
2495 NEXT JJ
2500 Q(II) = X / A(DI + II)
2505 NEXT II
2510 FOR II = N TO 2 STEP -1
2515 DI = DIAG(II) - II
2520 X = Q(II) / A(DI + II): Q(II) = X
2525 LL = DIAG(II - 1) - DI + 1
2530 FOR KK = LL TO II - 1
2535 Q(KK) = Q(KK) - X * A(DI + KK)
2540 NEXT KK
2545 NEXT II
2550 Q(1) = Q(1) / A(1)
2590 PRINT #2,
2610 PRINT #2, : PRINT #2, "THE NODAL VALUES OF THE DISPLACEMENTS ARE:-":
PRINT #2,
2620 FOR II = 1 TO NJ
2630 PRINT #2, "NODE="; II, "U("; II; ")="; Q(II * 2 - 1),
2640 PRINT #2, "V("; II; ")="; Q(II * 2)
2645 UG(II * 2 - 1) = Q(II * 2 - 1): UG(II * 2) = Q(II * 2)
2650 NEXT II
2660 FOR EL = 1 TO ES
2670 I = I(EL)
2680 J = J(EL)
2690 K = K(EL)
2700 L = L(EL)
2750 SU(1) = Q(2 * I - 1)
2760 SU(3) = Q(2 * J - 1)
2770 SU(5) = Q(2 * K - 1)
2780 SU(7) = Q(2 * L - 1)
2830 SU(2) = Q(2 * I)
2840 SU(4) = Q(2 * J)
2850 SU(6) = Q(2 * K)
2860 SU(8) = Q(2 * L)
2910 CO(1, 1) = X(I)
```

```
2920 CO(1, 2) = Y(I)
2930 CO(2, 1) = X(J)
2940 CO(2, 2) = Y(J)
2950 CO(3, 1) = X(K)
2960 CO(3, 2) = Y(K)
2970 CO(4, 1) = X(L)
2980 CO(4, 2) = Y(L)
3070 FOR LL = 1 TO 3
3080 IF LL = 1 THEN XI = -1
3090 IF LL = 2 THEN XI = 0
3100 IF LL = 3 THEN XI = 1
3110 FOR MM = 1 TO 3
3120 IF MM = 1 THEN ET = -1
3130 IF MM = 2 THEN ET = 0
3140 IF MM = 3 THEN ET = 1
3150 GOSUB 3330
3160 FOR II = 1 TO 3
3170 SR(II) = 0
3180 FOR JJ = 1 TO 8
3190 SR(II) = SR(II) + DB(II, JJ) * SU(JJ)
3200 NEXT JJ
3210 NEXT II
3220 PRINT #2, : PRINT #2, "THE STRESSES IN ELEMENT NO."; EL; "   "; I; "-"; J; "-"; K;
"-"; L; "  AT XI="; XI; " & ETA="; ET
3240 PRINT #2, "SIGMAX="; SR(1) / T, "SIGMAY="; SR(2) / T, "TAU X-Y="; SR(3) / T,
3241 SHEAR = .5 * SQR((SR(1) - SR(2)) ^ 2 + 4 * SR(3) ^ 2) / T
3242 PRINCIPAL = .5 * (SR(1) + SR(2)) / T + SHEAR
3243 MINIMUM = .5 * (SR(1) + SR(2)) / T - SHEAR
3244 PRINT #2, "SIGMA1="; PRINCIPAL, "SIGMA2="; MINIMUM
3246 IF XI = 0 OR ET = 0 THEN SHEAR = 0: GOTO 3270
3247 IF SR(1) < SR(2) THEN SHEAR = -SHEAR
3250 IF XI = -1 AND ET = -1 THEN SHEAR(I) = SHEAR(I) + SHEAR: PRINCIPAL(I) =
PRINCIPAL(I) + PRINCIPAL: MINIMUM(I) = MINIMUM(I) + MINIMUM
3255 IF XI = 1 AND ET = -1 THEN SHEAR(J) = SHEAR(J) + SHEAR: PRINCIPAL(J) =
PRINCIPAL(J) + PRINCIPAL: MINIMUM(J) = MINIMUM(J) + MINIMUM
3260 IF XI = 1 AND ET = 1 THEN SHEAR(K) = SHEAR(K) + SHEAR: PRINCIPAL(K) =
PRINCIPAL(K) + PRINCIPAL: MINIMUM(K) = MINIMUM(K) + MINIMUM
3265 IF XI = -1 AND ET = 1 THEN SHEAR(L) = SHEAR(L) + SHEAR: PRINCIPAL(L) =
PRINCIPAL(L) + PRINCIPAL: MINIMUM(L) = MINIMUM(L) + MINIMUM
3270 NEXT MM
3280 NEXT LL
3290 NEXT EL
3293 GOSUB 11000
3295 FOR II = 1 TO NJ
3297 VG(II) = SHEAR(II) / VEC(II)
3300 NEXT II
3310 GOSUB 6000
3320 END
3330 FOR II = 1 TO 2
3340 FOR JJ = 1 TO 2
3345 JA(II, JJ) = 0
3350 NEXT JJ
3360 NEXT II
```

467

3380 JA(1, 1) = ((ET - 1) * (CO(1, 1) - CO(2, 1)) + (ET + 1) * (CO(3, 1) - CO(4, 1))) / 4
3390 JA(1, 2) = ((ET - 1) * (CO(1, 2) - CO(2, 2)) + (ET + 1) * (CO(3, 2) - CO(4, 2))) / 4
3400 JA(2, 1) = ((XI - 1) * (CO(1, 1) - CO(4, 1)) + (XI + 1) * (CO(3, 1) - CO(2, 1))) / 4
3410 JA(2, 2) = ((XI - 1) * (CO(1, 2) - CO(4, 2)) + (XI + 1) * (CO(3, 2) - CO(2, 2))) / 4
3510 CN = JA(1, 1)
3520 JA(1, 1) = JA(2, 2)
3530 JA(2, 2) = CN
3540 JA(2, 1) = -JA(2, 1)
3550 JA(1, 2) = -JA(1, 2)
3560 DT = JA(1, 1) * JA(2, 2) - JA(2, 1) * JA(1, 2)
3570 FOR II = 1 TO 2
3580 FOR JJ = 1 TO 2
3590 JA(II, JJ) = JA(II, JJ) / DT
3600 NEXT JJ
3610 NEXT II
3620 B(1, 1) = (JA(1, 1) * (ET - 1) + JA(1, 2) * (XI - 1)) / 4
3630 B(1, 2) = (JA(1, 1) * (1 - ET) - JA(1, 2) * (XI + 1)) / 4
3640 B(1, 3) = (JA(1, 1) * (ET + 1) + JA(1, 2) * (XI + 1)) / 4
3650 B(1, 4) = (-JA(1, 1) * (ET + 1) + JA(1, 2) * (1 - XI)) / 4
3660 B(2, 1) = (JA(2, 1) * (ET - 1) + JA(2, 2) * (XI - 1)) / 4
3670 B(2, 2) = (JA(2, 1) * (1 - ET) - JA(2, 2) * (XI + 1)) / 4
3680 B(2, 3) = (JA(2, 1) * (ET + 1) + JA(2, 2) * (XI + 1)) / 4
3690 B(2, 4) = (-JA(2, 1) * (ET + 1) + JA(2, 2) * (1 - XI)) / 4
3840 FOR II = 1 TO 4
3850 BX(1, 2 * II - 1) = B(1, II)
3860 BX(2, 2 * II) = B(2, II)
3870 BX(3, 2 * II - 1) = B(2, II)
3880 BX(3, 2 * II) = B(1, II)
3890 NEXT II
3900 FOR II = 1 TO 3
3910 FOR JJ = 1 TO 8
3920 DB(II, JJ) = 0
3930 NEXT JJ
3940 NEXT II
3950 FOR II = 1 TO 3
3960 FOR JJ = 1 TO 8
3970 FOR KK = 1 TO 3
3980 DB(II, JJ) = DB(II, JJ) + D(II, KK) * BX(KK, JJ)
3990 NEXT KK
4000 NEXT JJ
4010 NEXT II
4020 RETURN
5000 REM OUTPUT OF INPUT
5010 PRINT #2, : PRINT #2, "PLANE STRESS & PLANE STRAIN USING A 4 NODE QUADRILERAL"
5020 PRINT #2, : PRINT #2, "NUMBER OF NODES="; NJ
5030 PRINT #2, "NUMBER OF ELEMENTS="; ES
5040 PRINT #2, "NUMBER OF NODES WITH ZERO DISPLACEMENTS="; NF
5050 PRINT #2, "ELASTIC MODULUS="; E
5060 PRINT #2, "POISSON'S RATIO="; NU
5070 PRINT #2, "PLATE THICKNESS="; T
5080 IF PS = 0 THEN PRINT #2, "THIS IS A PLANE STRAIN PROBLEM"
5090 IF PS = 1 THEN PRINT #2, "THIS IS A PLANE STRESS PROBLEM"

```
5100 PRINT #2, : PRINT #2, "THE NODAL COORDINATES (X & Y) IN GLOBAL
COORDINATES ARE:": PRINT #2,
5110 FOR II = 1 TO NJ
5120 PRINT #2, "NODE="; II, "X("; II; ")="; X(II), "Y("; II; ")="; Y(II)
5130 NEXT II
5140 PRINT #2, : PRINT #2, "THE NODAL POSITIONS, ETC. OF THE ZERO
DISPLACEMENTS ARE:": PRINT #2,
5150 FOR II = 1 TO NF
5160 SU = INT((NS(II * 2 - 1) + 1) / 2)
5170 IF SU = 0 THEN SU = INT(NS(II * 2) / 2)
5180 PRINT #2, "NODE="; SU
5190 N9 = NS(II * 2 - 1)
5200 IF N9 > 0 THEN PRINT #2, "DISPLACEMENT IN X DIRECTION IS ZERO AT NODE
"; SU
5210 N9 = NS(II * 2)
5220 IF N9 > 0 THEN PRINT #2, "DISPLACEMENT IN Y DIRECTION IS ZERO AT NODE
"; SU
5260 NEXT II
5270 PRINT #2, : PRINT #2, "ELEMENT TOPOLOGY": PRINT #2,
5280 FOR EL = 1 TO ES
5290 I = I(EL): J = J(EL): K = K(EL): L = L(EL)
5300 PRINT #2, "ELEMENT NO. "; EL, "NODES="; I; "-"; J; "-"; K; "-"; L
5330 NEXT EL
5340 PRINT #2, : PRINT #2, "NODAL POSITIONS & VALUES OF THE CONCENTRATED
LOADS": PRINT #2,
5350 FOR I = 1 TO NC
5360 POW = POW(I)
5370 PRINT #2, "NODE="; POW
5380 PRINT #2, "LOAD IN X DIRECTION="; Q(POW * 2 - 1)
5390 PRINT #2, "LOAD IN Y DIRECTION="; Q(POW * 2)
5400 NEXT I
5430 RETURN
6000 LF = 10: Z9 = 99: DIM YY(4), XX(4), FF(LF + 1)
6001 GOSUB 6006
6002 FOR II = 1 TO NJ: VG(II) = PRINCIPAL(II) / VEC(II): NEXT II: Z9 = 999: GOSUB 6006
6003 FOR II = 1 TO NJ: VG(II) = MINIMUM(II) / VEC(II): NEXT II: Z9 = 9999: GOSUB 6006
6004 SCREEN 0, 0
6005 END
6006 MAX = -1E+12: MAXX = -1E+12: MAXY = -1E+12: MAXU = -1E+12: MINU =
1E+12
6007 SCREEN 9
6010 CLS
6011 IF Z9 = 99 THEN PRINT "LINES OF CONSTANT MAXIMUM SHEAR STRESS"
6012 IF Z9 = 999 THEN PRINT "LINES OF CONSTANT MAXIMUM PRINCIPAL STRESS"
6013 IF Z9 = 9999 THEN PRINT "LINES OF CONSTANT MINIMUM PRINCIPAL STRESS"
6020 FOR I = 1 TO NJ
6030 IF VG(I) > MAXU THEN MAXU = VG(I)
6040 IF VG(I) < MINU THEN MINU = VG(I)
6045 IF ABS(X(I)) > MAXX THEN MAXX = ABS(X(I))
6050 IF ABS(Y(I)) > MAXY THEN MAXY = ABS(Y(I))
6055 IF MAXX > MAX THEN MAX = MAXX
6060 IF MAXY > MAX THEN MAX = MAXY
6065 NEXT I
```

```
6070 LF = 10: XS = 200 / MAX
6080 DIF = MAXU - MINU: STP = DIF / LF: AG = 1
6090 FOR I = MINU TO MAXU STEP STP
6100 FF(AG) = I: IF ABS(FF(AG)) < .000001 THEN FF(AG) = 0
6110 AG = AG + 1: NEXT I
6120 FOR N = 1 TO LF
6130 FOR EL = 1 TO ES
6135 AA = 0: BB = 0: CC = 0: DD = 0
6140 I = I(EL): J = J(EL): K = K(EL): L = L(EL)
6145 IF FF(N) = (INT(VG(I) * 1000)) / 1000 THEN GOTO 6155
6147 IF FF(N) = (INT(VG(J) * 1000)) / 1000 THEN GOTO 6155
6149 IF FF(N) = (INT(VG(K) * 1000)) / 1000 THEN GOTO 6155
6151 IF FF(N) = (INT(VG(L) * 1000)) / 1000 THEN GOTO 6155
6153 GOTO 6160
6155 FF(N) = FF(N) + .000001
6160 IF FF(N) > VG(I) AND FF(N) < VG(J) THEN GOSUB 7000
6165 IF FF(N) < VG(I) AND FF(N) > VG(J) THEN GOSUB 7000
6170 IF FF(N) > VG(I) AND FF(N) < VG(L) THEN GOSUB 7100
6175 IF FF(N) < VG(I) AND FF(N) > VG(L) THEN GOSUB 7100
6180 IF FF(N) > VG(J) AND FF(N) < VG(K) THEN GOSUB 7200
6185 IF FF(N) < VG(J) AND FF(N) > VG(K) THEN GOSUB 7200
6190 IF FF(N) > VG(K) AND FF(N) < VG(L) THEN GOSUB 7300
6195 IF FF(N) < VG(K) AND FF(N) > VG(L) THEN GOSUB 7300
6220 IF AA = 1 AND BB = 1 THEN LINE (40 + XS * (XX(1)), 280 - XS * (YY(1)))-(40 + XS
* (XX(2)), 280 - XS * (YY(2)))
6230 IF AA = 1 AND CC = 1 THEN LINE (40 + XS * (XX(1)), 280 - XS * (YY(1)))-(40 + XS
* (XX(3)), 280 - XS * (YY(3)))
6240 IF AA = 1 AND DD = 1 THEN LINE (40 + XS * (XX(1)), 280 - XS * (YY(1)))-(40 + XS
* (XX(4)), 280 - XS * (YY(4)))
6250 IF BB = 1 AND CC = 1 THEN LINE (40 + XS * (XX(2)), 280 - XS * (YY(2)))-(40 + XS
* (XX(3)), 280 - XS * (YY(3)))
6260 IF BB = 1 AND DD = 1 THEN LINE (40 + XS * (XX(2)), 280 - XS * (YY(2)))-(40 + XS
* (XX(4)), 280 - XS * (YY(4)))
6270 IF CC = 1 AND DD = 1 THEN LINE (40 + XS * (XX(3)), 280 - XS * (YY(3)))-(40 + XS
* (XX(4)), 280 - XS * (YY(4)))
6290 NEXT EL
6300 NEXT N
6310 A$ = INKEY$: IF A$ = "" THEN GOTO 6310
6320 IF A$ = "Y" OR A$ = "y" THEN SCREEN 0: RETURN
6330 GOTO 6310
6400 END
7000 DF = FF(N) - VG(I): DE = VG(J) - VG(I)
7010 DX = X(J) - X(I): DY = Y(J) - Y(I)
7020 IF DF = 0 AND DE = 0 THEN GOTO 7050
7030 XX(1) = (DF * DX / DE) + X(I): YY(1) = (DF * DY / DE) + Y(I)
7040 AA = 1: RETURN
7050 XX(1) = X(I): YY(1) = Y(I): AA = 1: RETURN
7100 DF = FF(N) - VG(I): DE = VG(L) - VG(I)
7110 DX = X(L) - X(I): DY = Y(L) - Y(I)
7120 IF DF = 0 AND DE = 0 THEN GOTO 7150
7130 XX(2) = (DF * DX / DE) + X(I): YY(2) = (DF * DY / DE) + Y(I)
7140 BB = 1: RETURN
7150 XX(2) = X(I): YY(2) = Y(I): BB = 1: RETURN
```

```
7200 DF = FF(N) - VG(J): DE = VG(K) - VG(J)
7210 DX = X(K) - X(J): DY = Y(K) - Y(J)
7220 IF DF = 0 AND DE = 0 THEN GOTO 7250
7230 XX(3) = (DF * DX / DE) + X(J): YY(3) = (DF * DY / DE) + Y(J)
7240 CC = 1: RETURN
7250 XX(3) = X(J): YY(3) = Y(J): CC = 1: RETURN
7300 DF = FF(N) - VG(K): DE = VG(L) - VG(K)
7310 DX = X(L) - X(K): DY = Y(L) - Y(K)
7320 IF DF = 0 AND DE = 0 THEN GOTO 7350
7330 XX(4) = (DF * DX / DE) + X(K): YY(4) = (DF * DY / DE) + Y(K)
7340 DD = 1: RETURN
7350 XX(4) = X(K): YY(4) = Y(K): DD = 1: RETURN
8000 REM DATA
9995 REM WARNING
9996 PRINT : PRINT "INCORRECT DATA": PRINT
9997 RETURN
10000 IF Z9 = 77 OR Z9 = 99 THEN 10295
10010 CLS : XZ = 1E+12: YZ = 1E+12: XM = -1E+12: YM = -1E+12
10020 PRINT "TO CONTINUE, TYPE Y"
10060 A$ = INKEY$: IF A$ = "" THEN 10060
10070 IF A$ = "Y" OR A$ = "y" THEN 10090
10080 GOTO 10060
10090 FOR EL = 1 TO ES
10100 I = I(EL): J = J(EL): K = K(EL): L = L(EL)
10105 IF X(I) < XZ THEN XZ = X(I)
10110 IF X(J) < XZ THEN XZ = X(J)
10115 IF X(K) < XZ THEN XZ = X(K)
10120 IF X(L) < XZ THEN XZ = X(L)
10130 IF Y(I) < YZ THEN YZ = Y(I)
10135 IF Y(J) < YZ THEN YZ = Y(J)
10140 IF Y(K) < YZ THEN YZ = Y(K)
10145 IF Y(L) < YZ THEN YZ = Y(L)
10155 IF X(I) > XM THEN XM = X(I)
10160 IF X(J) > XM THEN XM = X(J)
10165 IF X(K) > XM THEN XM = X(K)
10170 IF X(L) > XM THEN XM = X(L)
10180 IF Y(I) > YM THEN YM = Y(I)
10185 IF Y(J) > YM THEN YM = Y(J)
10190 IF Y(K) > YM THEN YM = Y(K)
10195 IF Y(L) > YM THEN YM = Y(L)
10200 NEXT EL
10210 XS = 200 / (XM - XZ): YS = 100 / (YM - YZ)
10295 IF Z9 = 0 OR Z9 = 77 THEN SCREEN 9
10296 IF Z9 = 0 THEN PRINT "THE MESH WILL NOW BE DRAWN"
10298 IF Z9 = 77 THEN PRINT "DEFLECTED FORM OF MESH"
10300 FOR EL = 1 TO ES
10310 I = I(EL): J = J(EL): K = K(EL): L = L(EL)
10315 FOR IV = 1 TO 4
10320 IF IV = 1 THEN II = I: JJ = J
10325 IF IV = 2 THEN II = J: JJ = K
10330 IF IV = 3 THEN II = K: JJ = L
10335 IF IV = 4 THEN II = L: JJ = I
10340 IF Z9 = 99 THEN 10450
```

```
10345 XI = (X(II) - XZ) * XS
10350 XJ = (X(JJ) - XZ) * XS
10355 YI = (Y(II) - YZ) * XS
10360 YJ = (Y(JJ) - YZ) * XS
10365 GOTO 10480
10450 XI = X(II) - XZ * XS: XJ = X(JJ) - XZ * XS
10455 YI = Y(II) - YZ * XS: YJ = Y(JJ) - YZ * XS
10480 LINE (XI + 40, 280 - YI)-(XJ + 40, 280 - YJ)
10490 NEXT IV
10495 NEXT EL
10510 IF Z9 = 77 THEN 10600
10515 PRINT "TO CONTIMUE, TYPE Y"
10520 A$ = INKEY$: IF A$ = "" THEN 10520
10530 IF A$ = "Y" OR A$ = "y" THEN 10560
10540 GOTO 10520
10560 SCREEN 0: WIDTH 80
10600 RETURN
11000 REM DEFLECTED FORM OF MESH
11010 CLS : PRINT "TO CONTINUE, TYPE Y"
11020 A$ = INKEY$: IF A$ = "" THEN 11020
11030 IF A$ = "Y" OR A$ = "y" THEN 11050
11040 GOTO 11020
11050 REM
11070 Z9 = 77
11080 GOSUB 10000
11090 UM = -1E+12
11100 FOR II = 1 TO NN
11110 IF ABS(UG(II)) > UM THEN UM = ABS(UG(II))
11120 NEXT II
11130 SC = 20
11140 FOR II = 1 TO NJ
11150 X(II) = X(II) * XS + SC * UG(II * 2 - 1) / UM
11160 Y(II) = Y(II) * XS + SC * UG(II * 2) / UM
11170 NEXT II
11180 Z9 = 99: GOSUB 10000
11190 FOR II = 1 TO NJ
11200 X(II) = XG(II)
11210 Y(II) = YG(II)
11220 NEXT II
11230 RETURN
```

Computer Program for In-Plane Stresses in Plates "STRESS8N"

```
50 CLS
60 REM COPYRIGHT OF DR.C.T.F.ROSS
70 REM 6 HURSTVILLE DRIVE, WATERLOOVILLE
80 REM PORTSMOUTH. ENGLAND
90 PRINT : PRINT "PLANE STRESS & PLANE STRAIN (8 NODE QUADRILATERAL)": PRINT
100 PRINT : PRINT "PROGRAM BY DR.C.T.F.ROSS": PRINT
110 PRINT : PRINT "NOT TO BE REPRODUCED": PRINT
111 INPUT "TYPE IN THE NAME OF YOUR INPUT FILE"; FILENAME$
112 PRINT "THE NAME OF YOUR INPUT FILE IS "; FILENAME$
113 PRINT "IF THIS NAME IS CORRECT, TYPE Y; ELSE N"
114 A$ = INKEY$: IF A$ = "" THEN GOTO 114
115 IF A$ = "Y" OR A$ = "y" THEN GOTO 119
116 IF A$ = "N" OR A$ = "n" THEN GOTO 111
117 END
119 OPEN FILENAME$ FOR INPUT AS #1
120 INPUT "TYPE IN THE NAME OF YOUR OUTPUT FILE "; OUT$
121 PRINT "THE NAME OF YOUR OUTPUT FILE IS "; OUT$
122 PRINT "IF THIS NAME IS CORRECT, TYPE Y, ELSE N"
123 B$ = INKEY$: IF B$ = "" THEN GOTO 123
124 IF B$ = "Y" OR B$ = "y" THEN GOTO 127
125 IF B$ = "N" OR B$ = "n" THEN GOTO 120
126 GOTO 123
127 OPEN OUT$ FOR OUTPUT AS #2
129 REM TYPE IN NO. OF NODES
130 INPUT #1, NJ: NN = 2 * NJ
140 IF NJ < 8 OR NJ > 299 THEN PRINT "NO. OF NODES="; NJ: GOSUB 9995: STOP
145 NNODE = 8: NDF = 2
150 REM TYPE IN THE NO. OF ELEMENTS
160 INPUT #1, ES
170 IF ES < 1 OR ES > 199 THEN PRINT "NO. OF ELEMENTS="; ES: GOSUB 9995: STOP
180 REM TYPE IN THE NO. OF NODES WITH ZERO DISPLACEMENTS
190 INPUT #1, NF
195 IF NF < 2 OR NF > NJ THEN PRINT "NO. OF NODES WITH ZERO
DISPLACEMENTS="; NF: GOSUB 9995: STOP
200 DIM I(ES), J(ES), K(ES), L(ES), M(ES), N(ES), O(ES), P(ES)
210 DIM SK(16, 16), KS(16, 16), DN(2, 8), D(3, 3), DB(3, 16), UG(NN), VG(NJ), VEC(NJ),
SHEAR(NJ)
220 DIM X(NJ), Y(NJ), XG(NJ), YG(NJ), NS(NF * 2)
230 DIM B(2, 8), BX(3, 16), BT(16, 3), JA(2, 2)
240 DIM CO(8, 2), SU(16), SR(3), DIAG(NN)
245 DIM PRINCIPAL(NJ), MINIMUM(NJ)
250 REM TYPE IN ELASTIC MODULUS & POISSON'S RATIO
260 INPUT #1, E, NU
270 IF E < 1 OR NU < 0 OR NU > .5 THEN PRINT "YOUNG'S MODULUS="; E; "
POISSON'S RATIO="; NU: GOSUB 9995: STOP
280 REM TYPE IN PLATE THICKNESS
282 INPUT #1, T
283 IF T < = 0 THEN PRINT "PLATE THICKNESS="; T: GOSUB 9995: STOP
290 REM IF PLANE STRESS, TYPE 1; BUT IF PLANE STRAIN, TYPE 0
```

```
300 INPUT #1, PS
310 IF PS = 0 OR PS = 1 THEN GOTO 330
320 PRINT "NEITHER PLANE STRESS NOR PLANE STRAIN. PS="; PS: GOSUB 9995: STOP
330 IF PS = 0 THEN GOTO 380
340 CN = E * T / (1 - NU ^ 2)
350 ZU = NU
360 MU = (1 - NU) / 2
370 GOTO 410
380 CN = E * T * (1 - NU) / ((1 + NU) * (1 - 2 * NU))
390 ZU = NU / (1 - NU)
400 MU = (1 - 2 * NU) * (2 * (1 - NU))
410 D(1, 1) = CN
420 D(2, 2) = CN
430 D(1, 2) = ZU * CN
440 D(2, 1) = D(1, 2)
450 D(3, 3) = MU * CN
460 REM TYPE IN THE NODAL COORDINATES
465 PRINT "  X        Y"
470 FOR II = 1 TO NJ
480 REM TYPE IN X(II) & Y(II)
490 INPUT #1, X(II), Y(II)
495 PRINT X(II), Y(II)
500 IF X(II) < -1000000! OR X(II) > 1000000! THEN PRINT "X("; II; ")="; X(II): GOSUB
9995: STOP
510 IF Y(II) < -1000000! OR Y(II) > 1000000! THEN PRINT "Y("; II; ")="; Y(II): GOSUB
9995: STOP
520 XG(II) = X(II): YG(II) = Y(II)
540 NEXT II
545 PRINT "TO CONTINUE, PRESS 'ANY' KEY"
547 C$ = INKEY$: IF C$ = "" THEN GOTO 547
550 REM TYPE IN THE NODAL POSITIONS OF THE ZERO DISPLACEMENTS, ETC.
560 FOR II = 1 TO NF * 2: NS(II) = 0: NEXT II
570 FOR II = 1 TO NF
580 REM TYPE NODAL POSITION 'SU', & IF XDISP & YDISP ARE ZERO
590 INPUT #1, SU, SUX, SVY
600 IF SU < 1 OR SU > NJ THEN PRINT "NO. OF NODE WITH ZERO
DISPLACEMENTS="; SU: GOSUB 9995: STOP
602 REM IF THE HORIZONTAL DEFLECTION IS ZERO. TYPE 1; ELSE TYPE 0
603 IF SUX = 1 THEN Y$ = "Y" ELSE IF SUX = 0 THEN Y$ = "N"
604 IF Y$ = "Y" OR Y$ = "y" THEN NS(II * 2 - 1) = SU * 2 - 1: GOTO 607
605 IF Y$ = "N" OR Y$ = "n" THEN GOTO 607
606 PRINT "SUX="; SUX; " AT NODE "; SU: GOSUB 9995: STOP
607 REM IF THE VERTICAL DEFLECTION IS ZERO. TYPE 1; ELSE TYPE 0
608 IF SVY = 1 THEN Y$ = "Y" ELSE IF SVY = 0 THEN Y$ = "N"
609 IF Y$ = "Y" OR Y$ = "y" THEN NS(II * 2) = SU * 2: GOTO 615
610 IF Y$ = "N" OR Y$ = "n" THEN GOTO 615
612 PRINT "SVY="; SVY; " AT NODE "; SU: GOSUB 9995: STOP
615 NEXT II
620 REM CALCULATION OF DIAGONAL POINTER VECTOR
625 PRINT "I   J   K   L   M   N   O   P"
630 FOR EL = 1 TO ES
640 REM TYPE IN THE NODAL NUMBERS IN AN ANTI-CLOCKWISE DIRECTION,
STARTING FROM A CORNER NODE
```

```
650 INPUT #1, I(EL), J(EL), K(EL), L(EL), M(EL), N(EL), O(EL), P(EL)
660 I = I(EL): J = J(EL): K = K(EL): L = L(EL)
670 M = M(EL): N = N(EL): O = O(EL): P = P(EL)
675 PRINT I; J; K; L; M; N; O; P
680 IF I < 1 OR I > NJ THEN PRINT "INODE="; I: GOSUB 9995: STOP
690 IF J < 1 OR J > NJ THEN PRINT "JNODE="; J: GOSUB 9995: STOP
700 IF K < 1 OR K > NJ THEN PRINT "KNODE="; K: GOSUB 9995: STOP
710 IF L < 1 OR L > NJ THEN PRINT "LNODE="; L: GOSUB 9995: STOP
720 IF M < 1 OR M > NJ THEN PRINT "MNODE="; M: GOSUB 9995: STOP
730 IF N < 1 OR N > NJ THEN PRINT "NNODE="; N: GOSUB 9995: STOP
740 IF O < 1 OR O > NJ THEN PRINT "ONODE="; O: GOSUB 9995: STOP
750 IF P < 1 OR P > NJ THEN PRINT "PNODE="; P: GOSUB 9995: STOP
755 IF I = J OR I = K OR I = L OR I = M OR I = N OR I = O OR I = P THEN GOSUB
9000: GOSUB 9995: STOP
760 IF J = K OR J = L OR J = M OR J = N OR J = O OR J = P THEN GOSUB 9000: GOSUB
9995: STOP
765 IF K = L OR K = M OR K = N OR K = O OR K = P THEN GOSUB 9000: GOSUB 9995:
STOP
770 IF L = M OR L = N OR L = O OR L = P THEN GOSUB 9000: GOSUB 9995: STOP
775 IF M = N OR M = O OR M = P THEN GOSUB 9000: GOSUB 9995: STOP
780 IF N = O OR N = P THEN GOSUB 9000: GOSUB 9995: STOP
785 IF O = P THEN GOSUB 9000: GOSUB 9995: STOP
790 FOR II = 1 TO NNODE: IA = P
795 IF II = 1 THEN IA = I
800 IF II = 2 THEN IA = J
805 IF II = 3 THEN IA = K
810 IF II = 4 THEN IA = L
815 IF II = 5 THEN IA = M
820 IF II = 6 THEN IA = N
825 IF II = 7 THEN IA = O
830 FOR IJ = 1 TO NDF
835 II1 = NDF * (IA - 1) + IJ
840 FOR JJ = 1 TO NNODE: IB = P
845 IF JJ = 1 THEN IB = I
850 IF JJ = 2 THEN IB = J
855 IF JJ = 3 THEN IB = K
860 IF JJ = 4 THEN IB = L
865 IF JJ = 5 THEN IB = M
870 IF JJ = 6 THEN IB = N
875 IF JJ = 7 THEN IB = O
880 FOR JI = 1 TO NDF
885 JJ1 = NDF * (IB - 1) + JI
890 IIJJ = JJ1 - II1 + 1
895 IF IIJJ > DIAG(JJ1) THEN DIAG(JJ1) = IIJJ
900 NEXT JI, JJ, IJ, II
940 NEXT EL
945 PRINT " TO CONTINUE, PRESS 'ANY' KEY"
947 C$ = INKEY$: IF C$ = "" THEN GOTO 947
950 DIAG(1) = 1: FOR II = 1 TO NN: DIAG(II) = DIAG(II - 1) + DIAG(II): NEXT II
960 NT = DIAG(NN)
970 DIM A(NT), Q(NN), POW(NJ)
980 REM TYPE IN THE NO. OF NODES WITH CONCENTRATED LOADS
990 INPUT #1, NC
```

```
1000 IF NC < 1 OR NC > NJ THEN PRINT "NO. OF CONCENTRATED LOADS="; NC:
GOSUB 9995: STOP
1010 FOR II = 1 TO NN: Q(II) = 0: NEXT II
1020 FOR II = 1 TO NJ: POW(II) = 0: NEXT II
1030 FOR II = 1 TO NC
1040 REM TYPE IN THE NODAL POSITION OF THE LOAD, & XLOAD & YLOAD
1050 INPUT #1, POW, XLOAD, YLOAD: POW(II) = POW
1055 IF POW < 1 OR POW > NJ THEN PRINT "NODAL POSITION OF CONCENTRATED
LOAD="; POW: GOSUB 9995: STOP
1060 REM TYPE IN THE LOAD IN THE 'X' DIRECTION FOR NODE POW
1070 Q(2 * POW - 1) = XLOAD
1080 REM TYPE IN THE LOAD IN THE 'Y' DIRECTION FOR NODE POW
1090 Q(2 * POW) = YLOAD
1160 NEXT II
1180 GOSUB 5000
1185 Z9 = 0: GOSUB 10000
1190 FOR EL = 1 TO ES
1200 PRINT : PRINT "ELEMENT NO."; EL; " UNDER COMPUTATION"
1210 I = I(EL)
1220 J = J(EL)
1230 K = K(EL)
1240 L = L(EL)
1250 M = M(EL)
1260 N = N(EL)
1270 O = O(EL)
1280 P = P(EL)
1282 VEC(I) = VEC(I) + 1: VEC(K) = VEC(K) + 1
1285 VEC(M) = VEC(M) + 1: VEC(O) = VEC(O) + 1
1287 VEC(J) = 1: VEC(L) = 1: VEC(N) = 1: VEC(P) = 1
1290 CO(1, 1) = X(I)
1300 CO(1, 2) = Y(I)
1310 CO(2, 1) = X(J)
1320 CO(2, 2) = Y(J)
1330 CO(3, 1) = X(K)
1340 CO(3, 2) = Y(K)
1350 CO(4, 1) = X(L)
1360 CO(4, 2) = Y(L)
1370 CO(5, 1) = X(M)
1380 CO(5, 2) = Y(M)
1390 CO(6, 1) = X(N)
1400 CO(6, 2) = Y(N)
1410 CO(7, 1) = X(O)
1420 CO(7, 2) = Y(O)
1430 CO(8, 1) = X(P)
1440 CO(8, 2) = Y(P)
1450 FOR II = 1 TO 16
1460 FOR JJ = 1 TO 16
1470 SK(II, JJ) = 0
1480 NEXT JJ
1490 NEXT II
1500 FOR LL = 1 TO 3
1510 IF LL = 1 THEN GOTO 1540
1520 IF LL = 2 THEN GOTO 1570
```

```
1530 IF LL = 3 THEN GOTO 1600
1540 XI = -.774597
1550 AI = 5 / 9
1560 GOTO 1620
1570 XI = 0
1580 AI = 8 / 9
1590 GOTO 1620
1600 XI = .774597
1610 AI = 5 / 9
1620 FOR MM = 1 TO 3
1630 IF MM = 1 THEN GOTO 1660
1640 IF MM = 2 THEN GOTO 1690
1650 IF MM = 3 THEN GOTO 1720
1660 ET = -.774597
1670 AJ = 5 / 9
1680 GOTO 1740
1690 ET = 0
1700 AJ = 8 / 9
1710 GOTO 1740
1720 ET = .774597
1730 AJ = 5 / 9
1740 GOSUB 3330
1750 FOR II = 1 TO 3
1760 FOR JJ = 1 TO 16
1770 BT(JJ, II) = BX(II, JJ)
1780 NEXT JJ, II
1790 FOR II = 1 TO 16
1800 FOR JJ = 1 TO 16
1810 KS(II, JJ) = 0
1820 NEXT JJ
1830 NEXT II
1840 FOR II = 1 TO 16
1850 FOR JJ = 1 TO 16
1860 FOR KK = 1 TO 3
1870 KS(II, JJ) = KS(II, JJ) + BT(II, KK) * DB(KK, JJ) * AI * AJ * DT
1880 NEXT KK
1890 NEXT JJ
1900 NEXT II
1910 FOR IJ = 1 TO 16
1920 FOR JI = 1 TO 16
1930 SK(IJ, JI) = SK(IJ, JI) + KS(IJ, JI)
1940 NEXT JI
1950 NEXT IJ
1960 NEXT MM
1970 NEXT LL
1980 I1 = 2 * I - 2: J1 = 2 * J - 2: K1 = 2 * K - 2: L1 = 2 * L - 2
1985 M1 = 2 * M - 2: N1 = 2 * N - 2: O1 = 2 * O - 2: P1 = 2 * P - 2
1990 FOR II = 1 TO 2
1995 FOR JJ = 1 TO 2
2000 MM = I1 + II: MN = I1 + JJ
2005 NM = J1 + II: N9 = J1 + JJ
2010 IF JJ > II THEN 2016
2012 A(DIAG(MM) + MN - MM) = A(DIAG(MM) + MN - MM) + SK(II, JJ)
```

2014 A(DIAG(NM) + N9 - NM) = A(DIAG(NM) + N9 - NM) + SK(II + 2, JJ + 2)
2016 IF I1 > J1 THEN 2022
2018 A(DIAG(NM) + MN - NM) = A(DIAG(NM) + MN - NM) + SK(II + 2, JJ)
2020 GOTO 2024
2022 A(DIAG(MM) + N9 - MM) = A(DIAG(MM) + N9 - MM) + SK(II, JJ + 2)
2024 NM = K1 + II: N9 = K1 + JJ
2026 IF JJ > II THEN 2030
2028 A(DIAG(NM) + N9 - NM) = A(DIAG(NM) + N9 - NM) + SK(II + 4, JJ + 4)
2030 IF I1 > K1 THEN 2036
2032 A(DIAG(NM) + MN - NM) = A(DIAG(NM) + MN - NM) + SK(II + 4, JJ)
2034 GOTO 2038
2036 A(DIAG(MM) + N9 - MM) = A(DIAG(MM) + N9 - MM) + SK(II, JJ + 4)
2038 NM = L1 + II: N9 = L1 + JJ
2040 IF JJ > II THEN 2044
2042 A(DIAG(NM) + N9 - NM) = A(DIAG(NM) + N9 - NM) + SK(II + 6, JJ + 6)
2044 IF I1 > L1 THEN 2050
2046 A(DIAG(NM) + MN - NM) = A(DIAG(NM) + MN - NM) + SK(II + 6, JJ)
2048 GOTO 2052
2050 A(DIAG(MM) + N9 - MM) = A(DIAG(MM) + N9 - MM) + SK(II, JJ + 6)
2052 NM = M1 + II: N9 = M1 + JJ
2054 IF JJ > II THEN 2058
2056 A(DIAG(NM) + N9 - NM) = A(DIAG(NM) + N9 - NM) + SK(II + 8, JJ + 8)
2058 IF I1 > M1 THEN 2064
2060 A(DIAG(NM) + MN - NM) = A(DIAG(NM) + MN - NM) + SK(II + 8, JJ)
2062 GOTO 2068
2064 A(DIAG(MM) + N9 - MM) = A(DIAG(MM) + N9 - MM) + SK(II, JJ + 8)
2068 NM = N1 + II: N9 = N1 + JJ
2070 IF JJ > II THEN 2074
2072 A(DIAG(NM) + N9 - NM) = A(DIAG(NM) + N9 - NM) + SK(II + 10, JJ + 10)
2074 IF I1 > N1 THEN 2082
2078 A(DIAG(NM) + MN - NM) = A(DIAG(NM) + MN - NM) + SK(II + 10, JJ)
2080 GOTO 2084
2082 A(DIAG(MM) + N9 - MM) = A(DIAG(MM) + N9 - MM) + SK(II, JJ + 10)
2084 NM = O1 + II: N9 = O1 + JJ
2086 IF JJ > II THEN 2130
2088 A(DIAG(NM) + N9 - NM) = A(DIAG(NM) + N9 - NM) + SK(II + 12, JJ + 12)
2130 IF I1 > O1 THEN 2136
2132 A(DIAG(NM) + MN - NM) = A(DIAG(NM) + MN - NM) + SK(II + 12, JJ)
2134 GOTO 2138
2136 A(DIAG(MM) + N9 - MM) = A(DIAG(MM) + N9 - MM) + SK(II, JJ + 12)
2138 NM = P1 + II: N9 = P1 + JJ
2140 IF JJ > II THEN 2144
2142 A(DIAG(NM) + N9 - NM) = A(DIAG(NM) + N9 - NM) + SK(II + 14, JJ + 14)
2144 IF I1 > P1 THEN 2150
2146 A(DIAG(NM) + MN - NM) = A(DIAG(NM) + MN - NM) + SK(II + 14, JJ)
2148 GOTO 2152
2150 A(DIAG(MM) + N9 - MM) = A(DIAG(MM) + N9 - MM) + SK(II, JJ + 14)
2152 MM = J1 + II: MN = J1 + JJ
2154 NM = K1 + II: N9 = K1 + JJ
2156 IF J1 > K1 THEN 2162
2158 A(DIAG(NM) + MN - NM) = A(DIAG(NM) + MN - NM) + SK(II + 4, JJ + 2)
2160 GOTO 2164
2162 A(DIAG(MM) + N9 - MM) = A(DIAG(MM) + N9 - MM) + SK(II + 2, JJ + 4)

```
2164 NM = L1 + II: N9 = L1 + JJ
2166 IF J1 > L1 THEN 2172
2168 A(DIAG(NM) + MN - NM) = A(DIAG(NM) + MN - NM) + SK(II + 6, JJ + 2)
2170 GOTO 2174
2172 A(DIAG(MM) + N9 - MM) = A(DIAG(MM) + N9 - MM) + SK(II + 2, JJ + 6)
2174 NM = M1 + II: N9 = M1 + JJ
2176 IF J1 > M1 THEN 2182
2178 A(DIAG(NM) + MN - NM) = A(DIAG(NM) + MN - NM) + SK(II + 8, JJ + 2)
2180 GOTO 2184
2182 A(DIAG(MM) + N9 - MM) = A(DIAG(MM) + N9 - MM) + SK(II + 2, JJ + 8)
2184 NM = N1 + II: N9 = N1 + JJ
2186 IF J1 > N1 THEN 2192
2188 A(DIAG(NM) + MN - NM) = A(DIAG(NM) + MN - NM) + SK(II + 10, JJ + 2)
2190 GOTO 2194
2192 A(DIAG(MM) + N9 - MM) = A(DIAG(MM) + N9 - MM) + SK(II + 2, JJ + 10)
2194 NM = O1 + II: N9 = O1 + JJ
2196 IF J1 > O1 THEN 2202
2198 A(DIAG(NM) + MN - NM) = A(DIAG(NM) + MN - NM) + SK(II + 12, JJ + 2)
2200 GOTO 2204
2202 A(DIAG(MM) + N9 - MM) = A(DIAG(MM) + N9 - MM) + SK(II + 2, JJ + 12)
2204 NM = P1 + II: N9 = P1 + JJ
2206 IF J1 > P1 THEN 2212
2208 A(DIAG(NM) + MN - NM) = A(DIAG(NM) + MN - NM) + SK(II + 14, JJ + 2)
2210 GOTO 2216
2212 A(DIAG(MM) + N9 - MM) = A(DIAG(MM) + N9 - MM) + SK(II + 2, JJ + 14)
2216 MM = K1 + II: MN = K1 + JJ
2218 NM = L1 + II: N9 = L1 + JJ
2220 IF K1 > L1 THEN 2226
2222 A(DIAG(NM) + MN - NM) = A(DIAG(NM) + MN - NM) + SK(II + 6, JJ + 4)
2224 GOTO 2228
2226 A(DIAG(MM) + N9 - MM) = A(DIAG(MM) + N9 - MM) + SK(II + 4, JJ + 6)
2228 NM = M1 + II: N9 = M1 + JJ
2230 IF K1 > M1 THEN 2236
2232 A(DIAG(NM) + MN - NM) = A(DIAG(NM) + MN - NM) + SK(II + 8, JJ + 4)
2234 GOTO 2238
2236 A(DIAG(MM) + N9 - MM) = A(DIAG(MM) + N9 - MM) + SK(II + 4, JJ + 8)
2238 NM = N1 + II: N9 = N1 + JJ
2240 IF K1 > N1 THEN 2248
2242 A(DIAG(NM) + MN - NM) = A(DIAG(NM) + MN - NM) + SK(II + 10, JJ + 4)
2246 GOTO 2250
2248 A(DIAG(MM) + N9 - MM) = A(DIAG(MM) + N9 - MM) + SK(II + 4, JJ + 10)
2250 NM = O1 + II: N9 = O1 + JJ
2252 IF K1 > O1 THEN 2258
2254 A(DIAG(NM) + MN - NM) = A(DIAG(NM) + MN - NM) + SK(II + 12, JJ + 4)
2256 GOTO 2260
2258 A(DIAG(MM) + N9 - MM) = A(DIAG(MM) + N9 - MM) + SK(II + 4, JJ + 12)
2260 NM = P1 + II: N9 = P1 + JJ
2262 IF K1 > P1 THEN 2268
2264 A(DIAG(NM) + MN - NM) = A(DIAG(NM) + MN - NM) + SK(II + 14, JJ + 4)
2266 GOTO 2270
2268 A(DIAG(MM) + N9 - MM) = A(DIAG(MM) + N9 - MM) + SK(II + 4, JJ + 14)
2270 MM = L1 + II: MN = L1 + JJ
2274 NM = M1 + II: N9 = M1 + JJ
```

```
2276 IF L1 > M1 THEN 2282
2278 A(DIAG(NM) + MN - NM) = A(DIAG(NM) + MN - NM) + SK(II + 8, JJ + 6)
2280 GOTO 2284
2282 A(DIAG(MM) + N9 - MM) = A(DIAG(MM) + N9 - MM) + SK(II + 6, JJ + 8)
2284 NM = N1 + II: N9 = N1 + JJ
2286 IF L1 > N1 THEN 2292
2288 A(DIAG(NM) + MN - NM) = A(DIAG(NM) + MN - NM) + SK(II + 10, JJ + 6)
2290 GOTO 2294
2292 A(DIAG(MM) + N9 - MM) = A(DIAG(MM) + N9 - MM) + SK(II + 6, JJ + 10)
2294 NM = O1 + II: N9 = O1 + JJ
2296 IF L1 > O1 THEN 2302
2298 A(DIAG(NM) + MN - NM) = A(DIAG(NM) + MN - NM) + SK(II + 12, JJ + 6)
2300 GOTO 2304
2302 A(DIAG(MM) + N9 - MM) = A(DIAG(MM) + N9 - MM) + SK(II + 6, JJ + 12)
2304 NM = P1 + II: N9 = P1 + JJ
2306 IF L1 > P1 THEN 2312
2308 A(DIAG(NM) + MN - NM) = A(DIAG(NM) + MN - NM) + SK(II + 14, JJ + 6)
2310 GOTO 2316
2312 A(DIAG(MM) + N9 - MM) = A(DIAG(MM) + N9 - MM) + SK(II + 6, JJ + 14)
2316 MM = M1 + II: MN = M1 + JJ
2318 NM = N1 + II: N9 = N1 + JJ
2320 IF M1 > N1 THEN 2326
2322 A(DIAG(NM) + MN - NM) = A(DIAG(NM) + MN - NM) + SK(II + 10, JJ + 8)
2324 GOTO 2328
2326 A(DIAG(MM) + N9 - MM) = A(DIAG(MM) + N9 - MM) + SK(II + 8, JJ + 10)
2328 NM = O1 + II: N9 = O1 + JJ
2330 IF M1 > O1 THEN 2336
2332 A(DIAG(NM) + MN - NM) = A(DIAG(NM) + MN - NM) + SK(II + 12, JJ + 8)
2334 GOTO 2338
2336 A(DIAG(MM) + N9 - MM) = A(DIAG(MM) + N9 - MM) + SK(II + 8, JJ + 12)
2338 NM = P1 + II: N9 = P1 + JJ
2340 IF M1 > P1 THEN 2346
2342 A(DIAG(NM) + MN - NM) = A(DIAG(NM) + MN - NM) + SK(II + 14, JJ + 8)
2344 GOTO 2350
2346 A(DIAG(MM) + N9 - MM) = A(DIAG(MM) + N9 - MM) + SK(II + 8, JJ + 14)
2350 MM = N1 + II: MN = N1 + JJ
2352 NM = O1 + II: N9 = O1 + JJ
2354 IF N1 > O1 THEN 2360
2356 A(DIAG(NM) + MN - NM) = A(DIAG(NM) + MN - NM) + SK(II + 12, JJ + 10)
2358 GOTO 2362
2360 A(DIAG(MM) + N9 - MM) = A(DIAG(MM) + N9 - MM) + SK(II + 10, JJ + 12)
2362 NM = P1 + II: N9 = P1 + JJ
2364 IF N1 > P1 THEN 2372
2368 A(DIAG(NM) + MN - NM) = A(DIAG(NM) + MN - NM) + SK(II + 14, JJ + 10)
2370 GOTO 2376
2372 A(DIAG(MM) + N9 - MM) = A(DIAG(MM) + N9 - MM) + SK(II + 10, JJ + 14)
2376 MM = O1 + II: MN = O1 + JJ
2378 NM = P1 + II: N9 = P1 + JJ
2380 IF O1 > P1 THEN 2386
2382 A(DIAG(NM) + MN - NM) = A(DIAG(NM) + MN - NM) + SK(II + 14, JJ + 12)
2384 GOTO 2390
2386 A(DIAG(MM) + N9 - MM) = A(DIAG(MM) + N9 - MM) + SK(II + 12, JJ + 14)
2390 NEXT JJ, II
```

```
2395 NEXT EL
2400 FOR II = 1 TO NF * 2
2402 N9 = NS(II)
2404 IF N9 = 0 THEN 2410
2406 Q(N9) = 0: DI = DIAG(N9)
2408 A(DI) = A(DI) * 1E+12 + 1E+12
2410 NEXT II
2415 CLOSE #1
2412 PRINT : PRINT "THE SIMULTANEOUS EQUATIONS ARE NOW BEING SOLVED":
PRINT
2414 N = NN
2416 A(1) = SQR(A(1))
2418 FOR II = 2 TO N
2420 DI = DIAG(II) - II
2422 LL = DIAG(II - 1) - DI + 1
2424 FOR JJ = LL TO II
2426 X = A(DI + JJ)
2428 DJ = DIAG(JJ) - JJ
2430 IF JJ = 1 THEN 2442
2432 LB = DIAG(JJ - 1) - DJ + 1
2434 IF LL > LB THEN LB = LL
2436 FOR KK = LB TO JJ - 1
2438 X = X - A(DI + KK) * A(DJ + KK)
2440 NEXT KK
2442 A(DI + JJ) = X / A(DJ + JJ)
2444 NEXT JJ
2446 A(DI + II) = SQR(X)
2448 NEXT II
2450 Q(1) = Q(1) / A(1)
2452 FOR II = 2 TO N
2454 DI = DIAG(II) - II
2456 LL = DIAG(II - 1) - DI + 1
2458 X = Q(II)
2460 FOR JJ = LL TO II - 1
2462 X = X - A(DI + JJ) * Q(JJ)
2464 NEXT JJ
2466 Q(II) = X / A(DI + II)
2468 NEXT II
2470 FOR II = N TO 2 STEP -1
2472 DI = DIAG(II) - II
2474 X = Q(II) / A(DI + II): Q(II) = X
2476 LL = DIAG(II - 1) - DI + 1
2478 FOR KK = LL TO II - 1
2480 Q(KK) = Q(KK) - X * A(DI + KK)
2482 NEXT KK
2484 NEXT II
2486 Q(1) = Q(1) / A(1)
2590 PRINT #2,
2610 PRINT #2, : PRINT #2, "THE NODAL VALUES OF THE DISPLACEMENTS ARE:-":
PRINT #2,
2620 FOR II = 1 TO NJ
2630 PRINT #2, "NODE="; II, "U("; II; ")="; Q(II * 2 - 1),
2640 PRINT #2, "V("; II; ")="; Q(II * 2)
```

```
2645 UG(II * 2 - 1) = Q(II * 2 - 1): UG(II * 2) = Q(II * 2)
2650 NEXT II
2660 FOR EL = 1 TO ES
2670 I = I(EL)
2680 J = J(EL)
2690 K = K(EL)
2700 L = L(EL)
2710 M = M(EL)
2720 N = N(EL)
2730 O = O(EL)
2740 P = P(EL)
2750 SU(1) = Q(2 * I - 1)
2760 SU(3) = Q(2 * J - 1)
2770 SU(5) = Q(2 * K - 1)
2780 SU(7) = Q(2 * L - 1)
2790 SU(9) = Q(2 * M - 1)
2800 SU(11) = Q(2 * N - 1)
2810 SU(13) = Q(2 * O - 1)
2820 SU(15) = Q(2 * P - 1)
2830 SU(2) = Q(2 * I)
2840 SU(4) = Q(2 * J)
2850 SU(6) = Q(2 * K)
2860 SU(8) = Q(2 * L)
2870 SU(10) = Q(2 * M)
2880 SU(12) = Q(2 * N)
2890 SU(14) = Q(2 * O)
2900 SU(16) = Q(2 * P)
2910 CO(1, 1) = X(I)
2920 CO(1, 2) = Y(I)
2930 CO(2, 1) = X(J)
2940 CO(2, 2) = Y(J)
2950 CO(3, 1) = X(K)
2960 CO(3, 2) = Y(K)
2970 CO(4, 1) = X(L)
2980 CO(4, 2) = Y(L)
2990 CO(5, 1) = X(M)
3000 CO(5, 2) = Y(M)
3010 CO(6, 1) = X(N)
3020 CO(6, 2) = Y(N)
3030 CO(7, 1) = X(O)
3040 CO(7, 2) = Y(O)
3050 CO(8, 1) = X(P)
3060 CO(8, 2) = Y(P)
3070 FOR LL = 1 TO 3
3080 IF LL = 1 THEN XI = -1
3090 IF LL = 2 THEN XI = 0
3100 IF LL = 3 THEN XI = 1
3110 FOR MM = 1 TO 3
3120 IF MM = 1 THEN ET = -1
3130 IF MM = 2 THEN ET = 0
3140 IF MM = 3 THEN ET = 1
3150 GOSUB 3330
3160 FOR II = 1 TO 3
```

```
3170 SR(II) = 0
3180 FOR JJ = 1 TO 16
3190 SR(II) = SR(II) + DB(II, JJ) * SU(JJ)
3200 NEXT JJ
3210 NEXT II
3220 PRINT #2, : PRINT #2, "THE STRESSES IN ELEMENT NO."; EL; "    "; I; "-"; J; "-";
K; "-"; L; "-"; M; "-"; N; "-"; O; "-"; P; "   AT XI="; XI; " & ETA="; ET
3240 PRINT #2, "SIGMAX="; SR(1) / T, "SIGMAY="; SR(2) / T, "TAU X-Y="; SR(3) / T,
3241 SHEAR = .5 * SQR((SR(1) - SR(2)) ^ 2 + 4 * SR(3) ^ 2) / T
3242 PRINCIPAL = .5 * (SR(1) + SR(2)) / T + SHEAR
3243 MINIMUM = .5 * (SR(1) + SR(2)) / T - SHEAR
3244 PRINT #2, "SIGMA1="; PRINCIPAL, "SIGMA2="; MINIMUM
3245 IF XI = 0 OR ET = 0 THEN SHEAR = 0: GOTO 3270
3246 IF SR(1) < SR(2) THEN SHEAR = -SHEAR
3248 IF XI = -1 AND ET = -1 THEN SHEAR(I) = SHEAR(I) + SHEAR: PRINCIPAL(I) =
PRINCIPAL(I) + PRINCIPAL: MINIMUM(I) = MINIMUM(I) + MINIMUM
3250 IF XI = 1 AND ET = -1 THEN SHEAR(K) = SHEAR(K) + SHEAR: PRINCIPAL(K) =
PRINCIPAL(K) + PRINCIPAL: MINIMUM(K) = MINIMUM(K) + MINIMUM
3252 IF XI = 1 AND ET = 1 THEN SHEAR(M) = SHEAR(M) + SHEAR: PRINCIPAL(M) =
PRINCIPAL(M) + PRINCIPAL: MINIMUM(M) = MINIMUM(M) + MINIMUM
3254 IF XI = -1 AND ET = 1 THEN SHEAR(O) = SHEAR(O) + SHEAR: PRINCIPAL(O) =
PRINCIPAL(O) + PRINCIPAL: MINIMUM(O) = MINIMUM(O) + MINIMUM
3270 NEXT MM
3280 NEXT LL
3290 NEXT EL
3292 GOSUB 11000
3295 FOR II = 1 TO NJ STEP 2
3297 VG(II) = SHEAR(II) / VEC(II)
3300 NEXT II
3310 GOSUB 6000
3320 END
3330 FOR II = 1 TO 2
3340 FOR JJ = 1 TO 2
3345 JA(II, JJ) = 0
3350 NEXT JJ
3360 NEXT II
3370 DN(1, 1) = (1 - ET) * (2 * XI + ET) / 4
3390 DN(1, 3) = (1 - ET) * (2 * XI - ET) / 4
3400 DN(1, 5) = (1 + ET) * (2 * XI + ET) / 4
3410 DN(1, 7) = (1 + ET) * (2 * XI - ET) / 4
3420 DN(1, 2) = -XI * (1 - ET)
3430 DN(1, 4) = (1 - ET ^ 2) / 2
3440 DN(1, 6) = -XI * (1 + ET)
3450 DN(1, 8) = -(1 - ET ^ 2) / 2
3460 DN(2, 1) = (1 - XI) * (XI + 2 * ET) / 4
3470 DN(2, 3) = (1 + XI) * (-XI + 2 * ET) / 4
3480 DN(2, 5) = (1 + XI) * (XI + 2 * ET) / 4
3490 DN(2, 7) = (1 - XI) * (-XI + 2 * ET) / 4
3500 DN(2, 2) = -(1 - XI ^ 2) / 2
3510 DN(2, 4) = -ET * (1 + XI)
3520 DN(2, 6) = (1 - XI ^ 2) / 2
3530 DN(2, 8) = -ET * (1 - XI)
3540 FOR II = 1 TO 2
```

```
3550 FOR JJ = 1 TO 2
3560 FOR KK = 1 TO 8
3570 JA(II, JJ) = JA(II, JJ) + DN(II, KK) * CO(KK, JJ)
3580 NEXT KK
3590 NEXT JJ
3600 NEXT II
3610 CN = JA(1, 1)
3620 JA(1, 1) = JA(2, 2)
3630 JA(2, 2) = CN
3640 JA(2, 1) = -JA(2, 1)
3650 JA(1, 2) = -JA(1, 2)
3660 DT = JA(1, 1) * JA(2, 2) - JA(1, 2) * JA(2, 1)
3670 FOR II = 1 TO 2
3680 FOR JJ = 1 TO 2
3690 JA(II, JJ) = JA(II, JJ) / DT
3700 NEXT JJ
3710 NEXT II
3720 FOR II = 1 TO 2
3730 FOR JJ = 1 TO 8
3740 B(II, JJ) = 0
3750 NEXT JJ
3760 NEXT II
3770 FOR II = 1 TO 2
3780 FOR JJ = 1 TO 8
3790 FOR KK = 1 TO 2
3800 B(II, JJ) = B(II, JJ) + JA(II, KK) * DN(KK, JJ)
3810 NEXT KK
3820 NEXT JJ
3830 NEXT II
3840 FOR II = 1 TO 8
3850 BX(1, 2 * II - 1) = B(1, II)
3860 BX(2, 2 * II) = B(2, II)
3870 BX(3, 2 * II - 1) = B(2, II)
3880 BX(3, 2 * II) = B(1, II)
3890 NEXT II
3900 FOR II = 1 TO 3
3910 FOR JJ = 1 TO 16
3920 DB(II, JJ) = 0
3930 NEXT JJ
3940 NEXT II
3950 FOR II = 1 TO 3
3960 FOR JJ = 1 TO 16
3970 FOR KK = 1 TO 3
3980 DB(II, JJ) = DB(II, JJ) + D(II, KK) * BX(KK, JJ)
3990 NEXT KK
4000 NEXT JJ
4010 NEXT II
4020 RETURN
5000 REM OUTPUT OF INPUT
5010 PRINT #2, : PRINT #2, "PLANE STRESS & PLANE STRAIN USING AN 8 NODE
QUADRILERAL"
5020 PRINT #2, : PRINT #2, "NUMBER OF NODES="; NJ
5030 PRINT #2, "NUMBER OF ELEMENTS="; ES
```

```
5040 PRINT #2, "NUMBER OF NODES WITH ZERO DISPLACEMENTS="; NF
5050 PRINT #2, "ELASTIC MODULUS="; E
5060 PRINT #2, "POISSON'S RATIO="; NU
5070 PRINT #2, "PLATE THICKNESS="; T
5080 IF PS = 0 THEN PRINT #2, "THIS IS A PLANE STRAIN PROBLEM"
5090 IF PS = 1 THEN PRINT #2, "THIS IS A PLANE STRESS PROBLEM"
5100 PRINT #2, : PRINT #2, "THE NODAL COORDINATES (X & Y) IN GLOBAL
COORDINATES ARE:": PRINT #2,
5110 FOR II = 1 TO NJ
5120 PRINT #2, "NODE="; II, "X("; II; ")="; X(II), "Y("; II; ")="; Y(II)
5130 NEXT II
5140 PRINT #2, : PRINT #2, "THE NODAL POSITIONS, ETC. OF THE ZERO
DISPLACEMENTS ARE:": PRINT #2,
5150 FOR II = 1 TO NF
5160 SU = INT((NS(II * 2 - 1) + 1) / 2)
5170 IF SU = 0 THEN SU = INT(NS(II * 2) / 2)
5180 PRINT #2, "NODE="; SU
5190 N9 = NS(II * 2 - 1)
5200 IF N9 > 0 THEN PRINT #2, "DISPLACEMENT IN X DIRECTION IS ZERO AT NODE
"; SU
5210 N9 = NS(II * 2)
5220 IF N9 > 0 THEN PRINT #2, "DISPLACEMENT IN Y DIRECTION IS ZERO AT NODE
"; SU
5260 NEXT II
5270 PRINT #2, : PRINT #2, "ELEMENT TOPOLOGY": PRINT #2,
5280 FOR EL = 1 TO ES
5290 I = I(EL): J = J(EL): K = K(EL): L = L(EL): M = M(EL): N = N(EL): O = O(EL): P
= P(EL)
5300 PRINT #2, "ELEMENT NO. "; EL, "NODES="; I; "-"; J; "-"; K; "-"; L; "-"; M; "-"; N;
"-"; O; "-"; P
5330 NEXT EL
5340 PRINT #2, : PRINT #2, "NODAL POSITIONS & VALUES OF THE CONCENTRATED
LOADS": PRINT #2,
5350 FOR I = 1 TO NC
5360 POW = POW(I)
5370 PRINT #2, "NODE="; POW
5380 PRINT #2, "LOAD IN X DIRECTION="; Q(POW * 2 - 1)
5390 PRINT #2, "LOAD IN Y DIRECTION="; Q(POW * 2)
5400 NEXT I
5430 RETURN
6000 LF = 10: Z9 = 99: DIM YY(4), XX(4), FF(LF + 1)
6001 GOSUB 6006
6002 FOR II = 1 TO NJ STEP 2: VG(II) = PRINCIPAL(II) / VEC(II): NEXT II: Z9 = 999:
GOSUB 6006
6003 FOR II = 1 TO NJ STEP 2: VG(II) = MINIMUM(II) / VEC(II): NEXT II: Z9 = 9999:
GOSUB 6006
6004 SCREEN 0, 0
6005 END
6006 MAX = MAXX = -1E+12: MAXU = -1E+12: MINU = 1E+12
6007 SCREEN 9
6010 CLS
6011 IF Z9 = 99 THEN PRINT "LINES OF CONSTANT MAXIMUM SHEAR STRESS"
6012 IF Z9 = 999 THEN PRINT "LINES OF CONSTANT MAXIMUM PRINCIPAL STRESS"
```

```
6013 IF Z9 = 9999 THEN PRINT "LINES OF CONSTANT MINIMUM PRINCIPAL STRESS"
6020 FOR I = 1 TO NJ STEP 2
6030 IF VG(I) > MAXU THEN MAXU = VG(I)
6040 IF VG(I) < MINU THEN MINU = VG(I)
6045 IF ABS(X(I)) > MAXX THEN MAXX = ABS(X(I))
6050 IF ABS(Y(I)) > MAXY THEN MAXY = ABS(Y(I))
6055 IF MAXX > MAX THEN MAX = MAXX
6060 IF MAXY > MAX THEN MAX = MAXY
6065 NEXT I
6070 LF = 10: XS = 200 / MAX
6080 DIF = MAXU - MINU: STP = DIF / LF: AG = 1
6090 FOR I = MINU TO MAXU STEP STP
6100 FF(AG) = I: IF ABS(FF(AG)) < .000001 THEN FF(AG) = 0
6110 AG = AG + 1: NEXT I
6120 FOR N = 1 TO LF
6130 FOR EL = 1 TO ES
6135 AA = 0: BB = 0: CC = 0: DD = 0
6140 I = I(EL): J = K(EL): K = M(EL): L = O(EL)
6145 IF FF(N) = (INT(VG(I) * 1000)) / 1000 THEN GOTO 6155
6147 IF FF(N) = (INT(VG(J) * 1000)) / 1000 THEN GOTO 6155
6149 IF FF(N) = (INT(VG(K) * 1000)) / 1000 THEN GOTO 6155
6151 IF FF(N) = (INT(VG(L) * 1000)) / 1000 THEN GOTO 6155
6153 GOTO 6160
6155 FF(N) = FF(N) + .000001
6160 IF FF(N) > VG(I) AND FF(N) < VG(J) THEN GOSUB 7000
6165 IF FF(N) < VG(I) AND FF(N) > VG(J) THEN GOSUB 7000
6170 IF FF(N) > VG(I) AND FF(N) < VG(L) THEN GOSUB 7100
6175 IF FF(N) < VG(I) AND FF(N) > VG(L) THEN GOSUB 7100
6180 IF FF(N) > VG(J) AND FF(N) < VG(K) THEN GOSUB 7200
6185 IF FF(N) < VG(J) AND FF(N) > VG(K) THEN GOSUB 7200
6190 IF FF(N) > VG(K) AND FF(N) < VG(L) THEN GOSUB 7300
6195 IF FF(N) < VG(K) AND FF(N) > VG(L) THEN GOSUB 7300
6220 IF AA = 1 AND BB = 1 THEN LINE (40 + XS * (XX(1)), 280 - XS * (YY(1)))-(40 + XS
* (XX(2)), 280 - XS * (YY(2)))
6230 IF AA = 1 AND CC = 1 THEN LINE (40 + XS * (XX(1)), 280 - XS * (YY(1)))-(40 + XS
* (XX(3)), 280 - XS * (YY(3)))
6240 IF AA = 1 AND DD = 1 THEN LINE (40 + XS * (XX(1)), 280 - XS * (YY(1)))-(40 + XS
* (XX(4)), 280 - XS * (YY(4)))
6250 IF BB = 1 AND CC = 1 THEN LINE (40 + XS * (XX(2)), 280 - XS * (YY(2)))-(40 + XS
* (XX(3)), 280 - XS * (YY(3)))
6260 IF BB = 1 AND DD = 1 THEN LINE (40 + XS * (XX(2)), 280 - XS * (YY(2)))-(40 + XS
* (XX(4)), 280 - XS * (YY(4)))
6270 IF CC = 1 AND DD = 1 THEN LINE (40 + XS * (XX(3)), 280 - XS * (YY(3)))-(40 + XS
* (XX(4)), 280 - XS * (YY(4)))
6290 NEXT EL
6300 NEXT N
6310 A$ = INKEY$: IF A$ = "" THEN GOTO 6310
6320 IF A$ = "Y" OR A$ = "y" THEN SCREEN 0: RETURN
6330 GOTO 6310
6400 END
7000 DF = FF(N) - VG(I): DE = VG(J) - VG(I)
7010 DX = X(J) - X(I): DY = Y(J) - Y(I)
7020 IF DF = 0 AND DE = 0 THEN GOTO 7050
```

```
7030 XX(1) = (DF * DX / DE) + X(I): YY(1) = (DF * DY / DE) + Y(I)
7040 AA = 1: RETURN
7050 XX(1) = X(I): YY(1) = Y(I): AA = 1: RETURN
7100 DF = FF(N) - VG(I): DE = VG(L) - VG(I)
7110 DX = X(L) - X(I): DY = Y(L) - Y(I)
7120 IF DF = 0 AND DE = 0 THEN GOTO 7150
7130 XX(2) = (DF * DX / DE) + X(I): YY(2) = (DF * DY / DE) + Y(I)
7140 BB = 1: RETURN
7150 XX(2) = X(I): YY(2) = Y(I): BB = 1: RETURN
7200 DF = FF(N) - VG(J): DE = VG(K) - VG(J)
7210 DX = X(K) - X(J): DY = Y(K) - Y(J)
7220 IF DF = 0 AND DE = 0 THEN GOTO 7250
7230 XX(3) = (DF * DX / DE) + X(J): YY(3) = (DF * DY / DE) + Y(J)
7240 CC = 1: RETURN
7250 XX(3) = X(J): YY(3) = Y(J): CC = 1: RETURN
7300 DF = FF(N) - VG(K): DE = VG(L) - VG(K)
7310 DX = X(L) - X(K): DY = Y(L) - Y(K)
7320 IF DF = 0 AND DE = 0 THEN GOTO 7350
7330 XX(4) = (DF * DX / DE) + X(K): YY(4) = (DF * DY / DE) + Y(K)
7340 DD = 1: RETURN
7350 XX(4) = X(K): YY(3) = Y(K): DD = 1: RETURN
9000 PRINT "AT LEAST TWO OF THE NODES HAVE THE SAME VALUES.";
9010 PRINT " THIS IS NOT PERMISSIBLE"
9020 PRINT "I="; I; "J="; J; "K="; K; "L="; L; "M="; M; "N="; N; "O="; O; "P="; P
9030 RETURN
9995 REM WARNING
9996 PRINT : PRINT "INCORRECT DATA": PRINT
9997 RETURN
10000 IF Z9 = 77 OR Z9 = 99 THEN 10295
10010 CLS : XZ = 1E+12: YZ = 1E+12: XM = -1E+12: YM = -1E+12
10020 PRINT "TO CONTINUE, TYPE Y"
10060 A$ = INKEY$: IF A$ = "" THEN 10060
10070 IF A$ = "Y" OR A$ = "y" THEN 10090
10080 GOTO 10060
10090 FOR EL = 1 TO ES
10100 I = I(EL): J = J(EL): K = K(EL): L = L(EL): M = M(EL): N = N(EL): O = O(EL): P
= P(EL)
10104 IF X(I) < XZ THEN XZ = X(I)
10106 IF X(J) < XZ THEN XZ = X(J)
10108 IF X(K) < XZ THEN XZ = X(K)
10110 IF X(L) < XZ THEN XZ = X(L)
10112 IF X(M) < XZ THEN XZ = X(M)
10114 IF X(N) < XZ THEN XZ = X(N)
10116 IF X(O) < XZ THEN XZ = X(O)
10120 IF X(P) < XZ THEN XZ = X(P)
10122 IF Y(I) < YZ THEN YZ = Y(I)
10124 IF Y(J) < YZ THEN YZ = Y(J)
10126 IF Y(K) < YZ THEN YZ = Y(K)
10128 IF Y(L) < YZ THEN YZ = Y(L)
10130 IF Y(M) < YZ THEN YZ = Y(M)
10132 IF Y(N) < YZ THEN YZ = Y(N)
10134 IF Y(O) < YZ THEN YZ = Y(O)
10140 IF Y(P) < YZ THEN YZ = Y(P)
```

```
10150 IF X(I) > XM THEN XM = X(I)
10152 IF X(J) > XM THEN XM = X(J)
10154 IF X(K) > XM THEN XM = X(K)
10156 IF X(L) > XM THEN XM = X(L)
10158 IF X(M) > XM THEN XM = X(M)
10160 IF X(N) > XM THEN XM = X(N)
10162 IF X(O) > XM THEN XM = X(O)
10170 IF X(P) > XM THEN XM = X(P)
10172 IF Y(I) > YM THEN YM = Y(I)
10174 IF Y(J) > YM THEN YM = Y(J)
10176 IF Y(K) > YM THEN YM = Y(K)
10178 IF Y(L) > YM THEN YM = Y(L)
10180 IF Y(M) > YM THEN YM = Y(M)
10182 IF Y(N) > YM THEN YM = Y(N)
10184 IF Y(O) > YM THEN YM = Y(O)
10186 IF Y(P) > YM THEN YM = Y(P)
10200 NEXT EL
10210 XS = 400 / (XM - XZ): YS = 200 / (YM - YZ)
10220 IF XS > YS THEN XS = YS
10295 IF Z9 = 0 OR Z9 = 77 THEN SCREEN 9
10296 IF Z9 = 0 THEN PRINT "THE MESH WILL NOW BE DRAWN"
10298 IF Z9 = 77 THEN PRINT "DEFLECTED FORM OF MESH"
10300 FOR EL = 1 TO ES
10310 I = I(EL): J = J(EL): K = K(EL): L = L(EL): M = M(EL): N = N(EL): O = O(EL): P
= P(EL)
10315 FOR IV = 1 TO 8
10318 IF IV = 1 THEN II = I: JJ = J
10320 IF IV = 2 THEN II = J: JJ = K
10322 IF IV = 3 THEN II = K: JJ = L
10324 IF IV = 4 THEN II = L: JJ = M
10326 IF IV = 5 THEN II = M: JJ = N
10330 IF IV = 6 THEN II = N: JJ = O
10332 IF IV = 7 THEN II = O: JJ = P
10340 IF IV = 8 THEN II = P: JJ = I
10342 IF Z9 = 99 THEN 10450
10345 XI = (X(II) - XZ) * XS
10350 XJ = (X(JJ) - XZ) * XS
10355 YI = (Y(II) - YZ) * XS
10360 YJ = (Y(JJ) - YZ) * XS
10365 GOTO 10480
10450 XI = X(II) - XZ * XS: XJ = X(JJ) - XZ * XS
10455 YI = Y(II) - YZ * XS: YJ = Y(JJ) - YZ * XS
10480 LINE (XI + 40, 280 - YI)-(XJ + 40, 280 - YJ)
10490 NEXT IV
10495 NEXT EL
10510 IF Z9 = 77 THEN 10600
10515 PRINT "TO CONTINUE, TYPE Y"
10520 A$ = INKEY$: IF A$ = "" THEN 10520
10530 IF A$ = "Y" OR A$ = "y" THEN 10560
10540 GOTO 10520
10560 SCREEN 0: WIDTH 80
10600 RETURN
11000 REM DEFLECTED FORM OF MESH
```

```
11010 CLS : PRINT "TO CONTINUE, TYPE Y"
11020 A$ = INKEY$: IF A$ = "" THEN 11020
11030 IF A$ = "Y" OR A$ = "y" THEN 11050
11040 GOTO 11020
11050 REM
11070 Z9 = 77
11080 GOSUB 10000
11090 UM = -1E+12
11100 FOR II = 1 TO NN
11110 IF ABS(UG(II)) > UM THEN UM = ABS(UG(II))
11120 NEXT II
11130 SC = 20
11140 FOR II = 1 TO NJ
11150 X(II) = X(II) * XS + SC * UG(II * 2 - 1) / UM
11160 Y(II) = Y(II) * XS + SC * UG(II * 2) / UM
11170 NEXT II
11180 Z9 = 99: GOSUB 10000
11190 FOR II = 1 TO NJ
11200 X(II) = XG(II)
11210 Y(II) = YG(II)
11220 NEXT II
11230 RETURN
```

Computer Program for Bending Stresses in Plates "PLATEBEF"

```
100 REM PROGRAM BY DR.C.T.F.ROSS
110 REM 6 HURSTVILLE DRIVE,
120 REM WATERLOOVILLE, PORTSMOUTH.
130 REM PO7 7NB ENGLAND
140 CLS : RESTORE
142 PRINT : PRINT "PLATE BENDING STRESSES": PRINT
PRINT "PROGRAM BY Dr.C.T.F.ROSS": PRINT
143 INPUT "TYPE IN THE NAME OF YOUR INPUT FILE "; FILENAME$
PRINT "IF YOU ARE SATISFIED WITH THIS FILENAME, TYPE Y; ELSE N"
144 A$ = INKEY$: IF A$ = "" THEN GOTO 144
IF A$ = "Y" OR A$ = "y" THEN GOTO 150
IF A$ = "N" OR A$ = "n" THEN GOTO 143
GOTO 144
150 OPEN FILENAME$ FOR INPUT AS #1
152 INPUT "TYPE IN THE NAME OF YOUR OUTPUT FILE "; OUT$
PRINT "IF YOU ARE SATISFIED WITH THIS FILENAME, TYPE Y; ELSE N"
154 B$ = INKEY$: IF B$ = "" THEN GOTO 154
IF B$ = "Y" OR B$ = "y" THEN GOTO 160
IF B$ = "N" OR B$ = "n" THEN GOTO 152
GOTO 154
160 OPEN OUT$ FOR OUTPUT AS #2
180 REM NO. OF NODES
190 INPUT #1, NN: NJ = NN
REM NO. OF ELEMENTS
INPUT #1, MS
200 M3 = 3 * MS
210 N2 = 3 * NN
220 REM NO. OF NODES WITH ZERO DISPLACEMENTS
230 INPUT #1, NF
240 DIM IJ(M3), NS(3 * NF), UG(N2), VG(N2)
250 DIM V(3), VEC(NN), SHEAR(NN)
260 DIM D(3, 3), SR(3), MOMXY(3), QG(9)
270 DIM PRINCIPAL(NN), MINIMUM(NN)
DIM B(3, 10), T(10, 9), CB(3, 10), BC(3, 9), BCT(9, 3), DB(3, 9)
DIM SK(9, 9)
DIM ZETA1(3), ZETA2(3), ZETA3(3), XA(3), YA(3), ZA(3), SY21(3), SY34(3)
DIM ZETAP(3, 3), ZETAN(3, 3), UNIT(3, 3), XL(9), COORD(9), UL(9)
DIM X(NN), Y(NN), XG(NN), YG(NN), NP(400), POSS(400)
DIM WZ(NN)
REM DETAILS OF ZERO DISPLACEMENTS
FOR II = 1 TO NF * 3: NS(II) = 0: NEXT II
FOR II = 1 TO NF
REM TYPE IN NODE WITH ZERO DISPLACEMENTS
INPUT #1, POSS(II), W0, THY, THX
REM IF W=0 THEN TYPE1; ELSE TYPE 0. SIMILARLY FOR DW/DX & DW/DY
IF W0 < > 0 THEN NS(II * 3 - 2) = POSS(II) * 3 - 2
IF THY < > 0 THEN NS(II * 3 - 1) = POSS(II) * 3 - 1
IF THX < > 0 THEN NS(II * 3) = POSS(II) * 3
NEXT II
```

```
REM TYPE IN NODAL COORDINATES (GLOBAL)
FOR II = 1 TO NN
REM X(II),Y(II)
INPUT #1, X(II), Y(II)
PRINT X(II), Y(II)
XG(II) = X(II): YG(II) = Y(II)
NEXT II
REM PLATE THICKNESS
INPUT #1, T: TH = T
280 REM ELASTIC MODULUS
290 INPUT #1, E, NU
300 REM POISSON'S RATIO
440 REM ELEMENT TOPOLOGY
450 MX = 0
460 FOR ME = 1 TO MS
470 REM I,J & K NODES
480 INPUT #1, IN, JN, KN
PRINT IN, JN, KN
490 I8 = 3 * ME
500 IJ(I8 - 2) = IN
510 IJ(I8 - 1) = JN
520 IJ(I8) = KN
530 IF ABS(JN - IN) > MX THEN MX = ABS(JN - IN)
540 IF ABS(KN - JN) > MX THEN MX = ABS(KN - JN)
550 IF ABS(KN - IN) > MX THEN MX = ABS(KN - IN)
560 NEXT ME
570 NW = (MX + 1) * 3
580 NT = NW + N2
590 DIM A(NT, NW), Q(NT), CV(NT)
670 REM NO. OF NODES WITH EXTERNAL LOADS
680 INPUT #1, NC
IF NC = 0 THEN GOTO 805
700 REM NODAL POSITIONS & VALUES OF EXTERNAL LOADS
710 FOR II = 1 TO NC
720 REM NODAL POSITION
730 INPUT #1, NP(II), WC
770 REM VERTICAL LOAD IN Z DIRECTION (I.E. +VE UPWARDS)
790 Q(3 * NP(II) - 2) = WC
800 NEXT II
805 REM TYPE IN PRESSURE NORMAL TO THE SURFACE_+VE 'INTERNAL'
INPUT #1, PRESSURE
GOSUB 2550
920 Z9 = 0: GOSUB 10000
930 FOR ME = 1 TO MS
940 PRINT : PRINT "ELEMENT NO."; ME; " UNDER COMPUTATION"
950 I8 = 3 * ME
960 IN = IJ(I8 - 2)
970 JN = IJ(I8 - 1)
KN = IJ(I8)
 I = IN: J = JN: K = KN
 VEC(IN) = VEC(IN) + 1
 VEC(JN) = VEC(JN) + 1
 VEC(KN) = VEC(KN) + 1
```

```
FOR II = 1 TO 9: FOR JJ = 1 TO 9
SK(II, JJ) = 0
NEXT JJ: QG(II) = 0: NEXT II
FOR II = 1 TO 3: FOR JJ = 1 TO 3
D(II, JJ) = 0
NEXT JJ, II
D(1, 1) = E * TH ^ 3 / (12 * (1 - NU * NU)): D(2, 2) = D(1, 1)
D(1, 2) = NU * D(1, 1): D(2, 1) = D(1, 2)
D(3, 3) = D(1, 1) * (1 - NU)
FOR IJ = 1 TO 3
IF IJ = 1 THEN L1 = .5: L2 = .5
IF IJ = 2 THEN L1 = 0: L2 = .5
IF IJ = 3 THEN L1 = .5: L2 = 0
L3 = 1 - L1 - L2
GOSUB 5000
FOR II = 1 TO 9
FOR JJ = 1 TO 9
FOR KK = 1 TO 3
SK(II, JJ) = SK(II, JJ) + BCT(II, KK) * DB(KK, JJ) * DEL / 6
NEXT KK
NEXT JJ, II
NEXT IJ
QG(1) = PRESSURE * DEL / 6
QG(4) = PRESSURE * DEL / 6
QG(7) = PRESSURE * DEL / 6
1190 I1 = 3 * IN - 3
1200 J1 = 3 * JN - 3
1210 K1 = 3 * KN - 3
1220 FOR JJ = 1 TO 3
1230 IF JJ = 1 THEN NR = I1
1240 IF JJ = 2 THEN NR = J1
1250 IF JJ = 3 THEN NR = K1
1260 FOR J9 = 1 TO 3
1270 NR = NR + 1: II = (JJ - 1) * 3 + J9
1280 FOR KK = 1 TO 3
1290 IF KK = 1 THEN N9 = I1
1300 IF KK = 2 THEN N9 = J1
1310 IF KK = 3 THEN N9 = K1
1320 FOR K9 = 1 TO 3
1330 LL = (KK - 1) * 3 + K9
1340 NK = N9 + K9 + 1 - NR
1350 IF NK < = 0 THEN GOTO 1370
1360 A(NR, NK) = A(NR, NK) + SK(II, LL)
1370 NEXT K9
1380 NEXT KK
Q(NR) = Q(NR) + QG(II)
1390 NEXT J9
1400 NEXT JJ
1410 NEXT ME
1420 FOR II = 1 TO NF * 3
1430 N9 = NS(II)
IF N9 = 0 THEN GOTO 1470
FOR JJ = 1 TO NW
```

```
A(N9, JJ) = 0
IF N9 - JJ + 1 < 1 THEN GOTO 1450
A(N9 - JJ + 1, JJ) = 0
1450 NEXT JJ
A(N9, 1) = 1
1460 Q(N9) = 0
1470 NEXT II
1480 FOR II = 1 TO N2
1490 CV(II) = Q(II)
1500 NEXT II
PRINT #2, "SOLVING"
1510 PRINT : PRINT "THE SIMULTANEOUS EQUATIONS ARE NOW BEING SOLVED":
PRINT
1520 GOSUB 2330
1530 PRINT #2, "THE NODAL DISPLACEMENTS U & V ARE:"
1540 FOR I = 1 TO NJ
1550 FOR J = 1 TO 3
1560 II = 3 * I - 3 + J
1570 UG(II) = CV(II)
IF J = 1 THEN PRINT #2, "W("; I; ")="; UG(II),
IF J = 2 THEN PRINT #2, "DW/DX("; I; ")="; UG(II),
IF J = 3 THEN PRINT #2, "DW/DY("; I; ")="; UG(II)
1600 NEXT J
1610 NEXT I
1630 FOR ME = 1 TO MS
1640 I8 = 3 * ME
1650 IN = IJ(I8 - 2)
1660 JN = IJ(I8 - 1)
1670 KN = IJ(I8)
I = IN: J = JN: K = KN
1680 PRINT #2, "ELEMENT NO."; ME; "   "; IN; "-"; JN; "-"; KN
FOR II = 1 TO 9: UL(II) = 0: NEXT II
FOR II = 1 TO 3
IF II = 1 THEN I1 = IN * 3 - 3
IF II = 2 THEN I1 = JN * 3 - 3
IF II = 3 THEN I1 = KN * 3 - 3
FOR JJ = 1 TO 3
J1 = II * 3 - 3 + JJ
UL(J1) = UG(I1 + JJ)
NEXT JJ, II
L1 = .3333: L2 = .3333: L3 = 1 - L1 - L2
FOR II = 1 TO 3: FOR JJ = 1 TO 3
D(II, JJ) = 0
NEXT JJ, II
D(1, 1) = E * TH ^ 3 / (12 * (1 - NU ^ 2))
D(2, 2) = D(1, 1)
D(1, 2) = NU * D(1, 1)
D(2, 1) = D(1, 2)
D(3, 3) = D(1, 1) * (1 - NU)
GOSUB 5000
FOR II = 1 TO 3: MOMXY(II) = 0
FOR JJ = 1 TO 9
MOMXY(II) = MOMXY(II) + DB(II, JJ) * UL(JJ)
```

```
NEXT JJ
IF II = 1 THEN PRINT #2, "MX="; -MOMXY(II)
IF II = 2 THEN PRINT #2, "MY="; -MOMXY(II)
IF II = 3 THEN PRINT #2, "M-XY="; -MOMXY(II)
NEXT II
FOR II = 1 TO 3
SR(II) = 6 * MOMXY(II) / TH ^ 2
NEXT II
SX = SR(1): SY = SR(2): SXY = SR(3)
2130 SHEAR = .5 * SQR((SX - SY) ^ 2 + 4 * SXY ^ 2)
2140 PRINCIPAL = .5 * (SX + SY) + SHEAR
2150 MINIMUM = .5 * (SX + SY) - SHEAR
2160 PRINT #2, "SIGMA1="; PRINCIPAL, "SIGMA2="; MINIMUM
2170 IF SX < SY THEN SHEAR = -SHEAR
2180 SHEAR(IN) = SHEAR(IN) + SHEAR / 3
2190 PRINCIPAL(IN) = PRINCIPAL(IN) + PRINCIPAL / 3
2200 MINIMUM(IN) = MINIMUM(IN) + MINIMUM / 3
2210 SHEAR(JN) = SHEAR(JN) + SHEAR / 3
2220 PRINCIPAL(JN) = PRINCIPAL(JN) + PRINCIPAL / 3
2230 MINIMUM(JN) = MINIMUM(JN) + MINIMUM / 3
2240 SHEAR(KN) = SHEAR(KN) + SHEAR / 3
2250 PRINCIPAL(KN) = PRINCIPAL(KN) + PRINCIPAL / 3
2260 MINIMUM(KN) = MINIMUM(KN) + MINIMUM / 3
2270 NEXT ME
NN = NJ
2280 FOR II = 1 TO NN
2290 VG(II) = SHEAR(II) / VEC(II)
2300 NEXT II
2310 GOSUB 11000
2320 GOTO 2950
2330 FOR II = 1 TO N2
2340 IK = II
2350 FOR JJ = 2 TO NW
2360 IK = IK + 1
2370 KN = A(II, JJ) / A(II, 1)
2380 JK = 0
2390 FOR KK = JJ TO NW
2400 JK = JK + 1
2410 A(IK, JK) = A(IK, JK) - KN * A(II, KK)
2420 NEXT KK
2430 A(II, JJ) = KN
2440 CV(IK) = CV(IK) - KN * CV(II)
2450 NEXT JJ
2460 CV(II) = CV(II) / A(II, 1)
2470 NEXT II
2480 FOR IZ = 2 TO N2
2490 II = N2 - IZ + 1
2500 FOR KK = 2 TO NW
2510 JJ = II + KK - 1
2520 CV(II) = CV(II) - A(II, KK) * CV(JJ)
2530 NEXT KK: NEXT IZ
2540 RETURN
2550 REM OUTPUT OF INPUT
```

```
2560 PRINT #2, : PRINT #2,
2570 PRINT #2, "PLATE BENDING STRESSES USING A 3 NODE TRIANGLE"
2580 PRINT #2,
2590 PRINT #2, "NO. OF ELEMENTS ="; MS
2600 PRINT #2, "NO. OF NODES ="; NN
2610 PRINT #2, "NO. OF NODES WITH ZERO DISPLACEMENTS ="; NF
PRINT #2, "PLATE THICKNESS ="; T
2620 PRINT #2, "; ELASTIC; MODULUS = "; E; " ";
2630 PRINT #2, "POISSON'S RATIO ="; NU
2670 PRINT #2, "ELEMENT TOPOLOGY"
2680 FOR II = 1 TO MS
2690 I8 = II * 3
2700 IN = IJ(I8 - 2)
2710 JN = IJ(I8 - 1)
2720 KN = IJ(I8)
2730 PRINT #2, "I,J & K ARE: "; IN; "-"; JN; "-"; KN
2740 NEXT II
2750 PRINT #2, "NODAL COORDINATES"
2760 FOR II = 1 TO NN
2770 PRINT #2, "X("; II; ")="; X(II); "   Y("; II; ")="; Y(II)
2780 NEXT II
2790 PRINT #2, "EXTERNAL LOADS"
2800 PRINT #2, "NO. OF NODES WITH CONCENTRATED LOADS ="; NC
IF NC = 0 THEN GOTO 2855
2810 FOR II = 1 TO NC
2820 PRINT #2, "NODE ="; NP(II)
PRINT #2, "LOAD IN Z DIRECTION ="; Q(NP(II) * 3 - 2)
2850 NEXT II
2855 PRINT #2, "PRESSURE ="; PRESSURE
2860 PRINT #2, "DETAILS OF ZERO DISPLACEMENTS"
2870 FOR II = 1 TO NF
2880 PRINT #2, "NODE ="; POSS(II)
2890 N9 = NS(II * 3 - 2)
2900 IF N9 < > 0 THEN PRINT #2, "W IS ZERO AT NODE "; POSS(II)
2910 N9 = NS(II * 3 - 1)
2920 IF N9 < > 0 THEN PRINT #2, "DW/DX IS ZERO AT NODE "; POSS(II)
N9 = NS(II * 3)
IF N9 < > 0 THEN PRINT #2, "DW/DY IS ZERO AT NODE "; POSS(II)
2930 NEXT II
2940 RETURN
2950 LF = 10: Z9 = 99
2960 DIM YY(3), XX(3), FF(LF + 1)
2970 GOSUB 3010
2980 FOR II = 1 TO NN: VG(II) = PRINCIPAL(II) / VEC(II): NEXT II: Z9 = 999: GOSUB 3010
2990 FOR II = 1 TO NN: VG(II) = MINIMUM(II) / VEC(II): NEXT II: Z9 = 9999: GOSUB 3010
3000 SCREEN 0, 0: END
3010 MAX = -1E+12: MAXX = -1E+12: MAXY = -1E+12: MAXU = -1E+12: MINU =
1E+12
3020 SCREEN 9
3030 IF Z9 = 99 THEN PRINT "LINES OF CONSTANT MAXIMUM SHEAR STRESS"
3040 IF Z9 = 999 THEN PRINT "LINES OF CONSTANT MAXIMUM PRINCIPAL STRESS"
3050 IF Z9 = 9999 THEN PRINT "LINES OF CONSTANT MINIMUM PRINCIPAL STRESS"
3060 FOR I = 1 TO NN
```

```
3070 IF VG(I) > MAXU THEN MAXU = VG(I)
3080 IF VG(I) < MINU THEN MINU = VG(I)
3090 IF ABS(X(I)) > MAXX THEN MAXX = ABS(X(I))
3100 IF ABS(Y(I)) > MAXY THEN MAXY = ABS(Y(I))
3110 IF MAXX > MAX THEN MAX = MAXX
3120 IF MAXY > MAX THEN MAX = MAXY
3130 NEXT I
3140 LF = 10
3150 DIF = MAXU - MINU: STP = DIF / LF: AG = 1
3160 FOR I = MINU TO MAXU STEP STP
3170 FF(AG) = I: IF ABS(FF(AG)) < .0000001 THEN FF(AG) = 0
3180 AG = AG + 1: NEXT I
3190 FOR N = 1 TO LF
3200 FOR ME = 1 TO MS
3210 I8 = 3 * ME: AA = 0: BB = 0: CC = 0
3220 I = IJ(I8 - 2): J = IJ(I8 - 1): K = IJ(I8)
3230 IF FF(N) = (INT(VG(I) * 1000)) / 1000 THEN GOTO 3270
3240 IF FF(N) = (INT(VG(J) * 1000)) / 1000 THEN GOTO 3270
3250 IF FF(N) = (INT(VG(K) * 1000)) / 1000 THEN GOTO 3270
3260 GOTO 3280
3270 FF(N) = FF(N) + .0000001
3280 IF FF(N) > VG(I) AND FF(N) < VG(J) THEN GOSUB 3430
3290 IF FF(N) < VG(I) AND FF(N) > VG(J) THEN GOSUB 3430
3300 IF FF(N) > VG(I) AND FF(N) < VG(K) THEN GOSUB 3490
3310 IF FF(N) < VG(I) AND FF(N) > VG(K) THEN GOSUB 3490
3320 IF FF(N) > VG(J) AND FF(N) < VG(K) THEN GOSUB 3550
3330 IF FF(N) < VG(J) AND FF(N) > VG(K) THEN GOSUB 3550
3340 IF AA = 1 AND BB = 1 THEN LINE (40 + XS * (XX(1)), 280 - XS * (YY(1)))-(40 + XS
* (XX(2)), 280 - XS * (YY(2)))
3350 IF AA = 1 AND CC = 1 THEN LINE (40 + XS * (XX(1)), 280 - XS * (YY(1)))-(40 + XS
* (XX(3)), 280 - XS * (YY(3)))
3360 IF BB = 1 AND CC = 1 THEN LINE (40 + XS * (XX(2)), 280 - XS * (YY(2)))-(40 + XS
* (XX(3)), 280 - XS * (YY(3)))
3370 NEXT ME
3380 NEXT N
3390 A$ = INKEY$: IF A$ = "" THEN GOTO 3390
3400 IF A$ = "Y" OR A$ = "y" THEN SCREEN 0: WIDTH 80: RETURN
3410 GOTO 3390
3420 END
3430 DF = FF(N) - VG(I): DE = VG(J) - VG(I)
3440 DX = X(J) - X(I): DY = Y(J) - Y(I)
3450 IF DF = 0 AND DE = 0 THEN GOTO 3480
3460 XX(1) = (DF * DX / DE) + X(I): YY(1) = (DF * DY / DE) + Y(I)
3470 AA = 1: RETURN
3480 XX(1) = X(I): YY(1) = Y(I): AA = 1: RETURN
3490 DF = FF(N) - VG(I): DE = VG(K) - VG(I)
3500 DX = X(K) - X(I): DY = Y(K) - Y(I)
3510 IF DF = 0 AND DE = 0 THEN GOTO 3540
3520 XX(2) = (DF * DX / DE) + X(I): YY(2) = (DF * DY / DE) + Y(I)
3530 BB = 1: RETURN
3540 XX(2) = X(I): YY(2) = Y(I): BB = 1: RETURN
3550 DF = FF(N) - VG(J): DE = VG(K) - VG(J)
3560 DX = X(K) - X(J): DY = Y(K) - Y(J)
```

```
3570 IF DF = 0 AND DE = 0 THEN GOTO 3600
3580 XX(3) = (DF * DX / DE) + X(J): YY(3) = (DF * DY / DE) + Y(J)
3590 CC = 1: RETURN
3600 XX(3) = X(J): YY(3) = Y(J): CC = 1: RETURN
5000 XJI = X(J) - X(I)
5010 YJI = Y(J) - Y(I)
5020 XKJ = X(K) - X(J)
5030 YKJ = Y(K) - Y(J)
5040 XIK = X(I) - X(K)
5050 YIK = Y(I) - Y(K)
5070 X1 = SQR(XJI * XJI + YJI * YJI)
5080 X2 = SQR(XKJ * XKJ + YKJ * YKJ)
5090 X3 = SQR(XIK * XIK + YIK * YIK)
5100 DEL = XJI * YKJ - YJI * XKJ
5110 XJI = XJI / X1
5120 YJI = YJI / X1
5130 XKJ = XKJ / X2
5140 YKJ = YKJ / X2
5150 XIK = XIK / X3
5160 YIK = YIK / X3
5170 AX = XJI * XJI
5180 AY = YJI * YJI
5190 AXY = 1.414 * XJI * YJI
5200 BX = XKJ * XKJ
5210 BY = YKJ * YKJ
5220 BXY = 1.414 * XKJ * YKJ
5230 CX = XIK * XIK
5240 CY = YIK * YIK
5250 CXY = 1.414 * XIK * YIK
5260 DET = AX * (BY * CXY - BXY * CY) - AY * (BX * CXY - BXY * CX) + AXY * (BX *
CY - BY * CX)
5270 C(1, 1) = BY * CXY - BXY * CY
5280 C(1, 2) = -AY * CXY + AXY * CY
5290 C(1, 3) = AY * BXY - AXY * BY
5300 C(2, 1) = -BX * CXY + BXY * CX
5310 C(2, 2) = AX * CXY - AXY * CX
5320 C(2, 3) = -AX * BXY + AXY * BX
5330 C(3, 1) = BX * CY - BY * CX
5340 C(3, 2) = -AX * CY + AY * CX
5350 C(3, 3) = AX * BY - AY * BX
5360 FOR II = 1 TO 3: FOR JJ = 1 TO 3: C(II, JJ) = C(II, JJ) / DET: NEXT JJ
5370 FOR JJ = 1 TO 10
5380 B(II, JJ) = 0
5390 NEXT JJ
5400 NEXT II
5410 B(1, 1) = 27 * L1 - 9
5420 B(1, 2) = 27 * L2 - 9
5430 B(1, 4) = 27 * L2 - 54 * L1 + 9
5440 B(1, 5) = 27 * L1 - 54 * L2 + 9
5450 B(1, 6) = 27 * L3
5460 B(1, 9) = 27 * L3
5470 B(1, 10) = -54 * L3
5480 B(2, 2) = 27 * L2 - 9
```

```
5490 B(2, 3) = 27 * L3 - 9
5500 B(2, 5) = 27 * L1
5510 B(2, 10) = -54 * L1
5520 B(2, 6) = 27 * L3 - 54 * L2 + 9
5530 B(2, 7) = 27 * L2 - 54 * L3 + 9
5540 B(2, 8) = 27 * L1
5550 B(3, 1) = 27 * L1 - 9
5560 B(3, 3) = 27 * L3 - 9
5570 B(3, 4) = 27 * L2
5580 B(3, 7) = 27 * L2
5590 B(3, 8) = 27 * L1 - 54 * L3 + 9
5600 B(3, 9) = 27 * L3 - 54 * L1 + 9
5610 B(3, 10) = -54 * L2
5620 FOR JJ = 1 TO 10
5630 B(1, JJ) = -B(1, JJ) / (X1 * X1)
5640 B(2, JJ) = -B(2, JJ) / (X2 * X2)
5650 B(3, JJ) = -B(3, JJ) / (X3 * X3)
5660 NEXT JJ
5670 A = 2 * XJI * X1
5680 B = 2 * YJI * X1
5690 C = 2 * XKJ * X2
5700 DD = 2 * YKJ * X2
5710 EE = 2 * XIK * X3
5720 F = 2 * YIK * X3
5750 FOR II = 1 TO 10
5760 FOR JJ = 1 TO 9
5770 T(II, JJ) = 0
5780 NEXT JJ
5790 NEXT II
5800 T(1, 1) = 1
5810 T(2, 4) = 1
5820 T(3, 7) = 1
5830 T(4, 1) = 20 / 27
5840 T(4, 2) = 2 * A / 27
5850 T(4, 3) = 2 * B / 27
5860 T(4, 4) = 7 / 27
5870 T(4, 5) = -A / 27
5880 T(4, 6) = -B / 27
5890 T(5, 1) = 7 / 27
5900 T(5, 2) = A / 27
5910 T(5, 3) = B / 27
5920 T(5, 4) = 20 / 27
5930 T(5, 5) = -2 * A / 27
5940 T(5, 6) = -2 * B / 27
5950 T(6, 4) = 20 / 27
5960 T(6, 5) = 2 * C / 27
5970 T(6, 6) = 2 * DD / 27
5980 T(6, 7) = 7 / 27
5990 T(6, 8) = -C / 27
6000 T(6, 9) = -DD / 27
6010 T(7, 4) = 7 / 27
6020 T(7, 5) = C / 27
6030 T(7, 6) = DD / 27
```

```
6040 T(7, 7) = 20 / 27
6050 T(7, 8) = -2 * C / 27
6060 T(7, 9) = -2 * DD / 27
6070 T(8, 1) = 7 / 27
6080 T(8, 2) = -EE / 27
6090 T(8, 3) = -F / 27
6100 T(8, 7) = 20 / 27
6110 T(8, 8) = 2 * EE / 27
6120 T(8, 9) = 2 * F / 27
6130 T(9, 1) = 20 / 27
6140 T(9, 2) = -2 * EE / 27
6150 T(9, 3) = -2 * F / 27
6160 T(9, 7) = 7 / 27
6170 T(9, 8) = EE / 27
6180 T(9, 9) = F / 27
6190 FOR II = 1 TO 9
6200 FOR JJ = 4 TO 9
6210 T(10, II) = T(10, II) + T(JJ, II) / 4
6220 NEXT JJ, II
6230 FOR II = 1 TO 9
6240 FOR JJ = 1 TO 3
6250 T(10, II) = T(10, II) - T(JJ, II) / 6
6260 NEXT JJ, II
6270 FOR II = 1 TO 3
6280 FOR JJ = 1 TO 10
6290 CB(II, JJ) = 0
6300 FOR KK = 1 TO 3
6310 CB(II, JJ) = CB(II, JJ) + C(II, KK) * B(KK, JJ)
6320 NEXT KK, JJ, II
6330 FOR II = 1 TO 3
6340 FOR JJ = 1 TO 9
6350 BC(II, JJ) = 0
6360 FOR KK = 1 TO 10
6370 BC(II, JJ) = BC(II, JJ) + CB(II, KK) * T(KK, JJ)
6380 NEXT KK: BCT(JJ, II) = BC(II, JJ): NEXT JJ, II
6390 FOR II = 1 TO 3
6400 FOR JJ = 1 TO 9
6410 DB(II, JJ) = 0
6420 FOR KK = 1 TO 3
6430 DB(II, JJ) = DB(II, JJ) + D(II, KK) * BC(KK, JJ)
6440 NEXT KK, JJ, II
6500 FS(1) = (9 * L1 ^ 3 - 9 * L1 ^ 2 + 2 * L1) / 2
6510 FS(2) = (9 * L2 ^ 3 - 9 * L2 ^ 2 + 2 * L2) / 2
6520 FS(3) = (9 * L3 ^ 3 - 9 * L3 ^ 2 + 2 * L3) / 2
6530 FS(4) = 13.5 * L1 ^ 2 * L2 - 4.5 * L1 * L2
6540 FS(5) = 13.5 * L1 * L2 ^ 2 - 4.5 * L1 * L2
6550 FS(6) = 13.5 * L2 ^ 2 * L3 - 4.5 * L2 * L3
6560 FS(7) = 13.5 * L2 * L3 ^ 2 - 4.5 * L2 * L3
6570 FS(8) = 13.5 * L3 ^ 2 * L1 - 4.5 * L3 * L1
6580 FS(9) = 13.5 * L3 * L1 ^ 2 - 4.5 * L3 * L1
6590 FS(10) = 27 * L1 * L2 * L3
6600 FOR II = 1 TO 9
6610 N(II) = 0
```

```
6620 FOR JJ = 1 TO 10
6630 N(II) = N(II) + FS(JJ) * T(JJ, II)
6640 NEXT JJ, II
6690 RETURN
10000 IF Z9 = 77 OR Z9 = 99 THEN GOTO 10295
10010 XZ = 1E+12: YZ = 1E+12
10030 XM = -1E+12: YM = -1E+12
10050 PRINT : PRINT "TO CONTINUE, TYPE Y"
10060 A$ = INKEY$: IF A$ = "" THEN GOTO 10060
10070 IF A$ = "Y" OR A$ = "y" THEN GOTO 10090
10080 GOTO 10060
10090 FOR ME = 1 TO MS
10100 I8 = 3 * ME
10110 I = IJ(I8 - 2)
10120 J = IJ(I8 - 1)
10130 K = IJ(I8)
10140 IF X(I) < XZ THEN XZ = X(I)
10150 IF X(J) < XZ THEN XZ = X(J)
10160 IF X(K) < XZ THEN XZ = X(K)
10170 IF Y(I) < YZ THEN YZ = Y(I)
10180 IF Y(J) < YZ THEN YZ = Y(J)
10190 IF Y(K) < YZ THEN YZ = Y(K)
10200 IF X(I) > XM THEN XM = X(I)
10210 IF X(J) > XM THEN XM = X(J)
10220 IF X(K) > XM THEN XM = X(K)
10230 IF Y(I) > YM THEN YM = Y(I)
10240 IF Y(J) > YM THEN YM = Y(J)
10250 IF Y(K) > YM THEN YM = Y(K)
10260 NEXT ME
10270 XS = 350 / (XM - XZ)
10280 YS = 175 / (YM - YZ)
10290 IF XS > YS THEN XS = YS
10295 SCREEN 9
10300 FOR ME = 1 TO MS
10310 I8 = 3 * ME
10320 II = IJ(I8 - 2)
10330 JJ = IJ(I8 - 1)
10340 KK = IJ(I8)
10350 FOR IV = 1 TO 3
10360 IF IV = 1 THEN I = II: J = JJ
10370 IF IV = 2 THEN I = JJ: J = KK
10380 IF IV = 3 THEN I = KK: J = II
10390 IF Z9 = 99 THEN GOTO 10450
10400 XI = (X(I) - XZ) * XS
10410 XJ = (X(J) - XZ) * XS
10420 YI = (Y(I) - YZ) * XS
10430 YJ = (Y(J) - YZ) * XS
10440 GOTO 10480
10450 XI = X(I) - XZ * XS: XJ = X(J) - XZ * XS
10460 YI = Y(I) - YZ * XS: YJ = Y(J) - YZ * XS
10470 IF Z9 = 77 THEN GOTO 10490
10480 LINE (XI + 40, 280 - YI)-(XJ + 40, 280 - YJ)
10490 NEXT IV
```

```
10500 NEXT ME
10510 IF Z9 = 77 THEN GOTO 10600
10515 PRINT "TO CONTINUE, TYPE Y"
10520 A$ = INKEY$: IF A$ = "" THEN GOTO 10520
10530 IF A$ = "Y" OR A$ = "y" THEN GOTO 10560
10540 GOTO 10520
10560 SCREEN 0, 0: WIDTH 80
10600 RETURN
11000 REM DEFLECTED FORM OF MESH
11020 Z9 = 77
11030 REM GOSUB 10000
11040 UM = -1E + 12
11050 FOR II = 1 TO N2
11060 IF ABS(UG(II)) > UM THEN UM = ABS(UG(II))
11070 NEXT II
11080 SC = 20
11090 FOR II = 1 TO NN
11100 X(II) = X(II) * XS: REM + SC * UG(II * 6 - 1) / UM
11110 Y(II) = Y(II) * XS: REM + SC * UG(II * 6 - 2) / UM
11120 NEXT II
11130 Z9 = 99
11140 GOSUB 10000
11150 FOR II = 1 TO NN
11160 X(II) = XG(II)
11170 Y(II) = YG(II)
11180 NEXT II
11190 RETURN
25000 END
```

Appendix 16

Computer Program for Stresses in Doubly-Curved Shells "SHELLSTF"

```
100 REM PROGRAM BY DR.C.T.F.ROSS
110 REM 6 HURSTVILLE DRIVE,
120 REM WATERLOOVILLE, PORTSMOUTH.
130 REM PO7 7NB ENGLAND
140 CLS
150 PRINT : PRINT "STRESSES IN DOUBLY-CURVED SHELLS": PRINT
160 PRINT "PROGRAM BY DR.C.T.F.ROSS"
162 INPUT "TYPE IN THE NAME OF YOUR INPUT FILE "; FILENAME$
164 PRINT "THE NAME OF YOUR INPUT FILE IS "; FILENAME$
166 PRINT "IF YOU ARE SATISFIED WITH THIS NAME, TYPE Y;ELSE N"
167 A$ = INKEY$: IF A$ = "" THEN GOTO 167
169 IF A$ = "Y" OR A$ = "y" THEN GOTO 178
170 IF A$ = "N" OR A$ = "n" THEN GOTO 162
175 END
178 OPEN FILENAME$ FOR INPUT AS #1
180 INPUT "TYPE IN THE NAME OF YOUR OUTPUT FILE "; OUT$
PRINT "IF YOU ARE SATISFIED WITH THIS FILENAME, TYPE Y; ELSE N"
182 B$ = INKEY$: IF B$ = "" THEN GOTO 182
IF B$ = "Y" OR B$ = "y" THEN GOTO 186
IF B$ = "N" OR B$ = "n" THEN GOTO 180
GOTO 182
186 OPEN OUT$ FOR OUTPUT AS #2
188 REM NO. OF NODES
190 INPUT #1, NN: NJ = NN
REM NO. OF ELEMENTS
INPUT #1, MS
NDF = 6
200 M3 = 3 * MS
210 N2 = 6 * NN
220 REM NO. OF NODES WITH ZERO DISPLACEMENTS
230 INPUT #1, NF
240 DIM SKP(6, 6), SKB(9, 9), STIFF(9, 9), IJ(M3), NS(6 * NF), UG(N2), VG(N2)
250 DIM B1(3), C1(3), U(3), V(3), VEC(NN), SHEAR(NN)
260 DIM BM(3, 6), BD(3, 6), D(3, 3), SR(3), MOMXY(3), QL(18), QG(18)
270 DIM PRINCIPAL(NN), MINIMUM(NN), DIAG(N2)
DIM B(3, 10), T(10, 9), CB(3, 10), BC(3, 9), BCT(9, 3), DB(3, 9), FS(10), n(9)
DIM SK(18, 18), DC(18, 18), TEMP(18, 18)
DIM ZETA1(3), ZETA2(3), ZETA3(3), XA(3), YA(3), ZA(3), SY21(3), SY34(3)
DIM ZETAP(3, 3), ZETAN(3, 3), UNIT(3, 3), XL(9), COORD(9), VL(9), UL(18)
DIM X(NN), Y(NN), Z(NN), XG(NN), YG(NN), ZG(NN), NP(400), POSS(400)
DIM WZ(NN)
REM DETAILS OF ZERO DISPLACEMENTS
FOR II = 1 TO NF * 6: NS(II) = 0: NEXT II
FOR II = 1 TO NF
REM TYPE IN NODE WITH ZERO DISPLACEMENTS
REM IF U0=0 THEN TYPE1; ELSE TYPE 0. SIMILARLY FOR V0,W0,
REM THETAX0,THETAY0 & THETAZ0
INPUT #1, POSS(II), U0, V0, W0, THX, THY, THZ
IF U0 < > 0 THEN NS(II * 6 - 5) = POSS(II) * 6 - 5
```

502

```
IF V0 < > 0 THEN NS(II * 6 - 4) = POSS(II) * 6 - 4
IF W0 < > 0 THEN NS(II * 6 - 3) = POSS(II) * 6 - 3
IF THX < > 0 THEN NS(II * 6 - 2) = POSS(II) * 6 - 2
IF THY < > 0 THEN NS(II * 6 - 1) = POSS(II) * 6 - 1
IF THZ < > 0 THEN NS(II * 6) = POSS(II) * 6
NEXT II
REM TYPE IN NODAL COORDINATES (GLOBAL)
FOR II = 1 TO NN
REM X(II),Y(II),Z(II)
INPUT #1, X(II), Y(II), Z(II)
PRINT X(II), Y(II), Z(II)
XG(II) = X(II): YG(II) = Y(II): ZG(II) = Z(II)
NEXT II
REM PLATE THICKNESS
INPUT #1, T: TH = T
280 REM ELASTIC MODULUS & POISSON'S RATIO
290 INPUT #1, E, NU
370 CN = E * T / (1 - NU ^ 2)
380 ZU = NU
390 MU = (1 - NU) / 2
440 REM ELEMENT TOPOLOGY
450 MX = 0
460 FOR ME = 1 TO MS
470 REM I,J & K NODES
480 INPUT #1, IN, JN, KN
PRINT IN, JN, KN
490 I8 = 3 * ME
500 IJ(I8 - 2) = IN
510 IJ(I8 - 1) = JN
520 IJ(I8) = KN
FOR II = 1 TO 3
IA = KN
IF II = 1 THEN IA = IN
IF II = 2 THEN IA = JN
FOR IJ = 1 TO NDF
II1 = NDF * (IA - 1) + IJ
FOR JJ = 1 TO 3
IB = KN
IF JJ = 1 THEN IB = IN
IF JJ = 2 THEN IB = JN
FOR JI = 1 TO NDF
JJ1 = NDF * (IB - 1) + JI
IIJJ = JJ1 - II1 + 1
IF IIJJ > DIAG(JJ1) THEN DIAG(JJ1) = IIJJ
NEXT JI, JJ, IJ, II
560 NEXT ME
DIAG(1) = 1: FOR II = 1 TO N2: DIAG(II) = DIAG(II - 1) + DIAG(II): NEXT II
580 NT = DIAG(N2)
590 DIM A(NT), Q(N2), CV(N2)
670 REM NO. OF NODES WITH EXTERNAL LOADS
680 INPUT #1, NC
IF NC = 0 THEN GOTO 805
700 REM NODAL POSITIONS & VALUES OF EXTERNAL LOADS
```

```
710 FOR II = 1 TO NC
720 REM NODAL POSITION
730 INPUT #1, NP(II), XWC, YWC.ZWC
740 REM HORIZONTAL LOAD IN X0 DIRECTION IS XWC
760 Q(6 * NP(II) - 5) = XWC
REM HORIZONTAL LOAD IN Y0 DIRECTION IS YWC
Q(6 * NP(II) - 4) = YWC
770 REM VERTICAL LOAD IN Z0 DIRECTION (I.E. +VE UPWARDS) IS ZWC
790 Q(6 * NP(II) - 3) = ZWC
800 NEXT II
805 REM TYPE IN PRESSURE NORMAL TO THE SURFACE_+VE 'INTERNAL'
INPUT #1, PRESSURE
GOSUB 2550
920 Z9 = 0: GOSUB 10000
930 FOR ME = 1 TO MS
940 PRINT : PRINT "ELEMENT NO."; ME; " UNDER COMPUTATION"
950 I8 = 3 * ME
960 IN = IJ(I8 - 2)
970 JN = IJ(I8 - 1)
KN = IJ(I8)
I = IN: J = JN: K = KN
VEC(IN) = VEC(IN) + 1
VEC(JN) = VEC(JN) + 1
VEC(KN) = VEC(KN) + 1
COORD(1) = X(I): COORD(4) = X(J): COORD(7) = X(K)
COORD(2) = Y(I): COORD(5) = Y(J): COORD(8) = Y(K)
COORD(3) = Z(I): COORD(6) = Z(J): COORD(9) = Z(K)
FOR II = 1 TO 9: XL(II) = 0: NEXT II
FOR II = 1 TO 18: FOR JJ = 1 TO 18
SK(II, JJ) = 0: TEMP(II, JJ) = 0: DC(II, JJ) = 0
NEXT JJ: QL(II) = 0: QG(II) = 0: NEXT II
FOR II = 1 TO 6: FOR JJ = 1 TO 6
SKP(II, JJ) = 0
NEXT JJ, II
FOR II = 1 TO 9: FOR JJ = 1 TO 9
STIFF(II, JJ) = 0:  SKB(II, JJ) = 0
NEXT JJ, II
FOR II = 1 TO 3: FOR JJ = 1 TO 3
D(II, JJ) = 0
NEXT JJ, II
D(1, 1) = E * TH ^ 3 / (12 * (1 - NU * NU)): D(2, 2) = D(1, 1)
D(1, 2) = NU * D(1, 1): D(2, 1) = D(1, 2)
D(3, 3) = D(1, 1) * (1 - NU)
GOSUB 7000
FOR II = 1 TO 9: FOR JJ = 1 TO 9
XL(II) = XL(II) + DC(II, JJ) * COORD(JJ)
NEXT JJ, II
FOR IJ = 1 TO 3
IF IJ = 1 THEN L1 = .5: L2 = .5
IF IJ = 2 THEN L1 = 0: L2 = .5
IF IJ = 3 THEN L1 = .5: L2 = 0
L3 = 1 - L1 - L2
GOSUB 5000
```

504

```
FOR II = 1 TO 9
FOR JJ = 1 TO 9
FOR KK = 1 TO 3
SKB(II, JJ) = SKB(II, JJ) + BCT(II, KK) * DB(KK, JJ) * DEL / 6
NEXT KK
NEXT JJ, II
NEXT IJ
FOR II = 1 TO 9: FOR JJ = 1 TO 9
STIFF(II, JJ) = SKB(II, JJ)
NEXT JJ, II
FOR II = 1 TO 3: FOR JJ = 1 TO 9
STIFF(3 * II - 1, JJ) = SKB(3 * II, JJ)
STIFF(3 * II, JJ) = -SKB(3 * II - 1, JJ)
NEXT JJ, II
FOR II = 1 TO 9: FOR JJ = 1 TO 9
SKB(II, JJ) = STIFF(II, JJ): NEXT JJ, II
FOR II = 1 TO 3: FOR JJ = 1 TO 9
STIFF(JJ, 3 * II - 1) = SKB(JJ, 3 * II)
STIFF(JJ, 3 * II) = -SKB(JJ, 3 * II - 1)
NEXT JJ, II
CN = E * TH / (1 - NU ^ 2): ZU = NU: MU = (1 - NU) / 2
B1(1) = XL(5) - XL(8)
B1(2) = XL(8) - XL(2)
B1(3) = XL(2) - XL(5)
C1(1) = XL(7) - XL(4)
C1(2) = XL(1) - XL(7)
C1(3) = XL(4) - XL(1)
DL = XL(1) * B1(1) + XL(4) * B1(2) + XL(7) * B1(3)
FOR II = 1 TO 3
I2 = 2 * II - 2
FOR JJ = 1 TO 3
J2 = 2 * JJ - 2
SKP(I2 + 1, J2 + 1) = .5 * CN * (B1(II) * B1(JJ) + MU * C1(II) * C1(JJ)) / DL
SKP(I2 + 2, J2 + 2) = .5 * CN * (C1(II) * C1(JJ) + MU * B1(II) * B1(JJ)) / DL
SKP(I2 + 1, J2 + 2) = .5 * CN * (ZU * B1(II) * C1(JJ) + MU * C1(II) * B1(JJ)) / DL
SKP(I2 + 2, J2 + 1) = .5 * CN * (ZU * B1(JJ) * C1(II) + MU * C1(JJ) * B1(II)) / DL
NEXT JJ, II
FOR KK = 1 TO 3: FOR LL = 1 TO 3
SK(KK + 2, LL + 2) = STIFF(KK, LL)
SK(KK + 2, 8 + LL) = STIFF(KK, LL + 3)
SK(KK + 2, 14 + LL) = STIFF(KK, LL + 6)
SK(8 + KK, 8 + LL) = STIFF(KK + 3, LL + 3)
SK(8 + KK, 14 + LL) = STIFF(KK + 3, LL + 6)
SK(14 + KK, 14 + LL) = STIFF(KK + 6, LL + 6)
NEXT LL, KK
FOR KK = 1 TO 2: FOR LL = 1 TO 2
SK(KK, LL) = SKP(KK, LL)
SK(KK, 6 + LL) = SKP(KK, LL + 2)
SK(KK, 12 + LL) = SKP(KK, LL + 4)
SK(KK + 6, LL + 6) = SKP(KK + 2, LL + 2): SK(KK + 6, 12 + LL) = SKP(KK + 2, LL + 4)
SK(KK + 12, 12 + LL) = SKP(KK + 4, LL + 4)
NEXT LL, KK
```

```
2957 GOSUB 8100
 FOR II = 1 TO 18
 FOR JJ = 1 TO 18
 TEMP(II, JJ) = 0: NEXT JJ
FOR JJ = 1 TO 18
 FOR KK = 1 TO 18
 TEMP(II, JJ) = TEMP(II, JJ) + SK(II, KK) * DC(KK, JJ)
 NEXT KK, JJ, II
 FOR II = 1 TO 18
 FOR JJ = 1 TO 18
 SK(II, JJ) = 0: NEXT JJ
FOR JJ = 1 TO 18
 FOR KK = 1 TO 18
 SK(II, JJ) = SK(II, JJ) + DC(KK, II) * TEMP(KK, JJ)
 NEXT KK, JJ, II
 IF SK(6, 6) > 20 THEN GOTO 1100
 SK(6, 6) = .005 * E * DL * TH: SK(6, 12) = -.5 * SK(6, 6): SK(12, 6) = SK(6, 12)
 SK(12, 12) = SK(6, 6): SK(12, 18) = SK(6, 12): SK(18, 12) = SK(12, 18)
 SK(18, 18) = SK(6, 6): SK(6, 18) = SK(6, 12): SK(18, 6) = SK(6, 18)
1100 REM CONTINUE
 QL(3) = PRESSURE * DL / 6
 QL(9) = PRESSURE * DL / 6
 QL(15) = PRESSURE * DL / 6
 FOR II = 1 TO 18
 QG(II) = 0
 FOR JJ = 1 TO 18
 QG(II) = QG(II) + DC(JJ, II) * QL(JJ)
 NEXT JJ, II
1190 I1 = 6 * IN - 6
1200 J1 = 6 * JN - 6
1210 K1 = 6 * KN - 6
 FOR III = 1 TO 3
 IIC = K1
 IF III = 1 THEN IIC = I1
 IF III = 2 THEN IIC = J1
 FOR IJ = 1 TO NDF
 IC = IIC + IJ
 IDIAG = DIAG(IC)
 II = 6 * (III - 1) + IJ
 FOR JJJ = 1 TO 3
 JJR = K1
 IF JJJ = 1 THEN JJR = I1
 IF JJJ = 2 THEN JJR = J1
 FOR JI = 1 TO NDF
 JJ = 6 * (JJJ - 1) + JI
 IR = JJR + JI
 IF IR > IC THEN GOTO 1390
 LOCAT = IDIAG - (IC - IR)
 A(LOCAT) = A(LOCAT) + SK(II, JJ)
 1390 NEXT JI: NEXT JJJ: NEXT IJ: NEXT III
 FOR JJ = 1 TO 3
 IF JJ = 1 THEN NR = I1
 IF JJ = 2 THEN NR = J1
```

```
IF JJ = 3 THEN NR = K1
FOR J9 = 1 TO NDF
NR = NR + 1: II = (JJ - 1) * NDF + J9
Q(NR) = Q(NR) + QG(II)
NEXT J9, JJ
1410 NEXT ME
1420 FOR II = 1 TO NF * 6
1430 N9 = NS(II)
IF N9 = 0 THEN GOTO 1470
DI = DIAG(NS(II))
A(DI) = A(DI) * 1E+12
1460 Q(N9) = 0
1470 NEXT II
1480 FOR II = 1 TO N2
1490 CV(II) = Q(II)
1500 NEXT II
PRINT #2, "SOLVING"
1510 PRINT : PRINT "THE SIMULTANEOUS EQUATIONS ARE NOW BEING SOLVED":
PRINT
1520 GOSUB 2330
1530 PRINT #2, "THE NODAL DISPLACEMENTS U & V ARE:"
1540 FOR I = 1 TO NJ
1550 FOR J = 1 TO 6
1560 II = 6 * I - 6 + J
1570 UG(II) = CV(II)
1580 IF J = 1 THEN PRINT #2, "U("; I; ")="; UG(II),
1590 IF J = 2 THEN PRINT #2, "V("; I; ")="; UG(II),
IF J = 3 THEN PRINT #2, "W("; I; ")="; UG(II)
IF J = 4 THEN PRINT #2, "THETAX("; I; ")="; UG(II),
IF J = 5 THEN PRINT #2, "THETAY("; I; ")="; UG(II),
IF J = 6 THEN PRINT #2, "THETAZ("; I; ")="; UG(II)
1600 NEXT J
1610 NEXT I
1620 CN = E / (1 - NU ^ 2)
1630 FOR ME = 1 TO MS
1640 I8 = 3 * ME
1650 IN = IJ(I8 - 2)
1660 JN = IJ(I8 - 1)
1670 KN = IJ(I8)
I = IN: J = JN: K = KN
1680 PRINT #2, "ELEMENT NO."; ME; "    "; IN; "-"; JN; "-"; KN
FOR II = 1 TO 9: XL(II) = 0: NEXT II
FOR II = 1 TO 18: UL(II) = 0: NEXT II
GOSUB 7000
FOR II = 1 TO 3
IF II = 1 THEN I1 = IN * 6 - 6
IF II = 2 THEN I1 = JN * 6 - 6
IF II = 3 THEN I1 = KN * 6 - 6
FOR JJ = 1 TO 6
J1 = II * 6 - 6 + JJ
VG(J1) = UG(I1 + JJ)
NEXT JJ, II
FOR II = 1 TO 18
```

```
UL(II) = 0
FOR JJ = 1 TO 18
UL(II) = UL(II) + DC(II, JJ) * VG(JJ)
NEXT JJ, II
1690 U(1) = UL(1)
1700 U(2) = UL(7)
1710 U(3) = UL(13)
1720 V(1) = UL(2)
1730 V(2) = UL(8)
1740 V(3) = UL(14)
COORD(1) = X(I): COORD(4) = X(J): COORD(7) = X(K)
COORD(2) = Y(I): COORD(5) = Y(J): COORD(8) = Y(K)
COORD(3) = Z(I): COORD(6) = Z(J): COORD(9) = Z(K)
FOR II = 1 TO 9
XL(II) = 0
FOR JJ = 1 TO 9
XL(II) = XL(II) + DC(II, JJ) * COORD(JJ)
NEXT JJ, II
1750 B1(1) = XL(5) - XL(8)
1760 B1(2) = XL(8) - XL(2)
1770 B1(3) = XL(2) - XL(5)
1780 C1(1) = XL(7) - XL(4)
1790 C1(2) = XL(1) - XL(7)
1800 C1(3) = XL(4) - XL(1)
1810 DL = XL(1) * B1(1) + XL(4) * B1(2) + XL(7) * B1(3)
FOR II = 1 TO 3: FOR JJ = 1 TO 3
D(II, JJ) = 0
NEXT JJ, II
1830 D(1, 1) = CN
1840 D(2, 2) = CN
1850 D(1, 2) = ZU * CN
1860 D(2, 1) = ZU * CN
1870 D(3, 3) = MU * CN
1880 FOR II = 1 TO 3
1890 BM(1, II) = B1(II) / DL
1900 BM(2, II + 3) = C1(II) / DL
1910 BM(3, II) = C1(II) / DL
1920 BM(3, II + 3) = B1(II) / DL
1930 NEXT II
1940 FOR II = 1 TO 3
1950 FOR JJ = 1 TO 6
1960 BD(II, JJ) = 0
1970 FOR KK = 1 TO 3
1980 BD(II, JJ) = BD(II, JJ) + D(II, KK) * BM(KK, JJ)
1990 NEXT KK
2000 NEXT JJ: NEXT II
2010 FOR II = 1 TO 3
2020 SR(II) = 0
2030 FOR JJ = 1 TO 6
2040 IF JJ > 3 THEN GOTO 2070
2050 SR(II) = SR(II) + BD(II, JJ) * U(JJ)
2060 GOTO 2080
2070 SR(II) = SR(II) + BD(II, JJ) * V(JJ - 3)
```

```
2080 NEXT JJ
2090 IF II = 1 THEN PRINT #2, "SIGMA X(MEMBRANE)="; SR(II), : SX = SR(II)
2100 IF II = 2 THEN PRINT #2, "SIGMA Y(MEMBRANE)="; SR(II), : SY = SR(II)
2110 IF II = 3 THEN PRINT #2, "TAU X-Y(MEMBRANE)="; SR(II), : SXY = SR(II)
2120 NEXT II
L1 = .3333: L2 = .3333: L3 = 1 - L1 - L2
FOR II = 1 TO 3: FOR JJ = 1 TO 3
D(II, JJ) = 0
NEXT JJ, II
D(1, 1) = E * TH ^ 3 / (12 * (1 - NU ^ 2))
D(2, 2) = D(1, 1)
D(1, 2) = NU * D(1, 1)
D(2, 1) = D(1, 2)
D(3, 3) = D(1, 1) * (1 - NU)
GOSUB 5000
VL(1) = UL(3)
VL(4) = UL(9)
VL(7) = UL(15)
VL(2) = -UL(5)
VL(5) = -UL(11)
VL(8) = -UL(17)
VL(3) = UL(4)
VL(6) = UL(10)
VL(9) = UL(16)
FOR II = 1 TO 3: MOMXY(II) = 0
FOR JJ = 1 TO 9
MOMXY(II) = MOMXY(II) + DB(II, JJ) * VL(JJ)
NEXT JJ
IF II = 1 THEN PRINT #2, "MX="; -MOMXY(II)
IF II = 2 THEN PRINT #2, "MY="; -MOMXY(II)
IF II = 3 THEN PRINT #2, "M-XY="; -MOMXY(II)
NEXT II
FOR II = 1 TO 3
SR(II) = SR(II) - 6 * MOMXY(II) / TH ^ 2
NEXT II
SX = SR(1): SY = SR(2): SXY = SR(3)
2130 SHEAR = .5 * SQR((SX - SY) ^ 2 + 4 * SXY ^ 2)
2140 PRINCIPAL = .5 * (SX + SY) + SHEAR
2150 MINIMUM = .5 * (SX + SY) - SHEAR
2160 PRINT #2, "SIGMA1="; PRINCIPAL, "SIGMA2="; MINIMUM
2170 IF SX < SY THEN SHEAR = -SHEAR
2180 SHEAR(IN) = SHEAR(IN) + SHEAR / 3
2190 PRINCIPAL(IN) = PRINCIPAL(IN) + PRINCIPAL / 3
2200 MINIMUM(IN) = MINIMUM(IN) + MINIMUM / 3
2210 SHEAR(JN) = SHEAR(JN) + SHEAR / 3
2220 PRINCIPAL(JN) = PRINCIPAL(JN) + PRINCIPAL / 3
2230 MINIMUM(JN) = MINIMUM(JN) + MINIMUM / 3
2240 SHEAR(KN) = SHEAR(KN) + SHEAR / 3
2250 PRINCIPAL(KN) = PRINCIPAL(KN) + PRINCIPAL / 3
2260 MINIMUM(KN) = MINIMUM(KN) + MINIMUM / 3
2270 NEXT ME
NN = NJ
2280 FOR II = 1 TO NN
```

```
2290 VG(II) = SHEAR(II) / VEC(II)
2300 NEXT II
2310 GOSUB 11000
2320 GOTO 2950
2330 A(1) = SQR(A(1))
FOR II = 2 TO N2
DI = DIAG(II) - II
LL = DIAG(II - 1) - DI + 1
FOR JJ = LL TO II
X = A(DI + JJ)
DJ = DIAG(JJ) - JJ
IF JJ = 1 THEN GOTO 2340
LB = DIAG(JJ - 1) - DJ + 1
IF LL > LB THEN LB = LL
FOR KK = LB TO JJ - 1
X = X - A(DI + KK) * A(DJ + KK)
NEXT KK
2340 A(DI + JJ) = X / A(DJ + JJ)
NEXT JJ
A(DI + II) = SQR(X)
NEXT II
CV(1) = CV(1) / A(1)
FOR II = 2 TO N2
DI = DIAG(II) - II
LL = DIAG(II - 1) - DI + 1
X = CV(II)
FOR JJ = LL TO II - 1
X = X - A(DI + JJ) * CV(JJ)
NEXT JJ
CV(II) = X / A(DI + II)
NEXT II
FOR II = N2 TO 2 STEP -1
DI = DIAG(II) - II
X = CV(II) / A(DI + II)
CV(II) = X
LL = DIAG(II - 1) - DI + 1
FOR KK = LL TO II - 1
CV(KK) = CV(KK) - X * A(DI + KK)
NEXT KK
NEXT II
CV(1) = CV(1) / A(1)
2540 RETURN
2550 REM OUTPUT OF INPUT
2560 PRINT #2, : PRINT #2,
2570 PRINT #2, "STRESSES IN DOUBLY-CURVED SHELLS USING A 3 NODE TRIANGLE"
2580 PRINT #2,
2590 PRINT #2, "NO. OF ELEMENTS ="; MS
2600 PRINT #2, "NO. OF NODES ="; NN
2610 PRINT #2, "NO. OF NODES WITH ZERO DISPLACEMENTS ="; NF
PRINT #2, "PLATE THICKNESS ="; T
2620 PRINT #2, "; ELASTIC; MODULUS = "; E; " ";
2630 PRINT #2, "POISSON'S RATIO ="; NU
2670 PRINT #2, "ELEMENT TOPOLOGY"
```

```
2680 FOR II = 1 TO MS
2690 I8 = II * 3
2700 IN = IJ(I8 - 2)
2710 JN = IJ(I8 - 1)
2720 KN = IJ(I8)
2730 PRINT #2, "I,J & K ARE: "; IN; "-"; JN; "-"; KN
2740 NEXT II
2750 PRINT #2, "NODAL COORDINATES"
2760 FOR II = 1 TO NN
2770 PRINT #2, "X("; II; ")="; X(II); "   Y("; II; ")="; Y(II);
PRINT #2, "   Z("; II; ")="; Z(II)
2780 NEXT II
2790 PRINT #2, "EXTERNAL LOADS"
2800 PRINT #2, "NO. OF NODES WITH CONCENTRATED LOADS="; NC
IF NC = 0 THEN GOTO 2855
2810 FOR II = 1 TO NC
2820 PRINT #2, "NODE="; NP(II)
2830 PRINT #2, "LOAD IN X0 DIRECTION="; Q(NP(II) * 6 - 5)
2840 PRINT #2, "LOAD IN Y0 DIRECTION="; Q(NP(II) * 6 - 4)
PRINT #2, "LOAD IN Z0 DIRECTION="; Q(NP(II) * 6 - 3)
2850 NEXT II
2855 PRINT #2, "PRESSURE="; PRESSURE
2860 PRINT #2, "DETAILS OF ZERO DISPLACEMENTS"
2870 FOR II = 1 TO NF
2880 PRINT #2, "NODE="; POSS(II)
2890 N9 = NS(II * 6 - 5)
2900 IF N9 < > 0 THEN PRINT #2, "U0 IS ZERO AT NODE "; POSS(II)
2910 N9 = NS(II * 6 - 4)
2920 IF N9 < > 0 THEN PRINT #2, "V0 IS ZERO AT NODE "; POSS(II)
N9 = NS(II * 6 - 3)
IF N9 < > 0 THEN PRINT #2, "W0 IS ZERO AT NODE "; POSS(II)
N9 = NS(II * 6 - 2)
IF N9 < > 0 THEN PRINT #2, "THETAX0 IS ZERO AT NODE "; POSS(II)
N9 = NS(II * 6 - 1)
IF N9 < > 0 THEN PRINT #2, "THETAY0 IS ZERO AT NODE "; POSS(II)
N9 = NS(II * 6)
IF N9 < > 0 THEN PRINT #2, "THETAZ0 IS ZERO AT NODE "; POSS(II)
2930 NEXT II
2940 RETURN
2950 LF = 10: Z9 = 99
2960 DIM YY(3), XX(3), FF(LF + 1)
2970 GOSUB 3010
2980 FOR II = 1 TO NN: VG(II) = PRINCIPAL(II) / VEC(II): NEXT II: Z9 = 999: GOSUB 3010
2990 FOR II = 1 TO NN: VG(II) = MINIMUM(II) / VEC(II): NEXT II: Z9 = 9999: GOSUB 3010
3000 SCREEN 0: WIDTH 80: END
3010 MAX = -1E+12: MAXX = -1E+12: MAXY = -1E+12: MAXU = -1E+12: MINU = 1E+12
3020 SCREEN 9
3030 IF Z9 = 99 THEN PRINT "LINES OF CONSTANT MAXIMUM SHEAR STRESS"
3040 IF Z9 = 999 THEN PRINT "LINES OF CONSTANT MAXIMUM PRINCIPAL STRESS"
3050 IF Z9 = 9999 THEN PRINT "LINES OF CONSTANT MINIMUM PRINCIPAL STRESS"
3060 FOR I = 1 TO NN
3070 IF VG(I) > MAXU THEN MAXU = VG(I)
```

```
3080 IF VG(I) < MINU THEN MINU = VG(I)
3090 IF ABS(X(I)) > MAXX THEN MAXX = ABS(X(I))
3100 IF ABS(Y(I)) > MAXY THEN MAXY = ABS(Y(I))
3110 IF MAXX > MAX THEN MAX = MAXX
3120 IF MAXY > MAX THEN MAX = MAXY
3130 NEXT I
3140 LF = 10
3150 DIF = MAXU - MINU: STP = DIF / LF: AG = 1
3160 FOR I = MINU TO MAXU STEP STP
3170 FF(AG) = I: IF ABS(FF(AG)) < .0000001 THEN FF(AG) = 0
3180 AG = AG + 1: NEXT I
3190 FOR n = 1 TO LF
3200 FOR ME = 1 TO MS
3210 I8 = 3 * ME: AA = 0: BB = 0: CC = 0
3220 I = IJ(I8 - 2): J = IJ(I8 - 1): K = IJ(I8)
3230 IF FF(n) = (INT(VG(I) * 1000)) / 1000 THEN GOTO 3270
3240 IF FF(n) = (INT(VG(J) * 1000)) / 1000 THEN GOTO 3270
3250 IF FF(n) = (INT(VG(K) * 1000)) / 1000 THEN GOTO 3270
3260 GOTO 3280
3270 FF(n) = FF(n) + .0000001
3280 IF FF(n) > VG(I) AND FF(n) < VG(J) THEN GOSUB 3430
3290 IF FF(n) < VG(I) AND FF(n) > VG(J) THEN GOSUB 3430
3300 IF FF(n) > VG(I) AND FF(n) < VG(K) THEN GOSUB 3490
3310 IF FF(n) < VG(I) AND FF(n) > VG(K) THEN GOSUB 3490
3320 IF FF(n) > VG(J) AND FF(n) < VG(K) THEN GOSUB 3550
3330 IF FF(n) < VG(J) AND FF(n) > VG(K) THEN GOSUB 3550
3340 IF AA = 1 AND BB = 1 THEN LINE (40 + XS * (XX(1)), 280 - XS * (YY(1)))-(40 + XS
* (XX(2)), 280 - XS * (YY(2)))
3350 IF AA = 1 AND CC = 1 THEN LINE (40 + XS * (XX(1)), 280 - XS * (YY(1)))-(40 + XS
* (XX(3)), 280 - XS * (YY(3)))
3360 IF BB = 1 AND CC = 1 THEN LINE (40 + XS * (XX(2)), 280 - XS * (YY(2)))-(40 + XS
* (XX(3)), 280 - XS * (YY(3)))
3370 NEXT ME
3380 NEXT n
3390 A$ = INKEY$: IF A$ = "" THEN GOTO 3390
3400 IF A$ = "Y" OR A$ = "y" THEN CLS : RETURN
3410 GOTO 3390
3420 END
3430 DF = FF(n) - VG(I): DE = VG(J) - VG(I)
3440 DX = X(J) - X(I): DY = Y(J) - Y(I)
3450 IF DF = 0 AND DE = 0 THEN GOTO 3480
3460 XX(1) = (DF * DX / DE) + X(I): YY(1) = (DF * DY / DE) + Y(I)
3470 AA = 1: RETURN
3480 XX(1) = X(I): YY(1) = Y(I): AA = 1: RETURN
3490 DF = FF(n) - VG(I): DE = VG(K) - VG(I)
3500 DX = X(K) - X(I): DY = Y(K) - Y(I)
3510 IF DF = 0 AND DE = 0 THEN GOTO 3540
3520 XX(2) = (DF * DX / DE) + X(I): YY(2) = (DF * DY / DE) + Y(I)
3530 BB = 1: RETURN
3540 XX(2) = X(I): YY(2) = Y(I): BB = 1: RETURN
3550 DF = FF(n) - VG(J): DE = VG(K) - VG(J)
3560 DX = X(K) - X(J): DY = Y(K) - Y(J)
3570 IF DF = 0 AND DE = 0 THEN GOTO 3600
```

```
3580 XX(3) = (DF * DX / DE) + X(J): YY(3) = (DF * DY / DE) + Y(J)
3590 CC = 1: RETURN
3600 XX(3) = X(J): YY(3) = Y(J): CC = 1: RETURN
5000 XJI = XL(4) - XL(1)
5010 YJI = XL(5) - XL(2)
5020 XKJ = XL(7) - XL(4)
5030 YKJ = XL(8) - XL(5)
5040 XIK = XL(1) - XL(7)
5050 YIK = XL(2) - XL(8)
5070 X1 = SQR(XJI * XJI + YJI * YJI)
5080 X2 = SQR(XKJ * XKJ + YKJ * YKJ)
5090 X3 = SQR(XIK * XIK + YIK * YIK)
5100 DEL = XJI * YKJ - YJI * XKJ
5110 XJI = XJI / X1
5120 YJI = YJI / X1
5130 XKJ = XKJ / X2
5140 YKJ = YKJ / X2
5150 XIK = XIK / X3
5160 YIK = YIK / X3
5170 AX = XJI * XJI
5180 AY = YJI * YJI
5190 AXY = 1.414 * XJI * YJI
5200 BX = XKJ * XKJ
5210 BY = YKJ * YKJ
5220 BXY = 1.414 * XKJ * YKJ
5230 CX = XIK * XIK
5240 CY = YIK * YIK
5250 CXY = 1.414 * XIK * YIK
5260 DET = AX * (BY * CXY - BXY * CY) - AY * (BX * CXY - BXY * CX) + AXY * (BX * CY - BY * CX)
5270 C(1, 1) = BY * CXY - BXY * CY
5280 C(1, 2) = -AY * CXY + AXY * CY
5290 C(1, 3) = AY * BXY - AXY * BY
5300 C(2, 1) = -BX * CXY + BXY * CX
5310 C(2, 2) = AX * CXY - AXY * CX
5320 C(2, 3) = -AX * BXY + AXY * BX
5330 C(3, 1) = BX * CY - BY * CX
5340 C(3, 2) = -AX * CY + AY * CX
5350 C(3, 3) = AX * BY - AY * BX
5360 FOR II = 1 TO 3: FOR JJ = 1 TO 3: C(II, JJ) = C(II, JJ) / DET: NEXT JJ
5370 FOR JJ = 1 TO 10
5380 B(II, JJ) = 0
5390 NEXT JJ
5400 NEXT II
5410 B(1, 1) = 27 * L1 - 9
5420 B(1, 2) = 27 * L2 - 9
5430 B(1, 4) = 27 * L2 - 54 * L1 + 9
5440 B(1, 5) = 27 * L1 - 54 * L2 + 9
5450 B(1, 6) = 27 * L3
5460 B(1, 9) = 27 * L3
5470 B(1, 10) = -54 * L3
5480 B(2, 2) = 27 * L2 - 9
5490 B(2, 3) = 27 * L3 - 9
```

```
5500 B(2, 5) = 27 * L1
5510 B(2, 10) = -54 * L1
5520 B(2, 6) = 27 * L3 - 54 * L2 + 9
5530 B(2, 7) = 27 * L2 - 54 * L3 + 9
5540 B(2, 8) = 27 * L1
5550 B(3, 1) = 27 * L1 - 9
5560 B(3, 3) = 27 * L3 - 9
5570 B(3, 4) = 27 * L2
5580 B(3, 7) = 27 * L2
5590 B(3, 8) = 27 * L1 - 54 * L3 + 9
5600 B(3, 9) = 27 * L3 - 54 * L1 + 9
5610 B(3, 10) = -54 * L2
5620 FOR JJ = 1 TO 10
5630 B(1, JJ) = -B(1, JJ) / (X1 * X1)
5640 B(2, JJ) = -B(2, JJ) / (X2 * X2)
5650 B(3, JJ) = -B(3, JJ) / (X3 * X3)
5660 NEXT JJ
5670 A = 2 * XJI * X1
5680 B = 2 * YJI * X1
5690 C = 2 * XKJ * X2
5700 DD = 2 * YKJ * X2
5710 EE = 2 * XIK * X3
5720 F = 2 * YIK * X3
5750 FOR II = 1 TO 10
5760 FOR JJ = 1 TO 9
5770 T(II, JJ) = 0
5780 NEXT JJ
5790 NEXT II
5800 T(1, 1) = 1
5810 T(2, 4) = 1
5820 T(3, 7) = 1
5830 T(4, 1) = 20 / 27
5840 T(4, 2) = 2 * A / 27
5850 T(4, 3) = 2 * B / 27
5860 T(4, 4) = 7 / 27
5870 T(4, 5) = -A / 27
5880 T(4, 6) = -B / 27
5890 T(5, 1) = 7 / 27
5900 T(5, 2) = A / 27
5910 T(5, 3) = B / 27
5920 T(5, 4) = 20 / 27
5930 T(5, 5) = -2 * A / 27
5940 T(5, 6) = -2 * B / 27
5950 T(6, 4) = 20 / 27
5960 T(6, 5) = 2 * C / 27
5970 T(6, 6) = 2 * DD / 27
5980 T(6, 7) = 7 / 27
5990 T(6, 8) = -C / 27
6000 T(6, 9) = -DD / 27
6010 T(7, 4) = 7 / 27
6020 T(7, 5) = C / 27
6030 T(7, 6) = DD / 27
6040 T(7, 7) = 20 / 27
```

```
6050 T(7, 8) = -2 * C / 27
6060 T(7, 9) = -2 * DD / 27
6070 T(8, 1) = 7 / 27
6080 T(8, 2) = -EE / 27
6090 T(8, 3) = -F / 27
6100 T(8, 7) = 20 / 27
6110 T(8, 8) = 2 * EE / 27
6120 T(8, 9) = 2 * F / 27
6130 T(9, 1) = 20 / 27
6140 T(9, 2) = -2 * EE / 27
6150 T(9, 3) = -2 * F / 27
6160 T(9, 7) = 7 / 27
6170 T(9, 8) = EE / 27
6180 T(9, 9) = F / 27
6190 FOR II = 1 TO 9
6200 FOR JJ = 4 TO 9
6210 T(10, II) = T(10, II) + T(JJ, II) / 4
6220 NEXT JJ, II
6230 FOR II = 1 TO 9
6240 FOR JJ = 1 TO 3
6250 T(10, II) = T(10, II) - T(JJ, II) / 6
6260 NEXT JJ, II
6270 FOR II = 1 TO 3
6280 FOR JJ = 1 TO 10
6290 CB(II, JJ) = 0
6300 FOR KK = 1 TO 3
6310 CB(II, JJ) = CB(II, JJ) + C(II, KK) * B(KK, JJ)
6320 NEXT KK, JJ, II
6330 FOR II = 1 TO 3
6340 FOR JJ = 1 TO 9
6350 BC(II, JJ) = 0
6360 FOR KK = 1 TO 10
6370 BC(II, JJ) = BC(II, JJ) + CB(II, KK) * T(KK, JJ)
6380 NEXT KK: BCT(JJ, II) = BC(II, JJ): NEXT JJ, II
6390 FOR II = 1 TO 3
6400 FOR JJ = 1 TO 9
6410 DB(II, JJ) = 0
6420 FOR KK = 1 TO 3
6430 DB(II, JJ) = DB(II, JJ) + D(II, KK) * BC(KK, JJ)
6440 NEXT KK, JJ, II
6500 FS(1) = (9 * L1 ^ 3 - 9 * L1 ^ 2 + 2 * L1) / 2
6510 FS(2) = (9 * L2 ^ 3 - 9 * L2 ^ 2 + 2 * L2) / 2
6520 FS(3) = (9 * L3 ^ 3 - 9 * L3 ^ 2 + 2 * L3) / 2
6530 FS(4) = 13.5 * L1 ^ 2 * L2 - 4.5 * L1 * L2
6540 FS(5) = 13.5 * L1 * L2 ^ 2 - 4.5 * L1 * L2
6550 FS(6) = 13.5 * L2 ^ 2 * L3 - 4.5 * L2 * L3
6560 FS(7) = 13.5 * L2 * L3 ^ 2 - 4.5 * L2 * L3
6570 FS(8) = 13.5 * L3 ^ 2 * L1 - 4.5 * L3 * L1
6580 FS(9) = 13.5 * L3 * L1 ^ 2 - 4.5 * L3 * L1
6590 FS(10) = 27 * L1 * L2 * L3
6600 FOR II = 1 TO 9
6610 n(II) = 0
6620 FOR JJ = 1 TO 10
```

6630 n(II) = n(II) + FS(JJ) * T(JJ, II)
6640 NEXT JJ, II
6690 RETURN
7000 REM CALCULATION OF DIRECTIONAL COSINES
7010 XA(1) = X(I)
7020 YA(1) = Y(I)
7030 ZA(1) = Z(I)
7040 XA(2) = X(J)
7050 YA(2) = Y(J)
7060 ZA(2) = Z(J)
7070 XA(3) = X(K)
7080 YA(3) = Y(K)
7090 ZA(3) = Z(K)
7100 REM DIRECTION COSINES
7110 SY21(1) = XA(2) - XA(1)
7120 SY21(2) = YA(2) - YA(1)
7130 SY21(3) = ZA(2) - ZA(1)
7140 SY31(1) = XA(3) - XA(1)
7150 SY31(2) = YA(3) - YA(1)
7160 SY31(3) = ZA(3) - ZA(1)
7170 L21 = SQR(SY21(1) ^ 2 + SY21(2) ^ 2 + SY21(3) ^ 2)
7180 FOR II = 1 TO 3
7190 ZETA1(II) = SY21(II) / L21
7200 NEXT II
7210 FOR II = 1 TO 3
7220 FOR JJ = 1 TO 3
7230 UNIT(II, JJ) = 0
7240 NEXT JJ
7250 UNIT(II, II) = 1
7260 NEXT II
7270 FOR II = 1 TO 3
7280 FOR JJ = 1 TO 3
7290 ZETAP(II, JJ) = 0
7300 NEXT JJ
7310 FOR JJ = 1 TO 3
7320 ZETAP(II, JJ) = ZETAP(II, JJ) + ZETA1(II) * ZETA1(JJ)
7330 ZETAN(II, JJ) = -ZETAP(II, JJ)
7340 NEXT JJ
7350 NEXT II
7360 FOR II = 1 TO 3
7370 FOR JJ = 1 TO 3
7380 ZETAP(II, JJ) = UNIT(II, JJ) + ZETAN(II, JJ)
7390 NEXT JJ
7400 NEXT II
7410 FOR II = 1 TO 3
7420 SY34(II) = 0
7430 NEXT II
7440 FOR II = 1 TO 3
7450 FOR JJ = 1 TO 3
7460 SY34(II) = SY34(II) + ZETAP(II, JJ) * SY31(JJ)
7470 NEXT JJ
7480 NEXT II
7490 L21 = SQR(SY34(1) ^ 2 + SY34(2) ^ 2 + SY34(3) ^ 2)

```
7500 FOR II = 1 TO 3
7510 ZETA2(II) = SY34(II) / L21
7520 NEXT II
7530 AXY = (XA(1) * (YA(2) - YA(3)) - YA(1) * (XA(2) - XA(3)) + (XA(2) * YA(3) - YA(2) *
XA(3))) / 2
7540 AYZ = (YA(1) * (ZA(2) - ZA(3)) - ZA(1) * (YA(2) - YA(3)) + (YA(2) * ZA(3) - ZA(2) *
YA(3))) / 2
7550 AZX = (ZA(1) * (XA(2) - XA(3)) - XA(1) * (ZA(2) - ZA(3)) + (ZA(2) * XA(3) - XA(2) *
ZA(3))) / 2
7560 TRIANG = SQR(AYZ ^ 2 + AZX ^ 2 + AXY ^ 2)
7570 ZETA3(1) = AYZ / TRIANG
7580 ZETA3(2) = AZX / TRIANG
7590 ZETA3(3) = AXY / TRIANG
7600 FOR II = 1 TO 3
7610 DC(1, II) = ZETA1(II)
7620 DC(2, II) = ZETA2(II)
7630 DC(3, II) = ZETA3(II)
7640 NEXT II
7650 FOR II = 1 TO 5
7660 MM = 1 + 3 * II
7670 MN = MM + 1
7680 NN = MN + 1
7690 DC(MM, MM) = DC(1, 1)
7700 DC(MM, MN) = DC(1, 2)
7710 DC(MM, NN) = DC(1, 3)
7720 DC(MN, MM) = DC(2, 1)
7730 DC(MN, MN) = DC(2, 2)
7740 DC(MN, NN) = DC(2, 3)
7750 DC(NN, MM) = DC(3, 1)
7760 DC(NN, MN) = DC(3, 2)
7770 DC(NN, NN) = DC(3, 3)
7780 NEXT II
7790 REM DIRCOS
7800 RETURN
8100 FOR II = 1 TO 18
8110 FOR JJ = 1 TO II
8120 SK(II, JJ) = SK(JJ, II)
8140 NEXT JJ, II
8150 RETURN
10000 IF Z9 = 77 OR Z9 = 99 THEN GOTO 10295
10010 XZ = 1E+12: YZ = 1E+12
10030 XM = -1E+12: YM = -1E+12: ZMAX = -1E+12: ZMIN = 1E+12
10050 PRINT : PRINT "TO CONTINUE, TYPE Y"
10060 A$ = INKEY$: IF A$ = "" THEN GOTO 10060
10070 IF A$ = "Y" OR A$ = "y" THEN GOTO 10090
10080 GOTO 10060
10090 FOR ME = 1 TO MS
10100 I8 = 3 * ME
10110 I = IJ(I8 - 2)
10120 J = IJ(I8 - 1)
10130 K = IJ(I8)
10140 IF X(I) < XZ THEN XZ = X(I)
10150 IF X(J) < XZ THEN XZ = X(J)
```

```
10160 IF X(K) < XZ THEN XZ = X(K)
10170 IF Y(I) < YZ THEN YZ = Y(I)
10180 IF Y(J) < YZ THEN YZ = Y(J)
10190 IF Y(K) < YZ THEN YZ = Y(K)
10200 IF X(I) > XM THEN XM = X(I)
10210 IF X(J) > XM THEN XM = X(J)
10220 IF X(K) > XM THEN XM = X(K)
10230 IF Y(I) > YM THEN YM = Y(I)
10240 IF Y(J) > YM THEN YM = Y(J)
10250 IF Y(K) > YM THEN YM = Y(K)
IF Z(I) > ZMAX THEN ZMAX = Z(I)
IF Z(J) > ZMAX THEN ZMAX = Z(J)
IF Z(K) > ZMAX THEN ZMAX = Z(K)
IF Z(I) < ZMIN THEN ZMIN = Z(I)
IF Z(J) < ZMIN THEN ZMIN = Z(J)
IF Z(K) < ZMIN THEN ZMIN = Z(K)
10260 NEXT ME
ZSCALE = 50 / (ZMAX - ZMIN)
10270 XS = 350 / (XM - XZ)
10280 YS = 175 / (YM - YZ)
10290 IF XS > YS THEN XS = YS
10295 SCREEN 9
10300 FOR ME = 1 TO MS
10310 I8 = 3 * ME
10320 II = IJ(I8 - 2)
10330 JJ = IJ(I8 - 1)
10340 KK = IJ(I8)
10350 FOR IV = 1 TO 3
10360 IF IV = 1 THEN I = II: J = JJ
10370 IF IV = 2 THEN I = JJ: J = KK
10380 IF IV = 3 THEN I = KK: J = II
10390 IF Z9 = 99 THEN GOTO 10450
10400 XI = (X(I) - XZ) * XS
10410 XJ = (X(J) - XZ) * XS
10420 YI = (Y(I) - YZ) * XS
10430 YJ = (Y(J) - YZ) * XS
ZI = (Z(I) - ZMIN) * XS: ZJ = (Z(J) - ZMIN) * XS
10440 GOTO 10480
10450 XI = X(I) - XZ * XS: XJ = X(J) - XZ * XS
10460 YI = Y(I) - YZ * XS: YJ = Y(J) - YZ * XS
ZI = Z(I) - ZMIN * XS: ZJ = Z(J) - ZMIN * XS
10470 IF Z9 = 77 THEN GOTO 10490
10480 LINE (XI + 40, 280 - YI)-(XJ + 40, 280 - YJ)
10490 NEXT IV
10500 NEXT ME
10510 IF Z9 = 77 THEN GOTO 10600
10515 PRINT "TO CONTINUE, TYPE Y"
10520 A$ = INKEY$: IF A$ = "" THEN GOTO 10520
10530 IF A$ = "Y" OR A$ = "y" THEN GOTO 10560
10540 GOTO 10520
10560 SCREEN 0: WIDTH 80
10600 RETURN
11000 REM DEFLECTED FORM OF MESH
```

```
11020 Z9 = 77
11030 REM GOSUB 10000
11040 UM = -1E+12
11050 FOR II = 1 TO N2
11060 IF ABS(UG(II)) > UM THEN UM = ABS(UG(II))
11070 NEXT II
11080 SC = 20
11090 FOR II = 1 TO NN
11100 X(II) = X(II) * XS: REM + SC * UG(II * 6 - 1) / UM
11110 Y(II) = Y(II) * XS: REM + SC * UG(II * 6 - 2) / UM
Z(II) = Z(II) * XS + SC * UG(II * 6 - 3) / UM
11120 NEXT II
11130 Z9 = 99
11140 GOSUB 10000
11150 FOR II = 1 TO NN
11160 X(II) = XG(II)
11170 Y(II) = YG(II)
Z(II) = ZG(II)
11180 NEXT II
11190 RETURN
25000 REM END
```

Computer Program for Vibration of In-Plane Plates "VIBPLANF"

```
100 GOSUB 3900
110 IF A$ < > "Y" THEN PRINT : PRINT "GO BACK & READ IT !": GOTO 25000
120 REM COPYRIGHT OF DR.C.T.F.ROSS
130 REM 6 HURSTVILLE DRIVE,  WATERLOOVILLE
140 REM PORTSMOUTH. ENGLAND
142 PRINT : PRINT "FREE VIBRATION OF IN-PLANE PLATES": PRINT
143 INPUT "TYPE IN THE NAME OF YOUR INPUT FILE "; FILENAME$
PRINT "IF YOU ARE SATISFIED WITH THIS FILENAME, TYPE Y; ELSE N"
144 A$ = INKEY$: IF A$ = "" THEN GOTO 144
146 IF A$ = "Y" OR A$ = "y" THEN GOTO 148
IF A$ = "N" OR A$ = "n" THEN GOTO 143
GOTO 144
148 OPEN FILENAME$ FOR INPUT AS #1
150 INPUT "TYPE IN THE NAME OF YOUR OUTPUT FILE "; OUT$
PRINT "IF YOU ARE SATISFIED WITH THIS FILENAME, TYPE Y; ELSE N"
152 B$ = INKEY$: IF B$ = "" THEN GOTO 152
153 IF B$ = "Y" OR B$ = "y" THEN GOTO 160
IF B$ = "N" OR B$ = "n" THEN GOTO 150
GOTO 152
160 OPEN OUT$ FOR OUTPUT AS #2
168 REM PRINT:PRINT"TYPE IN NUMBER OF NODAL POINTS"
170 INPUT #1, NN
180 IF NN < 3 OR NN > 199 THEN GOSUB 4130: STOP
200 NJ = NN
210 REM PRINT"TYPE IN NUMBERS OF ELEMENTS"
220 INPUT #1, LS
230 IF LS < 1 OR LS > 199 THEN GOSUB 4130: STOP
260 REM PRINT"TYPE IN THE NUMBER OF NODES WITH ZERO DISPLACEMENTS"
270 INPUT #1, NF
280 IF NF < 2 OR NF > NN THEN GOSUB 4130: STOP
310 N2 = NN * 2
320 NP = N2 + 1
350 DIM A(N2, N2), XX(N2, NP), X(200)
360 DIM SK(6, 6), SM(6, 6), NS(NF * 2), Y(NN)
365 DIM UG(N2), XG(NN), YG(NN), I9(LS), J9(LS), K9(LS)
367 DIM AM(10), VC(200, 10), Z(200), IG(200), VE(200), FR(200)
370 GOTO 1900
380 PRINT "THE EIGENVALUES ARE BEING DETERMINED": PRINT
390 GOSUB 510
410 FOR I = 1 TO M1
420 PRINT #2, "EIGENVALUE="; AM(I)
430 PRINT #2, "FREQUENCY="; SQR(1 / AM(I)) / (2 * 3.141593)
440 PRINT #2, "EIGENVECTOR IS"
450 PRINT #2,
460 FOR J = 1 TO N
470 PRINT #2, VC(J, I); "   ";
480 PRINT #2, : NEXT J
490 PRINT #2, : NEXT I
495 GOSUB 11000
```

```
500 GOTO 25000
510 MN = N
520 NN = N
530 GOSUB 1390
540 M = 1
550 FOR I = 1 TO NN
560 VC(I, M) = X(I)
570 XX(I, M) = VC(I, M)
580 NEXT I
590 AM(M) = XM
600 IF M1 < 2 THEN RETURN
610 FOR M = 2 TO M1
620 FOR I = 1 TO NN
630 K4 = ABS(XX(I, M - 1) - 1)
640 IF K4 < .00001 THEN IR = I
650 NEXT I
660 IG(M - 1) = IR
670 FOR I = 1 TO NN
680 XX(MN - I + 1, MN - M + 3) = A(IR, I)
690 NEXT I
700 FOR I = 1 TO NN
710 FOR J = 1 TO NN
720 Z1 = MN - J + 1
730 Z2 = MN - M + 3
740 A(I, J) = A(I, J) - XX(I, M - 1) * XX(Z1, Z2)
750 NEXT J
760 NEXT I
770 FOR I = 1 TO NN
780 IF I = IR THEN GOTO 870
790 IF I > IR THEN K1 = I - 1
800 IF I < IR THEN K1 = I
810 FOR J = 1 TO NN
820 IF J = IR THEN GOTO 860
830 IF J > IR THEN K2 = J - 1
840 IF J < IR THEN K2 = J
850 A(K1, K2) = A(I, J)
860 NEXT J
870 NEXT I
880 NN = NN - 1
890 M3 = NN
900 IF M < > MN THEN GOTO 940
910 XM = A(1, 1)
920 X(1) = 1
930 GOTO 950
940 GOSUB 1390
950 FOR I = 1 TO NN
960 XX(I, M) = X(I)
970 NEXT I
980 AM(M) = XM
990 M4 = M - 1
1000 M5 = 1000 - M4
1010 FOR M8 = M5 TO 999
1020 M6 = M3 + 1
```

```
1030 M2 = 1000 - M8
1040 M7 = IG(M2) + 1
1050 IF M6 < M7 THEN GOTO 1120
1060 N9 = 1000 - M7
1070 N8 = 1000 - M6
1080 FOR I3 = N8 TO N9
1090 I = 1000 - I3
1100 X(I) = X(I - 1)
1110 NEXT I3
1120 J = IG(M2)
1130 X(J) = 0
1140 SUM = 0
1150 FOR I = 1 TO M6
1160 Z3 = MN - I + 1
1170 Z4 = MN - M2 + 2
1180 SUM = SUM + XX(Z3, Z4) * X(I)
1190 NEXT I
1200 XK = (AM(M2) - XM) / SUM
1210 FOR I = 1 TO M6
1220 X(I) = XX(I, M2) - XK * X(I)
1230 NEXT I
1240 SUM = 0
1250 FOR I = 1 TO M6
1260 IF ABS(SUM) < ABS(X(I)) THEN SUM = X(I)
1270 NEXT I
1280 FOR I = 1 TO M6
1290 X(I) = X(I) / SUM
1300 NEXT I
1310 M3 = M3 + 1
1320 IF M2 < > 1 THEN GOTO 1360
1330 FOR I = 1 TO M3
1340 VC(I, M) = X(I)
1350 NEXT I
1360 NEXT M8
1370 NEXT M
1380 RETURN
1390 Y1 = 100000!
1400 FOR I = 1 TO NN
1410 X(I) = 1
1420 NEXT I
1430 XM = -100000!
1440 FOR I = 1 TO NN
1450 SG = 0
1460 FOR J = 1 TO NN
1470 SG = SG + A(I, J) * X(J)
1480 NEXT J
1490 Z(I) = SG
1500 NEXT I
1510 XM = 0
1520 FOR I = 1 TO NN
1530 IF ABS(XM) < ABS(Z(I)) THEN XM = Z(I)
1540 NEXT I
1550 FOR I = 1 TO NN
```

```
1560 X(I) = Z(I) / XM
1570 NEXT I
1580 IF ABS((Y1 - XM) / XM) > D THEN GOTO 1600
1590 GOTO 1620
1600 Y1 = XM
1610 GOTO 1440
1620 X3 = 0
1630 FOR I = 1 TO NN
1640 IF ABS(X3) < ABS(X(I)) THEN X3 = X(I)
1650 NEXT I
1660 FOR I = 1 TO NN
1670 X(I) = X(I) / X3
1680 NEXT I
1690 RETURN
1700 N9 = N - 1
1710 FOR NX = 1 TO N
1720 DI = A(NX, 1)
1730 IF DI = 0 THEN PRINT "MATRIX IS SINGULAR"
1740 FOR NY = 1 TO N9
1750 Y9 = NY + 1
1760 A(NX, NY) = A(NX, Y9) / DI
1770 NEXT NY
1780 A(NX, N) = 1 / DI
1790 FOR NZ = 1 TO N
1800 IF NZ = NX THEN GOTO 1870
1810 O = A(NZ, 1)
1820 FOR NY = 1 TO N9
1830 Y9 = NY + 1
1840 A(NZ, NY) = A(NZ, Y9) - A(NX, NY) * O
1850 NEXT NY
1860 A(NZ, N) = -A(NX, N) * O
1870 NEXT NZ
1880 NEXT NX
1890 RETURN
1900 REM START OF MAIN PART OF PROGRAM
1910 REM PRINT:PRINT"TYPE IN THE NODE NUMBERS & 'DIRECTIONS' OF THE ZERO
DISPLACEMENTS IN ASCENDING ORDER"
1920 SN = 0
1930 FOR I = 1 TO NF * 2: NS(I) = 0: NEXT I
1940 S2 = 0
1950 FOR I = 1 TO NF
1960 INPUT #1, SU, SUX, SUY
1970 IF SU < 1 OR SU > NN THEN GOSUB 4130: STOP
2000 IF SU > S2 THEN GOTO 2010
2005 GOTO 1960
2010 REM PRINT"IS THE HORIZONTAL DISPLACEMENT AT NODE ";SU;" ZERO ? TYPE
Y/N"
2020 Y$ = "FALSE"
IF SUX = 1 THEN Y$ = "Y"
IF SUX = 0 THEN Y$ = "N"
2030 IF Y$ = "Y" OR Y$ = "y" THEN NS(I * 2 - 1) = SU * 2 - 1: SN = SN + 1
2040 IF Y$ = "Y" OR Y$ = "N" OR Y$ = "y" OR Y$ = "n" THEN GOTO 2060
2050 GOSUB 4130: PRINT "SUX="; SUX; " AT NODE "; SU: STOP
```

```
2060 REM PRINT"IS THE VERTICAL DISPLACEMENT AT NODE ";SU;" ZERO ? TYPE Y/N"
2070 Y$ = "FALSE"
IF SUY = 1 THEN Y$ = "Y"
IF SUY = 0 THEN Y$ = "N"
2080 IF Y$ = "Y" OR Y$ = "y" THEN NS(I * 2) = SU * 2: SN = SN + 1
2090 IF Y$ = "Y" OR Y$ = "N" OR Y$ = "y" OR Y$ = "n" THEN GOTO 2110
2100 GOSUB 4130: PRINT "SUY="; SUY; " AT NODE "; SU: STOP
2110 NEXT I
2140 REM PRINT:PRINT"TYPE IN THE NODAL COORDINATES"
2150 FOR I = 1 TO NN
2160 REM PRINT"TYPE IN X COORD. OF NODE ";I
2170 INPUT #1, X(I), Y(I)
2175 XG(I) = X(I)
2180 REM PRINT"TYPE IN Y COORD. OF NODE ";I
2195 YG(I) = Y(I)
2200 NEXT I
2230 N = N2 - SN
2240 REM PRINT:PRINT"TYPE IN THE NUMBER OF FREQUENCIES - THIS MUST BE
< =";N
2245 INPUT #1, M1
2250 IF M1 < 1 OR M1 > N THEN GOSUB 4130: STOP
2260 INPUT #1, TH: REM "TYPE IN PLATE THICKNESS";TH
2265 INPUT #1, E, NU, RH: REM  "TYPE IN YOUNG'S MODULUS";E
2270 REM  "TYPE IN POISSON'S RATIO";NU
2275 REM  "TYPE IN DENSITY";RH
2295 CN = E * TH / (1 - NU ^ 2): ZU = NU: MU = (1 - NU) / 2
2310 GOSUB 4160
2320 REM PRINT:PRINT"TYPE IN MEMBER DETAILS"
2330 FOR LE = 1 TO LS
2340 REM PRINT:PRINT"TYPE IN DETAILS OF MEMBER NUMBER ";LE
2350 REM PRINT"TYPE IN I NODE"
2360 INPUT #1, I, J, K
2380 IF I < 1 OR I > NN THEN GOSUB 4130: STOP
2400 REM PRINT"TYPE IN J NODE"
2420 IF J < 1 OR J > NN THEN GOSUB 4130: STOP
2425 REM PRINT"TYPE IN K NODE"
2435 IF K < 1 OR K > NN THEN GOSUB 4130: STOP
2450 IF I = J OR I = K OR J = K THEN GOSUB 4130: STOP
2460 I9(LE) = I: J9(LE) = J: K9(LE) = K
2575 NEXT LE
2576 Z9 = 0: GOSUB 10000
2577 FOR LE = 1 TO LS
2578 I = I9(LE): J = J9(LE): K = K9(LE)
2580 PRINT : PRINT "ELEMENT NUMBER "; LE; " UNDER COMPUTATION": PRINT
2590 B(1) = Y(J) - Y(K)
2600 B(2) = Y(K) - Y(I)
2610 B(3) = Y(I) - Y(J)
2620 C(1) = X(K) - X(J)
2630 C(2) = X(I) - X(K)
2640 C(3) = X(J) - X(I)
2650 DL = X(I) * B(1) + X(J) * B(2) + X(K) * B(3)
2660 FOR II = 1 TO 3
2670 I2 = 2 * II - 2
```

```
2680 FOR JJ = 1 TO 3
2690 J2 = 2 * JJ - 2
2700 SK(I2 + 1, J2 + 1) = .5 * CN * (B(II) * B(JJ) + MU * C(II) * C(JJ)) / DL
2705 SK(I2 + 2, J2 + 2) = .5 * CN * (C(II) * C(JJ) + MU * B(II) * B(JJ)) / DL
2710 SK(I2 + 1, J2 + 2) = .5 * CN * (ZU * B(II) * C(JJ) + MU * C(II) * B(JJ)) / DL
2715 SK(I2 + 2, J2 + 1) = .5 * CN * (ZU * B(JJ) * C(II) + MU * C(JJ) * B(II)) / DL
2720 NEXT JJ, II
2750 FOR JJ = 1 TO 6
2755 FOR II = JJ TO 6
2760 SK(II, JJ) = SK(JJ, II)
2765 NEXT II, JJ
2780 FOR II = 1 TO 6
2790 FOR JJ = 1 TO 6
2800 SM(II, JJ) = 0
2810 NEXT JJ
2820 NEXT II
2830 SM(1, 1) = 2
2840 SM(2, 2) = 2
2850 SM(3, 3) = 2
2860 SM(4, 4) = 2: SM(5, 5) = 2: SM(6, 6) = 2
2870 SM(1, 3) = 1: SM(3, 1) = 1
2875 SM(1, 5) = 1: SM(5, 1) = 1
2877 SM(2, 4) = 1: SM(4, 2) = 1
2878 SM(2, 6) = 1: SM(6, 2) = 1
2900 SM(3, 5) = 1: SM(5, 3) = 1
2904 SM(4, 6) = 1: SM(6, 4) = 1
2910 FOR II = 1 TO 6
2920 FOR JJ = 1 TO 6
2950 KONST = .5 * RH * DL * TH / 12
2960 SM(II, JJ) = SM(II, JJ) * KONST
2970 NEXT JJ
2980 NEXT II
2990 I1 = 2 * I - 2
3000 J1 = 2 * J - 2
3005 K1 = 2 * K - 2
3010 FOR II = 1 TO 2
3020 FOR JJ = 1 TO 2
3030 MM = I1 + II
3040 MN = I1 + JJ
3050 NM = J1 + II
3060 N9 = J1 + JJ
3070 A(MM, MN) = A(MM, MN) + SK(II, JJ)
3080 XX(MM, MN) = XX(MM, MN) + SM(II, JJ)
3090 A(NM, MN) = A(NM, MN) + SK(II + 2, JJ)
3100 XX(NM, MN) = XX(NM, MN) + SM(II + 2, JJ)
3110 A(MM, N9) = A(MM, N9) + SK(II, JJ + 2)
3120 XX(MM, N9) = XX(MM, N9) + SM(II, JJ + 2)
3130 A(NM, N9) = A(NM, N9) + SK(II + 2, JJ + 2)
3140 XX(NM, N9) = XX(NM, N9) + SM(II + 2, JJ + 2)
3147 NM = K1 + II: N9 = K1 + JJ
3149 A(MM, N9) = A(MM, N9) + SK(II, JJ + 4)
3150 XX(MM, N9) = XX(MM, N9) + SM(II, JJ + 4)
3151 A(NM, MN) = A(NM, MN) + SK(II + 4, JJ)
```

```
3152 XX(NM, MN) = XX(NM, MN) + SM(II + 4, JJ)
3160 MM = K1 + II: MN = K1 + JJ
3164 A(MM, MN) = A(MM, MN) + SK(II + 4, JJ + 4)
3166 XX(MM, MN) = XX(MM, MN) + SM(II + 4, JJ + 4)
3176 MM = J1 + II: MN = J1 + JJ
3178 NM = K1 + II: N9 = K1 + JJ
3179 A(MM, N9) = A(MM, N9) + SK(II + 2, JJ + 4)
3180 XX(MM, N9) = XX(MM, N9) + SM(II + 2, JJ + 4)
3182 A(NM, MN) = A(NM, MN) + SK(II + 4, JJ + 2)
3183 XX(NM, MN) = XX(NM, MN) + SM(II + 4, JJ + 2)
3185 NEXT JJ, II
3187 NEXT LE
3189 MM = 0
3190 FOR I = 1 TO NF * 2
3200 N7 = NS(I)
3210 IF N7 = 0 THEN GOTO 3410
3220 MM = MM + 1
3230 N7 = N7 - MM + 1
3240 N8 = N2 - MM
3250 IF N8 < N7 THEN GOTO 3410
3260 FOR II = N7 TO N8
3270 FOR JJ = N7 TO N8
3280 A(II, JJ) = A(II + 1, JJ + 1)
3290 XX(II, JJ) = XX(II + 1, JJ + 1)
3300 NEXT JJ
3310 NEXT II
3320 IF N7 = 1 THEN GOTO 3410
3330 FOR II = 1 TO N7 - 1
3340 FOR JJ = N7 TO N8
3350 A(II, JJ) = A(II, JJ + 1)
3360 XX(II, JJ) = XX(II, JJ + 1)
3370 A(JJ, II) = A(JJ + 1, II)
3380 XX(JJ, II) = XX(JJ + 1, II)
3390 NEXT JJ
3400 NEXT II
3410 NEXT I
3420 REM PRINT:PRINT"TYPE IN THE NUMBER OF CONCENTRATED MASSES":PRINT
3430 INPUT #1, NC
3440 IF NC < 0 OR NC > NN THEN GOSUB 4130: STOP
3490 IF NC = 0 THEN GOTO 3750
3500 DIM PO(NC), MC(NC)
3510 FOR II = 1 TO NC: PO(II) = 0: MC(II) = 0: NEXT II
3520 FOR II = 1 TO NC
3530 PRINT "TYPE IN NODE NUMBER OF MASS"
3540 INPUT #1, POSM, MASS
3550 IF POSM < 1 OR POSM > NN THEN GOSUB 4130: GOTO 3530
3580 PRINT "TYPE IN VALUE OF MASS"
3600 IF MASS < 0 THEN GOSUB 4130: GOTO 3580
3630 PO(II) = POSM
3640 MC(II) = MASS
3650 I1 = 2 * POSM - 1
3660 II9 = 2 * POSM - 1
3670 FOR JJ = 1 TO SN
```

```
3680 IF NS(JJ) < II9 THEN I1 = I1 - 1
3690 NEXT JJ
3700 J1 = I1 + 1
3710 XX(I1, I1) = XX(I1, I1) + MASS
3720 XX(J1, J1) = XX(J1, J1) + MASS
3730 NEXT II
3740 GOSUB 4410
3750 PRINT : PRINT "THE STIFFNESS MATRIX IS BEING INVERTED": PRINT
3760 GOSUB 1700
3770 FOR I = 1 TO N
3780 FOR J = 1 TO N
3790 VE(J) = 0
3800 FOR K = 1 TO N
3810 VE(J) = VE(J) + A(I, K) * XX(K, J)
3820 NEXT K
3830 NEXT J
3840 FOR J = 1 TO N
3850 A(I, J) = VE(J)
3860 NEXT J
3870 NEXT I
3880 D = .001
3890 GOTO 380
3900 CLS
3910 PRINT "COPYRIGHT OF DR.C.T.F.ROSS"
3920 PRINT
3930 PRINT "NOT TO BE REPRODUCED"
3940 PRINT
3950 PRINT "HAVE YOU READ THE APPROPRIATE MANUAL"
3980 PRINT
3990 PRINT "BY"
4000 PRINT
4010 PRINT "C.T.F.ROSS": PRINT
4020 PRINT "IF YES, TYPE Y; ELSE N"
4030 A$ = INKEY$: IF A$ = "" THEN GOTO 4030
4035 IF A$ = "y" THEN A$ = "Y"
4040 RETURN
4050 PRINT : PRINT
4060 PRINT "IF YOU ARE SATISFIED WITH THE INPUT OF YOUR LAST BATCH OF DATA,
TYPE Y; ELSE N": PRINT
4070 X$ = INKEY$: IF X$ = "" THEN GOTO 4070
4075 IF X$ = "y" THEN X$ = "Y"
4080 RETURN
4130 REM WARNING
4140 PRINT : PRINT "INCORRECT DATA": PRINT
4150 RETURN
4160 REM OUTPUT 1
4180 PRINT #2,
4190 PRINT #2, "FREE VIBRATIONS OF IN-PLANE PLATES"
4200 PRINT #2, : PRINT #2, "NUMBER OF NODAL POINTS="; NN
4210 PRINT #2, "NUMBER OF ELEMENTS"; LS
4220 PRINT #2, "NUMBER OF NODES WITH ZERO DISPLACEMENTS="; NF
4230 PRINT #2, : PRINT #2, "THE NODAL POSITIONS & 'DIRECTIONS' OF THE ZERO
DISPLACEMENTS ARE:-"
```

```
4240 FOR I = 1 TO NF
4250 SU = INT((NS(I * 2 - 1) + 1) / 2)
4260 IF SU = 0 THEN SU = INT((NS(I * 2)) / 2)
4270 PRINT #2, "NODE="; SU
4280 N9 = NS(I * 2 - 1)
4290 IF N9 > 0 THEN PRINT #2, "HORIZONTAL DISPLACEMENT AT NODE "; SU; " IS
ZERO"
4300 N9 = NS(I * 2)
4310 IF N9 > 0 THEN PRINT #2, "VERTICAL DISPLACEMENT AT NODE "; SU; " IS ZERO"
4320 NEXT I
4330 PRINT #2, : PRINT #2, "THE NODAL COORDINATES X0 & Y0 ARE:-"
4340 FOR I = 1 TO NN
4350 PRINT #2, "NODE="; I
4360 PRINT #2, "X("; I; ")="; X(I); "    Y("; I; ")="; Y(I)
4370 NEXT I
4380 PRINT #2, : PRINT #2, "NUMBER OF FREQUENCIES="; M1
4385 PRINT #2, "PLATE THICKNESS="; TH
4390 PRINT #2, "YOUNG'S MODULUS="; E
4395 PRINT #2, "POISSON'S RATIO="; NU
4397 PRINT #2, "DENSITY="; RH
4400 RETURN
4410 REM OUTPUT 2
4420 PRINT #2, : PRINT #2, "NUMBER OF CONCENTRATED MASSES="; NC
4430 IF NC = 0 THEN RETURN
4440 FOR I = 1 TO NC
4450 PRINT #2, "NODAL POSITION OF MASS="; PO(I)
4460 PRINT #2, "VALUE OF MASS="; MC(I)
4470 NEXT I
4490 RETURN
4500 REM OUTPUT 3
4520 PRINT #2, : PRINT #2, "DETAILS OF MEMBER "; I; "-"; J
4530 PRINT #2, "CROSS-SECTIONAL AREA="; AR
4540 PRINT #2, "ELASTIC MODULUS="; E
4550 PRINT #2, "DENSITY="; RH
4570 RETURN
4580 REM END
10000 PRINT "TO CONTINUE, PRESS Y"
IF Z9 = 99 THEN PRINT "EIGENMODE="; IJ: PRINT ; "FREQUENCY="; SQR(1 / AM(IJ))
/ (2 * 3.14159); "Hz"
10001 IF Z9 = 77 OR Z9 = 99 THEN GOTO 10300
10005 XZ = 1E+12
10010 YZ = 1E+12
10020 XM = -1E+12
10030 YM = -1E+12
10070 FOR EL = 1 TO LS
10080 I = I9(EL)
10090 J = J9(EL)
10100 K = K9(EL)
10130 IF X(I) < XZ THEN XZ = X(I)
10140 IF Y(I) < YZ THEN YZ = Y(I)
10150 IF X(J) < XZ THEN XZ = X(J)
10160 IF Y(J) < YZ THEN YZ = Y(J)
10165 IF X(K) < XZ THEN XZ = X(K)
```

```
10167 IF Y(K) < YZ THEN YZ = Y(K)
10170 IF X(I) > XM THEN XM = X(I)
10180 IF Y(I) > YM THEN YM = Y(I)
10190 IF X(J) > XM THEN XM = X(J)
10200 IF Y(J) > YM THEN YM = Y(J)
10205 IF X(K) > XM THEN XM = X(K)
10207 IF Y(K) > YM THEN YM = Y(K)
10210 NEXT EL
10250 XS = 350 / (XM - XZ)
10270 YS = 175 / (YM - YZ)
10280 IF XS > YS THEN XS = YS
10300 IF Z9 = 0 OR Z9 = 77 THEN SCREEN 9
10350 FOR EL = 1 TO LS
10360 I = I9(EL)
10370 J = J9(EL)
10375 K = K9(EL)
10377 FOR IV = 1 TO 3
10378 IF IV = 1 THEN II = I: JJ = J
10379 IF IV = 2 THEN II = J: JJ = K
10380 IF IV = 3 THEN II = K: JJ = I
10385 IF Z9 = 99 THEN GOTO 10440
10390 XI = (X(II) - XZ) * XS
10400 YI = (Y(II) - YZ) * XS
10410 XJ = (X(JJ) - XZ) * XS
10420 YJ = (Y(JJ) - YZ) * XS
10430 GOTO 10460
10440 XI = X(II) - XZ * XS: XJ = X(JJ) - XZ * XS
10450 YI = Y(II) - YZ * XS: YJ = Y(JJ) - YZ * XS
10460 LINE (XI + 40, 250 - YI)-(XJ + 40, 250 - YJ)
10465 NEXT IV
10470 NEXT EL
10650 IF Z9 = 77 THEN GOTO 10800
10660 A$ = INKEY$: IF A$ = "" THEN GOTO 10660
10670 IF A$ = "Y" OR A$ = "y" THEN GOTO 10720
10700 GOTO 10660
10720 SCREEN 0: WIDTH 80
10800 RETURN
11000 REM PLOT OF EIGENMODES
11010 FR(0) = 0
11020 FOR II = 1 TO N
11030 FR(II) = FR(II - 1)
11040 FR(II) = FR(II) + 1
11050 FOR JJ = 1 TO NF * 2
11070 IF FR(II) = NS(JJ) THEN GOTO 11040
11080 NEXT JJ
11090 NEXT II
11100 FOR IJ = 1 TO M1
11110 FOR II = 1 TO NJ
11120 X(II) = XG(II)
11130 Y(II) = YG(II)
11140 NEXT II
11150 Z9 = 77
11160 GOSUB 10000
```

```
11180 FOR II = 1 TO N
11190 UG(FR(II)) = VC(II, IJ)
11300 NEXT II
11310 SC = 10
11320 FOR II = 1 TO NJ
11330 X(II) = XG(II) * XS + SC * UG(II * 2 - 1)
11340 Y(II) = YG(II) * XS + SC * UG(II * 2)
11360 NEXT II
11380 Z9 = 99
11390 GOSUB 10000
11410 NEXT IJ
11430 RETURN
25000 REM END
```

Computer Program for Lateral Vibrations of Flat Plates "VIBPLATF"

```
100 GOSUB 3900
110 IF A$ < > "Y" THEN PRINT : PRINT "GO BACK & INPUT #1, IT !": GOTO 25000
120 REM COPYRIGHT OF DR.C.T.F.ROSS
130 REM 6 HURSTVILLE DRIVE,  WATERLOOVILLE
140 REM PORTSMOUTH. ENGLAND
142 PRINT : PRINT "FREE VIBRATION OF PLATES IN BENDING": PRINT
143 INPUT "TYPE IN THE NAME OF YOUR INPUT FILE "; FILENAME$
PRINT "IF YOU ARE SATISFIED WITH THIS FILENAME, TYPE Y; ELSE N"
144 A$ = INKEY$: IF A$ = "" THEN GOTO 144
145 IF A$ = "Y" OR A$ = "y" THEN GOTO 150
IF A$ = "N" OR A$ = "n" THEN GOTO 143
GOTO 144
150 OPEN FILENAME$ FOR INPUT AS #1
152 INPUT "TYPE IN THE NAME OF YOUR OUTPUT FILE "; OUT$
PRINT "IF YOU ARE SATISFIED WITH THIS FILENAME, TYPE Y; ELSE N"
153 B$ = INKEY$: IF B$ = "" THEN GOTO 153
154 IF B$ = "Y" OR B$ = "y" THEN GOTO 160
IF B$ = "N" OR B$ = "n" THEN GOTO 152
GOTO 153
160 OPEN OUT$ FOR OUTPUT AS #2
168 REM PRINT:PRINT"TYPE IN NUMBER OF NODAL POINTS"
170 INPUT #1, NN
180 IF NN < 3 OR NN > 199 THEN GOSUB 4130: STOP
200 NJ = NN
210 REM PRINT"TYPE IN NUMBERS OF ELEMENTS"
220 INPUT #1, LS
230 IF LS < 1 OR LS > 199 THEN GOSUB 4130: STOP
260 REM PRINT"TYPE IN THE NUMBER OF NODES WITH ZERO DISPLACEMENTS"
270 INPUT #1, NF
280 IF NF < 2 OR NF > NN THEN GOSUB 4130: STOP
310 N2 = NN * 3
320 NP = N2 + 1
350 DIM A(N2, N2), XX(N2, NP), X(200)
355 DIM D(3, 3), C(3, 3), B(3, 10), T(10, 9)
357 DIM CB(3, 10), BC(3, 9), BCT(9, 3), DB(3, 9), FS(10), N(9)
360 DIM SK(9, 9), SM(9, 9), NS(NF * 3), Y(NN)
365 DIM UG(N2), XG(NN), YG(NN), I9(LS), J9(LS), K9(LS)
367 DIM AM(21), VC(100, 21), Z(100), IG(100), VE(100), FR(100)
370 GOTO 1900
380 PRINT "THE EIGENVALUES ARE BEING DETERMINED": PRINT
390 GOSUB 510
410 FOR I = 1 TO M1
420 PRINT #2, "EIGENVALUE="; AM(I)
430 PRINT #2, "FREQUENCY="; SQR(1 / AM(I)) / (2 * 3.141593)
440 PRINT #2, "EIGENVECTOR IS"
450 PRINT #2,
460 FOR J = 1 TO N
470 PRINT #2, VC(J, I); "   ";
480 PRINT #2, : NEXT J
```

```
490 PRINT #2, : NEXT I
495 GOSUB 11000
500 GOTO 25000
510 MN = N
520 NN = N
530 GOSUB 1390
540 M = 1
550 FOR I = 1 TO NN
560 VC(I, M) = X(I)
570 XX(I, M) = VC(I, M)
580 NEXT I
590 AM(M) = XM
600 IF M1 < 2 THEN RETURN
610 FOR M = 2 TO M1
620 FOR I = 1 TO NN
630 K4 = ABS(XX(I, M - 1) - 1)
640 IF K4 < .00001 THEN IR = I
650 NEXT I
660 IG(M - 1) = IR
670 FOR I = 1 TO NN
680 XX(MN - I + 1, MN - M + 3) = A(IR, I)
690 NEXT I
700 FOR I = 1 TO NN
710 FOR J = 1 TO NN
720 Z1 = MN - J + 1
730 Z2 = MN - M + 3
740 A(I, J) = A(I, J) - XX(I, M - 1) * XX(Z1, Z2)
750 NEXT J
760 NEXT I
770 FOR I = 1 TO NN
780 IF I = IR THEN GOTO 870
790 IF I > IR THEN K1 = I - 1
800 IF I < IR THEN K1 = I
810 FOR J = 1 TO NN
820 IF J = IR THEN GOTO 860
830 IF J > IR THEN K2 = J - 1
840 IF J < IR THEN K2 = J
850 A(K1, K2) = A(I, J)
860 NEXT J
870 NEXT I
880 NN = NN - 1
890 M3 = NN
900 IF M < > MN THEN GOTO 940
910 XM = A(1, 1)
920 X(1) = 1
930 GOTO 950
940 GOSUB 1390
950 FOR I = 1 TO NN
960 XX(I, M) = X(I)
970 NEXT I
980 AM(M) = XM
990 M4 = M - 1
1000 M5 = 1000 - M4
```

```
1010 FOR M8 = M5 TO 999
1020 M6 = M3 + 1
1030 M2 = 1000 - M8
1040 M7 = IG(M2) + 1
1050 IF M6 < M7 THEN GOTO 1120
1060 N9 = 1000 - M7
1070 N8 = 1000 - M6
1080 FOR I3 = N8 TO N9
1090 I = 1000 - I3
1100 X(I) = X(I - 1)
1110 NEXT I3
1120 J = IG(M2)
1130 X(J) = 0
1140 SUM = 0
1150 FOR I = 1 TO M6
1160 Z3 = MN - I + 1
1170 Z4 = MN - M2 + 2
1180 SUM = SUM + XX(Z3, Z4) * X(I)
1190 NEXT I
1200 XK = (AM(M2) - XM) / SUM
1210 FOR I = 1 TO M6
1220 X(I) = XX(I, M2) - XK * X(I)
1230 NEXT I
1240 SUM = 0
1250 FOR I = 1 TO M6
1260 IF ABS(SUM) < ABS(X(I)) THEN SUM = X(I)
1270 NEXT I
1280 FOR I = 1 TO M6
1290 X(I) = X(I) / SUM
1300 NEXT I
1310 M3 = M3 + 1
1320 IF M2 < > 1 THEN GOTO 1360
1330 FOR I = 1 TO M3
1340 VC(I, M) = X(I)
1350 NEXT I
1360 NEXT M8
1370 NEXT M
1380 RETURN
1390 Y1 = 100000!
1400 FOR I = 1 TO NN
1410 X(I) = 1
1420 NEXT I
1430 XM = -100000!
1440 FOR I = 1 TO NN
1450 SG = 0
1460 FOR J = 1 TO NN
1470 SG = SG + A(I, J) * X(J)
1480 NEXT J
1490 Z(I) = SG
1500 NEXT I
1510 XM = 0
1520 FOR I = 1 TO NN
1530 IF ABS(XM) < ABS(Z(I)) THEN XM = Z(I)
```

```
1540 NEXT I
1550 FOR I = 1 TO NN
1560 X(I) = Z(I) / XM
1570 NEXT I
1580 IF ABS((Y1 - XM) / XM) > D THEN GOTO 1600
1590 GOTO 1620
1600 Y1 = XM
1610 GOTO 1440
1620 X3 = 0
1630 FOR I = 1 TO NN
1640 IF ABS(X3) < ABS(X(I)) THEN X3 = X(I)
1650 NEXT I
1660 FOR I = 1 TO NN
1670 X(I) = X(I) / X3
1680 NEXT I
1690 RETURN
1700 N9 = N - 1
1710 FOR NX = 1 TO N
1720 DI = A(NX, 1)
1730 IF DI = 0 THEN PRINT "MATRIX IS SINGULAR"
1740 FOR NY = 1 TO N9
1750 Y9 = NY + 1
1760 A(NX, NY) = A(NX, Y9) / DI
1770 NEXT NY
1780 A(NX, N) = 1 / DI
1790 FOR NZ = 1 TO N
1800 IF NZ = NX THEN GOTO 1870
1810 O = A(NZ, 1)
1820 FOR NY = 1 TO N9
1830 Y9 = NY + 1
1840 A(NZ, NY) = A(NZ, Y9) - A(NX, NY) * O
1850 NEXT NY
1860 A(NZ, N) = -A(NX, N) * O
1870 NEXT NZ
1880 NEXT NX
1890 RETURN
1900 REM START OF MAIN PART OF PROGRAM
1910 REM PRINT:PRINT"TYPE IN THE NODE NUMBERS & 'DIRECTIONS' OF THE ZERO
DISPLACEMENTS IN ASCENDING ORDER"
1920 SN = 0
1930 FOR I = 1 TO NF * 3: NS(I) = 0: NEXT I
1940 S2 = 0
1950 FOR I = 1 TO NF
1960 INPUT #1, SU, SUW, SUX, SUY
1970 IF SU < 1 OR SU > NN THEN GOSUB 4130: STOP
2000 IF SU > S2 THEN GOTO 2010
2005 GOTO 1960
2010 REM PRINT"IS THE VERTICAL DISPLACEMENT AT NODE ";SU;" ZERO ? TYPE Y/N"
2020 Y$ = "FALSE"
IF SUW = 1 THEN Y$ = "Y"
IF SUW = 0 THEN Y$ = "N"
2030 IF Y$ = "Y" OR Y$ = "y" THEN NS(I * 3 - 2) = SU * 3 - 2: SN = SN + 1
2040 IF Y$ = "Y" OR Y$ = "N" OR Y$ = "y" OR Y$ = "n" THEN GOTO 2060
```

2050 GOSUB 4130: PRINT "SUW="; SUW; " AT NODE "; SU: STOP
2060 REM PRINT"IS THE DW/DX DISPLACEMENT AT NODE ";SU;" ZERO ? TYPE Y/N"
2070 Y$ = "FALSE"
IF SUX = 1 THEN Y$ = "Y"
IF SUX = 0 THEN Y$ = "N"
2080 IF Y$ = "Y" OR Y$ = "y" THEN NS(I * 3 - 1) = SU * 3 - 1: SN = SN + 1
2090 IF Y$ = "Y" OR Y$ = "N" OR Y$ = "y" OR Y$ = "n" THEN GOTO 2093
2092 GOSUB 4130: PRINT 'SUX=";SUX;" AT NODE ";SU: STOP
2093 REM PRINT"IS THE DW/DY DISPLACEMENT AT NODE ";SU;" ZERO ? TYPE Y/N"
2094 Y$ = "FALSE"
IF SUY = 1 THEN Y$ = "Y"
IF SUY = 0 THEN Y$ = "N"
2095 IF Y$ = "Y" OR Y$ = "y" THEN NS(I * 3) = SU * 3: SN = SN + 1
2096 IF Y$ = "Y" OR Y$ = "N" OR Y$ = "y" OR Y$ = "n" THEN GOTO 2110
2109 GOSUB 4130: PRINT "SUY="; SUY; " AT NODE "; SU: STOP
2110 NEXT I
2140 REM PRINT:PRINT"TYPE IN THE NODAL COORDINATES"
2150 FOR I = 1 TO NN
2160 REM PRINT"TYPE IN X COORD. OF NODE ";I
2170 INPUT #1, X(I), Y(I)
2175 XG(I) = X(I)
2180 REM PRINT"TYPE IN Y COORD. OF NODE ";I
2195 YG(I) = Y(I)
2200 NEXT I
2230 N = N2 - SN
2240 REM PRINT:PRINT"TYPE IN THE NUMBER OF FREQUENCIES - THIS MUST BE
< =";N
2245 INPUT #1, M1
2250 IF M1 < 1 OR M1 > N THEN GOSUB 4130: STOP
2260 INPUT #1, TH: REM "TYPE IN PLATE THICKNESS";TH
2265 INPUT #1, E, NU, RH: REM "TYPE IN YOUNG'S MODULUS";E
2270 REM "TYPE IN POISSON'S RATIO";NU
2275 RH: REM "TYPE IN DENSITY";RH
2310 GOSUB 4160
2320 REM PRINT:PRINT"TYPE IN MEMBER DETAILS"
2330 FOR LE = 1 TO LS
2340 REM PRINT:PRINT"TYPE IN DETAILS OF MEMBER NUMBER ";LE
2350 REM PRINT"TYPE IN I NODE"
2360 INPUT #1, I, J, K
2380 IF I < 1 OR I > NN THEN GOSUB 4130: PRINT "I NODE="; I: STOP
2400 REM PRINT"TYPE IN J NODE"
2420 IF J < 1 OR J > NN THEN GOSUB 4130: PRINT "J NODE="; J: STOP
2425 REM PRINT"TYPE IN K NODE"
2435 IF K < 1 OR K > NN THEN GOSUB 4130: PRINT "K NODE="; K: STOP
2450 IF I = J OR I = K OR J = K THEN GOSUB 4130: PRINT "I,J,K NODES="; I; J; K: STOP
2460 I9(LE) = I: J9(LE) = J: K9(LE) = K
2575 NEXT LE
2576 Z9 = 0: GOSUB 10000
2577 FOR LE = 1 TO LS
2578 I = I9(LE): J = J9(LE): K = K9(LE)
2580 PRINT : PRINT "ELEMENT NUMBER "; LE; " UNDER COMPUTATION": PRINT
2581 FOR II = 1 TO 9: FOR JJ = 1 TO 9
2582 SK(II, JJ) = 0: SM(II, JJ) = 0

```
2583 NEXT JJ, II
2584 FOR II = 1 TO 3: FOR JJ = 1 TO 3
2585 D(II, JJ) = 0
2586 NEXT JJ, II
2588 D(1, 1) = E * TH ^ 3 / (12 * (1 - NU * NU)): D(2, 2) = D(1, 1)
2590 D(1, 2) = NU * D(1, 1): D(2, 1) = D(1, 2)
2592 D(3, 3) = D(1, 1) * (1 - NU)
2595 FOR IJ = 1 TO 3
2600 IF IJ = 1 THEN L1 = .5: L2 = .5
2610 IF IJ = 2 THEN L1 = 0: L2 = .5
2620 IF IJ = 3 THEN L1 = .5: L2 = 0
2630 L3 = 1 - L1 - L2
2640 GOSUB 5000
2650 FOR II = 1 TO 9
2660 FOR JJ = 1 TO 9
2670 FOR KK = 1 TO 3
2680 SK(II, JJ) = SK(II, JJ) + BCT(II, KK) * DB(KK, JJ) * DEL / 6
2690 NEXT KK
2700 SM(II, JJ) = SM(II, JJ) + N(II) * N(JJ) * DEL * RH * TH / 6
2710 NEXT JJ, II
2800 NEXT IJ
2990 I1 = 3 * I - 3
3000 J1 = 3 * J - 3
3005 K1 = 3 * K - 3
3010 FOR II = 1 TO 3
3020 FOR JJ = 1 TO 3
3030 MM = I1 + II
3040 MN = I1 + JJ
3050 NM = J1 + II
3060 N9 = J1 + JJ
3070 A(MM, MN) = A(MM, MN) + SK(II, JJ)
3080 XX(MM, MN) = XX(MM, MN) + SM(II, JJ)
3090 A(NM, MN) = A(NM, MN) + SK(II + 3, JJ)
3100 XX(NM, MN) = XX(NM, MN) + SM(II + 3, JJ)
3110 A(MM, N9) = A(MM, N9) + SK(II, JJ + 3)
3120 XX(MM, N9) = XX(MM, N9) + SM(II, JJ + 3)
3130 A(NM, N9) = A(NM, N9) + SK(II + 3, JJ + 3)
3140 XX(NM, N9) = XX(NM, N9) + SM(II + 3, JJ + 3)
3147 NM = K1 + II: N9 = K1 + JJ
3149 A(MM, N9) = A(MM, N9) + SK(II, JJ + 6)
3150 XX(MM, N9) = XX(MM, N9) + SM(II, JJ + 6)
3151 A(NM, MN) = A(NM, MN) + SK(II + 6, JJ)
3152 XX(NM, MN) = XX(NM, MN) + SM(II + 6, JJ)
3160 MM = K1 + II: MN = K1 + JJ
3164 A(MM, MN) = A(MM, MN) + SK(II + 6, JJ + 6)
3166 XX(MM, MN) = XX(MM, MN) + SM(II + 6, JJ + 6)
3176 MM = J1 + II: MN = J1 + JJ
3178 NM = K1 + II: N9 = K1 + JJ
3179 A(MM, N9) = A(MM, N9) + SK(II + 3, JJ + 6)
3180 XX(MM, N9) = XX(MM, N9) + SM(II + 3, JJ + 6)
3182 A(NM, MN) = A(NM, MN) + SK(II + 6, JJ + 3)
3183 XX(NM, MN) = XX(NM, MN) + SM(II + 6, JJ + 3)
3185 NEXT JJ, II
```

```
3187 NEXT LE
3189 MM = 0
3190 FOR I = 1 TO NF * 3
3200 N7 = NS(I)
3210 IF N7 = 0 THEN GOTO 3410
3220 MM = MM + 1
3230 N7 = N7 - MM + 1
3240 N8 = N2 - MM
3250 IF N8 < N7 THEN GOTO 3410
3260 FOR II = N7 TO N8
3270 FOR JJ = N7 TO N8
3280 A(II, JJ) = A(II + 1, JJ + 1)
3290 XX(II, JJ) = XX(II + 1, JJ + 1)
3300 NEXT JJ
3310 NEXT II
3320 IF N7 = 1 THEN GOTO 3410
3330 FOR II = 1 TO N7 - 1
3340 FOR JJ = N7 TO N8
3350 A(II, JJ) = A(II, JJ + 1)
3360 XX(II, JJ) = XX(II, JJ + 1)
3370 A(JJ, II) = A(JJ + 1, II)
3380 XX(JJ, II) = XX(JJ + 1, II)
3390 NEXT JJ
3400 NEXT II
3410 NEXT I
3420 REM PRINT:PRINT"TYPE IN THE NUMBER OF CONCENTRATED MASSES":PRINT
3430 INPUT #1, NC
3440 IF NC < 0 OR NC > NN THEN GOSUB 4130: STOP
3490 IF NC = 0 THEN GOTO 3750
3500 DIM PO(NC), MC(NC)
3510 FOR II = 1 TO NC: PO(II) = 0: MC(II) = 0: NEXT II
3520 FOR II = 1 TO NC
3530 PRINT "TYPE IN NODE NUMBER OF MASS"
3540 INPUT POSM, MASS
3550 IF POSM < 1 OR POSM > NN THEN GOSUB 4130: PRINT "NODAL POSITION OF
MASS="; POSM: GOTO 3530
3580 PRINT "TYPE IN VALUE OF MASS"
3600 IF MASS < 0 THEN GOSUB 4130: PRINT "VALUE OF MASS="; MASS: GOTO 3580
3630 PO(II) = POSM
3640 MC(II) = MASS
3650 I1 = 3 * POSM - 2
3660 II9 = 3 * POSM - 2
3670 FOR JJ = 1 TO SN
3680 IF NS(JJ) < II9 THEN I1 = I1 - 1
3690 NEXT JJ
3700 J1 = I1 + 1
3710 XX(I1, I1) = XX(I1, I1) + MASS
3720 XX(J1, J1) = XX(J1, J1) + MASS
3730 NEXT II
3740 GOSUB 4410
3750 PRINT : PRINT "THE STIFFNESS MATRIX IS BEING INVERTED": PRINT
3760 GOSUB 1700
3770 FOR I = 1 TO N
```

```
3780 FOR J = 1 TO N
3790 VE(J) = 0
3800 FOR K = 1 TO N
3810 VE(J) = VE(J) + A(I, K) * XX(K, J)
3820 NEXT K
3830 NEXT J
3840 FOR J = 1 TO N
3850 A(I, J) = VE(J)
3860 NEXT J
3870 NEXT I
3880 D = .001
3890 GOTO 380
3900 CLS
3910 PRINT "COPYRIGHT OF DR.C.T.F.ROSS"
3920 PRINT
3930 PRINT "NOT TO BE REPRODUCED"
3940 PRINT
3950 PRINT "HAVE YOU READ THE APPROPRIATE MANUAL"
3980 PRINT
3990 PRINT "BY"
4000 PRINT
4010 PRINT "C.T.F.ROSS": PRINT
4020 PRINT "IF YES, TYPE Y; ELSE N"
4030 A$ = INKEY$: IF A$ = "" THEN GOTO 4030
4035 IF A$ = "y" THEN A$ = "Y"
4040 RETURN
4050 PRINT : PRINT
4060 PRINT "IF YOU ARE SATISFIED WITH THE INPUT OF YOUR LAST BATCH OF DATA,
TYPE Y; ELSE N": PRINT
4070 X$ = INKEY$: IF X$ = "" THEN GOTO 4070
4075 IF X$ = "y" THEN X$ = "Y"
4080 RETURN
4130 REM WARNING
4140 PRINT : PRINT "INCORRECT DATA": PRINT
4150 RETURN
4160 REM OUTPUT 1
4180 PRINT #2,
4190 PRINT #2, "FREE VIBRATIONS OF PLATES IN FLEXURE"
4200 PRINT #2, : PRINT #2, "NUMBER OF NODAL POINTS="; NN
4210 PRINT #2, "NUMBER OF ELEMENTS"; LS
4220 PRINT #2, "NUMBER OF NODES WITH ZERO DISPLACEMENTS="; NF
4230 PRINT #2, : PRINT #2, "THE NODAL POSITIONS & 'DIRECTIONS' OF THE ZERO
DISPLACEMENTS ARE:-"
4240 FOR I = 1 TO NF
4250 SU = INT((NS(I * 3 - 2) + 3) / 3)
4260 IF SU = 0 THEN SU = INT((NS(I * 3 - 1) + 2) / 3)
4265 IF SU = 0 THEN SU = INT((NS(I * 3) + 1) / 3)
4270 PRINT #2, "NODE="; SU
4280 N9 = NS(I * 3 - 2)
4290 IF N9 > 0 THEN PRINT #2, "VERTICAL DISPLACEMENT AT NODE "; SU; " IS ZERO"
4300 N9 = NS(I * 3 - 1)
4310 IF N9 > 0 THEN PRINT #2, "THE DW/DX DISPLACEMENT AT NODE "; SU; " IS
ZERO"
```

```
4315 N9 = NS(I * 3)
4317 IF N9 > 0 THEN PRINT #2, "THE DW/DY DISPLACEMENT AT NODE "; SU; " IS
ZERO"
4320 NEXT I
4330 PRINT #2, : PRINT #2, "THE NODAL COORDINATES X0 & Y0 ARE:-"
4340 FOR I = 1 TO NN
4350 PRINT #2, "NODE="; I
4360 PRINT #2, "X("; I; ")="; X(I); "    Y("; I; ")="; Y(I)
4370 NEXT I
4380 PRINT #2, : PRINT #2, "NUMBER OF FREQUENCIES="; M1
4385 PRINT #2, "PLATE THICKNESS="; TH
4390 PRINT #2, "YOUNG'S MODULUS="; E
4395 PRINT #2, "POISSON'S RATIO="; NU
4397 PRINT #2, "DENSITY="; RH
4400 RETURN
4410 REM OUTPUT 2
4420 PRINT #2, : PRINT #2, "NUMBER OF CONCENTRATED MASSES="; NC
4430 IF NC = 0 THEN RETURN
4440 FOR I = 1 TO NC
4450 PRINT #2, "NODAL POSITION OF MASS="; PO(I)
4460 PRINT #2, "VALUE OF MASS="; MC(I)
4470 NEXT I
4490 RETURN
4500 REM OUTPUT 3
4520 PRINT #2, : PRINT #2, "DETAILS OF MEMBER "; I; "-"; J
4530 PRINT #2, "CROSS-SECTIONAL AREA="; AR
4540 PRINT #2, "ELASTIC MODULUS="; E
4550 PRINT #2, "DENSITY="; RH
4570 RETURN
4580 REM END
5000 XJI = X(J) - X(I)
5010 YJI = Y(J) - Y(I)
5020 XKJ = X(K) - X(J)
5030 YKJ = Y(K) - Y(J)
5040 XIK = X(I) - X(K)
5050 YIK = Y(I) - Y(K)
5070 X1 = SQR(XJI * XJI + YJI * YJI)
5080 X2 = SQR(XKJ * XKJ + YKJ * YKJ)
5090 X3 = SQR(XIK * XIK + YIK * YIK)
5100 DEL = XJI * YKJ - YJI * XKJ
5110 XJI = XJI / X1
5120 YJI = YJI / X1
5130 XKJ = XKJ / X2
5140 YKJ = YKJ / X2
5150 XIK = XIK / X3
5160 YIK = YIK / X3
5170 AX = XJI * XJI
5180 AY = YJI * YJI
5190 AXY = 1.414 * XJI * YJI
5200 BX = XKJ * XKJ
5210 BY = YKJ * YKJ
5220 BXY = 1.414 * XKJ * YKJ
5230 CX = XIK * XIK
```

```
5240 CY = YIK * YIK
5250 CXY = 1.414 * XIK * YIK
5260 DET = AX * (BY * CXY - BXY * CY) - AY * (BX * CXY - BXY * CX) + AXY * (BX *
CY - BY * CX)
5270 C(1, 1) = BY * CXY - BXY * CY
5280 C(1, 2) = -AY * CXY + AXY * CY
5290 C(1, 3) = AY * BXY - AXY * BY
5300 C(2, 1) = -BX * CXY + BXY * CX
5310 C(2, 2) = AX * CXY - AXY * CX
5320 C(2, 3) = -AX * BXY + AXY * BX
5330 C(3, 1) = BX * CY - BY * CX
5340 C(3, 2) = -AX * CY + AY * CX
5350 C(3, 3) = AX * BY - AY * BX
5360 FOR II = 1 TO 3: FOR JJ = 1 TO 3: C(II, JJ) = C(II, JJ) / DET: NEXT JJ
5370 FOR JJ = 1 TO 10
5380 B(II, JJ) = 0
5390 NEXT JJ
5400 NEXT II
5410 B(1, 1) = 27 * L1 - 9
5420 B(1, 2) = 27 * L2 - 9
5430 B(1, 4) = 27 * L2 - 54 * L1 + 9
5440 B(1, 5) = 27 * L1 - 54 * L2 + 9
5450 B(1, 6) = 27 * L3
5460 B(1, 9) = 27 * L3
5470 B(1, 10) = -54 * L3
5480 B(2, 2) = 27 * L2 - 9
5490 B(2, 3) = 27 * L3 - 9
5500 B(2, 5) = 27 * L1
5510 B(2, 10) = -54 * L1
5520 B(2, 6) = 27 * L3 - 54 * L2 + 9
5530 B(2, 7) = 27 * L2 - 54 * L3 + 9
5540 B(2, 8) = 27 * L1
5550 B(3, 1) = 27 * L1 - 9
5560 B(3, 3) = 27 * L3 - 9
5570 B(3, 4) = 27 * L2
5580 B(3, 7) = 27 * L2
5590 B(3, 8) = 27 * L1 - 54 * L3 + 9
5600 B(3, 9) = 27 * L3 - 54 * L1 + 9
5610 B(3, 10) = -54 * L2
5620 FOR JJ = 1 TO 10
5630 B(1, JJ) = -B(1, JJ) / (X1 * X1)
5640 B(2, JJ) = -B(2, JJ) / (X2 * X2)
5650 B(3, JJ) = -B(3, JJ) / (X3 * X3)
5660 NEXT JJ
5670 A = 2 * (X(J) - X(I))
5680 B = 2 * (Y(J) - Y(I))
5690 C = 2 * (X(K) - X(J))
5700 DD = 2 * (Y(K) - Y(J))
5710 EE = 2 * (X(I) - X(K))
5720 F = 2 * (Y(I) - Y(K))
5750 FOR II = 1 TO 10
5760 FOR JJ = 1 TO 9
5770 T(II, JJ) = 0
```

```
5780 NEXT JJ
5790 NEXT II
5800 T(1, 1) = 1
5810 T(2, 4) = 1
5820 T(3, 7) = 1
5830 T(4, 1) = 20 / 27
5840 T(4, 2) = 2 * A / 27
5850 T(4, 3) = 2 * B / 27
5860 T(4, 4) = 7 / 27
5870 T(4, 5) = -A / 27
5880 T(4, 6) = -B / 27
5890 T(5, 1) = 7 / 27
5900 T(5, 2) = A / 27
5910 T(5, 3) = B / 27
5920 T(5, 4) = 20 / 27
5930 T(5, 5) = -2 * A / 27
5940 T(5, 6) = -2 * B / 27
5950 T(6, 4) = 20 / 27
5960 T(6, 5) = 2 * C / 27
5970 T(6, 6) = 2 * DD / 27
5980 T(6, 7) = 7 / 27
5990 T(6, 8) = -C / 27
6000 T(6, 9) = -DD / 27
6010 T(7, 4) = 7 / 27
6020 T(7, 5) = C / 27
6030 T(7, 6) = DD / 27
6040 T(7, 7) = 20 / 27
6050 T(7, 8) = -2 * C / 27
6060 T(7, 9) = -2 * DD / 27
6070 T(8, 1) = 7 / 27
6080 T(8, 2) = -EE / 27
6090 T(8, 3) = -F / 27
6100 T(8, 7) = 20 / 27
6110 T(8, 8) = 2 * EE / 27
6120 T(8, 9) = 2 * F / 27
6130 T(9, 1) = 20 / 27
6140 T(9, 2) = -2 * EE / 27
6150 T(9, 3) = -2 * F / 27
6160 T(9, 7) = 7 / 27
6170 T(9, 8) = EE / 27
6180 T(9, 9) = F / 27
6190 FOR II = 1 TO 9
6200 FOR JJ = 4 TO 9
6210 T(10, II) = T(10, II) + T(JJ, II) / 4
6220 NEXT JJ, II
6230 FOR II = 1 TO 9
6240 FOR JJ = 1 TO 3
6250 T(10, II) = T(10, II) - T(JJ, II) / 6
6260 NEXT JJ, II
6270 FOR II = 1 TO 3
6280 FOR JJ = 1 TO 10
6290 CB(II, JJ) = 0
6300 FOR KK = 1 TO 3
```

```
6310 CB(II, JJ) = CB(II, JJ) + C(II, KK) * B(KK, JJ)
6320 NEXT KK, JJ, II
6330 FOR II = 1 TO 3
6340 FOR JJ = 1 TO 9
6350 BC(II, JJ) = 0
6360 FOR KK = 1 TO 10
6370 BC(II, JJ) = BC(II, JJ) + CB(II, KK) * T(KK, JJ)
6380 NEXT KK: BCT(JJ, II) = BC(II, JJ): NEXT JJ, II
6390 FOR II = 1 TO 3
6400 FOR JJ = 1 TO 9
6410 DB(II, JJ) = 0
6420 FOR KK = 1 TO 3
6430 DB(II, JJ) = DB(II, JJ) + D(II, KK) * BC(KK, JJ)
6440 NEXT KK, JJ, II
6500 FS(1) = (9 * L1 ^ 3 - 9 * L1 ^ 2 + 2 * L1) / 2
6510 FS(2) = (9 * L2 ^ 3 - 9 * L2 ^ 2 + 2 * L2) / 2
6520 FS(3) = (9 * L3 ^ 3 - 9 * L3 ^ 2 + 2 * L3) / 2
6530 FS(4) = 13.5 * L1 ^ 2 * L2 - 4.5 * L1 * L2
6540 FS(5) = 13.5 * L1 * L2 ^ 2 - 4.5 * L1 * L2
6550 FS(6) = 13.5 * L2 ^ 2 * L3 - 4.5 * L2 * L3
6560 FS(7) = 13.5 * L2 * L3 ^ 2 - 4.5 * L2 * L3
6570 FS(8) = 13.5 * L3 ^ 2 * L1 - 4.5 * L3 * L1
6580 FS(9) = 13.5 * L3 * L1 ^ 2 - 4.5 * L3 * L1
6590 FS(10) = 27 * L1 * L2 * L3
6600 FOR II = 1 TO 9
6610 N(II) = 0
6620 FOR JJ = 1 TO 10
6630 N(II) = N(II) + FS(JJ) * T(JJ, II)
6640 NEXT JJ, II
6690 RETURN
10000 PRINT "TO CONTINUE, PRESS Y"
IF Z9 = 99 THEN PRINT "EIGENMODE="; IJ: PRINT "FREQUENCY="; SQR(1 / AM(IJ)) /
(2 * 3.14159); "Hz"
10001 IF Z9 = 77 OR Z9 = 99 THEN GOTO 10300
10005 XZ = 1E+12
10010 YZ = 1E+12
10020 XM = -1E+12
10030 YM = -1E+12
10070 FOR EL = 1 TO LS
10080 I = I9(EL)
10090 J = J9(EL)
10100 K = K9(EL)
10130 IF X(I) < XZ THEN XZ = X(I)
10140 IF Y(I) < YZ THEN YZ = Y(I)
10150 IF X(J) < XZ THEN XZ = X(J)
10160 IF Y(J) < YZ THEN YZ = Y(J)
10165 IF X(K) < XZ THEN XZ = X(K)
10167 IF Y(K) < YZ THEN YZ = Y(K)
10170 IF X(I) > XM THEN XM = X(I)
10180 IF Y(I) > YM THEN YM = Y(I)
10190 IF X(J) > XM THEN XM = X(J)
10200 IF Y(J) > YM THEN YM = Y(J)
10205 IF X(K) > XM THEN XM = X(K)
```

```
10207 IF Y(K) > YM THEN YM = Y(K)
10210 NEXT EL
10250 XS = 350 / (XM - XZ)
10270 YS = 175 / (YM - YZ)
10280 IF XS > YS THEN XS = YS
10300 IF Z9 = 0 OR Z9 = 77 THEN SCREEN 9
10350 FOR EL = 1 TO LS
10360 I = I9(EL)
10370 J = J9(EL)
10375 K = K9(EL)
10377 FOR IV = 1 TO 3
10378 IF IV = 1 THEN II = I: JJ = J
10379 IF IV = 2 THEN II = J: JJ = K
10380 IF IV = 3 THEN II = K: JJ = I
10385 IF Z9 = 99 THEN GOTO 10440
10390 XI = (X(II) - XZ) * XS
10400 YI = (Y(II) - YZ) * XS
10410 XJ = (X(JJ) - XZ) * XS
10420 YJ = (Y(JJ) - YZ) * XS
10430 GOTO 10460
10440 XI = X(II) - XZ * XS: XJ = X(JJ) - XZ * XS
10450 YI = Y(II) - YZ * XS: YJ = Y(JJ) - YZ * XS
10460 LINE ((XI + YI) * .866 + 40, 150 - (YI - XI))-((XJ + YJ) * .866 + 40, 150 - (YJ - XJ))
10465 NEXT IV
10470 NEXT EL
10650 IF Z9 = 77 THEN GOTO 10800
10660 A$ = INKEY$: IF A$ = "" THEN GOTO 10660
10670 IF A$ = "Y" OR A$ = "y" THEN GOTO 10720
10700 GOTO 10660
10720 SCREEN 0: WIDTH 80
10800 RETURN
11000 REM PLOT OF EIGENMODES
11010 FR(0) = 0
11020 FOR II = 1 TO N
11030 FR(II) = FR(II - 1)
11040 FR(II) = FR(II) + 1
11050 FOR JJ = 1 TO NF * 3
11070 IF FR(II) = NS(JJ) THEN GOTO 11040
11080 NEXT JJ
11090 NEXT II
11100 FOR IJ = 1 TO M1
11110 FOR II = 1 TO NJ
11120 X(II) = XG(II)
11130 Y(II) = YG(II)
11140 NEXT II
11150 Z9 = 77
11160 GOSUB 10000
11180 FOR II = 1 TO N
11190 UG(FR(II)) = VC(II, IJ)
11300 NEXT II
11310 SC = 10
11320 FOR II = 1 TO NJ
11330 X(II) = XG(II) * XS: REM + SC * UG(II * 2 - 1)
```

```
11340 Y(II) = YG(II) * XS + SC * UG(II * 3 - 1)
11360 NEXT II
11380 Z9 = 99
11390 GOSUB 10000
11410 NEXT IJ
11430 RETURN
25000 END
```

Appendix 19

Computer Program for the Vibration of Doubly-Curved Shells "VIBSHELF"

```
100 GOSUB 3900
110 IF A$ < > "Y" THEN PRINT : PRINT "GO BACK & READ IT !": GOTO 25000
120 REM COPYRIGHT OF DR.C.T.F.ROSS
130 REM 6 HURSTVILLE DRIVE, WATERLOOVILLE
140 REM PORTSMOUTH. ENGLAND
150 PRINT : PRINT "FREE VIBRATION OF DOUBLY-CURVED SHELLS": PRINT
151 INPUT "TYPE IN THE NAME OF YOUR INPUT FILE "; FILENAME$
PRINT "IF YOU ARE SATISFIED WIT THIS FILENAME, TYPE Y; ELSE N"
152 A$ = INKEY$: IF A$ = "" THEN GOTO 152
IF A$ = "N" OR A$ = "n" THEN GOTO 151
IF A$ = "Y" OR A$ = "y" THEN GOTO 153
GOTO 152
153 INPUT "TYPE IN THE NAME OF YOUR OUTPUT FILE "; OUT$
PRINT "IF YOU ARE SATISFIED WITH THIS FILENAME, TYPE Y; ELSE N"
154 B$ = INKEY$: IF B$ = "" THEN GOTO 154
IF B$ = "N" OR B$ = "n" THEN GOTO 153
IF B$ = "Y" OR B$ = "y" THEN GOTO 160
GOTO 154
160 OPEN FILENAME$ FOR INPUT AS #1
162 OPEN OUT$ FOR OUTPUT AS #2
169 REM PRINT:PRINT"TYPE IN NUMBER OF NODAL POINTS"
170 INPUT #1, NN
180 IF NN < 3 OR NN > 199 THEN GOSUB 4130: STOP
200 NJ = NN
INPUT #1, N
210 REM PRINT"TYPE IN NUMBERS OF ELEMENTS"
220 INPUT #1, LS
230 IF LS < 1 OR LS > 199 THEN GOSUB 4130: STOP
260 REM PRINT"TYPE IN THE NUMBER OF NODES WITH ZERO DISPLACEMENTS"
270 INPUT #1, NF
280 IF NF < 0 OR NF > NN THEN GOSUB 4130: STOP
310 NJ6 = 199
320 NP = NJ6 + 1
350 DIM A(NJ6, NJ6), XX(NJ6, NP), X(200)
355 DIM D(3, 3), C(3, 3), B(3, 10), T(10, 9), STIFF(9, 9), MASS(9, 9)
357 DIM CB(3, 10), BC(3, 9), BCT(9, 3), DB(3, 9), FS(10), N(9)
360 DIM SK(18, 18), SM(18, 18), SKB(9, 9), SKP(6, 6), SMB(9, 9), SMP(6, 6), NS(NF), Y(NN)
365 DIM UG(NJ6), XG(NN), YG(NN), ZG(NN), I9(LS), J9(LS), K9(LS)
367 DIM DC(18, 18), TEMP(18, 18), B1(3), C1(3)
368 DIM AM(20), VC(100, 20), Z(200), IG(100), VE(100), FR(100)
369 DIM ZETA1(3), ZETA2(3), ZETA3(3), XA(3), YA(3), ZA(3), SY21(3), SY34(3)
370 DIM ZETAP(3, 3), ZETAN(3, 3), UNIT(3, 3), XL(9), COORD(9)
375 GOTO 1900
380 PRINT "THE EIGENVALUES ARE BEING DETERMINED": PRINT
390 GOSUB 510
410 FOR I = 1 TO M1
420 PRINT #2, "EIGENVALUE="; AM(I)
430 PRINT #2, "FREQUENCY="; SQR(1 / AM(I)) / (2 * 3.141593)
440 PRINT #2, "EIGENVECTOR IS"
```

```
450 PRINT #2,
460 FOR J = 1 TO N
470 PRINT #2, VC(J, I); "   ";
480 PRINT #2, : NEXT J
490 PRINT #2, : NEXT I
495 REM GOSUB 11000
500 GOTO 25000
510 MN = N
520 NN = N
530 GOSUB 1390
540 M = 1
550 FOR I = 1 TO NN
560 VC(I, M) = X(I)
570 XX(I, M) = VC(I, M)
580 NEXT I
590 AM(M) = XM
600 IF M1 < 2 THEN RETURN
610 FOR M = 2 TO M1
620 FOR I = 1 TO NN
630 K4 = ABS(XX(I, M - 1) - 1)
640 IF K4 < .00001 THEN IR = I
650 NEXT I
660 IG(M - 1) = IR
670 FOR I = 1 TO NN
680 XX(MN - I + 1, MN - M + 3) = A(IR, I)
690 NEXT I
700 FOR I = 1 TO NN
710 FOR J = 1 TO NN
720 Z1 = MN - J + 1
730 Z2 = MN - M + 3
740 A(I, J) = A(I, J) - XX(I, M - 1) * XX(Z1, Z2)
750 NEXT J
760 NEXT I
770 FOR I = 1 TO NN
780 IF I = IR THEN GOTO 870
790 IF I > IR THEN K1 = I - 1
800 IF I < IR THEN K1 = I
810 FOR J = 1 TO NN
820 IF J = IR THEN GOTO 860
830 IF J > IR THEN K2 = J - 1
840 IF J < IR THEN K2 = J
850 A(K1, K2) = A(I, J)
860 NEXT J
870 NEXT I
880 NN = NN - 1
890 M3 = NN
900 IF M < > MN THEN GOTO 940
910 XM = A(1, 1)
920 X(1) = 1
930 GOTO 950
940 GOSUB 1390
950 FOR I = 1 TO NN
960 XX(I, M) = X(I)
```

```
970 NEXT I
980 AM(M) = XM
990 M4 = M - 1
1000 M5 = 1000 - M4
1010 FOR M8 = M5 TO 999
1020 M6 = M3 + 1
1030 M2 = 1000 - M8
1040 M7 = IG(M2) + 1
1050 IF M6 < M7 THEN GOTO 1120
1060 N9 = 1000 - M7
1070 N8 = 1000 - M6
1080 FOR I3 = N8 TO N9
1090 I = 1000 - I3
1100 X(I) = X(I - 1)
1110 NEXT I3
1120 J = IG(M2)
1130 X(J) = 0
1140 SUM = 0
1150 FOR I = 1 TO M6
1160 Z3 = MN - I + 1
1170 Z4 = MN - M2 + 2
1180 SUM = SUM + XX(Z3, Z4) * X(I)
1190 NEXT I
1200 XK = (AM(M2) - XM) / SUM
1210 FOR I = 1 TO M6
1220 X(I) = XX(I, M2) - XK * X(I)
1230 NEXT I
1240 SUM = 0
1250 FOR I = 1 TO M6
1260 IF ABS(SUM) < ABS(X(I)) THEN SUM = X(I)
1270 NEXT I
1280 FOR I = 1 TO M6
1290 X(I) = X(I) / SUM
1300 NEXT I
1310 M3 = M3 + 1
1320 IF M2 < > 1 THEN GOTO 1360
1330 FOR I = 1 TO M3
1340 VC(I, M) = X(I)
1350 NEXT I
1360 NEXT M8
1370 NEXT M
1380 RETURN
1390 Y1 = 100000!
1400 FOR I = 1 TO NN
1410 X(I) = 1
1420 NEXT I
1430 XM = -100000!
1440 FOR I = 1 TO NN
1450 SG = 0
1460 FOR J = 1 TO NN
1470 SG = SG + A(I, J) * X(J)
1480 NEXT J
1490 Z(I) = SG
```

```
1500 NEXT I
1510 XM = 0
1520 FOR I = 1 TO NN
1530 IF ABS(XM) < ABS(Z(I)) THEN XM = Z(I)
1540 NEXT I
1550 FOR I = 1 TO NN
1560 X(I) = Z(I) / XM
1570 NEXT I
1580 IF ABS((Y1 - XM) / XM) > D THEN GOTO 1600
1590 GOTO 1620
1600 Y1 = XM
1610 GOTO 1440
1620 X3 = 0
1630 FOR I = 1 TO NN
1640 IF ABS(X3) < ABS(X(I)) THEN X3 = X(I)
1650 NEXT I
1660 FOR I = 1 TO NN
1670 X(I) = X(I) / X3
1680 NEXT I
1690 RETURN
1700 N9 = N - 1
1710 FOR NX = 1 TO N
1720 DI = A(NX, 1)
1730 IF DI = 0 THEN PRINT "MATRIX IS SINGULAR"
1740 FOR NY = 1 TO N9
1750 Y9 = NY + 1
1760 A(NX, NY) = A(NX, Y9) / DI
1770 NEXT NY
1780 A(NX, N) = 1 / DI
1790 FOR NZ = 1 TO N
1800 IF NZ = NX THEN GOTO 1870
1810 O = A(NZ, 1)
1820 FOR NY = 1 TO N9
1830 Y9 = NY + 1
1840 A(NZ, NY) = A(NZ, Y9) - A(NX, NY) * O
1850 NEXT NY
1860 A(NZ, N) = -A(NX, N) * O
1870 NEXT NZ
1880 NEXT NX
1890 RETURN
1900 REM START OF MAIN PART OF PROGRAM
1910 REM PRINT:PRINT"TYPE IN THE NODE NUMBERS & 'DIRECTIONS' OF THE ZERO
DISPLACEMENTS IN ASCENDING ORDER"
1930 FOR I = 1 TO NF: NS(I) = 0: NEXT I
1950 FOR I = 1 TO NF
1960 INPUT #1, NS(I)
2110 NEXT I
2140 REM PRINT:PRINT"TYPE IN THE NODAL COORDINATES"
2150 FOR I = 1 TO NN
2160 REM PRINT"TYPE IN X COORD. OF NODE ";I
2170 INPUT #1, X(I), Y(I), Z(I)
2175 XG(I) = X(I)
2180 REM PRINT"TYPE IN Y COORD. OF NODE ";I
```

```
2195 YG(I) = Y(I)
2197 REM PRINT"TYPE IN Z COORD. OF NODE ";I
2199 ZG(I) = Z(I)
2200 NEXT I
2245 INPUT #1, M1
2250 IF M1 < 1 OR M1 > N THEN GOSUB 4130: STOP
2260 INPUT #1, TH, E, NU, RH: REM "TYPE IN PLATE THICKNESS";TH,E,NU,RH
2310 GOSUB 4160
2320 REM PRINT:PRINT"TYPE IN MEMBER DETAILS"
2330 FOR LE = 1 TO LS
2340 REM PRINT:PRINT"TYPE IN DETAILS OF MEMBER NUMBER ";LE
2350 REM PRINT"TYPE IN I NODE"
2360 INPUT #1, I, J, K
2400 REM PRINT"TYPE IN J NODE"
2425 REM PRINT"TYPE IN K NODE"
2460 I9(LE) = I: J9(LE) = J: K9(LE) = K
GOSUB 4500
NEXT LE
IDS = 0
IMAX = 0
FOR LE = 1 TO LS
ISUP = 0
JSUP = 0
KSUP = 0
I = I9(LE): J = J9(LE): K = K9(LE)
GOSUB 14000
IF I < > 0 THEN GOTO 2465
INPUT #1, XI, YI, ZI: PRINT #2, "X0,Y0 & Z0 FOR I NODE="; XI, YI, ZI: GOTO 2466
2465 XI = X(I): YI = Y(I): ZI = Z(I)
2466 IF J < > 0 THEN GOTO 2467
INPUT #1, XJ, YJ, ZJ: PRINT #2, "X0,Y0 & Z0 FOR J NODE="; XJ, YJ, ZJ: GOTO 2468
2467 XJ = X(J): YJ = Y(J): ZJ = Z(J)
2468 IF K < > 0 THEN GOTO 2469
INPUT #1, XK, YK, ZK: PRINT #2, "X0,Y0 & Z0 FOR K NODE="; XK, YK, ZK: GOTO 2470
2469 XK = X(K): YK = Y(K): ZK = Z(K)
2470 REM
2480 COORD(1) = 0: COORD(4) = XJ - XI: COORD(7) = XK - XI
2482 COORD(2) = 0: COORD(5) = YJ - YI: COORD(8) = YK - YI
2483 COORD(3) = 0: COORD(6) = ZJ - ZI: COORD(9) = ZK - ZI
2485 PRINT : PRINT "ELEMENT NUMBER "; LE; " UNDER COMPUTATION": PRINT
2487 FOR II = 1 TO 9: XL(II) = 0: NEXT II
2490 FOR II = 1 TO 18: FOR JJ = 1 TO 18
2500 SK(II, JJ) = 0: SM(II, JJ) = 0: TEMP(II, JJ) = 0: DC(II, JJ) = 0
2510 NEXT JJ, II
2520 FOR II = 1 TO 6: FOR JJ = 1 TO 6
2530 SKP(II, JJ) = 0: SMP(II, JJ) = 0
2540 NEXT JJ, II
2581 FOR II = 1 TO 9: FOR JJ = 1 TO 9
2582 STIFF(II, JJ) = 0: MASS(II, JJ) = 0: SKB(II, JJ) = 0: SMB(II, JJ) = 0
2583 NEXT JJ, II
2584 FOR II = 1 TO 3: FOR JJ = 1 TO 3
2585 D(II, JJ) = 0
2586 NEXT JJ, II
```

```
2588 D(1, 1) = E * TH ^ 3 / (12 * (1 - NU * NU)): D(2, 2) = D(1, 1)
2590 D(1, 2) = NU * D(1, 1): D(2, 1) = D(1, 2)
2592 D(3, 3) = D(1, 1) * (1 - NU)
2593 GOSUB 7000
2595 FOR II = 1 TO 9: FOR JJ = 1 TO 9
2597 XL(II) = XL(II) + DC(II, JJ) * COORD(JJ)
2598 NEXT JJ, II
2600 FOR IJ = 1 TO 3
2605 IF IJ = 1 THEN L1 = .5: L2 = .5
2610 IF IJ = 2 THEN L1 = 0: L2 = .5
2620 IF IJ = 3 THEN L1 = .5: L2 = 0
2630 L3 = 1 - L1 - L2
2640 GOSUB 5000
2650 FOR II = 1 TO 9
2660 FOR JJ = 1 TO 9
2670 FOR KK = 1 TO 3
2680 SKB(II, JJ) = SKB(II, JJ) + BCT(II, KK) * DB(KK, JJ) * DEL / 6
2690 NEXT KK
2700 SMB(II, JJ) = SMB(II, JJ) + N(II) * N(JJ) * DEL * RH * TH / 6
2710 NEXT JJ, II
2800 NEXT IJ
2810 FOR II = 1 TO 9: FOR JJ = 1 TO 9
2820 STIFF(II, JJ) = SKB(II, JJ): MASS(II, JJ) = SMB(II, JJ)
2830 NEXT JJ, II
2840 FOR II = 1 TO 3: FOR JJ = 1 TO 9
2850 STIFF(3 * II - 1, JJ) = SKB(3 * II, JJ): MASS(3 * II - 1, JJ) = SMB(3 * II, JJ)
2860 STIFF(3 * II, JJ) = -SKB(3 * II - 1, JJ): MASS(3 * II, JJ) = -SMB(3 * II - 1, JJ)
NEXT JJ, II
FOR II = 1 TO 9: FOR JJ = 1 TO 9: SKB(II, JJ) = STIFF(II, JJ)
SMB(II, JJ) = MASS(II, JJ): NEXT JJ, II
FOR II = 1 TO 3: FOR JJ = 1 TO 9
2865 STIFF(JJ, 3 * II - 1) = SKB(JJ, 3 * II): MASS(JJ, 3 * II - 1) = SMB(JJ, 3 * II)
2867 STIFF(JJ, 3 * II) = -SKB(JJ, 3 * II - 1): MASS(JJ, 3 * II) = -SMB(JJ, 3 * II - 1)
2870 NEXT JJ, II
2875 CN = E * TH / (1 - NU ^ 2): ZU = NU: MU = (1 - NU) / 2
2880 B1(1) = XL(5) - XL(8)
2882 B1(2) = XL(8) - XL(2)
2884 B1(3) = XL(2) - XL(5)
2886 C1(1) = XL(7) - XL(4)
2888 C1(2) = XL(1) - XL(7)
2890 C1(3) = XL(4) - XL(1)
2892 DL = XL(1) * B1(1) + XL(4) * B1(2) + XL(7) * B1(3)
2900 GOSUB 7900
2902 FOR II = 1 TO 3
2904 I2 = 2 * II - 2
2906 FOR JJ = 1 TO 3
2910 J2 = 2 * JJ - 2
2912 SKP(I2 + 1, J2 + 1) = .5 * CN * (B1(II) * B1(JJ) + MU * C1(II) * C1(JJ)) / DL
2914 SKP(I2 + 2, J2 + 2) = .5 * CN * (C1(II) * C1(JJ) + MU * B1(II) * B1(JJ)) / DL
2916 SKP(I2 + 1, J2 + 2) = .5 * CN * (ZU * B1(II) * C1(JJ) + MU * C1(II) * B1(JJ)) / DL
2918 SKP(I2 + 2, J2 + 1) = .5 * CN * (ZU * B1(JJ) * C1(II) + MU * C1(JJ) * B1(II)) / DL
2920 NEXT JJ, II
2924 FOR KK = 1 TO 3: FOR LL = 1 TO 3
```

550

2926 SK(KK + 2, LL + 2) = STIFF(KK, LL): SM(KK + 2, LL + 2) = MASS(KK, LL)
2927 SK(KK + 2, 8 + LL) = STIFF(KK, LL + 3): SM(KK + 2, 8 + LL) = MASS(KK, LL + 3)
2928 SK(KK + 2, 14 + LL) = STIFF(KK, LL + 6): SM(KK + 2, 14 + LL) = MASS(KK, LL + 6)
2929 SK(8 + KK, 8 + LL) = STIFF(KK + 3, LL + 3): SM(8 + KK, 8 + LL) = MASS(KK + 3, LL + 3)
2930 SK(8 + KK, 14 + LL) = STIFF(KK + 3, LL + 6): SM(8 + KK, 14 + LL) = MASS(KK + 3, LL + 6)
2931 SK(14 + KK, 14 + LL) = STIFF(KK + 6, LL + 6): SM(14 + KK, 14 + LL) = MASS(KK + 6, LL + 6)
2933 NEXT LL, KK
2934 FOR KK = 1 TO 2: FOR LL = 1 TO 2
2936 SK(KK, LL) = SKP(KK, LL): SM(KK, LL) = SMP(KK, LL)
2938 SK(KK, 6 + LL) = SKP(KK, LL + 2): SM(KK, 6 + LL) = SMP(KK, LL + 2)
2940 SK(KK, 12 + LL) = SKP(KK, LL + 4): SM(KK, 12 + LL) = SMP(KK, LL + 4)
2942 SK(KK + 6, LL + 6) = SKP(KK + 2, LL + 2): SM(KK + 6, LL + 6) = SMP(KK + 2, LL + 2)
2944 SK(KK + 6, 12 + LL) = SKP(KK + 2, LL + 4): SM(KK + 6, 12 + LL) = SMP(KK + 2, LL + 4)
2946 SK(KK + 12, 12 + LL) = SKP(KK + 4, LL + 4): SM(KK + 12, 12 + LL) = SMP(KK + 4, LL + 4)
2950 NEXT LL, KK
2951 IF ABS(SK(6, 6)) > .1 THEN GOTO 2957
2952 SK(6, 6) = .5 * E * DL * TH * .00001: SK(6, 12) = -.5 * SK(6, 6): SK(12, 6) = SK(6, 12)
2954 SK(12, 12) = SK(6, 6): SK(6, 18) = SK(6, 12): SK(18, 6) = SK(6, 18)
2956 SK(18, 18) = SK(6, 6): SK(12, 18) = SK(6, 12): SK(18, 12) = SK(12, 18)
2957 GOSUB 8100
2958 FOR II = 1 TO 18
2960 FOR JJ = 1 TO 18
2962 TEMP(II, JJ) = 0: NEXT JJ
FOR JJ = 1 TO 18
2964 FOR KK = 1 TO 18
2966 TEMP(II, JJ) = TEMP(II, JJ) + SK(II, KK) * DC(KK, JJ)
2968 NEXT KK, JJ, II
2970 FOR II = 1 TO 18
2972 FOR JJ = 1 TO 18
2974 SK(II, JJ) = 0: NEXT JJ
FOR JJ = 1 TO 18
2976 FOR KK = 1 TO 18
2978 SK(II, JJ) = SK(II, JJ) + DC(KK, II) * TEMP(KK, JJ)
2979 NEXT KK, JJ, II
2980 FOR II = 1 TO 18
2981 FOR JJ = 1 TO 18
2982 TEMP(II, JJ) = 0: NEXT JJ
FOR JJ = 1 TO 18
2984 FOR KK = 1 TO 18
2986 TEMP(II, JJ) = TEMP(II, JJ) + SM(II, KK) * DC(KK, JJ)
2988 NEXT KK, JJ, II
2990 FOR II = 1 TO 18
2992 FOR JJ = 1 TO 18
2994 SM(II, JJ) = 0: NEXT JJ
FOR JJ = 1 TO 18

551

```
2996 FOR KK = 1 TO 18
2998 SM(II, JJ) = SM(II, JJ) + DC(KK, II) * TEMP(KK, JJ)
2999 NEXT KK, JJ, II
3010 FOR II = 1 TO 3
IF II = 1 THEN MM = 6 * I - 6
IF II = 2 THEN MM = 6 * J - 6
IF II = 3 THEN MM = 6 * K - 6
IF MM < 0 THEN GOTO 3150
FOR JJ = 1 TO 3
IF JJ = 1 THEN MN = 6 * I - 6
IF JJ = 2 THEN MN = 6 * J - 6
IF JJ = 3 THEN MN = 6 * K - 6
IF MN < 0 THEN GOTO 3100
NM = 6 * II - 6
K1 = MM - IDS
FOR IP1 = 1 TO 6
K1 = K1 + 1
L1 = MN - IDS
NM = NM + 1
N2 = 6 * JJ - 6
FOR JP1 = 1 TO 6
L1 = L1 + 1
N2 = N2 + 1
A(K1, L1) = A(K1, L1) + SK(NM, N2)
XX(K1, L1) = XX(K1, L1) + SM(NM, N2)
NEXT JP1
NEXT IP1
3100 NEXT JJ
3150 NEXT II
REM BRANCHING TO EIGENVALUE ECONOMISER
IF I > IMAX THEN IMAX = I
IF J > IMAX THEN IMAX = J
IF K > IMAX THEN IMAX = K
NINST = 6 * IMAX - IDS
IF ISUP = 0 THEN GOTO 3156
IF IAX = 0 THEN GOTO 3152
S = 6 * I - IDS - 2
GOSUB 14410
3152 IF IAY = 0 THEN GOTO 3154
S = 6 * I - IDS - 1
GOSUB 14410
3154 IF IAZ = 0 THEN GOTO 3156
S = 6 * I - IDS
GOSUB 14410
3156 IF JSUP = 0 THEN GOTO 3164
IF JAX = 0 THEN GOTO 3160
S = 6 * J - IDS - 2
GOSUB 14410
3160 IF JAY = 0 THEN GOTO 3162
S = 6 * J - IDS - 1
GOSUB 14410
3162 IF JAZ = 0 THEN GOTO 3164
S = 6 * J - IDS
```

```
GOSUB 14410
3164 IF KSUP = 0 THEN GOTO 3174
IF KAX = 0 THEN GOTO 3170
S = 6 * K - IDS - 2
GOSUB 14410
3170 IF KAY = 0 THEN GOTO 3172
S = 6 * K - IDS - 1
GOSUB 14410
3172 IF KAZ = 0 THEN GOTO 3174
S = 6 * K - IDS
GOSUB 14410
3174 REM END OF ECONOMISING TECHNIQUE
NEXT LE
IF NF = 0 THEN GOTO 3420
3190 FOR I = 1 TO NF
3200 N7 = NS(I)
A(N7, N7) = A(N7, N7) * 1E+08
3410 NEXT I
3420 REM PRINT:PRINT"TYPE IN THE NUMBER OF CONCENTRATED MASSES":PRINT
3430 INPUT #1, NC
3440 IF NC < 0 OR NC > NN THEN GOSUB 4130: STOP
3490 IF NC = 0 THEN GOTO 3750
3500 DIM PO(NC), MC(NC)
3510 FOR II = 1 TO NC: PO(II) = 0: MC(II) = 0: NEXT II
3520 FOR II = 1 TO NC
3530 PRINT "TYPE IN NODE NUMBER OF MASS"
3540 INPUT POSM
3550 IF POSM < 1 OR POSM > NN THEN GOSUB 4130: GOTO 3530
3580 PRINT "TYPE IN VALUE OF MASS"
3590 INPUT MASS
3600 IF MASS < 0 THEN GOSUB 4130: GOTO 3580
3630 PO(II) = POSM
3640 MC(II) = MASS
3650 J1 = 3 * POSM
XX(J1 - 2, J1 - 2) = XX(J1 - 2, J1 - 2) + MASS
XX(J1 - 1, J1 - 1) = XX(J1 - 1, J1 - 1) + MASS
3720 XX(J1, J1) = XX(J1, J1) + MASS
3730 NEXT II
3740 GOSUB 4410
3750 PRINT : PRINT "THE STIFFNESS MATRIX IS BEING INVERTED": PRINT
3760 GOSUB 1700
3770 FOR I = 1 TO N
3780 FOR J = 1 TO N
3790 VE(J) = 0
3800 FOR K = 1 TO N
3810 VE(J) = VE(J) + A(I, K) * XX(K, J)
3820 NEXT K
3830 NEXT J
3840 FOR J = 1 TO N
3850 A(I, J) = VE(J)
3860 NEXT J
3870 NEXT I
3880 D = .001
```

```
3890 GOTO 380
3900 CLS
3910 PRINT "COPYRIGHT OF DR.C.T.F.ROSS"
3920 PRINT
3930 PRINT "NOT TO BE REPRODUCED"
3940 PRINT
3950 PRINT "HAVE YOU READ THE APPROPRIATE MANUAL"
3980 PRINT
3990 PRINT "BY"
4000 PRINT
4010 PRINT "C.T.F.ROSS": PRINT
4020 PRINT "IF YES, TYPE Y; ELSE N"
4030 A$ = INKEY$: IF A$ = "" THEN GOTO 4030
4035 IF A$ = "y" THEN A$ = "Y"
4040 RETURN
4050 PRINT : PRINT
4060 PRINT "IF YOU ARE SATISFIED WITH THE INPUT OF YOUR LAST BATCH OF DATA,
TYPE Y; ELSE N": PRINT
4070 X$ = INKEY$: IF X$ = "" THEN GOTO 4070
4075 IF X$ = "y" THEN X$ = "Y"
4080 RETURN
4130 REM WARNING
4140 PRINT : PRINT "INCORRECT DATA": PRINT
4150 RETURN
4160 REM OUTPUT 1
4180 PRINT #2,
4190 PRINT #2, "FREE VIBRATIONS OF DOUBLY-CURVED SHELLS"
4200 PRINT #2, : PRINT #2, "NUMBER OF NON-ZERO NODAL POINTS="; NN
PRINT #2, "NUMBER OF FREE DISPLACEMENTS="; N
4210 PRINT #2, "NUMBER OF ELEMENTS"; LS
4220 PRINT #2, "NUMBER OF NODES WITH PIN-JOINTS="; NF
IF NF = 0 THEN GOTO 4330
4230 PRINT #2, : PRINT #2, "THE DISPLACEMENT POSITIONS OF THE 'FINISHED' ZERO
DISPLACEMENTS ARE:-"
4240 FOR II = 1 TO NF
PRINT #2, NS(II)
4320 NEXT II
4330 PRINT #2, : PRINT #2, "THE NODAL COORDINATES OF THE NON-ZERO NODES
ARE:-"
4340 FOR I = 1 TO NN
4350 PRINT #2, "NODE="; I
4360 PRINT #2, "X("; I; ")="; X(I); "   Y("; I; ")="; Y(I); "   Z("; I; ")="; Z(I)
4370 NEXT I
4380 PRINT #2, : PRINT #2, "NUMBER OF FREQUENCIES="; M1
4385 PRINT #2, "PLATE THICKNESS="; TH
4390 PRINT #2, "YOUNG'S MODULUS="; E
4395 PRINT #2, "POISSON'S RATIO="; NU
4397 PRINT #2, "DENSITY="; RH
4400 RETURN
4410 REM OUTPUT 2
4420 PRINT #2, : PRINT #2, "NUMBER OF CONCENTRATED MASSES="; NC
4430 IF NC = 0 THEN RETURN
4440 FOR II = 1 TO NC
```

4450 PRINT #2, "THE NODAL POSITION OF MASS="; PO(II)
4460 PRINT #2, "VALUE OF MASS="; MC(II)
4470 NEXT II
4490 RETURN
4500 REM OUTPUT 3
4520 PRINT #2, : PRINT #2, "TOPOLOGY OF ELEMENT "; LE; ": "; I; "-"; J; "-"; K
4570 RETURN
4580 REM END
5000 XJI = XL(4) - XL(1)
5010 YJI = XL(5) - XL(2)
5020 XKJ = XL(7) - XL(4)
5030 YKJ = XL(8) - XL(5)
5040 XIK = XL(1) - XL(7)
5050 YIK = XL(2) - XL(8)
5070 X1 = SQR(XJI * XJI + YJI * YJI)
5080 X2 = SQR(XKJ * XKJ + YKJ * YKJ)
5090 X3 = SQR(XIK * XIK + YIK * YIK)
5100 DEL = XJI * YKJ - YJI * XKJ
5110 XJI = XJI / X1
5120 YJI = YJI / X1
5130 XKJ = XKJ / X2
5140 YKJ = YKJ / X2
5150 XIK = XIK / X3
5160 YIK = YIK / X3
5170 AX = XJI * XJI
5180 AY = YJI * YJI
5190 AXY = 1.414 * XJI * YJI
5200 BX = XKJ * XKJ
5210 BY = YKJ * YKJ
5220 BXY = 1.414 * XKJ * YKJ
5230 CX = XIK * XIK
5240 CY = YIK * YIK
5250 CXY = 1.414 * XIK * YIK
5260 DET = AX * (BY * CXY - BXY * CY) - AY * (BX * CXY - BXY * CX) + AXY * (BX * CY - BY * CX)
5270 C(1, 1) = BY * CXY - BXY * CY
5280 C(1, 2) = -AY * CXY + AXY * CY
5290 C(1, 3) = AY * BXY - AXY * BY
5300 C(2, 1) = -BX * CXY + BXY * CX
5310 C(2, 2) = AX * CXY - AXY * CX
5320 C(2, 3) = -AX * BXY + AXY * BX
5330 C(3, 1) = BX * CY - BY * CX
5340 C(3, 2) = -AX * CY + AY * CX
5350 C(3, 3) = AX * BY - AY * BX
5360 FOR II = 1 TO 3: FOR JJ = 1 TO 3: C(II, JJ) = C(II, JJ) / DET: NEXT JJ
5370 FOR JJ = 1 TO 10
5380 B(II, JJ) = 0
5390 NEXT JJ
5400 NEXT II
5410 B(1, 1) = 27 * L1 - 9
5420 B(1, 2) = 27 * L2 - 9
5430 B(1, 4) = 27 * L2 - 54 * L1 + 9
5440 B(1, 5) = 27 * L1 - 54 * L2 + 9

555

```
5450 B(1, 6) = 27 * L3
5460 B(1, 9) = 27 * L3
5470 B(1, 10) = -54 * L3
5480 B(2, 2) = 27 * L2 - 9
5490 B(2, 3) = 27 * L3 - 9
5500 B(2, 5) = 27 * L1
5510 B(2, 10) = -54 * L1
5520 B(2, 6) = 27 * L3 - 54 * L2 + 9
5530 B(2, 7) = 27 * L2 - 54 * L3 + 9
5540 B(2, 8) = 27 * L1
5550 B(3, 1) = 27 * L1 - 9
5560 B(3, 3) = 27 * L3 - 9
5570 B(3, 4) = 27 * L2
5580 B(3, 7) = 27 * L2
5590 B(3, 8) = 27 * L1 - 54 * L3 + 9
5600 B(3, 9) = 27 * L3 - 54 * L1 + 9
5610 B(3, 10) = -54 * L2
5620 FOR JJ = 1 TO 10
5630 B(1, JJ) = -B(1, JJ) / (X1 * X1)
5640 B(2, JJ) = -B(2, JJ) / (X2 * X2)
5650 B(3, JJ) = -B(3, JJ) / (X3 * X3)
5660 NEXT JJ
5670 A = 2 * XJI * X1
5680 B = 2 * YJI * X1
5690 C = 2 * XKJ * X2
5700 DD = 2 * YKJ * X2
5710 EE = 2 * XIK * X3
5720 F = 2 * YIK * X3
5750 FOR II = 1 TO 10
5760 FOR JJ = 1 TO 9
5770 T(II, JJ) = 0
5780 NEXT JJ
5790 NEXT II
5800 T(1, 1) = 1
5810 T(2, 4) = 1
5820 T(3, 7) = 1
5830 T(4, 1) = 20 / 27
5840 T(4, 2) = 2 * A / 27
5850 T(4, 3) = 2 * B / 27
5860 T(4, 4) = 7 / 27
5870 T(4, 5) = -A / 27
5880 T(4, 6) = -B / 27
5890 T(5, 1) = 7 / 27
5900 T(5, 2) = A / 27
5910 T(5, 3) = B / 27
5920 T(5, 4) = 20 / 27
5930 T(5, 5) = -2 * A / 27
5940 T(5, 6) = -2 * B / 27
5950 T(6, 4) = 20 / 27
5960 T(6, 5) = 2 * C / 27
5970 T(6, 6) = 2 * DD / 27
5980 T(6, 7) = 7 / 27
5990 T(6, 8) = -C / 27
```

```
6000 T(6, 9) = -DD / 27
6010 T(7, 4) = 7 / 27
6020 T(7, 5) = C / 27
6030 T(7, 6) = DD / 27
6040 T(7, 7) = 20 / 27
6050 T(7, 8) = -2 * C / 27
6060 T(7, 9) = -2 * DD / 27
6070 T(8, 1) = 7 / 27
6080 T(8, 2) = -EE / 27
6090 T(8, 3) = -F / 27
6100 T(8, 7) = 20 / 27
6110 T(8, 8) = 2 * EE / 27
6120 T(8, 9) = 2 * F / 27
6130 T(9, 1) = 20 / 27
6140 T(9, 2) = -2 * EE / 27
6150 T(9, 3) = -2 * F / 27
6160 T(9, 7) = 7 / 27
6170 T(9, 8) = EE / 27
6180 T(9, 9) = F / 27
T(10, 1) = -1 / 6: T(10, 4) = -1 / 6: T(10, 7) = -1 / 6
6190 FOR II = 1 TO 9
6200 FOR JJ = 4 TO 9
6210 T(10, II) = T(10, II) + T(JJ, II) / 4
6220 NEXT JJ, II
6270 FOR II = 1 TO 3
6280 FOR JJ = 1 TO 10
6290 CB(II, JJ) = 0
6300 FOR KK = 1 TO 3
6310 CB(II, JJ) = CB(II, JJ) + C(II, KK) * B(KK, JJ)
6320 NEXT KK, JJ, II
6330 FOR II = 1 TO 3
6340 FOR JJ = 1 TO 9
6350 BC(II, JJ) = 0
6360 FOR KK = 1 TO 10
6370 BC(II, JJ) = BC(II, JJ) + CB(II, KK) * T(KK, JJ)
6380 NEXT KK: BCT(JJ, II) = BC(II, JJ): NEXT JJ, II
6390 FOR II = 1 TO 3
6400 FOR JJ = 1 TO 9
6410 DB(II, JJ) = 0
6420 FOR KK = 1 TO 3
6430 DB(II, JJ) = DB(II, JJ) + D(II, KK) * BC(KK, JJ)
6440 NEXT KK, JJ, II
6500 FS(1) = (9 * L1 ^ 3 - 9 * L1 ^ 2 + 2 * L1) / 2
6510 FS(2) = (9 * L2 ^ 3 - 9 * L2 ^ 2 + 2 * L2) / 2
6520 FS(3) = (9 * L3 ^ 3 - 9 * L3 ^ 2 + 2 * L3) / 2
6530 FS(4) = 13.5 * L1 ^ 2 * L2 - 4.5 * L1 * L2
6540 FS(5) = 13.5 * L1 * L2 ^ 2 - 4.5 * L1 * L2
6550 FS(6) = 13.5 * L2 ^ 2 * L3 - 4.5 * L2 * L3
6560 FS(7) = 13.5 * L2 * L3 ^ 2 - 4.5 * L2 * L3
6570 FS(8) = 13.5 * L3 ^ 2 * L1 - 4.5 * L3 * L1
6580 FS(9) = 13.5 * L3 * L1 ^ 2 - 4.5 * L3 * L1
6590 FS(10) = 27 * L1 * L2 * L3
6600 FOR II = 1 TO 9
```

```
6610 N(II) = 0
6620 FOR JJ = 1 TO 10
6630 N(II) = N(II) + FS(JJ) * T(JJ, II)
6640 NEXT JJ, II
6690 RETURN
7000 REM CALCULATION OF DIRECTIONAL COSINES
7010 XA(1) = XI
7020 YA(1) = YI
7030 ZA(1) = ZI
7040 XA(2) = XJ
7050 YA(2) = YJ
7060 ZA(2) = ZJ
7070 XA(3) = XK
7080 YA(3) = YK
7090 ZA(3) = ZK
7100 REM DIRECTION COSINES
7110 SY21(1) = XA(2) - XA(1)
7120 SY21(2) = YA(2) - YA(1)
7130 SY21(3) = ZA(2) - ZA(1)
7140 SY31(1) = XA(3) - XA(1)
7150 SY31(2) = YA(3) - YA(1)
7160 SY31(3) = ZA(3) - ZA(1)
7170 L21 = SQR(SY21(1) ^ 2 + SY21(2) ^ 2 + SY21(3) ^ 2)
7180 FOR II = 1 TO 3
7190 ZETA1(II) = SY21(II) / L21
7200 NEXT II
7210 FOR II = 1 TO 3
7220 FOR JJ = 1 TO 3
7230 UNIT(II, JJ) = 0
7240 NEXT JJ
7250 UNIT(II, II) = 1
7260 NEXT II
7270 FOR II = 1 TO 3
7280 FOR JJ = 1 TO 3
7290 ZETAP(II, JJ) = 0
7300 NEXT JJ
7310 FOR JJ = 1 TO 3
7320 ZETAP(II, JJ) = ZETAP(II, JJ) + ZETA1(II) * ZETA1(JJ)
7330 ZETAN(II, JJ) = -ZETAP(II, JJ)
7340 NEXT JJ
7350 NEXT II
7360 FOR II = 1 TO 3
7370 FOR JJ = 1 TO 3
7380 ZETAP(II, JJ) = UNIT(II, JJ) + ZETAN(II, JJ)
7390 NEXT JJ
7400 NEXT II
7410 FOR II = 1 TO 3
7420 SY34(II) = 0
7430 NEXT II
7440 FOR II = 1 TO 3
7450 FOR JJ = 1 TO 3
7460 SY34(II) = SY34(II) + ZETAP(II, JJ) * SY31(JJ)
7470 NEXT JJ
```

```
7480 NEXT II
7490 L21 = SQR(SY34(1) ^ 2 + SY34(2) ^ 2 + SY34(3) ^ 2)
7500 FOR II = 1 TO 3
7510 ZETA2(II) = SY34(II) / L21
7520 NEXT II
7530 AXY = (XA(1) * (YA(2) - YA(3)) - YA(1) * (XA(2) - XA(3)) + (XA(2) * YA(3) - YA(2) *
XA(3))) / 2
7540 AYZ = (YA(1) * (ZA(2) - ZA(3)) - ZA(1) * (YA(2) - YA(3)) + (YA(2) * ZA(3) - ZA(2) *
YA(3))) / 2
7550 AZX = (ZA(1) * (XA(2) - XA(3)) - XA(1) * (ZA(2) - ZA(3)) + (ZA(2) * XA(3) - XA(2) *
ZA(3))) / 2
7560 TRIANG = SQR(AYZ ^ 2 + AZX ^ 2 + AXY ^ 2)
7570 ZETA3(1) = AYZ / TRIANG
7580 ZETA3(2) = AZX / TRIANG
7590 ZETA3(3) = AXY / TRIANG
7600 FOR II = 1 TO 3
7610 DC(1, II) = ZETA1(II)
7620 DC(2, II) = ZETA2(II)
7630 DC(3, II) = ZETA3(II)
7640 NEXT II
7650 FOR II = 1 TO 5
7660 MM = 1 + 3 * II
7670 MN = MM + 1
7680 NN = MN + 1
7690 DC(MM, MM) = DC(1, 1)
7700 DC(MM, MN) = DC(1, 2)
7710 DC(MM, NN) = DC(1, 3)
7720 DC(MN, MM) = DC(2, 1)
7730 DC(MN, MN) = DC(2, 2)
7740 DC(MN, NN) = DC(2, 3)
7750 DC(NN, MM) = DC(3, 1)
7760 DC(NN, MN) = DC(3, 2)
7770 DC(NN, NN) = DC(3, 3)
7780 NEXT II
7790 REM DIRCOS
7800 RETURN
7900 REM CALCULATION OF (SMP)
7910 KN = RH * DL * TH / 24
7920 FOR II = 1 TO 6: FOR JJ = 1 TO 6
7930 SMP(II, JJ) = 0
7940 NEXT JJ
7950 SMP(II, II) = 2 * KN
7960 NEXT II
7970 SMP(1, 3) = KN: SMP(3, 1) = KN
7975 SMP(1, 5) = KN: SMP(5, 1) = KN
7980 SMP(2, 4) = KN: SMP(4, 2) = KN
7985 SMP(2, 6) = KN: SMP(6, 2) = KN
7990 SMP(3, 5) = KN: SMP(5, 3) = KN
8000 SMP(4, 6) = KN: SMP(6, 4) = KN
8010 RETURN
8100 FOR II = 1 TO 18
8110 FOR JJ = 1 TO II
8120 SK(II, JJ) = SK(JJ, II)
```

```
8130 SM(II, JJ) = SM(JJ, II)
8140 NEXT JJ, II
8150 RETURN
14000 REM EIGENVALUE ECONOMISER (PART1)
14010 IF I > = 0 THEN 14110
14020 I = -I
14030 ISUP = 1
14040 IX = 0
14050 IY = 0
14060 IZ = 0
14070 IAX = 1
14080 IAY = 1
14090 IAZ = 1
14100 GOTO 14210
14110 IF I < 999 THEN 14210
14120 I = I / 1000
14130 ISUP = 1
14140 IX = 1
14150 IY = 1
14160 IZ = 1
14170 IAX = 1
14180 IAY = 1
14190 IAZ = 1
14200 GOTO 14310
14210 IF J > = 0 THEN 14310
14220 J = -J
14230 JSUP = 1
14240 JX = 0
14250 JY = 0
14260 JZ = 0
14270 JAX = 1
14280 JAY = 1
14290 JAZ = 1
14300 GOTO 14392
14310 IF J < 999 THEN 14392
14320 J = J / 1000
14330 JSUP = 1
14340 JX = 1
14350 JY = 1
14360 JZ = 1
14370 JAX = 1
14380 JAY = 1
14390 JAZ = 1
14392 IF K > = 0 THEN GOTO 14398
K = -K
KSUP = 1
KAX = 1
KAY = 1
KAZ = 1
14398 REM END OF PART 1
14400 RETURN
14410 REM EIGENVALUE ECONOMISER (PART2)
14420 FOR II = 1 TO NINST
```

```
14430 IF II = S THEN 14490
14440 FOR JJ = 1 TO NINST
14450 IF JJ = S THEN 14480
14460 A(II, JJ) = A(II, JJ) - A(II, S) * A(S, JJ) / A(S, S)
14470 XX(II, JJ) = XX(II, JJ) - XX(II, S) * A(S, JJ) / A(S, S) - XX(S, JJ) * A(II, S) / A(S, S) +
XX(S, S) * A(S, JJ) * A(II, S) / (A(S, S) ^ 2)
14480 NEXT JJ
14490 NEXT II
14500 NINST = NINST - 1
14510 IDS = IDS + 1
14520 IF S > NINST THEN RETURN
14530 FOR II = 1 TO S - 1
14540 FOR JJ = S TO NINST
14550 A(II, JJ) = A(II, JJ + 1)
14560 A(JJ, II) = A(JJ + 1, II)
14570 XX(II, JJ) = XX(II, JJ + 1)
14580 XX(JJ, II) = XX(JJ + 1, II)
14590 NEXT JJ
14600 NEXT II
14610 FOR II = S TO NINST
14620 FOR JJ = S TO NINST
14630 A(II, JJ) = A(II + 1, JJ + 1): XX(II, JJ) = XX(II + 1, JJ + 1)
14640 NEXT JJ
14650 NEXT II
14660 FOR II = 1 TO NINST + 1
14670 A(NINST + 1, II) = 0
14680 A(II, NINST + 1) = 0
14690 XX(NINST + 1, II) = 0
14700 XX(II, NINST + 1) = 0
14710 NEXT II
14720 RETURN
25000 END
```

Computer Program for Stresses in Solids "TETRAF"

```
50 CLS : RESTORE
60 REM COPYRIGHT OF DR.C.T.F.ROSS
70 REM 6 HURSTVILLE DRIVE, WATERLOOVILLE
80 REM PORTSMOUTH. ENGLAND
90 PRINT : PRINT "3D STRESS (4 NODE TETRAHEDRON)": PRINT
100 PRINT : PRINT "PROGRAM BY DR.C.T.F.ROSS": PRINT
110 PRINT : PRINT "NOT TO BE REPRODUCED": PRINT
112 INPUT "TYPE IN THE NAME OF YOUR INPUT FILE "; FILENAME$
PRINT "IF YOU ARE SATISFIED WITH THIS FILENAME, TYPE Y; ELSE N"
113 A$ = INKEY$: IF A$ = "" THEN GOTO 113
IF A$ = "Y" OR A$ = "y" THEN GOTO 114
IF A$ = "N" OR A$ = "n" THEN GOTO 112
GOTO 113
114 OPEN FILENAME$ FOR INPUT AS #1
115 INPUT "TYPE IN THE NAME OF YOUR OUTPUT FILE "; OUT$
PRINT "IF YOU ARE SATISFIED WITH THIS FILENAME, TYPE Y; ELSE N"
116 B$ = INKEY$: IF B$ = "" THEN GOTO 116
IF B$ = "Y" OR B$ = "y" THEN GOTO 118
IF B$ = "N" OR B$ = "n" THEN GOTO 115
GOTO 116
118 OPEN OUT$ FOR OUTPUT AS #2
120 REM TYPE IN NO. OF NODES
130 INPUT #1, NJ: NN = 3 * NJ
135 NN3 = NN
140 IF NJ < 8 OR NJ > 199 THEN GOSUB 9995: STOP
145 NNODE = 8: NDF = 3
150 REM TYPE IN THE NO. OF 'CUBIC' ELEMENTS
160 INPUT #1, ES
170 IF ES < 1 OR ES > 199 THEN GOSUB 9995: STOP
180 REM TYPE IN THE NO. OF NODES WITH ZERO DISPLACEMENTS
190 INPUT #1, NF
195 IF NF < 3 OR NF > NJ THEN GOSUB 9995: STOP
200 DIM I(ES), J(ES), K(ES), L(ES)
205 DIM M(ES), N(ES), O(ES), P(ES)
210 DIM SK(12, 12), D(6, 6), BX(6, 12), BT(12, 6), DB(6, 12)
215 DIM VG(NJ), VEC(NJ), SHEAR(NJ)
220 DIM X(NJ), Y(NJ), Z(NJ), XG(NJ), YG(NJ), ZG(NJ), NS(NF * 3)
240 DIM CO(8, 3), SU(24), SR(6), DIAG(NN3)
245 DIM PRINCIPAL(NJ), MINIMUM(NJ)
250 REM TYPE IN ELASTIC MODULUS & POISSON'S RATIO
260 INPUT #1, E, NU
270 IF E < 1 OR NU < 0 OR NU > .5 THEN GOSUB 9995: STOP
340 CN = E * (1 - NU) / ((1 + NU) * (1 - 2 * NU))
350 ZU = CN * NU / (1 - NU)
360 MU = CN * (1 - 2 * NU) / (2 * (1 - NU))
410 D(1, 1) = CN
420 D(2, 2) = CN: D(3, 3) = CN
430 D(1, 2) = ZU: D(1, 3) = ZU: D(2, 3) = ZU
440 D(2, 1) = D(1, 2): D(3, 1) = D(1, 3): D(3, 2) = D(2, 3)
```

```
450 D(4, 4) = MU: D(5, 5) = MU: D(6, 6) = MU
460 REM TYPE IN THE NODAL COORDINATES
470 FOR II = 1 TO NJ
480 REM TYPE IN X(II),Y(II) & Z(II)
490 INPUT #1, X(II), Y(II), Z(II)
500 IF X(II) < -1000000! OR X(II) > 1000000! THEN GOSUB 9995: STOP
510 IF Y(II) < -1000000! OR Y(II) > 1000000! THEN GOSUB 9995: STOP
515 IF Z(II) < -1000000! OR Z(II) > 1000000! THEN GOSUB 9995: STOP
520 XG(II) = X(II): YG(II) = Y(II): ZG(II) = Z(II)
540 NEXT II
550 REM TYPE IN THE NODAL POSITIONS OF THE ZERO DISPLACEMENTS, ETC.
560 FOR II = 1 TO NF * 3: NS(II) = 0: NEXT II
570 FOR II = 1 TO NF
580 REM TYPE NODAL POSITION 'SU'
590 INPUT #1, SU, SUX, SUY, SUZ
600 IF SU < 1 OR SU > NJ THEN GOSUB 9995: STOP
601 Y$ = "ERROR"
602 IF SUX = 1 THEN Y$ = "Y"
603 IF SUX = 0 THEN Y$ = "N"
604 IF Y$ = "Y" OR Y$ = "y" THEN NS(II * 3 - 2) = SU * 3 - 2: GOTO 607
605 IF Y$ = "N" OR Y$ = "n" THEN GOTO 607
606 GOSUB 9995: STOP
607 Y$ = "ERROR"
608 IF SUY = 1 THEN Y$ = "Y" ELSE IF SUY = 0 THEN Y$ = "N"
609 IF Y$ = "Y" OR Y$ = "y" THEN NS(II * 3 - 1) = SU * 3 - 1: GOTO 613
610 IF Y$ = "N" OR Y$ = "n" THEN GOTO 613
612 GOSUB 9995: STOP
613 Y$ = "ERROR"
614 IF SUZ = 1 THEN Y$ = "Y" ELSE IF SUZ = 0 THEN Y$ = "N"
615 IF Y$ = "Y" THEN NS(II * 3) = SU * 3: GOTO 623
616 IF Y$ = "N" THEN GOTO 623
620 GOSUB 9995: STOP
623 NEXT II
625 REM CALCULATION OF DIAGONAL POINTER VECTOR
630 FOR EL = 1 TO ES
640 REM TYPE IN THE NODAL NUMBERS IN AN ANTI-CLOCKWISE DIRECTION
650 INPUT #1, I(EL), J(EL), K(EL), L(EL), M(EL), N(EL), O(EL), P(EL)
660 I = I(EL): J = J(EL): K = K(EL): L = L(EL)
670 M = M(EL): N = N(EL): O = O(EL): P = P(EL)
680 IF I < 1 OR I > NJ THEN GOSUB 9995: STOP
690 IF J < 1 OR J > NJ THEN GOSUB 9995: STOP
700 IF K < 1 OR K > NJ THEN GOSUB 9995: STOP
710 IF L < 1 OR L > NJ THEN GOSUB 9995: STOP
755 IF I = J OR I = K OR I = L THEN GOSUB 9995: STOP
760 IF J = K OR J = L THEN GOSUB 9995: STOP
765 IF K = L THEN GOSUB 9995: STOP
770 FOR TETRA = 1 TO 5
775 II = I: JJ = L: KK = N: LL = O
777 IF TETRA = 1 THEN II = I: JJ = L: KK = J: LL = N
778 IF TETRA = 2 THEN II = I: JJ = L: KK = K: LL = O
780 IF TETRA = 3 THEN II = N: JJ = O: KK = M: LL = I
785 IF TETRA = 4 THEN II = N: JJ = O: KK = P: LL = L
790 FOR III = 1 TO 4
```

```
795 IF III = 1 THEN IA = II
800 IF III = 2 THEN IA = JJ
805 IF III = 3 THEN IA = KK
806 IF III = 4 THEN IA = LL
810 FOR IJ = 1 TO NDF
815 II1 = NDF * (IA - 1) + IJ
820 FOR JJJ = 1 TO 4
825 IF JJJ = 1 THEN IB = II
830 IF JJJ = 2 THEN IB = JJ
835 IF JJJ = 3 THEN IB = KK
836 IF JJJ = 4 THEN IB = LL
840 FOR JI = 1 TO NDF
845 JJ1 = NDF * (IB - 1) + JI
850 IIJJ = JJ1 - II1 + 1
855 IF IIJJ > DIAG(JJ1) THEN DIAG(JJ1) = IIJJ
860 NEXT JI, JJJ, IJ, III
870 NEXT TETRA
940 NEXT EL
950 DIAG(1) = 1: FOR II = 1 TO NN3: DIAG(II) = DIAG(II - 1) + DIAG(II): NEXT II
960 NT = DIAG(NN3)
970 DIM A(NT), Q(NN3), UG(NN3), POW(NJ)
980 REM TYPE IN THE NO. OF NODES WITH CONCENTRATED LOADS
990 INPUT #1, NC
1000 IF NC < 1 OR NC > NJ THEN GOSUB 9995: STOP
1010 FOR III = 1 TO NN3: Q(III) = 0: NEXT III
1020 FOR III = 1 TO NJ: POW(III) = 0: NEXT III
1030 FOR III = 1 TO NC
1040 REM TYPE IN THE NODAL POSITION OF THE LOAD
1050 INPUT #1, POW, XLOAD, YLOAD, ZLOAD: POW(III) = POW
1060 REM LOAD IN THE 'X' DIRECTION AT NODE POW
1070 Q(3 * POW - 2) = XLOAD
1080 REM LOAD IN THE 'Y' DIRECTION AT NODE POW
1090 Q(3 * POW - 1) = YLOAD
1100 REM LOAD IN THE 'Z' DIRECTION AT NODE POW
1110 Q(3 * POW) = ZLOAD
1160 NEXT III
1180 GOSUB 5000
1185 Z9 = 0: GOSUB 10000
1190 FOR EL = 1 TO ES
1200 PRINT : PRINT "'CUBIC' ELEMENT NO."; EL; " UNDER COMPUTATION"
1210 I = I(EL)
1220 J = J(EL)
1230 K = K(EL)
1240 L = L(EL)
1241 M = M(EL)
1242 N = N(EL)
1243 O = O(EL)
1244 P = P(EL)
1250 FOR TETRA = 1 TO 5
1255 IF TETRA = 1 THEN II = I: JJ = L: KK = J: LL = N
1260 IF TETRA = 2 THEN II = I: JJ = L: KK = K: LL = O
1265 IF TETRA = 3 THEN II = N: JJ = O: KK = M: LL = I
1270 IF TETRA = 4 THEN II = N: JJ = O: KK = P: LL = L
```

```
1280 IF TETRA = 5 THEN II = I: JJ = L: KK = N: LL = O
1285 PRINT #2, "TERAHEDRON NO."; II; "-"; JJ; "-"; KK; "-"; LL
1290 X1 = X(II)
1300 Y1 = Y(II): Z1 = Z(II)
1310 X2 = X(JJ)
1320 Y2 = Y(JJ): Z2 = Z(JJ)
1330 X3 = X(KK)
1340 Y3 = Y(KK): Z3 = Z(KK)
1350 X4 = X(LL)
1360 Y4 = Y(LL): Z4 = Z(LL)
1370 GOSUB 1500
1375 PRINT #2, "VOLUME="; VOL
1380 FOR III = 1 TO 12
1390 FOR JJJ = 1 TO 12
1400 SK(III, JJJ) = 0
1410 NEXT JJJ, III
1420 FOR III = 1 TO 12
1430 FOR JJJ = 1 TO 12
1435 SK(III, JJJ) = 0
1440 FOR KKK = 1 TO 6
1450 SK(III, JJJ) = SK(III, JJJ) + BT(III, KKK) * DB(KKK, JJJ)
1460 NEXT KKK: SK(III, JJJ) = SK(III, JJJ) * VOL: NEXT JJJ, III
1470 GOSUB 1980
1480 NEXT TETRA: NEXT EL
1490 GOTO 2300
1500 FOR III = 1 TO 6
1510 FOR JJJ = 1 TO 12
1520 BX(III, JJJ) = 0
1530 NEXT JJJ, III
1540 X14 = X1 - X4: Y14 = Y1 - Y4: Z14 = Z1 - Z4
1550 X24 = X2 - X4: Y24 = Y2 - Y4: Z24 = Z2 - Z4
1560 X34 = X3 - X4: Y34 = Y3 - Y4: Z34 = Z3 - Z4
1570 DETJ = X14 * (Y24 * Z34 - Z24 * Y34)
1580 DETJ = DETJ + Y14 * (Z24 * X34 - X24 * Z34)
1590 DETJ = DETJ + Z14 * (X24 * Y34 - Y24 * X34)
1600 B11 = Y24 * Z34 - Y34 * Z24
1610 B21 = X34 * Z24 - X24 * Z34
1620 B31 = X24 * Y34 - X34 * Y24
1630 B12 = Y34 * Z14 - Y14 * Z34
1640 B22 = X14 * Z34 - X34 * Z14
1650 B32 = X34 * Y14 - X14 * Y34
1660 B13 = Y14 * Z24 - Y24 * Z14
1670 B23 = X24 * Z14 - X14 * Z24
1680 B33 = X14 * Y24 - X24 * Y14
1690 BX(1, 1) = B11: BX(1, 4) = B12
1700 BX(1, 7) = B13: B1B = -(B11 + B12 + B13): BX(1, 10) = B1B
1710 BX(2, 2) = B21: BX(2, 5) = B22
1720 BX(2, 8) = B23: B2B = -(B21 + B22 + B23): BX(2, 11) = B2B
1730 BX(3, 3) = B31: BX(3, 6) = B32
1740 BX(3, 9) = B33: B3B = -(B31 + B32 + B33): BX(3, 12) = B3B
1750 BX(4, 2) = B31: BX(4, 3) = B21: BX(4, 5) = B32
1760 BX(4, 6) = B22: BX(4, 8) = B33: BX(4, 9) = B23
1770 BX(4, 11) = B3B: BX(4, 12) = B2B
```

```
1780 BX(5, 1) = B31: BX(5, 3) = B11: BX(5, 4) = B32
1790 BX(5, 6) = B12: BX(5, 7) = B33: BX(5, 9) = B13
1800 BX(5, 10) = B3B: BX(5, 12) = B1B
1810 BX(6, 1) = B21: BX(6, 2) = B12: BX(6, 4) = B22
1820 BX(6, 5) = B12: BX(6, 7) = B23: BX(6, 8) = B13
1830 BX(6, 10) = B2B: BX(6, 11) = B1B
1840 FOR III = 1 TO 6: FOR JJJ = 1 TO 12
1845 BX(III, JJJ) = BX(III, JJJ) / DETJ
1850 NEXT JJJ, III
1860 VOL = ABS(DETJ) / 6
1870 FOR III = 1 TO 6
1872 FOR JJJ = 1 TO 12
1875 DB(III, JJJ) = 0
1877 FOR KKK = 1 TO 6
1880 DB(III, JJJ) = DB(III, JJJ) + D(III, KKK) * BX(KKK, JJJ)
1890 NEXT KKK, JJJ, III
1900 FOR III = 1 TO 12
1910 FOR JJJ = 1 TO 6
1920 BT(III, JJJ) = BX(JJJ, III)
1930 NEXT JJJ, III
1970 RETURN
1980 I1 = 3 * II - 3: J1 = 3 * JJ - 3
1990 K1 = 3 * KK - 3: L1 = 3 * LL - 3
2020 FOR III = 1 TO 4
2040 IF III = 1 THEN IIC = I1
2050 IF III = 2 THEN IIC = J1
2060 IF III = 3 THEN IIC = K1
2070 IF III = 4 THEN IIC = L1
2110 FOR IJ = 1 TO NDF
2120 IC = IIC + IJ
2130 IDIAG = DIAG(IC)
2140 I3 = 3 * (III - 1) + IJ
2150 FOR JJJ = 1 TO 4
2170 IF JJJ = 1 THEN JJR = I1
2180 IF JJJ = 2 THEN JJR = J1
2190 IF JJJ = 3 THEN JJR = K1
2200 IF JJJ = 4 THEN JJR = L1
2240 FOR JI = 1 TO 3
2250 J3 = 3 * (JJJ - 1) + JI
2260 IR = JJR + JI
2270 IF IR > IC THEN GOTO 2290
2280 LOCAT = IDIAG - (IC - IR)
2285 A(LOCAT) = A(LOCAT) + SK(I3, J3)
2290 NEXT JI: NEXT JJJ, IJ, III
2295 RETURN
2300 FOR II = 1 TO NF * 3
2310 N9 = NS(II)
2320 IF N9 = 0 THEN GOTO 2350
2330 Q(N9) = 0: DI = DIAG(N9)
2340 A(DI) = A(DI) * 1E+10
2350 NEXT II
2360 PRINT : PRINT "THE SIMULTANEOUS EQUATIONS ARE NOW BEING SOLVED":
PRINT
```

```
2370 N = NN3
2375 A(1) = SQR(A(1))
2380 FOR II = 2 TO N
2385 DI = DIAG(II) - II
2390 LL = DIAG(II - 1) - DI + 1
2395 FOR JJ = LL TO II
2400 X = A(DI + JJ)
2405 DJ = DIAG(JJ) - JJ
2410 IF JJ = 1 THEN 2440
2415 LB = DIAG(JJ - 1) - DJ + 1
2420 IF LL > LB THEN LB = LL
2425 FOR KK = LB TO JJ - 1
2430 X = X - A(DI + KK) * A(DJ + KK)
2435 NEXT KK
2440 A(DI + JJ) = X / A(DJ + JJ)
2445 NEXT JJ
2450 A(DI + II) = SQR(X)
2455 NEXT II
2460 Q(1) = Q(1) / A(1)
2465 FOR II = 2 TO N
2470 DI = DIAG(II) - II
2475 LL = DIAG(II - 1) - DI + 1
2480 X = Q(II)
2485 FOR JJ = LL TO II - 1
2490 X = X - A(DI + JJ) * Q(JJ)
2495 NEXT JJ
2500 Q(II) = X / A(DI + II)
2505 NEXT II
2510 FOR II = N TO 2 STEP -1
2515 DI = DIAG(II) - II
2520 X = Q(II) / A(DI + II): Q(II) = X
2525 LL = DIAG(II - 1) - DI + 1
2530 FOR KK = LL TO II - 1
2535 Q(KK) = Q(KK) - X * A(DI + KK)
2540 NEXT KK
2545 NEXT II
2550 Q(1) = Q(1) / A(1)
2590 PRINT #2,
2610 PRINT #2, : PRINT #2, "THE NODAL VALUES OF THE DISPLACEMENTS ARE:-":
PRINT #2,
2620 FOR III = 1 TO NJ
2630 PRINT #2, "NODE="; III, "U("; III; ")="; Q(III * 3 - 2),
2640 PRINT #2, "V("; III; ")="; Q(III * 3 - 1)
2642 PRINT #2, "W("; III; ")="; Q(III * 3)
2650 NEXT III
FOR III = 1 TO NJ * 3
UG(III) = Q(III)
NEXT III
2660 FOR EL = 1 TO ES
2670 I = I(EL)
2680 J = J(EL)
2690 K = K(EL)
2700 L = L(EL)
```

```
2705 M = M(EL)
2710 N = N(EL)
2715 O = O(EL)
2720 P = P(EL)
2730 PRINT #2, "'CUBE' NO."; I; "-"; J; "-"; K; "-"; L; "-"; M; "-";
2740 PRINT #2, N; "-"; O; "-"; P
2750 FOR TETRA = 1 TO 5
2760 IF TETRA = 1 THEN II = I: JJ = L: KK = J: LL = N
2770 IF TETRA = 2 THEN II = I: JJ = L: KK = K: LL = O
2780 IF TETRA = 3 THEN II = N: JJ = O: KK = M: LL = I
2790 IF TETRA = 4 THEN II = N: JJ = O: KK = P: LL = L
2800 IF TETRA = 5 THEN II = I: JJ = L: KK = N: LL = O
2810 X1 = X(II): Y1 = Y(II): Z1 = Z(II)
2820 X2 = X(JJ): Y2 = Y(JJ): Z2 = Z(JJ)
2830 X3 = X(KK): Y3 = Y(KK): Z3 = Z(KK)
2840 X4 = X(LL): Y4 = Y(LL): Z4 = Z(LL)
2850 GOSUB 1500
2860 SU(1) = Q(II * 3 - 2): SU(2) = Q(II * 3 - 1): SU(3) = Q(II * 3)
2870 SU(4) = Q(JJ * 3 - 2): SU(5) = Q(JJ * 3 - 1): SU(6) = Q(JJ * 3)
2880 SU(7) = Q(KK * 3 - 2): SU(8) = Q(KK * 3 - 1): SU(9) = Q(KK * 3)
2890 SU(10) = Q(LL * 3 - 2): SU(11) = Q(LL * 3 - 1): SU(12) = Q(LL * 3)
3160 FOR III = 1 TO 6
3170 SR(III) = 0
3180 FOR JJJ = 1 TO 12
3190 SR(III) = SR(III) + DB(III, JJJ) * SU(JJJ)
3200 NEXT JJJ
3210 NEXT III
3220 PRINT #2, "THE STRESSES IN TETRAHEDRON NO. "; II; "-"; JJ; "-"; KK;
3230 PRINT #2, "-"; LL; " ARE:-"
3240 PRINT #2, "SIGMAX="; SR(1), "SIGMAY="; SR(2), "SIGMAZ="; SR(3)
3242 PRINT #2, "TAU Y-Z="; SR(4), "TAU X-Z="; SR(5), "TAU X-Y="; SR(6)
3250 NEXT TETRA
3290 NEXT EL
3293 GOSUB 11000
3320 END
5000 REM OUTPUT OF INPUT
5010 PRINT #2, : PRINT #2, "3D STRESS USING A 4 NODE TETRAHEDRON"
5020 PRINT #2, : PRINT #2, "NUMBER OF NODES="; NJ
5030 PRINT #2, "NUMBER OF ELEMENTS="; ES
5040 PRINT #2, "NUMBER OF NODES WITH ZERO DISPLACEMENTS="; NF
5050 PRINT #2, "ELASTIC MODULUS="; E
5060 PRINT #2, "POISSON'S RATIO="; NU
5100 PRINT #2, : PRINT #2, "THE NODAL COORDINATES (X,Y & Z) IN GLOBAL
COORDINATES ARE:": PRINT #2,
5110 FOR II = 1 TO NJ
5120 PRINT #2, "NODE="; II, "X("; II; ")="; X(II), "Y("; II; ")="; Y(II), "Z("; II; ")="; Z(II)
5130 NEXT II
5140 PRINT #2, : PRINT #2, "THE NODAL POSITIONS, ETC. OF THE ZERO
DISPLACEMENTS ARE:": PRINT #2,
5150 FOR II = 1 TO NF
5160 SU = INT((NS(II * 3 - 2) + 2) / 3)
5170 IF SU = 0 THEN SU = INT((NS(II * 3 - 1) + 1) / 3)
5175 IF SU = 0 THEN SU = INT(NS(II * 3) / 3)
```

```
5180 PRINT #2, "NODE="; SU
5190 N9 = NS(II * 3 - 2)
5200 IF N9 > 0 THEN PRINT #2, "DISPLACEMENT IN X DIRECTION IS ZERO AT NODE
"; SU
5210 N9 = NS(II * 3 - 1)
5220 IF N9 > 0 THEN PRINT #2, "DISPLACEMENT IN Y DIRECTION IS ZERO AT NODE
"; SU
5230 N9 = NS(II * 3)
5240 IF N9 > 0 THEN PRINT #2, "DISPLACEMENT IN Z DIRECTION IS ZERO AT NODE
"; SU
5260 NEXT II
5270 PRINT #2, : PRINT #2, "ELEMENT TOPOLOGY": PRINT #2,
5280 FOR EL = 1 TO ES
5290 I = I(EL): J = J(EL): K = K(EL): L = L(EL)
5295 M = M(EL): N = N(EL): O = O(EL): P = P(EL)
5300 PRINT #2, "ELEMENT NO. "; EL, "NODES="; I; "-"; J; "-"; K; "-"; L;
5302 PRINT #2, "-"; M; "-"; N; "-"; O; "-"; P
5330 NEXT EL
PRINT #2, : PRINT #2, "NO. OF NODES WITH CONCENTRATED LOADS="; NC
5340 PRINT #2, : PRINT #2, "NODAL POSITIONS & VALUES OF THE CONCENTRATED
LOADS": PRINT #2,
5350 FOR I = 1 TO NC
5360 POW = POW(I)
5370 PRINT #2, "NODE="; POW
5380 PRINT #2, "LOAD IN X DIRECTION="; Q(POW * 3 - 2)
5390 PRINT #2, "LOAD IN Y DIRECTION="; Q(POW * 3 - 1)
5395 PRINT #2, "LOAD IN Z DIRECTION="; Q(POW * 3)
5400 NEXT I
5430 RETURN
9995 REM WARNING
9996 PRINT : PRINT "INCORRECT DATA": PRINT
9997 RETURN
10000 IF Z9 = 99 THEN 10300
10010 CLS : XZ = 1E+12: YZ = 1E+12: XM = -1E+12: YM = -1E+12
ZZ = 1E+12: ZM = -1E+12
10020 PRINT "TO CONTINUE, TYPE Y"
10060 A$ = INKEY$: IF A$ = "" THEN 10060
10070 IF A$ = "Y" OR A$ = "y" THEN 10090
10080 GOTO 10060
10090 PRINT "TYPE IN THE VIEWING ANGLE IN DEGREES": INPUT "ALPHA="; ALPHA:
INPUT "BETA="; BETA
PI = 3.1415926536#: ALPHA = PI * ALPHA / 180: BETA = -PI * BETA / 180
FOR EL = 1 TO ES
FOR TWO = 1 TO 2
10100 IF TWO = 1 THEN I = I(EL): J = J(EL): K = K(EL): L = L(EL)
IF TWO = 2 THEN I = M(EL): J = N(EL): K = O(EL): L = P(EL)
10105 IF X(I) < XZ THEN XZ = X(I)
10110 IF X(J) < XZ THEN XZ = X(J)
10115 IF X(K) < XZ THEN XZ = X(K)
10120 IF X(L) < XZ THEN XZ = X(L)
10130 IF Y(I) < YZ THEN YZ = Y(I)
10135 IF Y(J) < YZ THEN YZ = Y(J)
10140 IF Y(K) < YZ THEN YZ = Y(K)
```

```
10145 IF Y(L) < YZ THEN YZ = Y(L)
10155 IF X(I) > XM THEN XM = X(I)
10160 IF X(J) > XM THEN XM = X(J)
10165 IF X(K) > XM THEN XM = X(K)
10170 IF X(L) > XM THEN XM = X(L)
10180 IF Y(I) > YM THEN YM = Y(I)
10185 IF Y(J) > YM THEN YM = Y(J)
10190 IF Y(K) > YM THEN YM = Y(K)
10195 IF Y(L) > YM THEN YM = Y(L)
IF Z(I) < ZZ THEN ZZ = Z(I)
IF Z(J) < ZZ THEN ZZ = Z(J)
IF Z(K) < ZZ THEN ZZ = Z(K)
IF Z(L) < ZZ THEN ZZ = Z(L)
IF Z(I) > ZM THEN ZM = Z(I)
IF Z(J) > ZM THEN ZM = Z(J)
IF Z(K) > ZM THEN ZM = Z(K)
IF Z(L) > ZM THEN ZM = Z(L)
NEXT TWO
10200 NEXT EL
10210 XS = 400 / (XM - XZ)
IF YM - YZ = 0 OR ZM - ZZ = 0 THEN GOTO 10280
YS = 200 / (YM - YZ): ZS = 200 / (ZM - ZZ)
IF YS > ZS THEN YS = ZS
IF XS > YS THEN XS = YS
10280 CONX = 1E+12
CONZ = 1E+12
JUMP = 0
10295 IF Z9 = 0 OR Z9 = 77 THEN SCREEN 9
10296 IF Z9 = 0 THEN PRINT "THE MESH WILL NOW BE DRAWN"
10298 IF Z9 = 77 THEN PRINT "DEFLECTED FORM OF MESH"
10300 FOR EL = 1 TO ES
10310 NI = I(EL): NNJ = J(EL): NK = K(EL): NL = L(EL)
NM = M(EL): NNN = N(EL): NO = O(EL): NP = P(EL)
FOR TETRA = 1 TO 5
I = NI: J = NL: K = NNN: L = NO
IF TETRA = 1 THEN I = NI: J = NL: K = NNJ: L = NNN
IF TETRA = 2 THEN I = NI: J = NL: K = NK: L = NO
IF TETRA = 3 THEN I = NNN: J = NO: K = NM: L = NI
IF TETRA = 4 THEN I = NNN: J = NO: K = NP: L = NL
10315 FOR IV = 1 TO 6
10320 IF IV = 1 THEN II = I: JJ = J
10325 IF IV = 2 THEN II = J: JJ = K
10330 IF IV = 3 THEN II = K: JJ = L
10335 IF IV = 4 THEN II = L: JJ = I
IF IV = 5 THEN II = I: JJ = K
IF IV = 6 THEN II = J: JJ = L
10340 IF Z9 = 99 THEN 10450
10345 XI = (X(II) - XZ) * XS
10350 XJ = (X(JJ) - XZ) * XS
10355 YI = (Y(II) - YZ) * XS
10360 YJ = (Y(JJ) - YZ) * XS
ZI = (Z(II) - ZZ) * XS
ZJ = (Z(JJ) - ZZ) * XS
```

```
10365 GOTO 10460
10450 XI = X(II) - XZ * XS: XJ = X(JJ) - XZ * XS
10455 YI = Y(II) - YZ * XS: YJ = Y(JJ) - YZ * XS
ZI = Z(II) - ZZ * XS: ZJ = Z(JJ) - ZZ * XS
10460 XII = -XI * SIN(BETA) + YI * COS(BETA)
XIJ = -XJ * SIN(BETA) + YJ * COS(BETA)
ZII = -XI * SIN(ALPHA) * COS(BETA) - YI * SIN(ALPHA) * SIN(BETA) + ZI * COS(ALPHA)
ZIJ = -XJ * SIN(ALPHA) * COS(BETA) - YJ * SIN(ALPHA) * SIN(BETA) + ZJ * COS(ALPHA)
IF Z9 = 99 THEN GOTO 10480
IF XII < CONX THEN CONX = XII
IF XIJ < CONX THEN CONX = XIJ
IF ZII < CONZ THEN CONZ = ZII
IF ZIJ < CONZ THEN CONZ = ZIJ
IF JUMP = 0 THEN GOTO 10490
10480 LINE (XII - CONX + 40, 300 + CONZ - ZII)-(XIJ - CONX + 40, 300 + CONZ - ZIJ)
10490 NEXT IV
NEXT TETRA
10495 NEXT EL
JUMP = JUMP + 1
IF JUMP = 1 THEN GOTO 10300
10510 IF Z9 = 77 THEN 10600
10515 PRINT "TO CONTIMUE, TYPE Y"
10520 A$ = INKEY$: IF A$ = "" THEN 10520
10530 IF A$ = "Y" OR A$ = "y" THEN 10560
10540 GOTO 10520
10560 SCREEN 0: WIDTH 80
10600 RETURN
11000 REM DEFLECTED FORM OF MESH
11010 CLS : PRINT "TO CONTINUE, TYPE Y"
11020 A$ = INKEY$: IF A$ = "" THEN 11020
11030 IF A$ = "Y" OR A$ = "y" THEN 11050
11040 GOTO 11020
11050 REM
11070 Z9 = 77
11080 GOSUB 10000
11090 UM = -1E+12
11100 FOR II = 1 TO NJ * 3
11110 IF ABS(UG(II)) > UM THEN UM = ABS(UG(II))
11120 NEXT II
11130 SC = 20
11140 FOR II = 1 TO NJ
11150 X(II) = X(II) * XS + SC * UG(II * 3 - 2) / UM
11160 Y(II) = Y(II) * XS + SC * UG(II * 3 - 1) / UM
Z(II) = Z(II) * XS + SC * UG(II * 3) / UM
11170 NEXT II
11180 Z9 = 99: GOSUB 10000
11190 FOR II = 1 TO NJ
11200 X(II) = XG(II)
11210 Y(II) = YG(II)
Z(II) = ZG(II)
11220 NEXT II
11230 RETURN
```

Computer Program for Field Problems "FIELD3SF"

```
90 CLS
100 REM PROGRAM BY DR.C.T.F.ROSS
110 REM 6 HURSTVILLE DRIVE, WATERLOOVILLE.
120 REM PORTSMOUTH. ENGLAND.
130 PRINT : PRINT "PROGRAM FOR SOLVING 2D EQUATIONS OF THE LAPLACE &
POISSON TYPE": PRINT
140 PRINT : PRINT "BY DR.C.T.F.ROSS - NOT TO BE COPIED": PRINT
142 INPUT "TYPE IN THE NAME OF YOUR INPUT FILE "; FILENAME$
143 PRINT "IF YOU ARE SATISFIED WITH THIS FILENAME, TYPE Y; ELSE N"
145 A$ = INKEY$: IF A$ = "" THEN GOTO 145
146 IF A$ = "Y" OR A$ = "y" THEN GOTO 149
147 IF A$ = "N" OR A$ = "n" THEN GOTO 142
148 GOTO 145
149 OPEN FILENAME$ FOR INPUT AS #1
150 INPUT "TYPE IN THE NAME OF YOUR OUTPUT FILE "; OUT$
151 PRINT "IF YOU ARE SATISFIED WITH THIS FILENAME, TYPE Y; ELSE N"
152 B$ = INKEY$: IF B$ = "" THEN GOTO 152
153 IF B$ = "Y" OR B$ = "y" THEN GOTO 158
155 IF B$ = "N" OR B$ = "n" THEN GOTO 150
156 GOTO 152
158 OPEN OUT$ FOR OUTPUT AS #2
160 PRINT #2, "FIELD PROBLEMS (3 NODE TRIANGULAR ELEMENT)"
165 INPUT #1, MS: PRINT #2, "NO. OF ELEMENTS="; MS
168 M3 = 3 * MS
170 INPUT #1, NN: PRINT #2, "NO, OF NODES="; NN
175 INPUT #1, NF: PRINT #2, "NO. OF KNOWN  NODAL BOUNDARY VALUES="; NF
177 IF NF < 0 OR NF > NN THEN GOSUB 4990: PRINT "NO. OF NODAL BOUNDARY
VALUES="; NF: STOP
180 REM TYPE IN NO. OF DIFFERENT VALUES OF 'K'
190 INPUT #1, ZK: PRINT #2, "NO. OF DIFFERENT VALUES OF 'K'="; ZK
200 IF ZK < 1 OR ZK > MS THEN GOSUB 4900: PRINT "NO. OF DIFFERENT VALUES OF
K="; ZK: STOP
230 DIM SK(3, 3), IJ(M3), NS(NF), NQ(NF), UG(NN), C8(ZK)
240 DIM B(3), C(3), UL(3), BM(2, 3), SL(2), SQ(3), DIAG(NN)
250 INPUT #1, Q7
280 IF Q7 = 0 THEN PRINT #2, "'Q' VARIES OVER THE WHOLE SYSTEM": GOTO 335
290 IF Q7 = 1 THEN PRINT #2, "'Q' IS CONSTANT": GOTO 320
300 GOSUB 4990: STOP
320 INPUT #1, Q1: PRINT #2, "'Q'="; Q1
335 IF ZK > 1 THEN GOTO 350
340 INPUT #1, C8: PRINT #2, "K="; C8
342 IF C8 <= 0 THEN GOSUB 4990: PRINT "K="; C8: STOP
345 GOTO 360
350 FOR II = 1 TO ZK
352 REM TYPE IN THE VALUE OF 'K' FOR K-TYPE II
355 INPUT #1, C8(II): PRINT #2, "K-TYPE="; II, "K="; C8(II)
356 IF C8(II) <= 0 THEN GOSUB 4990: PRINT "K-TYPE="; C8(II): STOP
357 NEXT II
360 INPUT #1, XH
```

```
370 IF XH = 1 THEN PRINT #2, "'HL' HAS VALUES"
375 IF XH = 0 THEN PRINT #2, "HL=0"
380 IF XH = 1 OR XH = 0 THEN GOTO 410
390 GOSUB 4990: STOP
410 INPUT #1, XQ
415 IF XQ = 0 THEN PRINT #2, "QG=0": GOTO 460
420 IF XQ = 1 THEN PRINT #2, "'QG' HAS VALUES": GOTO 460
430 GOSUB 4990: PRINT "XQ="; XQ: STOP
460 REM TYPE IN ELEMENT TOPOLOGY
465 PRINT #2, "I,J & K NODES ARE"
470 FOR II = 0 TO NN: DIAG(II) = 0: NEXT II
475 PRINT "   i     j      k"
480 FOR ME = 1 TO MS
490 REM I,J & K NODES FOR ELEMENT ME
500 INPUT #1, IN, JN, KN
502 PRINT IN, JN, KN
505 PRINT #2, "ELEMENT="; ME; " I,J & K ="; IN; "-"; JN; "-"; KN
510 I8 = 3 * ME
520 IJ(I8 - 2) = IN
530 IJ(I8 - 1) = JN
540 IJ(I8) = KN
542 FOR II = 1 TO 3
543 IA = KN
544 IF II = 1 THEN IA = IN
545 IF II = 2 THEN IA = JN
546 II1 = IA
548 FOR JJ = 1 TO 3
549 IB = KN
550 IF JJ = 1 THEN IB = IN
552 IF JJ = 2 THEN IB = JN
554 JJ1 = IB
556 IIJJ = JJ1 - II1 + 1
558 IF IIJJ > DIAG(JJ1) THEN DIAG(JJ1) = IIJJ
560 NEXT JJ, II
570 NEXT ME
572 DIAG(1) = 1: FOR II = 2 TO NN: DIAG(II) = DIAG(II - 1) + DIAG(II): NEXT II
574 NT = DIAG(NN)
610 DIM A(NT), Q(NN), CV(NN), X(NN), Y(NN)
615 PRINT #2, "NODAL COORDINATES X & Y ARE"
617 PRINT "   X        Y"
620 REM TYPE IN THE NODAL COORDINATES (GLOBAL)
630 FOR II = 1 TO NN
640 INPUT #1, X(II), Y(II)
645 PRINT X(II), Y(II)
650 PRINT #2, "X("; II; ")="; X(II), "Y("; II; ")="; Y(II)
680 NEXT II
690 IF NF = 0 THEN PRINT #2, "THERE ARE NO BOUNDARY VALUES": GOTO 785
695 PRINT #2, "THE BOUNDARY VALUES ARE"
700 REM TYPE IN THE BOUNDARY CONDITIONS
710 FOR IC = 1 TO NF
720 REM TYPE IN THE NODAL POSITION IC
730 INPUT #1, N9, PH: PRINT #2, "NODE="; N9,
740 PRINT #2, "BOUNDARY VALUE="; PH
```

```
750 NS(IC) = N9
760 NQ(IC) = PH
780 NEXT IC
785 GOSUB 10000
790 FOR ME = 1 TO MS
800 PRINT : PRINT "ELEMENT NO. "; ME; " UNDER COMPUTATION"
810 I8 = 3 * ME
820 IN = IJ(I8 - 2)
830 JN = IJ(I8 - 1)
840 KN = IJ(I8)
841 PRINT IN, JN, KN
842 IF ZK = 1 THEN GOTO 850
844 REM TYPE IN K-TYPE
846 INPUT #1, ZZ: PRINT #2, "ELEMENT="; IN; "-"; JN; "-"; KN, "K-TYPE="; ZZ
848 C8 = C8(ZZ)
850 B(1) = Y(JN) - Y(KN)
860 B(2) = Y(KN) - Y(IN)
870 B(3) = Y(IN) - Y(JN)
880 C(1) = X(KN) - X(JN)
890 C(2) = X(IN) - X(KN)
900 C(3) = X(JN) - X(IN)
910 DL = X(IN) * B(1) + X(JN) * B(2) + X(KN) * B(3)
920 DL = ABS(DL)
930 IF Q7 = 1 THEN GOTO 960
940 REM TYPE IN THE VALUE OF 'Q' FOR ELEMENT NO. ME
950 INPUT #1, Q1: PRINT #2, "ELEMENT="; IN; "-"; JN; "-"; KN, "'Q'="; Q1
960 FOR I = 1 TO 3
970 FOR J = 1 TO 3
980 SK(I, J) = .5 * C8 * (B(I) * B(J) + C(I) * C(J)) / DL
990 NEXT J
1000 SQ(I) = .5 * Q1 * DL / 3
1010 NEXT I
1020 I1 = IN - 1
1030 J1 = JN - 1
1040 K1 = KN - 1
1050 FOR II = 1 TO 3
1060 IF II = 1 THEN NCT = I1
1070 IF II = 2 THEN NCT = J1
1080 IF II = 3 THEN NCT = K1
1082 NC = NCT + 1
1085 IDIAG = DIAG(NC)
1090 FOR JJ = 1 TO 3
1100 IF JJ = 1 THEN NRT = I1
1110 IF JJ = 2 THEN NRT = J1
1120 IF JJ = 3 THEN NRT = K1
1125 NR = NRT + 1
1130 IF NR > NC THEN GOTO 1160
1140 LOCATION = IDIAG - (NC - NR)
1150 A(LOCATION) = A(LOCATION) + SK(II, JJ)
1160 NEXT JJ
1170 Q(NC) = Q(NC) + SQ(II)
1180 NEXT II
1190 IF XH = 0 AND XQ = 0 THEN GOTO 1250
```

1200 REM IF THIS ELEMENT HAS A BOUNDARY ON WHICH EITHER HL OR QG HAS A VALUE, TYPEY; ELSE TYPE N
1210 INPUT #1, B$
1220 IF B$ = "Y" OR B$ = "y" THEN PRINT #2, "'HL' OR 'QG'HAS A VALUE": GOTO 1240
1225 IF B$ = "N" OR B$ = "n" THEN PRINT #2, "'NEITHER HL' NOR 'QG'HAS A VALUE": GOTO 1240
1230 GOSUB 4990: PRINT "B$="; B$: STOP
1240 IF B$ = "Y" OR B$ = "y" THEN GOSUB 1750
1250 NEXT ME
1260 IF NF = 0 THEN GOTO 1320
1270 FOR II = 1 TO NF
1280 N9 = NS(II)
1295 DI = DIAG(N9)
1290 Q(N9) = 1E+12 * NQ(II) * A(DI)
1300 A(DI) = A(DI) * 1E+12
1310 NEXT II
1320 FOR II = 1 TO NN
1330 CV(II) = Q(II)
1340 NEXT II
1350 PRINT : PRINT "THE SIMULTANEOUS EQUATIONS ARE NOW BEING SOLVED": PRINT
1360 GOSUB 1500
1370 PRINT #2, : PRINT #2, "THE NODAL VALUES OF THE FUNCTION ARE AS FOLLOWS:-": PRINT
1380 FOR I = 1 TO NN
1390 UG(I) = CV(I)
1400 PRINT #2, "THE VALUE OF THE FUNCTION AT NODE "; I; "="; UG(I)
1430 NEXT I
1440 REM IF THE SLOPES OF THE FUNCTIONS ARE REQUIRED, TYPE Y; ELSE TYPE N
1450 INPUT #1, C$
1460 IF C$ = "Y" OR C$ = "y" THEN PRINT #2, "THE SLOPES ARE:": GOTO 1480
1465 IF C$ = "N" OR C$ = "n" THEN PRINT #2, "THE SLOPES ARE NOT REQUIRED:": GOTO 1480
1470 GOSUB 4990: PRINT "C$="; C$: STOP
1480 IF C$ = "Y" OR C$ = "y" THEN GOSUB 2220
1490 GOTO 2990
1500 N = NN
1510 A(1) = SQR(A(1))
1520 FOR II = 2 TO N
1530 DI = DIAG(II) - II
1540 LL = DIAG(II - 1) - DI + 1
1550 FOR JJ = LL TO II
1560 X = A(DI + JJ)
1570 DJ = DIAG(JJ) - JJ
1580 IF JJ = 1 THEN GOTO 1640
1590 LB = DIAG(JJ - 1) - DJ + 1
1600 IF LL > LB THEN LB = LL
1610 FOR KK = LB TO JJ - 1
1620 X = X - A(DI + KK) * A(DJ + KK)
1630 NEXT KK
1640 A(DI + JJ) = X / A(DJ + JJ)
1650 NEXT JJ
1660 A(DI + II) = SQR(X)

```
1670 NEXT II
1680 CV(1) = CV(1) / A(1)
1690 FOR II = 2 TO N
1692 DI = DIAG(II) - II
1694 LL = DIAG(II - 1) - DI + 1
1696 X = CV(II)
1698 FOR JJ = LL TO II - 1
1700 X = X - A(DI + JJ) * CV(JJ)
1702 NEXT JJ
1704 CV(II) = X / A(DI + II)
1706 NEXT II
1708 FOR II = N TO 2 STEP -1
1710 DI = DIAG(II) - II
1712 X = CV(II) / A(DI + II)
1714 CV(II) = X
1715 LL = DIAG(II - 1) - DI + 1
1716 FOR KK = LL TO II - 1
1717 CV(KK) = CV(KK) - X * A(DI + KK)
1718 NEXT KK
1719 NEXT II
1720 CV(1) = CV(1) / A(1)
1725 RETURN
1750 IF XH = 0 THEN GOTO 2030
1760 INPUT #1, LH: PRINT #2, "NO. OF BOUNDARIES ON WHICH 'HL' HAS A VALUE=";
LH
1770 IF LH = 0 THEN GOTO 2030
1790 IF LH = 1 OR LH = 2 OR LH = 3 THEN GOTO 1820
1800 GOSUB 4990: STOP
1820 REM TYPE IN THE NODES DEFINING EACH BOUNDARY ON WHICH HL HAS A
VALUE, TOGETHER WITH THE VALUES OF HL & T(INFINITY)
1840 FOR II = 1 TO LH
1850 REM TYPE IN THE NODES DEFINING THE BOUNDARY
1860 INPUT #1, IL, JL, HL, IT
1870 PRINT #2, "BOUNDARY="; IL; "-"; JL, "'HL'="; HL, "T(INFINITY)="; IT
1910 XL = SQR((X(IL) - X(JL)) ^ 2 + (Y(IL) - Y(JL)) ^ 2)
1920 Q(IL) = Q(IL) + XL * HL * IT / 2
1930 Q(JL) = Q(JL) + XL * HL * IT / 2
1940 ILD = DIAG(IL): A(ILD) = A(ILD) + XL * HL / 3
1950 JLD = DIAG(JL): A(JLD) = A(JLD) + XL * HL / 3
1970 IF JL > IL THEN GOTO 2000
1975 NLOC = ILD - (IL - JL)
1980 A(NLOC) = A(NLOC) + XL * HL / 6
1990 GOTO 2020
2000 NLOC = JLD + IL - JL
2010 A(NLOC) = A(NLOC) + XL * HL / 6
2020 NEXT II
2030 IF XQ = 0 THEN GOTO 2210
2040 INPUT #1, GQ: PRINT #2, "NO. OF BOUNDARIES ON WHICH 'QG' HAS A VALUE=";
GQ
2050 IF GQ = 0 THEN GOTO 2210
2060 IF GQ = 1 OR GQ = 2 OR GQ = 3 THEN GOTO 2100
2080 GOSUB 4990: STOP
2100 REM TYPE IN NODES DEFINING EACH BOUNDARY OF THIS ELEMENT ON WHICH
```

QG HAS A VALUE, TOGETHER WITH THE VALUE OF QG
2120 FOR II = 1 TO GQ
2130 REM TYPE IN THE NODES DEFINING THE BOUNDARY
2140 READ IL, JL, QG
2150 PRINT #2, "BOUNDARY="; IL; "-"; JL, "'QG'="; QG
2170 XL = SQR((X(IL) - X(JL)) ^ 2 + (Y(IL) - Y(JL)) ^ 2)
2180 Q(IL) = Q(IL) - XL * QG / 2
2190 Q(JL) = Q(JL) - XL * QG / 2
2200 NEXT II
2210 RETURN
2220 FOR ME = 1 TO MS
2230 I8 = 3 * ME
2240 IN = IJ(I8 - 2)
2250 JN = IJ(I8 - 1)
2260 KN = IJ(I8)
2270 B(1) = Y(JN) - Y(KN)
2280 B(2) = Y(KN) - Y(IN)
2290 B(3) = Y(IN) - Y(JN)
2300 C(1) = X(KN) - X(JN)
2310 C(2) = X(IN) - X(KN)
2320 C(3) = X(JN) - X(IN)
2325 DL = ABS(X(IN) * B(1) + X(JN) * B(2) + X(KN) * B(3))
2330 UL(1) = UG(IN)
2340 UL(2) = UG(JN)
2350 UL(3) = UG(KN)
2360 FOR II = 1 TO 3
2370 BM(1, II) = B(II) / DL
2380 BM(2, II) = C(II) / DL
2390 NEXT II
2400 FOR II = 1 TO 2
2410 SL(II) = 0
2420 FOR JJ = 1 TO 3
2430 SL(II) = SL(II) + BM(II, JJ) * UL(JJ)
2440 NEXT JJ
2450 NEXT II
2460 PRINT #2, : PRINT #2, "THE SLOPES WITH RESPECT TO X & Y AXES FOR ELEMENT
NO."; IN; "-"; JN; "-"; KN; " ARE AS FOLLOWS:-"
2480 PRINT #2, SL(1), SL(2)
2510 NEXT ME
2520 RETURN
2990 REM GRAPHICAL DISPLAY OF RESULTS
3000 MAX = MAXX = MAXY = MAXU = 0: MINU = 1E+12: DIM YY(3), XX(3)
3001 SCREEN 9
3002 FOR I = 1 TO NN
3003 IF UG(I) > MAXU THEN MAXU = UG(I)
3004 IF UG(I) < MINU THEN MINU = UG(I)
3006 NEXT I
3010 FOR I = 1 TO NN
3020 IF ABS(X(I)) > MAXX THEN MAXX = ABS(X(I))
3030 IF ABS(Y(I)) > MAXY THEN MAXY = ABS(Y(I))
3032 IF MAXX > MAX THEN MAX = MAXX
3034 IF MAXY > MAX THEN MAX = MAXY
3040 NEXT I

```
3045 XS = 250 / MAX
3050 LF = 10
3060 DIF = MAXU - MINU: STP = DIF / LF: DIM FF(LF + 1): AG = 1
3065 FOR I = MINU TO MAXU STEP STP
3067 FF(AG) = I: IF ABS(FF(AG)) < .0000001 THEN FF(AG) = 0
3070 AG = AG + 1: NEXT I
3090 FOR N = 1 TO LF
3100 FOR ME = 1 TO MS
3110 I8 = 3 * ME: AA = 0: BB = 0: CC = 0
3120 I = IJ(I8 - 2): J = IJ(I8 - 1): K = IJ(I8)
3130 IF FF(N) = (INT(UG(I) * 1000)) / 1000 THEN GOTO 3155
3140 IF FF(N) = (INT(UG(J) * 1000)) / 1000 THEN GOTO 3155
3150 IF FF(N) = (INT(UG(K) * 1000)) / 1000 THEN GOTO 3155
3154 GOTO 3160
3155 FF(N) = FF(N) + .0000001
3160 IF FF(N) > UG(I) AND FF(N) < UG(J) THEN GOSUB 4000
3170 IF FF(N) < UG(I) AND FF(N) > UG(J) THEN GOSUB 4000
3180 IF FF(N) > UG(I) AND FF(N) < UG(K) THEN GOSUB 4100
3190 IF FF(N) < UG(I) AND FF(N) > UG(K) THEN GOSUB 4100
3200 IF FF(N) > UG(J) AND FF(N) < UG(K) THEN GOSUB 4200
3210 IF FF(N) < UG(J) AND FF(N) > UG(K) THEN GOSUB 4200
3220 IF AA = 1 AND BB = 1 THEN LINE (40 + XS * (XX(1)), 280 - XS * (YY(1)))-(40 + XS
* (XX(2)), 280 - XS * (YY(2)))
3230 IF AA = 1 AND CC = 1 THEN LINE (40 + XS * (XX(1)), 280 - XS * (YY(1)))-(40 + XS
* (XX(3)), 280 - XS * (YY(3)))
3240 IF BB = 1 AND CC = 1 THEN LINE (40 + XS * (XX(2)), 280 - XS * (YY(2)))-(40 + XS
* (XX(3)), 280 - XS * (YY(3)))
3250 NEXT ME
3260 NEXT N
3270 PRINT "TO CONTINUE, TYPE Y"
3280 A$ = INKEY$: IF A$ = "" THEN GOTO 3280
3290 IF A$ = "Y" OR A$ = "y" THEN GOTO 3310
3300 GOTO 3280
3310 SCREEN 0: WIDTH 80
3320 END
4000 DF = FF(N) - UG(I): DE = UG(J) - UG(I)
4010 DX = X(J) - X(I): DY = Y(J) - Y(I)
4015 IF DF = 0 AND DE = 0 THEN GOTO 4040
4020 XX(1) = (DF * DX / DE) + X(I): YY(1) = (DF * DY / DE) + Y(I)
4030 AA = 1: RETURN
4040 XX(1) = X(I): YY(1) = Y(I): AA = 1: RETURN
4100 DF = FF(N) - UG(I): DE = UG(K) - UG(I)
4110 DX = X(K) - X(I): DY = Y(K) - Y(I)
4115 IF DF = 0 AND DE = 0 THEN GOTO 4140
4120 XX(2) = (DF * DX / DE) + X(I): YY(2) = (DF * DY / DE) + Y(I)
4130 BB = 1: RETURN
4140 XX(2) = X(I): YY(2) = Y(I): BB = 1: RETURN
4200 DF = FF(N) - UG(J): DE = UG(K) - UG(J)
4210 DX = X(K) - X(J): DY = Y(K) - Y(J)
4215 IF DF = 0 AND DE = 0 THEN GOTO 4240
4220 XX(3) = (DF * DX / DE) + X(J): YY(3) = (DF * DY / DE) + Y(J)
4230 CC = 1: RETURN
4240 XX(3) = X(J): YY(3) = Y(J): CC = 1: RETURN
```

```
4300 STOP
4900 REM
4990 PRINT : PRINT "INCORRECT DATA": PRINT
4991 RETURN
8900 STOP
10000 REM GRAPH
10010 XZ = 1E+12: YZ = 1E+12
10030 XM = -1E+12: YM = -1E+12
10050 PRINT : PRINT "TO CONTINUE, TYPE Y"
10060 A$ = INKEY$: IF A$ = "" THEN GOTO 10060
10070 IF A$ = "Y" OR A$ = "y" THEN GOTO 10090
10080 GOTO 10060
10090 FOR ME = 1 TO MS
10100 I8 = 3 * ME
10110 I = IJ(I8 - 2)
10120 J = IJ(I8 - 1)
10130 K = IJ(I8)
10140 IF X(I) < XZ THEN XZ = X(I)
10150 IF X(J) < XZ THEN XZ = X(J)
10160 IF X(K) < XZ THEN XZ = X(K)
10170 IF Y(I) < YZ THEN YZ = Y(I)
10180 IF Y(J) < YZ THEN YZ = Y(J)
10190 IF Y(K) < YZ THEN YZ = Y(K)
10200 IF X(I) > XM THEN XM = X(I)
10210 IF X(J) > XM THEN XM = X(J)
10220 IF X(K) > XM THEN XM = X(K)
10230 IF Y(I) > YM THEN YM = Y(I)
10240 IF Y(J) > YM THEN YM = Y(J)
10250 IF Y(K) > YM THEN YM = Y(K)
10260 NEXT ME
10270 XS = 350 / (XM - XZ)
10280 YS = 175 / (YM - YZ)
10290 IF XS > YS THEN XS = YS
10295 SCREEN 9
10300 FOR ME = 1 TO MS
10310 I8 = 3 * ME
10320 II = IJ(I8 - 2)
10330 JJ = IJ(I8 - 1)
10340 KK = IJ(I8)
10350 FOR IV = 1 TO 3
10360 IF IV = 1 THEN I = II: J = JJ
10370 IF IV = 2 THEN I = JJ: J = KK
10380 IF IV = 3 THEN I = KK: J = II
10390 IF Z9 = 99 THEN GOTO 10450
10400 XI = (X(I) - XZ) * XS
10410 XJ = (X(J) - XZ) * XS
10420 YI = (Y(I) - YZ) * XS
10430 YJ = (Y(J) - YZ) * XS
10450 REM
10480 LINE (XI + 40, 250 - YI)-(XJ + 40, 250 - YJ)
10490 NEXT IV
10500 NEXT ME
10515 PRINT "TO CONTINUE, TYPE Y"
```

```
10520 A$ = INKEY$: IF A$ = "" THEN GOTO 10520
10530 IF A$ = "Y" OR A$ = "y" THEN GOTO 10560
10540 GOTO 10520
10560 SCREEN 0: WIDTH 80
10600 RETURN
```

Computer Program for Field Problems "FIELD4SF"

```
50 CLS
60 REM COPYRIGHT OF DR.C.T.F.ROSS
70 REM 6 HURSTVILLE DRIVE, WATERLOOVILLE
80 REM PORTSMOUTH. ENGLAND
90 PRINT : PRINT "FIELD PROBLEMS USING A 4 NODE ISOPARAMETRIC
QUADRILATERAL ELEMENT"
100 PRINT : PRINT "PROGRAM BY DR.C.T.F.ROSS": PRINT
101 PRINT : PRINT "NOT TO BE REPRODUCED": PRINT
103 INPUT "TYPE IN THE NAME OF YOUR INPUT FILE "; FILENAME$
104 PRINT "IF YOU ARE SATISFIED WITH YOUR INPUT FILENAME, TYPE Y; ELSE N"
105 A$ = INKEY$: IF A$ = "" THEN GOTO 105
106 IF A$ = "Y" OR A$ = "y" THEN GOTO 109
107 IF A$ = "N" OR A$ = "n" THEN GOTO 103
108 GOTO 105
109 OPEN FILENAME$ FOR INPUT AS #1
110 INPUT "TYPE IN THE NAME OF YOUR OUTPUT FILE "; OUT$
112 PRINT "IF YOU ARE SATISFIED WITH YOUR OUTPUT FILENAME, TYPE Y; ELSE N"
113 B$ = INKEY$: IF B$ = "" THEN GOTO 113
114 IF B$ = "Y" OR B$ = "y" THEN GOTO 118
115 IF B$ = "N" OR B$ = "n" THEN GOTO 110
116 GOTO 113
118 OPEN OUT$ FOR OUTPUT AS #2
119 PRINT #2, : PRINT #2, "FIELD PROBLEMS USING A 4 NODE ISOPARAMETRIC
QUADRILATERAL ELEMENT"
120 REM TYPE IN NO. OF NODES
130 INPUT #1, NJ: NN = NJ: PRINT #2, "NO. OF NODES="; NJ
140 IF NJ < 4 OR NJ > 299 THEN GOSUB 9995: STOP
150 REM TYPE IN THE NO. OF ELEMENTS
160 INPUT #1, ES: PRINT #2, "NO. OF ELEMENTS="; ES
170 IF ES < 1 OR ES > 199 THEN GOSUB 9995: STOP
180 REM TYPE IN THE NO. OF NODES WITH KNOWN BOUNDARY VALUES
190 INPUT #1, NF: PRINT #2, "NO. OF NODES WITH KNOWN BOUNDARY VALUES="; NF
191 IF NF < 0 OR NF > NJ THEN GOSUB 9995: STOP
193 REM TYPE IN NO. OF K-TYPES
194 INPUT #1, KZ: PRINT #2, "NO. OF K-TYPES="; KZ
195 IF KZ > = 1 OR KZ < ES THEN GOTO 200
196 GOSUB 9995: STOP
200 DIM I(ES), J(ES), K(ES), L(ES)
210 DIM SK(4, 4), KS(4, 4), D(2, 2), DB(2, 4), VG(NJ), VEC(NJ)
220 DIM X(NJ), Y(NJ), NS(NF), W(NF)
230 DIM B(2, 4), BT(4, 2), JA(2, 2), KX(KZ), KY(KZ)
240 DIM CO(4, 2), QS(4), SQ(4), DIAG(NN)
250 REM (D) MATRIX
255 IF KZ > 1 THEN GOTO 320
260 REM TYPE IN KX & KY
270 INPUT #1, KX, KY: PRINT #2, "KX="; KX, "KY="; KY
275 IF KX < 0 OR KY < 0 THEN GOSUB 9995: STOP
280 D(1, 1) = KX: D(2, 2) = KY
290 D(1, 2) = 0
```

```
300 D(2, 1) = 0
310 GOTO 460
320 REM TYPE IN K-TYPES,ETC
330 FOR II = 1 TO KZ
340 REM KX(II)=KX & KY(II)=KY FOR K-TYPES 'II'
350 INPUT #1, KX(II), KY(II)
360 PRINT #2, "K-TYPE="; II, "KX="; KX(II), "KY="; KY(II)
370 NEXT II
460 REM TYPE IN THE NODAL COORDINATES
465 PRINT #2, "THE NODAL COORDINATES ARE:"
470 FOR II = 1 TO NJ
480 REM TYPE IN X(II) & Y(II)
490 INPUT #1, X(II), Y(II): PRINT #2, "X("; II; ")="; X(II), "Y("; II; ")="; Y(II)
500 IF X(II) < -1000000! OR X(II) > 1000000! THEN GOSUB 9995: STOP
510 IF Y(II) < -1000000! OR Y(II) > 1000000! THEN GOSUB 9995: STOP
540 NEXT II
545 IF NF = 0 THEN GOTO 617
550 REM TYPE IN THE KNOWN BOUNDARY CONDITIONS
555 PRINT #2, "THE KNOWN BOUNDARY VALUES ARE:"
560 FOR II = 1 TO NF: NS(II) = 0: NEXT II
570 FOR II = 1 TO NF
580 REM TYPE NODAL POSITION 'SU'
590 INPUT #1, SU: PRINT #2, "NODE="; SU,
600 IF SU < 1 OR SU > NJ THEN GOSUB 9995: STOP
605 REM TYPE IN THE VALUE OF THE FUNCTION AT NODE II
610 INPUT #1, W(II): PRINT #2, "VALUE OF FUNCTION="; W(II)
612 NS(II) = SU
615 NEXT II
617 REM IF ANY BOUNDARY HAS A VALUE FOR HL, TYPE 1 ELSE TYPE 0
618 INPUT #1, XH: IF XH = 0 OR XH = 1 THEN GOTO 620
619 GOSUB 9995: STOP
620 IF XH = 1 THEN PRINT #2, "SOME BOUNDARIES HAVE VALUES FOR 'HL'"
621 IF XH = 0 THEN PRINT #2, "NO BOUNDARY HAS A VALUE FOR 'HL'"
622 INPUT #1, XQ: IF XQ = 0 OR XQ = 1 THEN GOTO 624
623 GOSUB 9995: STOP
624 IF XQ = 1 THEN PRINT #2, "SOME BOUNDARIES HAVE A VALUE FOR 'QG'" ELSE
PRINT #2, "NO BOUNDARY HAS A VALUE FOR 'QG'": REM IF 'Q' IS CONSTANT OR
ZERO OVER ENTIRE SURFACE, TYPE 1; ELSE TYPE 0
625 INPUT #1, Q7: IF Q7 = 0 OR Q7 = 1 THEN GOTO 627
626 GOSUB 9995: STOP
627 IF Q7 = 1 THEN PRINT #2, "'Q' IS CONSTANT OVER ENTIRE SURFACE" ELSE PRINT
#2, "'Q' VARIES OVER THE SURFACE": REM IF Q7=1 THEN TYPE IN 'Q'
628 IF Q7 = 1 THEN INPUT #1, Q: PRINT #2, "'Q'="; Q
632 PRINT #2, "ELEMENT TOPOLOGY"
635 FOR EL = 1 TO ES
640 REM TYPE IN THE NODAL NUMBERS IN AN ANTI-CLOCKWISE DIRECTION,
STARTING FROM A CORNER NODE
650 INPUT #1, I(EL), J(EL), K(EL), L(EL)
660 I = I(EL): J = J(EL): K = K(EL): L = L(EL)
675 PRINT #2, "ELEMENT "; I; "-"; J; "-"; K; "-"; L
680 IF I < 1 OR I > NJ THEN GOSUB 9995: STOP
690 IF J < 1 OR J > NJ THEN GOSUB 9995: STOP
700 IF K < 1 OR K > NJ THEN GOSUB 9995: STOP
```

```
710 IF L < 1 OR L > NJ THEN GOSUB 9995: STOP
755 IF I = J OR I = K OR I = L THEN GOSUB 9995: STOP
760 IF J = K OR J = L THEN GOSUB 9995: STOP
765 IF K = L THEN GOSUB 9995: STOP
770 FOR II = 1 TO 4
780 IA = L
790 IF II = 1 THEN IA = I
800 IF II = 2 THEN IA = J
810 IF II = 3 THEN IA = K
820 FOR JJ = 1 TO 4
830 IB = L
840 IF JJ = 1 THEN IB = I
850 IF JJ = 2 THEN IB = J
860 IF JJ = 3 THEN IB = K
870 IIJJ = IB - IA + 1
880 IF IIJJ > DIAG(IB) THEN DIAG(IB) = IIJJ
890 NEXT JJ, II
900 NEXT EL
905 DIAG(0) = 0
910 DIAG(1) = 1: FOR II = 2 TO NJ: DIAG(II) = DIAG(II - 1) + DIAG(II): NEXT II
920 NT = DIAG(NJ)
970 DIM A(NT), Q(NJ)
980 GOSUB 10000
1190 FOR EL = 1 TO ES
1200 PRINT : PRINT "ELEMENT NO."; EL; " UNDER COMPUTATION"
1210 I = I(EL)
1220 J = J(EL)
1230 K = K(EL)
1240 L = L(EL)
1272 IF KZ = 1 THEN GOTO 1281
1273 REM ZK=K-TYPE
1274 INPUT #1, ZK: PRINT #2, "K-TYPE="; ZK
1275 KX = KX(ZK): KY = KY(ZK)
1276 D(1, 1) = KX: D(2, 2) = KY
1277 D(1, 2) = 0: D(2, 1) = 0
1281 IF Q7 = 1 THEN GOTO 1284
1282 INPUT #1, Q: PRINT #2, "Q="; Q
1284 VEC(I) = VEC(I) + 1: VEC(K) = VEC(K) + 1
1285 VEC(J) = VEC(J) + 1: VEC(L) = VEC(L) + 1
1290 CO(1, 1) = X(I)
1300 CO(1, 2) = Y(I)
1310 CO(2, 1) = X(J)
1320 CO(2, 2) = Y(J)
1330 CO(3, 1) = X(K)
1340 CO(3, 2) = Y(K)
1350 CO(4, 1) = X(L)
1360 CO(4, 2) = Y(L)
1450 FOR II = 1 TO 4
1455 SQ(II) = 0
1460 FOR JJ = 1 TO 4
1470 SK(II, JJ) = 0
1480 NEXT JJ
1490 NEXT II
```

```
1500 FOR LL = 1 TO 2
1510 IF LL = 1 THEN XI = -.5773502700000001#
1520 IF LL = 2 THEN XI = .5773502700000001#
1620 FOR MM = 1 TO 2
1630 IF MM = 1 THEN ET = -.5773502700000001#
1640 IF MM = 2 THEN ET = .5773502700000001#
1740 GOSUB 3330
1790 FOR II = 1 TO 4
1800 FOR JJ = 1 TO 4
1810 KS(II, JJ) = 0
1820 NEXT JJ
1830 NEXT II
1840 FOR II = 1 TO 4
1850 FOR JJ = 1 TO 4
1860 FOR KK = 1 TO 2
1870 KS(II, JJ) = KS(II, JJ) + BT(II, KK) * DB(KK, JJ) * DT
1880 NEXT KK
1890 NEXT JJ
1900 NEXT II
1910 FOR IJ = 1 TO 4
1915 SQ(IJ) = SQ(IJ) + QS(IJ) * Q
1920 FOR JI = 1 TO 4
1930 SK(IJ, JI) = SK(IJ, JI) + KS(IJ, JI)
1940 NEXT JI
1950 NEXT IJ
1960 NEXT MM
1970 NEXT LL
1971 IF XH = 0 AND XQ = 0 THEN GOTO 1980
1972 REM IF EITHER 'HL' OR 'QG' HAS A VALUE ON THIS ELEMENT, TYPE Y; ELSE
TYPE N
1973 INPUT #1, B$
1974 IF B$ = "Y" OR B$ = "y" OR B$ = "N" OR B$ = "n" THEN GOTO 1976
1975 GOSUB 9995: STOP
1976 IF B$ = "N" OR B$ = "n" THEN GOTO 1980
1977 GOSUB 4380
1980 FOR JJ = 1 TO 4
1990 IF JJ = 1 THEN NR = I
2000 IF JJ = 2 THEN NR = J
2010 IF JJ = 3 THEN NR = K
2020 IF JJ = 4 THEN NR = L
2030 IDIAG = DIAG(NR)
2100 FOR KK = 1 TO 4
2110 IF KK = 1 THEN N9 = I
2120 IF KK = 2 THEN N9 = J
2130 IF KK = 3 THEN N9 = K
2140 IF KK = 4 THEN N9 = L
2220 IF N9 > NR THEN GOTO 2260
2230 LOCATION = IDIAG - (NR - N9)
2240 A(LOCATION) = A(LOCATION) + SK(JJ, KK)
2260 NEXT KK
2270 Q(NR) = Q(NR) + SQ(JJ)
2280 NEXT JJ
2290 NEXT EL
```

```
2295 IF NF = 0 THEN GOTO 2360
2300 FOR II = 1 TO NF
2310 N9 = NS(II)
2320 IF N9 = 0 THEN GOTO 2350
2325 DI = DIAG(N9)
2330 Q(N9) = 100000! * W(II) * A(DI)
2340 A(DI) = A(DI) * 100000!
2350 NEXT II
2360 PRINT : PRINT "THE SIMULTANEOUS EQUATIONS ARE NOW BEING SOLVED":
PRINT
2370 N = NJ
2380 A(1) = SQR(A(1))
2390 FOR II = 2 TO N
2400 DI = DIAG(II) - II
2405 LL = DIAG(II - 1) - DI + 1
2410 FOR JJ = LL TO II
2415 X = A(DI + JJ)
PRINT #2, JJ
2420 DJ = DIAG(JJ) - JJ
2425 IF JJ = 1 THEN GOTO 2455
2430 LB = DIAG(JJ - 1) - DJ + 1
2435 IF LL > LB THEN LB = LL
2440 FOR KK = LB TO JJ - 1
2445 X = X - A(DI + KK) * A(DJ + KK)
2450 NEXT KK
2455 A(DI + JJ) = X / A(DJ + JJ)
2460 NEXT JJ
2470 A(DI + II) = SQR(X)
2480 NEXT II
2490 Q(1) = Q(1) / A(1)
2495 FOR II = 2 TO N
2500 DI = DIAG(II) - II
2505 LL = DIAG(II - 1) - DI + 1
2510 X = Q(II)
2515 FOR JJ = LL TO II - 1
2520 X = X - A(DI + JJ) * Q(JJ)
2525 NEXT JJ
2530 Q(II) = X / A(DI + II)
2535 NEXT II
2540 FOR II = N TO 2 STEP -1
2545 DI = DIAG(II) - II
2550 X = Q(II) / A(DI + II)
2555 Q(II) = X
2560 LL = DIAG(II - 1) - DI + 1
2565 FOR KK = LL TO II - 1
2570 Q(KK) = Q(KK) - X * A(DI + KK)
2575 NEXT KK
2580 NEXT II
2585 Q(1) = Q(1) / A(1)
2590 PRINT #2,
2600 PRINT #2, " THE VALUES OF THE FUNCTIONS ARE:"
2610 FOR I = 1 TO NJ
2620 PRINT #2, "NODE="; I; " PHI("; I; ")="; Q(I)
```

```
2630 VG(I) = Q(I)
2640 NEXT I
2660 GOSUB 6000
2710 END
3330 FOR II = 1 TO 2
3340 FOR JJ = 1 TO 2
3345 JA(II, JJ) = 0
3350 NEXT JJ
3360 NEXT II
3370 JA(1, 1) = ((ET - 1) * (CO(1, 1) - CO(2, 1)) + (ET + 1) * (CO(3, 1) - CO(4, 1))) / 4
3380 JA(1, 2) = ((ET - 1) * (CO(1, 2) - CO(2, 2)) + (ET + 1) * (CO(3, 2) - CO(4, 2))) / 4
3390 JA(2, 1) = ((XI - 1) * (CO(1, 1) - CO(4, 1)) + (XI + 1) * (CO(3, 1) - CO(2, 1))) / 4
3400 JA(2, 2) = ((XI - 1) * (CO(1, 2) - CO(4, 2)) + (XI + 1) * (CO(3, 2) - CO(2, 2))) / 4
3610 CN = JA(1, 1)
3620 JA(1, 1) = JA(2, 2)
3630 JA(2, 2) = CN
3640 JA(2, 1) = -JA(2, 1)
3650 JA(1, 2) = -JA(1, 2)
3660 DT = JA(1, 1) * JA(2, 2) - JA(1, 2) * JA(2, 1)
3670 FOR II = 1 TO 2
3680 FOR JJ = 1 TO 2
3690 JA(II, JJ) = JA(II, JJ) / DT
3700 NEXT JJ
3705 NEXT II
3707 QS(1) = (1 - XI) * (1 - ET) * DT / 4
3708 QS(2) = (1 + XI) * (1 - ET) * DT / 4
3709 QS(3) = (1 + XI) * (1 + ET) * DT / 4
3710 QS(4) = (1 - XI) * (1 + ET) * DT / 4
3720 B(1, 1) = (JA(1, 1) * (ET - 1) + JA(1, 2) * (XI - 1)) / 4
3730 B(1, 2) = (JA(1, 1) * (1 - ET) - JA(1, 2) * (XI + 1)) / 4
3740 B(1, 3) = (JA(1, 1) * (ET + 1) + JA(1, 2) * (XI + 1)) / 4
3750 B(1, 4) = (-JA(1, 1) * (ET + 1) + JA(1, 2) * (1 - XI)) / 4
3760 B(2, 1) = (JA(2, 1) * (ET - 1) + JA(2, 2) * (XI - 1)) / 4
3770 B(2, 2) = (JA(2, 1) * (1 - ET) - JA(2, 2) * (XI + 1)) / 4
3780 B(2, 3) = (JA(2, 1) * (ET + 1) + JA(2, 2) * (XI + 1)) / 4
3790 B(2, 4) = (-JA(2, 1) * (ET + 1) + JA(2, 2) * (1 - XI)) / 4
3840 FOR II = 1 TO 2
3850 FOR JJ = 1 TO 4
3860 BT(JJ, II) = B(II, JJ)
3870 NEXT JJ
3880 NEXT II
3900 FOR II = 1 TO 2
3910 FOR JJ = 1 TO 4
3920 DB(II, JJ) = 0
3930 NEXT JJ
3940 NEXT II
3950 FOR II = 1 TO 2
3960 FOR JJ = 1 TO 4
3970 FOR KK = 1 TO 2
3980 DB(II, JJ) = DB(II, JJ) + D(II, KK) * B(KK, JJ)
3990 NEXT KK
4000 NEXT JJ
4010 NEXT II
```

```
4020 RETURN
4380 REM EFFECTS OF HL
4390 IF XH = 0 THEN GOTO 4950
4400 REM LH=NO. OF BOUNDARIES ON WHICH HL HAS A VALUE
4410 INPUT #1, LH: PRINT #2, "NO. OF BOUNDARIES ON WHICH HL HAS A VALUE=";
LH
4420 IF LH = 0 OR LH = 1 OR LH = 2 OR LH = 3 OR LH = 4 THEN GOTO 4440
4430 GOSUB 9995: STOP
4440 IF LH = 0 THEN GOTO 4950
4490 FOR II = 1 TO LH
4500 ZI = 0: ZJ = 0
4510 REM TYPE IN NODES DEFINING BOUNDARY
4520 INPUT #1, IN, JN: PRINT #2, "THE NODES DEFINING THE BOUNDARY ON WHICH
HL HAS A VALUE ARE "; IN; "-"; JN
4530 IF IN = I THEN ZI = 1
4535 IF IN = J THEN ZI = 2
4540 IF IN = K THEN ZI = 3
4545 IF IN = L THEN ZI = 4
4550 IF JN = I THEN ZJ = 1
4555 IF JN = J THEN ZJ = 2
4560 IF JN = K THEN ZJ = 3
4565 IF JN = L THEN ZJ = 4
4570 IF ZJ = 0 THEN GOSUB 9995: STOP
4580 IF ZI = ZJ THEN GOSUB 9995: STOP
4590 IF ZI = 0 OR ZJ = 0 THEN GOSUB 9995: STOP
4650 REM TYPE IN THE VALUES OF HL & T(INFINITY)
4660 INPUT #1, HL, IT: PRINT #2, "HL="; HL, "T(INFINITY)="; IT
4670 LL = SQR((X(JN) - X(IN)) ^ 2 + (Y(JN) - Y(IN)) ^ 2)
4680 SK(ZI, ZI) = SK(ZI, ZI) + HL * LL / 3
4690 SK(ZJ, ZJ) = SK(ZJ, ZJ) + HL * LL / 3
4700 SK(ZI, ZJ) = SK(ZI, ZJ) + HL * LL / 6
4710 SK(ZJ, ZI) = SK(ZJ, ZI) + HL * LL / 6
4720 SQ(ZI) = SQ(ZI) + HL * LL * IT / 2: SQ(ZJ) = SQ(ZJ) + HL * LL * IT / 2
4940 NEXT II
4950 REM EFFECTS OF 'QG'
4960 IF XQ = 0 THEN GOTO 5430
4970 REM GQ=NO. OF BOUNDARIES ON WHICH 'QG' HAS A VALUE
4980 INPUT #1, GQ: PRINT #2, "THE NO. OF BOUNDARIES ON WHICH 'QG' HAS A
VALUE="; GQ
4990 IF GQ = 0 OR GQ = 1 OR GQ = 2 OR GQ = 3 OR GQ = 4 THEN GOTO 5010
5000 GOSUB 9995: STOP
5010 IF GQ = 0 THEN GOTO 5430
5060 FOR II = 1 TO GQ
5070 ZI = 0: ZJ = 0
5080 REM TYPE IN THE NODES DEFINING THE BOUNDARY ON WHICH THERE IS A
VALUE OF 'QG'
5090 INPUT #1, IN, JN: PRINT #2, "THE NODES DEFINING THE BOUNDARY ON WHICH
'QG' HAS A VALUE ARE: "; IN, JN
5100 IF IN = I THEN ZI = 1
5110 IF IN = J THEN ZI = 2
5120 IF IN = K THEN ZI = 3
5130 IF IN = L THEN ZI = 4
5140 IF JN = I THEN ZJ = 1
```

```
5145 IF JN = J THEN ZJ = 2
5150 IF JN = K THEN ZJ = 3
5160 IF JN = L THEN ZJ = 4
5170 IF ZI = ZJ THEN GOSUB 9995: STOP
5180 IF ZI = 0 OR ZJ = 0 THEN GOSUB 9995: STOP
5220 REM TYPE IN THE VALUE OF QG
5230 INPUT #1, QG: PRINT #2, "QG="; QG
5240 LL = SQR((X(JN) - X(IN)) ^ 2 + (Y(JN) - Y(IN)) ^ 2)
5250 SQ(ZI) = SQ(ZI) - LL * QG / 2
5260 SQ(ZJ) = SQ(ZJ) - LL * QG / 2
5420 NEXT II
5430 RETURN
6000 MAX = MAXX = -1E+12: MAXU = -1E+12: MINU = 1E+12: DIM YY(4), XX(4)
6010 SCREEN 9
6020 FOR I = 1 TO NJ
6030 IF VG(I) > MAXU THEN MAXU = VG(I)
6040 IF VG(I) < MINU THEN MINU = VG(I)
6045 IF ABS(X(I)) > MAXX THEN MAXX = ABS(X(I))
6050 IF ABS(Y(I)) > MAXY THEN MAXY = ABS(Y(I))
6055 IF MAXX > MAX THEN MAX = MAXX
6060 IF MAXY > MAX THEN MAX = MAXY
6065 NEXT I
6070 LF = 10: XS = 250 / MAX
6080 DIF = MAXU - MINU: STP = DIF / LF: DIM FF(LF + 1): AG = 1
6090 FOR I = MINU TO MAXU STEP STP
6100 FF(AG) = I: IF ABS(FF(AG)) < .000001 THEN FF(AG) = 0
6110 AG = AG + 1: NEXT I
6120 FOR N = 1 TO LF
6130 FOR EL = 1 TO ES
6135 AA = 0: BB = 0: CC = 0: DD = 0
6140 I = I(EL): J = J(EL): K = K(EL): L = L(EL)
6145 IF FF(N) = (INT(VG(I) * 1000)) / 1000 THEN GOTO 6155
6147 IF FF(N) = (INT(VG(J) * 1000)) / 1000 THEN GOTO 6155
6149 IF FF(N) = (INT(VG(K) * 1000)) / 1000 THEN GOTO 6155
6151 IF FF(N) = (INT(VG(L) * 1000)) / 1000 THEN GOTO 6155
6153 GOTO 6160
6155 FF(N) = FF(N) + .000001
6160 IF FF(N) > VG(I) AND FF(N) < VG(J) THEN GOSUB 7000
6165 IF FF(N) < VG(I) AND FF(N) > VG(J) THEN GOSUB 7000
6170 IF FF(N) > VG(I) AND FF(N) < VG(L) THEN GOSUB 7100
6175 IF FF(N) < VG(I) AND FF(N) > VG(L) THEN GOSUB 7100
6180 IF FF(N) > VG(J) AND FF(N) < VG(K) THEN GOSUB 7200
6185 IF FF(N) < VG(J) AND FF(N) > VG(K) THEN GOSUB 7200
6190 IF FF(N) > VG(K) AND FF(N) < VG(L) THEN GOSUB 7300
6195 IF FF(N) < VG(K) AND FF(N) > VG(L) THEN GOSUB 7300
6220 IF AA = 1 AND BB = 1 THEN LINE (40 + XS * (XX(1)), 270 - XS * (YY(1)))-(40 + XS
* (XX(2)), 270 - XS * (YY(2)))
6230 IF AA = 1 AND CC = 1 THEN LINE (40 + XS * (XX(1)), 270 - XS * (YY(1)))-(40 + XS
* (XX(3)), 270 - XS * (YY(3)))
6240 IF AA = 1 AND DD = 1 THEN LINE (40 + XS * (XX(1)), 270 - XS * (YY(1)))-(40 + XS
* (XX(4)), 270 - XS * (YY(4)))
6250 IF BB = 1 AND CC = 1 THEN LINE (40 + XS * (XX(2)), 270 - XS * (YY(2)))-(40 + XS
* (XX(3)), 270 - XS * (YY(3)))
```

```
6260 IF BB = 1 AND DD = 1 THEN LINE (40 + XS * (XX(2)), 270 - XS * (YY(2)))-(40 + XS
     * (XX(4)), 270 - XS * (YY(4)))
6270 IF CC = 1 AND DD = 1 THEN LINE (40 + XS * (XX(3)), 270 - XS * (YY(3)))-(40 + XS
     * (XX(4)), 270 - XS * (YY(4)))
6290 NEXT EL
6300 NEXT N
6310 A$ = INKEY$: IF A$ = "" THEN GOTO 6310
6320 IF A$ = "Y" OR A$ = "y" THEN GOTO 6400
6330 GOSUB 9995: GOTO 6310
6400 RETURN
7000 DF = FF(N) - VG(I): DE = VG(J) - VG(I)
7010 DX = X(J) - X(I): DY = Y(J) - Y(I)
7020 IF DF = 0 AND DE = 0 THEN GOTO 7050
7030 XX(1) = (DF * DX / DE) + X(I): YY(1) = (DF * DY / DE) + Y(I)
7040 AA = 1: RETURN
7050 XX(1) = X(I): YY(1) = Y(I): AA = 1: RETURN
7100 DF = FF(N) - VG(I): DE = VG(L) - VG(I)
7110 DX = X(L) - X(I): DY = Y(L) - Y(I)
7120 IF DF = 0 AND DE = 0 THEN GOTO 7150
7130 XX(2) = (DF * DX / DE) + X(I): YY(2) = (DF * DY / DE) + Y(I)
7140 BB = 1: RETURN
7150 XX(2) = X(I): YY(2) = Y(I): BB = 1: RETURN
7200 DF = FF(N) - VG(J): DE = VG(K) - VG(J)
7210 DX = X(K) - X(J): DY = Y(K) - Y(J)
7220 IF DF = 0 AND DE = 0 THEN GOTO 7250
7230 XX(3) = (DF * DX / DE) + X(J): YY(3) = (DF * DY / DE) + Y(J)
7240 CC = 1: RETURN
7250 XX(3) = X(J): YY(3) = Y(J): CC = 1: RETURN
7300 DF = FF(N) - VG(K): DE = VG(L) - VG(K)
7310 DX = X(L) - X(K): DY = Y(L) - Y(K)
7320 IF DF = 0 AND DE = 0 THEN GOTO 7350
7330 XX(4) = (DF * DX / DE) + X(K): YY(4) = (DF * DY / DE) + Y(K)
7340 DD = 1: RETURN
7350 XX(4) = X(K): YY(3) = Y(K): DD = 1: RETURN
9995 REM WARNING
9996 PRINT : PRINT "INCORRECT DATA": PRINT
9997 RETURN
10000 REM GRAPH
10010 XZ = 1E+12: YZ = 1E+12
10030 XM = -1E+12: YM = -1E+12
10050 PRINT : PRINT "TO CONTINUE, TYPE Y"
10060 A$ = INKEY$: IF A$ = "" THEN GOTO 10060
10070 IF A$ = "Y" OR A$ = "y" THEN GOTO 10090
10080 GOTO 10060
10090 FOR ME = 1 TO ES
10110 I = I(ME)
10120 J = J(ME)
10130 K = K(ME)
10135 L = L(ME)
10140 IF X(I) < XZ THEN XZ = X(I)
10150 IF X(J) < XZ THEN XZ = X(J)
10160 IF X(K) < XZ THEN XZ = X(K)
10165 IF X(L) < XZ THEN XZ = X(L)
```

589

```
10170 IF Y(I) < YZ THEN YZ = Y(I)
10180 IF Y(J) < YZ THEN YZ = Y(J)
10190 IF Y(K) < YZ THEN YZ = Y(K)
10195 IF Y(L) < YZ THEN YZ = Y(L)
10200 IF X(I) > XM THEN XM = X(I)
10210 IF X(J) > XM THEN XM = X(J)
10220 IF X(K) > XM THEN XM = X(K)
10225 IF X(L) > XM THEN XM = X(L)
10230 IF Y(I) > YM THEN YM = Y(I)
10240 IF Y(J) > YM THEN YM = Y(J)
10250 IF Y(K) > YM THEN YM = Y(K)
10255 IF Y(L) > YM THEN YM = Y(L)
10260 NEXT ME
10270 XS = 350 / (XM - XZ)
10280 YS = 175 / (YM - YZ)
10290 IF XS > YS THEN XS = YS
10295 SCREEN 9
10300 FOR ME = 1 TO ES
10320 II = I(ME)
10330 JJ = J(ME)
10340 KK = K(ME)
10345 LL = L(ME)
10350 FOR IV = 1 TO 4
10360 IF IV = 1 THEN I = II: J = JJ
10370 IF IV = 2 THEN I = JJ: J = KK
10380 IF IV = 3 THEN I = KK: J = LL
10385 IF IV = 4 THEN I = LL: J = II
10400 XI = (X(I) - XZ) * XS
10410 XJ = (X(J) - XZ) * XS
10420 YI = (Y(I) - YZ) * XS
10430 YJ = (Y(J) - YZ) * XS
10480 LINE (XI + 40, 270 - YI)-(XJ + 40, 270 - YJ)
10490 NEXT IV
10500 NEXT ME
10510 IF Z9 = 77 THEN GOTO 10600
10515 PRINT "TO CONTINUE, TYPE Y"
10520 A$ = INKEY$: IF A$ = "" THEN GOTO 10520
10530 IF A$ = "Y" OR A$ = "y" THEN GOTO 10560
10540 GOTO 10520
10560 SCREEN 0: WIDTH 80
10600 RETURN
```

Computer Program for Field Problems "FIELD8SF"

```
50 CLS
60 REM COPYRIGHT OF DR.C.T.F.ROSS
70 REM 6 HURSTVILLE DRIVE, WATERLOOVILLE
80 REM PORTSMOUTH. ENGLAND
90 PRINT : PRINT "FIELD PROBLEMS USING AN 8 NODE ISOPARAMETRIC
QUADRILATERAL ELEMENT"
100 PRINT : PRINT "PROGRAM BY DR.C.T.F.ROSS": PRINT
101 PRINT : PRINT "NOT TO BE REPRODUCED": PRINT
102 INPUT "TYPE IN THE NAME OF YOUR INPUT FILE "; FILENAME$
103 PRINT "IF YOU ARE SATISFIED WITH YOUR INPUT FILENAME, TYPE Y; ELSE N"
104 A$ = INKEY$: IF A$ = "" THEN GOTO 104
105 IF A$ = "Y" OR A$ = "y" THEN GOTO 109
106 IF A$ = "N" OR A$ = "n" THEN GOTO 102
107 GOTO 104
109 OPEN FILENAME$ FOR INPUT AS #1
110 INPUT "TYPE IN THE NAME OF YOUR OUTPUT FILE "; OUT$
112 PRINT "IF YOU ARE SATISFIED WITH YOUR OUTPUT FILENAME, TYPE Y; ELSE N"
113 B$ = INKEY$: IF B$ = "" THEN GOTO 113
114 IF B$ = "Y" OR B$ = "y" THEN GOTO 118
115 IF B$ = "N" OR B$ = "n" THEN GOTO 110
116 GOTO 113
118 OPEN OUT$ FOR OUTPUT AS #2
119 PRINT #2, : PRINT #2, "FIELD PROBLEMS USING AN 8 NODE ISOPARAMETRIC
QUADRILATERAL ELEMENT"
120 REM TYPE IN NO. OF NODES
130 INPUT #1, NJ: NN = NJ: PRINT #2, "NO. OF NODES="; NJ
140 IF NJ < 8 OR NJ > 299 THEN GOSUB 9995: STOP
150 REM TYPE IN THE NO. OF ELEMENTS
160 INPUT #1, ES: PRINT #2, "NO. OF ELEMENTS="; ES
170 IF ES < 1 OR ES > 299 THEN GOSUB 9995: STOP
180 REM TYPE IN THE NO. OF NODES WITH KNOWN BOUNDARY VALUES
190 INPUT #1, NF: PRINT #2, "NO. OF NODES WITH KNOWN BOUNDARY VALUES="; NF
191 IF NF < 0 OR NF > NJ THEN GOSUB 9995: STOP
193 REM TYPE IN NO. OF K-TYPES
194 INPUT #1, KZ: PRINT #2, "NO. OF K-TYPES="; KZ
195 IF KZ > = 1 OR KZ < ES THEN GOTO 200
196 GOSUB 9995: STOP
200 DIM I(ES), J(ES), K(ES), L(ES), M(ES), N(ES), O(ES), P(ES)
210 DIM SK(8, 8), KS(8, 8), DN(2, 8), D(2, 2), DB(2, 8), VG(NJ), VEC(NJ)
220 DIM X(NJ), Y(NJ), NS(NF), W(NF)
230 DIM B(2, 8), BT(8, 2), JA(2, 2), KX(KZ), KY(KZ)
240 DIM CO(8, 2), QS(8), SQ(8), DIAG(NJ)
250 REM (D) MATRIX
255 IF KZ > 1 THEN GOTO 320
260 REM TYPE IN KX & KY
270 INPUT #1, KX, KY: PRINT #2, "KX="; KX, "KY="; KY
275 IF KX < 0 OR KY < 0 THEN GOSUB 9995: STOP
280 D(1, 1) = KX: D(2, 2) = KY
290 D(1, 2) = 0
```

300 D(2, 1) = 0
310 GOTO 460
320 REM TYPE IN K-TYPES,ETC
330 FOR II = 1 TO KZ
340 REM KX(II)=KX & KY(II)=KY FOR K-TYPES 'II'
350 INPUT #1, KX(II), KY(II)
360 PRINT #2, "K-TYPE="; II, "KX="; KX(II), "KY="; KY(II)
370 NEXT II
460 REM TYPE IN THE NODAL COORDINATES
465 PRINT #2, "THE NODAL COORDINATES ARE:"
470 FOR II = 1 TO NJ
480 REM TYPE IN X(II) & Y(II)
490 INPUT #1, X(II), Y(II): PRINT #2, "X("; II; ")="; X(II), "Y("; II; ")="; Y(II)
500 IF X(II) < -1000000! OR X(II) > 1000000! THEN GOSUB 9995: STOP
510 IF Y(II) < -1000000! OR Y(II) > 1000000! THEN GOSUB 9995: STOP
540 NEXT II
545 IF NF = 0 THEN GOTO 617
550 REM TYPE IN THE KNOWN BOUNDARY CONDITIONS
555 PRINT #2, "THE KNOWN BOUNDARY VALUES ARE:"
560 FOR II = 1 TO NF: NS(II) = 0: NEXT II
570 FOR II = 1 TO NF
580 REM TYPE NODAL POSITION 'SU'
590 INPUT #1, SU: PRINT #2, "NODE="; SU,
600 IF SU < 1 OR SU > NJ THEN GOSUB 9995: STOP
605 REM TYPE IN THE VALUE OF THE FUNCTION AT NODE II
610 INPUT #1, W(II): PRINT #2, "VALUE OF FUNCTION="; W(II)
612 NS(II) = SU
615 NEXT II
617 REM IF ANY BOUNDARY HAS A VALUE FOR HL, TYPE 1 ELSE TYPE 0
618 INPUT #1, XH: IF XH = 0 OR XH = 1 THEN GOTO 620
619 GOSUB 9995: STOP
620 IF XH = 1 THEN PRINT #2, "SOME BOUNDARIES HAVE VALUES FOR 'HL'"
621 IF XH = 0 THEN PRINT #2, "NO BOUNDARY HAS A VALUE FOR 'HL'"
622 INPUT #1, XQ: IF XQ = 0 OR XQ = 1 THEN GOTO 624
623 GOSUB 9995: STOP
624 IF XQ = 1 THEN PRINT #2, "SOME BOUNDARIES HAVE A VALUE FOR 'QG'" ELSE
PRINT #2, "NO BOUNDARY HAS A VALUE FOR 'QG'": REM IF 'Q' IS CONSTANT OR
ZERO OVER ENTIRE SURFACE, TYPE 1; ELSE TYPE 0
625 INPUT #1, Q7: IF Q7 = 0 OR Q7 = 1 THEN GOTO 627
626 GOSUB 9995: STOP
627 IF Q7 = 1 THEN PRINT #2, "'Q' IS CONSTANT OVER ENTIRE SURFACE" ELSE PRINT
#2, "'Q' VARIES OVER THE SURFACE": REM IF Q7=1 THEN TYPE IN 'Q'
628 IF Q7 = 1 THEN INPUT #1, Q: PRINT #2, "'Q'="; Q
632 PRINT #2, "ELEMENT TOPOLOGY"
635 FOR EL = 1 TO ES
640 REM TYPE IN THE NODAL NUMBERS IN AN ANTI-CLOCKWISE DIRECTION,
STARTING FROM A CORNER NODE
650 INPUT #1, I(EL), J(EL), K(EL), L(EL), M(EL), N(EL), O(EL), P(EL)
660 I = I(EL): J = J(EL): K = K(EL): L = L(EL)
670 M = M(EL): N = N(EL): O = O(EL): P = P(EL)
675 PRINT #2, "ELEMENT "; I; "-"; J; "-"; K; "-"; L; "-"; M; "-"; N; "-"; O; "-"; P
680 IF I < 1 OR I > NJ THEN GOSUB 9995: STOP
690 IF J < 1 OR J > NJ THEN GOSUB 9995: STOP

```
700 IF K < 1 OR K > NJ THEN GOSUB 9995: STOP
710 IF L < 1 OR L > NJ THEN GOSUB 9995: STOP
720 IF M < 1 OR M > NJ THEN GOSUB 9995: STOP
730 IF N < 1 OR N > NJ THEN GOSUB 9995: STOP
740 IF O < 1 OR O > NJ THEN GOSUB 9995: STOP
750 IF P < 1 OR P > NJ THEN GOSUB 9995: STOP
755 IF I = J OR I = K OR I = L OR I = M OR I = N OR I = O OR I = P THEN GOSUB
9995: STOP
760 IF J = K OR J = L OR J = M OR J = N OR J = O OR J = P THEN GOSUB 9995: STOP
765 IF K = L OR K = M OR K = N OR K = O OR K = P THEN GOSUB 9995: STOP
770 IF L = M OR L = N OR L = O OR L = P THEN GOSUB 9995: STOP
775 IF M = N OR M = O OR M = P THEN GOSUB 9995: STOP
780 IF N = O OR N = P THEN GOSUB 9995: STOP
785 IF O = P THEN GOSUB 9995: STOP
790 FOR II = 1 TO 8
800 IA = P
810 IF II = 1 THEN IA = I
820 IF II = 2 THEN IA = J
830 IF II = 3 THEN IA = K
840 IF II = 4 THEN IA = L
850 IF II = 5 THEN IA = M
860 IF II = 6 THEN IA = N
870 IF II = 7 THEN IA = O
880 FOR JJ = 1 TO 8
885 IB = P
890 IF JJ = 1 THEN IB = I
895 IF JJ = 2 THEN IB = J
900 IF JJ = 3 THEN IB = K
905 IF JJ = 4 THEN IB = L
910 IF JJ = 5 THEN IB = M
915 IF JJ = 6 THEN IB = N
920 IF JJ = 7 THEN IB = O
925 IIJJ = IB - IA + 1
930 IF IIJJ > DIAG(IB) THEN DIAG(IB) = IIJJ
935 NEXT JJ, II
940 NEXT EL
950 DIAG(1) = 1: FOR II = 2 TO NJ: DIAG(II) = DIAG(II - 1) + DIAG(II): NEXT II
960 NT = DIAG(NJ)
970 DIM A(NT), Q(NJ)
980 GOSUB 10000
1190 FOR EL = 1 TO ES
1200 PRINT : PRINT "ELEMENT NO."; EL; " UNDER COMPUTATION"
1210 I = I(EL)
1220 J = J(EL)
1230 K = K(EL)
1240 L = L(EL)
1250 M = M(EL)
1260 N = N(EL)
1265 O = O(EL)
1270 P = P(EL)
1272 IF KZ = 1 THEN GOTO 1281
1273 REM ZK=K-TYPE
1274 INPUT #1, ZK: PRINT #2, "K-TYPE="; ZK
```

```
1275 KX = KX(ZK): KY = KY(ZK)
1276 D(1, 1) = KX: D(2, 2) = KY
1277 D(1, 2) = 0: D(2, 1) = 0
1281 IF Q7 = 1 THEN GOTO 1284
1282 INPUT #1, Q: PRINT #2, "Q="; Q
1284 VEC(I) = VEC(I) + 1: VEC(K) = VEC(K) + 1
1285 VEC(M) = VEC(M) + 1: VEC(O) = VEC(O) + 1
1287 VEC(J) = 1: VEC(L) = 1: VEC(N) = 1: VEC(P) = 1
1290 CO(1, 1) = X(I)
1300 CO(1, 2) = Y(I)
1310 CO(2, 1) = X(J)
1320 CO(2, 2) = Y(J)
1330 CO(3, 1) = X(K)
1340 CO(3, 2) = Y(K)
1350 CO(4, 1) = X(L)
1360 CO(4, 2) = Y(L)
1370 CO(5, 1) = X(M)
1380 CO(5, 2) = Y(M)
1390 CO(6, 1) = X(N)
1400 CO(6, 2) = Y(N)
1410 CO(7, 1) = X(O)
1420 CO(7, 2) = Y(O)
1430 CO(8, 1) = X(P)
1440 CO(8, 2) = Y(P)
1450 FOR II = 1 TO 8
1455 SQ(II) = 0
1460 FOR JJ = 1 TO 8
1470 SK(II, JJ) = 0
1480 NEXT JJ
1490 NEXT II
1500 FOR LL = 1 TO 3
1510 IF LL = 1 THEN GOTO 1540
1520 IF LL = 2 THEN GOTO 1570
1530 IF LL = 3 THEN GOTO 1600
1540 XI = -.774597
1550 AI = 5 / 9
1560 GOTO 1620
1570 XI = 0
1580 AI = 8 / 9
1590 GOTO 1620
1600 XI = .774597
1610 AI = 5 / 9
1620 FOR MM = 1 TO 3
1630 IF MM = 1 THEN GOTO 1660
1640 IF MM = 2 THEN GOTO 1690
1650 IF MM = 3 THEN GOTO 1720
1660 ET = -.774597
1670 AJ = 5 / 9
1680 GOTO 1740
1690 ET = 0
1700 AJ = 8 / 9
1710 GOTO 1740
1720 ET = .774597
```

```
1730 AJ = 5 / 9
1740 GOSUB 3330
1790 FOR II = 1 TO 8
1800 FOR JJ = 1 TO 8
1810 KS(II, JJ) = 0
1820 NEXT JJ
1830 NEXT II
1840 FOR II = 1 TO 8
1850 FOR JJ = 1 TO 8
1860 FOR KK = 1 TO 2
1870 KS(II, JJ) = KS(II, JJ) + BT(II, KK) * DB(KK, JJ) * AI * AJ * DT
1880 NEXT KK
1890 NEXT JJ
1900 NEXT II
1910 FOR IJ = 1 TO 8
1915 SQ(IJ) = SQ(IJ) + QS(IJ) * AI * AJ * Q
1920 FOR JI = 1 TO 8
1930 SK(IJ, JI) = SK(IJ, JI) + KS(IJ, JI)
1940 NEXT JI
1950 NEXT IJ
1960 NEXT MM
1970 NEXT LL
1971 IF XH = 0 AND XQ = 0 THEN GOTO 1980
1972 REM IF EITHER 'HL' OR 'QG' HAS A VALUE ON THIS ELEMENT, TYPE Y; ELSE
TYPE N
1973 INPUT #1, B$
1974 IF B$ = "Y" OR B$ = "y" OR B$ = "N" OR B$ = "n" THEN GOTO 1976
1975 GOSUB 9995: STOP
1976 IF B$ = "N" OR B$ = "n" THEN GOTO 1980
1977 GOSUB 4380
1980 FOR JJ = 1 TO 8
1990 IF JJ = 1 THEN NR = I
2000 IF JJ = 2 THEN NR = J
2010 IF JJ = 3 THEN NR = K
2020 IF JJ = 4 THEN NR = L
2030 IF JJ = 5 THEN NR = M
2040 IF JJ = 6 THEN NR = N
2050 IF JJ = 7 THEN NR = O
2060 IF JJ = 8 THEN NR = P
2065 IDIAG = DIAG(NR)
2100 FOR KK = 1 TO 8
2110 IF KK = 1 THEN N9 = I
2120 IF KK = 2 THEN N9 = J
2130 IF KK = 3 THEN N9 = K
2140 IF KK = 4 THEN N9 = L
2150 IF KK = 5 THEN N9 = M
2160 IF KK = 6 THEN N9 = N
2170 IF KK = 7 THEN N9 = O
2180 IF KK = 8 THEN N9 = P
2220 IF N9 > NR THEN GOTO 2260
2230 LOCATION = IDIAG - (NR - N9)
2240 A(LOCATION) = A(LOCATION) + SK(JJ, KK)
2260 NEXT KK
```

2270 Q(NR) = Q(NR) + SQ(JJ)
2280 NEXT JJ
2290 NEXT EL
2295 IF NF = 0 THEN GOTO 2360
2300 FOR II = 1 TO NF
2310 N9 = NS(II)
2320 IF N9 = 0 THEN GOTO 2350
2325 IDIAG = DIAG(N9)
2330 Q(N9) = 100000! * W(II) * A(IDIAG)
2340 A(IDIAG) = A(IDIAG) * 100000!
2350 NEXT II
2360 PRINT : PRINT "THE SIMULTANEOUS EQUATIONS ARE NOW BEING SOLVED":
PRINT
2370 N = NN
2372 A(1) = SQR(A(1))
2375 FOR II = 2 TO N
2380 DI = DIAG(II) - II
2385 LL = DIAG(II - 1) - DI + 1
2390 FOR JJ = LL TO II
2395 X = A(DI + JJ)
2400 DJ = DIAG(JJ) - JJ
2405 IF JJ = 1 THEN GOTO 2435
2410 LB = DIAG(JJ - 1) - DJ + 1
2415 IF LL > LB THEN LB = LL
2420 FOR KK = LB TO JJ - 1
2425 X = X - A(DI + KK) * A(DJ + KK)
2430 NEXT KK
2435 A(DI + JJ) = X / A(DJ + JJ)
2440 NEXT JJ
2445 A(DI + II) = SQR(X)
2450 NEXT II
2455 Q(1) = Q(1) / A(1)
2460 FOR II = 2 TO N
2465 DI = DIAG(II) - II
2470 LL = DIAG(II - 1) - DI + 1
2475 X = Q(II)
2480 FOR JJ = LL TO II - 1
2485 X = X - A(DI + JJ) * Q(JJ)
2490 NEXT JJ
2495 Q(II) = X / A(DI + II)
2500 NEXT II
2505 FOR II = N TO 2 STEP -1
2510 DI = DIAG(II) - II
2515 X = Q(II) / A(DI + II)
2520 Q(II) = X
2525 LL = DIAG(II - 1) - DI + 1
2530 FOR KK = LL TO II - 1
2535 Q(KK) = Q(KK) - X * A(DI + KK)
2540 NEXT KK
2545 NEXT II
2550 Q(1) = Q(1) / A(1)
2590 PRINT #2,
2600 PRINT #2, " THE VALUES OF THE FUNCTIONS ARE:"

```
2610 FOR I = 1 TO NJ
2620 PRINT #2, "NODE="; I; "  PHI("; I; ")="; Q(I)
2630 VG(I) = Q(I)
2640 NEXT I
2660 GOSUB 6000
2710 END
3330 FOR II = 1 TO 2
3340 FOR JJ = 1 TO 2
3345 JA(II, JJ) = 0
3350 NEXT JJ
3360 NEXT II
3370 DN(1, 1) = (1 - ET) * (2 * XI + ET) / 4
3390 DN(1, 3) = (1 - ET) * (2 * XI - ET) / 4
3400 DN(1, 5) = (1 + ET) * (2 * XI + ET) / 4
3410 DN(1, 7) = (1 + ET) * (2 * XI - ET) / 4
3420 DN(1, 2) = -XI * (1 - ET)
3430 DN(1, 4) = (1 - ET ^ 2) / 2
3440 DN(1, 6) = -XI * (1 + ET)
3450 DN(1, 8) = -(1 - ET ^ 2) / 2
3460 DN(2, 1) = (1 - XI) * (XI + 2 * ET) / 4
3470 DN(2, 3) = (1 + XI) * (-XI + 2 * ET) / 4
3480 DN(2, 5) = (1 + XI) * (XI + 2 * ET) / 4
3490 DN(2, 7) = (1 - XI) * (-XI + 2 * ET) / 4
3500 DN(2, 2) = -(1 - XI ^ 2) / 2
3510 DN(2, 4) = -ET * (1 + XI)
3520 DN(2, 6) = (1 - XI ^ 2) / 2
3530 DN(2, 8) = -ET * (1 - XI)
3540 FOR II = 1 TO 2
3550 FOR JJ = 1 TO 2
3560 FOR KK = 1 TO 8
3570 JA(II, JJ) = JA(II, JJ) + DN(II, KK) * CO(KK, JJ)
3580 NEXT KK
3590 NEXT JJ
3600 NEXT II
3610 CN = JA(1, 1)
3620 JA(1, 1) = JA(2, 2)
3630 JA(2, 2) = CN
3640 JA(2, 1) = -JA(2, 1)
3650 JA(1, 2) = -JA(1, 2)
3660 DT = JA(1, 1) * JA(2, 2) - JA(1, 2) * JA(2, 1)
3670 FOR II = 1 TO 2
3680 FOR JJ = 1 TO 2
3690 JA(II, JJ) = JA(II, JJ) / DT
3700 NEXT JJ
3705 NEXT II
3707 QS(1) = -(1 - XI) * (1 - ET) * (XI + ET + 1) * DT / 4
3708 QS(2) = (1 - XI ^ 2) * (1 - ET) * DT / 2
3709 QS(3) = (1 + XI) * (1 - ET) * (XI - ET - 1) * DT / 4
3710 QS(4) = (1 - ET ^ 2) * (1 + XI) * DT / 2
3711 QS(5) = (1 + XI) * (1 + ET) * (XI + ET - 1) * DT / 4
3712 QS(6) = (1 - XI ^ 2) * (1 + ET) * DT / 2
3713 QS(7) = -(1 - XI) * (1 + ET) * (XI - ET + 1) * DT / 4
3714 QS(8) = (1 - ET ^ 2) * (1 - XI) * DT / 2
```

597

```
3725 FOR II = 1 TO 2
3730 FOR JJ = 1 TO 8
3740 B(II, JJ) = 0
3750 NEXT JJ
3760 NEXT II
3770 FOR II = 1 TO 2
3780 FOR JJ = 1 TO 8
3790 FOR KK = 1 TO 2
3800 B(II, JJ) = B(II, JJ) + JA(II, KK) * DN(KK, JJ)
3810 NEXT KK
3820 NEXT JJ
3830 NEXT II
3840 FOR II = 1 TO 2
3850 FOR JJ = 1 TO 8
3860 BT(JJ, II) = B(II, JJ)
3870 NEXT JJ
3880 NEXT II
3900 FOR II = 1 TO 2
3910 FOR JJ = 1 TO 8
3920 DB(II, JJ) = 0
3930 NEXT JJ
3940 NEXT II
3950 FOR II = 1 TO 2
3960 FOR JJ = 1 TO 8
3970 FOR KK = 1 TO 2
3980 DB(II, JJ) = DB(II, JJ) + D(II, KK) * B(KK, JJ)
3990 NEXT KK
4000 NEXT JJ
4010 NEXT II
4020 RETURN
4380 REM EFFECTS OF HL
4390 IF XH = 0 THEN GOTO 4950
4400 REM LH=NO. OF BOUNDARIES ON WHICH HL HAS A VALUE
4410 INPUT #1, LH: PRINT #2, "NO. OF BOUNDARIES ON WHICH HL HAS A VALUE=";
LH
4420 IF LH = 0 OR LH = 1 OR LH = 2 OR LH = 3 OR LH = 4 THEN GOTO 4440
4430 GOSUB 9995: STOP
4440 IF LH = 0 THEN GOTO 4950
4490 FOR II = 1 TO LH
4500 ZJ = 0
4510 REM JN=THE MID-SIDE NODE ON WHICH HL HAS A VALUE
4520 INPUT #1, JN: PRINT #2, "THE MID-SIDE NODE DEFINING THE BOUNDARY ON
WHICH HL HAS A VALUE IS "; JN
4530 IF JN = J THEN ZJ = 2
4540 IF JN = L THEN ZJ = 4
4550 IF JN = N THEN ZJ = 6
4560 IF JN = P THEN ZJ = 8
4570 IF ZJ = 0 THEN GOSUB 9995: STOP
4590 IF ZJ = 2 THEN IN = I: KN = K
4600 IF ZJ = 4 THEN IN = K: KN = M
4610 IF ZJ = 6 THEN IN = M: KN = O
4620 IF ZJ = 8 THEN IN = O: KN = I
4630 ZI = ZJ - 1: ZK = ZJ + 1
```

```
4640 IF ZK = 9 THEN ZK = 1
4650 REM TYPE IN THE VALUES OF HL & T(INFINITY)
4660 INPUT #1, HL, IT: PRINT #2, "HL="; HL, "T(INFINITY)="; IT
4670 X1 = X(IN): Y1 = Y(IN)
4680 X2 = X(JN): Y2 = Y(JN)
4690 X3 = X(KN): Y3 = Y(KN)
4700 X2 = X2 - X1: Y2 = Y2 - Y1
4710 X3 = X3 - X1: Y3 = Y3 - Y1
4720 X1 = 0: Y1 = 0
4730 IF X2 = 0 AND X3 = 0 THEN LL = Y3
4740 IF X2 = 0 AND X3 = 0 THEN GOTO 4810
4750 IF X3 - X2 = 0 THEN X2 = .99 * X3
4760 IF X2 = 0 THEN X2 = .01 * X3
4770 IF X3 = 0 THEN X3 = .01 * X2
4780 CC = (Y3 / X3 - Y2 / X2) / (X3 - X2)
4790 BB = Y2 / X2 - CC * X2
4800 LL = (BB ^ 2 / 2 + 1) * X3 + BB * CC * X3 ^ 2 + 2 * CC ^ 2 * X3 ^ 3 / 3
4810 LL = ABS(LL)
4820 SK(ZI, ZI) = SK(ZI, ZI) + 4 * HL * LL / 30
4830 SK(ZJ, ZJ) = SK(ZJ, ZJ) + 16 * HL * LL / 30
4840 SK(ZK, ZK) = SK(ZK, ZK) + 4 * HL * LL / 30
4850 SK(ZI, ZJ) = SK(ZI, ZJ) + 2 * HL * LL / 30
4860 SK(ZJ, ZI) = SK(ZJ, ZI) + 2 * HL * LL / 30
4870 SK(ZI, ZK) = SK(ZI, ZK) - HL * LL / 30
4880 SK(ZK, ZI) = SK(ZK, ZI) - HL * LL / 30
4890 SK(ZJ, ZK) = SK(ZJ, ZK) + 2 * HL * LL / 30
4900 SK(ZK, ZJ) = SK(ZK, ZJ) + 2 * HL * LL / 30
4910 SQ(ZI) = SQ(ZI) + HL * LL * IT / 6
4920 SQ(ZJ) = SQ(ZJ) + 4 * HL * LL * IT / 6
4930 SQ(ZK) = SQ(ZK) + HL * LL * IT / 6
4940 NEXT II
4950 REM EFFECTS OF 'QG'
4960 IF XQ = 0 THEN GOTO 5430
4970 REM GQ=NO. OF BOUNDARIES ON WHICH 'QG' HAS A VALUE
4980 INPUT #1, GQ: PRINT #2, "THE NO. OF BOUNDARIES ON WHICH 'QG' HAS A
VALUE="; GQ
4990 IF GQ = 0 OR GQ = 1 OR GQ = 2 OR GQ = 3 OR GQ = 4 THEN GOTO 5010
5000 GOSUB 9995: STOP
5010 IF GQ = 0 THEN GOTO 5430
5060 FOR II = 1 TO GQ
5070 ZJ = 0
5080 REM JN=THE MID-SIDE NODE ON THE BOUNDARY WHERE THERE IS A VALUE OF
'QG'
5090 INPUT #1, JN
5100 IF JN = J THEN ZJ = 2
5110 IF JN = L THEN ZJ = 4
5120 IF JN = N THEN ZJ = 6
5130 IF JN = P THEN ZJ = 8
5140 IF ZJ = 0 THEN GOSUB 9995: STOP
5160 IF ZJ = 2 THEN IN = I: KN = K
5170 IF ZJ = 4 THEN IN = K: KN = M
5180 IF ZJ = 6 THEN IN = M: KN = O
5190 IF ZJ = 8 THEN IN = O: KN = I
```

```
5200 ZI = ZJ - 1: ZK = ZJ + 1
5210 IF ZK = 9 THEN ZK = 1
5220 REM TYPE IN THE VALUE OF QG
5230 INPUT #1, QG: PRINT #2, "QG="; QG
5240 X1 = X(IN): Y1 = Y(IN)
5250 X2 = X(JN): Y2 = Y(JN)
5260 X3 = X(KN): Y3 = Y(KN)
5270 X2 = X2 - X1: Y2 = Y2 - Y1
5280 X3 = X3 - X1: Y3 = Y3 - Y12
5290 X1 = 0: Y1 = 0
5300 IF X2 = 0 AND X3 = 0 THEN LL = Y3
5310 IF X2 = 0 AND X3 = 0 THEN GOTO 5380
5320 IF X3 - X2 = 0 THEN X2 = .99 * X3
5330 IF X2 = 0 THEN X2 = .01 * X3
5340 IF X3 = 0 THEN X3 = .01 * X2
5350 CC = (Y3 / X3 - Y2 / X2) / (X3 - X2)
5360 BB = Y2 / X2 - CC * X2
5370 LL = (BB ^ 2 / 2 + 1) * X3 + BB * CC * X3 ^ 2 + 2 * CC ^ 2 * X3 ^ 3 / 3
5380 LL = ABS(LL)
5390 SQ(ZI) = SQ(ZI) - QG * LL / 6
5400 SQ(ZJ) = SQ(ZJ) - 4 * QG * LL / 6
5410 SQ(ZK) = SQ(ZK) - QG * LL / 6
5420 NEXT II
5430 RETURN
6000 MAX = MAXX = -1E+12: MAXU = -1E+12: MINU = 1E+12: DIM YY(4), XX(4)
6010 SCREEN 9
6020 FOR I = 1 TO NJ STEP 2
6030 IF VG(I) > MAXU THEN MAXU = VG(I)
6040 IF VG(I) < MINU THEN MINU = VG(I)
6045 IF ABS(X(I)) > MAXX THEN MAXX = ABS(X(I))
6050 IF ABS(Y(I)) > MAXY THEN MAXY = ABS(Y(I))
6055 IF MAXX > MAX THEN MAX = MAXX
6060 IF MAXY > MAX THEN MAX = MAXY
6065 NEXT I
6070 LF = 10: XS = 250 / MAX
6080 DIF = MAXU - MINU: STP = DIF / LF: DIM FF(LF + 1): AG = 1
6090 FOR I = MINU TO MAXU STEP STP
6100 FF(AG) = I: IF ABS(FF(AG)) < .000001 THEN FF(AG) = 0
6110 AG = AG + 1: NEXT I
6120 FOR N = 1 TO LF
6130 FOR EL = 1 TO ES
6135 AA = 0: BB = 0: CC = 0: DD = 0
6140 I = I(EL): J = K(EL): K = M(EL): L = O(EL)
6145 IF FF(N) = (INT(VG(I) * 1000)) / 1000 THEN GOTO 6155
6147 IF FF(N) = (INT(VG(J) * 1000)) / 1000 THEN GOTO 6155
6149 IF FF(N) = (INT(VG(K) * 1000)) / 1000 THEN GOTO 6155
6151 IF FF(N) = (INT(VG(L) * 1000)) / 1000 THEN GOTO 6155
6153 GOTO 6160
6155 FF(N) = FF(N) + .000001
6160 IF FF(N) > VG(I) AND FF(N) < VG(J) THEN GOSUB 7000
6165 IF FF(N) < VG(I) AND FF(N) > VG(J) THEN GOSUB 7000
6170 IF FF(N) > VG(I) AND FF(N) < VG(L) THEN GOSUB 7100
6175 IF FF(N) < VG(I) AND FF(N) > VG(L) THEN GOSUB 7100
```

```
6180 IF FF(N) > VG(J) AND FF(N) < VG(K) THEN GOSUB 7200
6185 IF FF(N) < VG(J) AND FF(N) > VG(K) THEN GOSUB 7200
6190 IF FF(N) > VG(K) AND FF(N) < VG(L) THEN GOSUB 7300
6195 IF FF(N) < VG(K) AND FF(N) > VG(L) THEN GOSUB 7300
6220 IF AA = 1 AND BB = 1 THEN LINE (40 + XS * (XX(1)), 270 - XS * (YY(1)))-(40 + XS
* (XX(2)), 270 - XS * (YY(2)))
6230 IF AA = 1 AND CC = 1 THEN LINE (40 + XS * (XX(1)), 270 - XS * (YY(1)))-(40 + XS
* (XX(3)), 270 - XS * (YY(3)))
6240 IF AA = 1 AND DD = 1 THEN LINE (40 + XS * (XX(1)), 270 - XS * (YY(1)))-(40 + XS
* (XX(4)), 270 - XS * (YY(4)))
6250 IF BB = 1 AND CC = 1 THEN LINE (40 + XS * (XX(2)), 270 - XS * (YY(2)))-(40 + XS
* (XX(3)), 270 - XS * (YY(3)))
6260 IF BB = 1 AND DD = 1 THEN LINE (40 + XS * (XX(2)), 270 - XS * (YY(2)))-(40 + XS
* (XX(4)), 270 - XS * (YY(4)))
6270 IF CC = 1 AND DD = 1 THEN LINE (40 + XS * (XX(3)), 270 - XS * (YY(3)))-(40 + XS
* (XX(4)), 270 - XS * (YY(4)))
6290 NEXT EL
6300 NEXT N
6310 A$ = INKEY$: IF A$ = "" THEN GOTO 6310
6320 IF A$ = "Y" OR A$ = "y" THEN GOTO 6400
6330 GOSUB 9995: GOTO 6310
6400 RETURN
7000 DF = FF(N) - VG(I): DE = VG(J) - VG(I)
7010 DX = X(J) - X(I): DY = Y(J) - Y(I)
7020 IF DF = 0 AND DE = 0 THEN GOTO 7050
7030 XX(1) = (DF * DX / DE) + X(I): YY(1) = (DF * DY / DE) + Y(I)
7040 AA = 1: RETURN
7050 XX(1) = X(I): YY(1) = Y(I): AA = 1: RETURN
7100 DF = FF(N) - VG(I): DE = VG(L) - VG(I)
7110 DX = X(L) - X(I): DY = Y(L) - Y(I)
7120 IF DF = 0 AND DE = 0 THEN GOTO 7150
7130 XX(2) = (DF * DX / DE) + X(I): YY(2) = (DF * DY / DE) + Y(I)
7140 BB = 1: RETURN
7150 XX(2) = X(I): YY(2) = Y(I): BB = 1: RETURN
7200 DF = FF(N) - VG(J): DE = VG(K) - VG(J)
7210 DX = X(K) - X(J): DY = Y(K) - Y(J)
7220 IF DF = 0 AND DE = 0 THEN GOTO 7250
7230 XX(3) = (DF * DX / DE) + X(J): YY(3) = (DF * DY / DE) + Y(J)
7240 CC = 1: RETURN
7250 XX(3) = X(J): YY(3) = Y(J): CC = 1: RETURN
7300 DF = FF(N) - VG(K): DE = VG(L) - VG(K)
7310 DX = X(L) - X(K): DY = Y(L) - Y(K)
7320 IF DF = 0 AND DE = 0 THEN GOTO 7350
7330 XX(4) = (DF * DX / DE) + X(K): YY(4) = (DF * DY / DE) + Y(K)
7340 DD = 1: RETURN
7350 XX(4) = X(K): YY(3) = Y(K): DD = 1: RETURN
9995 REM WARNING
9996 PRINT : PRINT "INCORRECT DATA": PRINT
9997 RETURN
10000 REM GRAPH
10010 XZ = 1E+12: YZ = 1E+12
10030 XM = -1E+12: YM = -1E+12
10050 PRINT : PRINT "TO CONTINUE, TYPE Y"
```

```
10060 A$ = INKEY$: IF A$ = "" THEN GOTO 10060
10070 IF A$ = "Y" OR A$ = "y" THEN GOTO 10090
10080 GOTO 10060
10090 FOR ME = 1 TO ES
10100 I = I(ME)
10105 J = J(ME)
10110 K = K(ME)
10115 L = L(ME)
10120 M = M(ME)
10125 N = N(ME)
10130 O = O(ME)
10135 P = P(ME)
10140 IF X(I) < XZ THEN XZ = X(I)
10150 IF X(J) < XZ THEN XZ = X(J)
10160 IF X(K) < XZ THEN XZ = X(K)
10165 IF X(L) < XZ THEN XZ = X(L)
10166 IF X(M) < XZ THEN XZ = X(M)
10167 IF X(N) < XZ THEN XZ = X(N)
10168 IF X(O) < XZ THEN XZ = X(O)
10169 IF X(P) < XZ THEN XZ = X(P)
10170 IF Y(I) < YZ THEN YZ = Y(I)
10180 IF Y(J) < YZ THEN YZ = Y(J)
10190 IF Y(K) < YZ THEN YZ = Y(K)
10195 IF Y(L) < YZ THEN YZ = Y(L)
10196 IF Y(M) < YZ THEN YZ = Y(M)
10197 IF Y(N) < YZ THEN YZ = Y(N)
10198 IF Y(O) < YZ THEN YZ = Y(O)
10199 IF Y(P) < YZ THEN YZ = Y(P)
10200 IF X(I) > XM THEN XM = X(I)
10210 IF X(J) > XM THEN XM = X(J)
10220 IF X(K) > XM THEN XM = X(K)
10225 IF X(L) > XM THEN XM = X(L)
10226 IF X(M) > XM THEN XM = X(M)
10227 IF X(N) > XM THEN XM = X(N)
10228 IF X(O) > XM THEN XM = X(O)
10229 IF X(P) > XM THEN XM = X(P)
10230 IF Y(I) > YM THEN YM = Y(I)
10240 IF Y(J) > YM THEN YM = Y(J)
10250 IF Y(K) > YM THEN YM = Y(K)
10255 IF Y(L) > YM THEN YM = Y(L)
10256 IF Y(M) > YM THEN YM = Y(M)
10257 IF Y(N) > YM THEN YM = Y(N)
10258 IF Y(O) > YM THEN YM = Y(O)
10259 IF Y(P) > YM THEN YM = Y(P)
10260 NEXT ME
10270 XS = 350 / (XM - XZ)
10280 YS = 175 / (YM - YZ)
10290 IF XS > YS THEN XS = YS
10295 SCREEN 9
10300 FOR ME = 1 TO ES
10310 II = I(ME)
10315 JJ = J(ME)
10320 KK = K(ME)
```

```
10325 LL = L(ME)
10330 MM = M(ME)
10335 NN = N(ME)
10340 OO = O(ME)
10345 PP = P(ME)
10350 FOR IV = 1 TO 8
10360 IF IV = 1 THEN I = II: J = JJ
10370 IF IV = 2 THEN I = JJ: J = KK
10380 IF IV = 3 THEN I = KK: J = LL
10385 IF IV = 4 THEN I = LL: J = MM
10386 IF IV = 5 THEN I = MM: J = NN
10387 IF IV = 6 THEN I = NN: J = OO
10388 IF IV = 7 THEN I = OO: J = PP
10389 IF IV = 8 THEN I = PP: J = II
10400 XI = (X(I) - XZ) * XS
10410 XJ = (X(J) - XZ) * XS
10420 YI = (Y(I) - YZ) * XS
10430 YJ = (Y(J) - YZ) * XS
10480 LINE (XI + 40, 270 - YI)-(XJ + 40, 270 - YJ)
10490 NEXT IV
10500 NEXT ME
10515 PRINT "TO CONTINUE, TYPE Y"
10520 A$ = INKEY$: IF A$ = "" THEN GOTO 10520
10530 IF A$ = "Y" OR A$ = "y" THEN GOTO 10560
10540 GOTO 10520
10560 SCREEN 0, 0
10600 RETURN
```

Appendix 24

Computer Program for the Solution of Helmholtz's Equation "ACOUSTIC"

```
100 GOSUB 3900
110 IF A$ < > "Y" THEN PRINT : PRINT "GO BACK & READ IT !": GOTO 25000
120 REM COPYRIGHT OF DR.C.T.F.ROSS
130 REM 6 HURSTVILLE DRIVE,  WATERLOOVILLE
140 REM PORTSMOUTH. ENGLAND
142 PRINT : PRINT "TWO-DIMENSIONAL ACOUSTIC VIBRATIONS": PRINT
143 INPUT "TYPE IN THE NAME OF YOUR INPUT FILE "; FILENAME$
PRINT "IF YOU ARE SATISFIED WITH THIS FILENAME, TYPE Y; ELSE N"
144 A$ = INKEY$: IF A$ = "" THEN GOTO 144
146 IF A$ = "Y" OR A$ = "y" THEN GOTO 148
IF A$ = "N" OR A$ = "n" THEN GOTO 143
GOTO 144
148 OPEN FILENAME$ FOR INPUT AS #1
150 INPUT "TYPE IN THE NAME OF YOUR OUTPUT FILE "; OUT$
PRINT "IF YOU ARE SATISFIED WITH THIS FILENAME, TYPE Y; ELSE N"
152 B$ = INKEY$: IF B$ = "" THEN GOTO 152
153 IF B$ = "Y" OR B$ = "y" THEN GOTO 160
IF B$ = "N" OR B$ = "n" THEN GOTO 150
GOTO 152
160 OPEN OUT$ FOR OUTPUT AS #2
168 REM PRINT:PRINT"TYPE IN NUMBER OF NODAL POINTS"
170 INPUT #1, NN
180 IF NN < 3 OR NN > 199 THEN GOSUB 4130: STOP
200 NJ = NN
N = NN
210 REM PRINT"TYPE IN NUMBERS OF ELEMENTS"
220 INPUT #1, LS
230 IF LS < 1 OR LS > 299 THEN GOSUB 4130: STOP
260 REM PRINT"TYPE IN THE NUMBER OF NODES WITH ZERO DISPLACEMENTS"
270 INPUT #1, NF             `
280 IF NF < 0 OR NF > NN THEN GOSUB 4130: STOP
310 N2 = NN
320 NP = N2 + 1
350 DIM A(N2, N2), XX(N2, NP), X(200)
360 DIM SK(3, 3), SM(3, 3), NS(NF), Y(NN)
365 DIM UG(N2), XG(NN), YG(NN), I9(LS), J9(LS), K9(LS)
367 DIM AM(50), VC(200, 50), Z(200), IG(200), VE(200), FR(200)
370 GOTO 1900
380 PRINT "THE EIGENVALUES ARE BEING DETERMINED": PRINT
390 GOSUB 510
410 FOR I = 1 TO M1
420 PRINT #2, "EIGENVALUE="; AM(I)
PRINT "FREQUENCY="; SQR(AM(I)) / (2 * 3.141593); "Hz"
430 PRINT #2, "FREQUENCY="; SQR(AM(I)) / (2 * 3.141593); "Hz"
440 PRINT #2, "EIGENVECTOR IS"
450 PRINT #2,
460 FOR J = 1 TO N
470 PRINT #2, VC(J, I); "   ";
480 PRINT #2, : NEXT J
```

```
490 PRINT #2, : NEXT I
495 GOSUB 11000
500 GOTO 25000
510 MN = N
520 NN = N
530 GOSUB 1390
540 M = 1
550 FOR I = 1 TO NN
560 VC(I, M) = X(I)
570 XX(I, M) = VC(I, M)
580 NEXT I
590 AM(M) = XM
600 IF M1 < 2 THEN RETURN
610 FOR M = 2 TO M1
620 FOR I = 1 TO NN
630 K4 = ABS(XX(I, M - 1) - 1)
640 IF K4 < .00001 THEN IR = I
650 NEXT I
660 IG(M - 1) = IR
670 FOR I = 1 TO NN
680 XX(MN - I + 1, MN - M + 3) = A(IR, I)
690 NEXT I
700 FOR I = 1 TO NN
710 FOR J = 1 TO NN
720 Z1 = MN - J + 1
730 Z2 = MN - M + 3
740 A(I, J) = A(I, J) - XX(I, M - 1) * XX(Z1, Z2)
750 NEXT J
760 NEXT I
770 FOR I = 1 TO NN
780 IF I = IR THEN GOTO 870
790 IF I > IR THEN K1 = I - 1
800 IF I < IR THEN K1 = I
810 FOR J = 1 TO NN
820 IF J = IR THEN GOTO 860
830 IF J > IR THEN K2 = J - 1
840 IF J < IR THEN K2 = J
850 A(K1, K2) = A(I, J)
860 NEXT J
870 NEXT I
880 NN = NN - 1
890 M3 = NN
900 IF M < > MN THEN GOTO 940
910 XM = A(1, 1)
920 X(1) = 1
930 GOTO 950
940 GOSUB 1390
950 FOR I = 1 TO NN
960 XX(I, M) = X(I)
970 NEXT I
980 AM(M) = XM
990 M4 = M - 1
1000 M5 = 1000 - M4
```

```
1010 FOR M8 = M5 TO 999
1020 M6 = M3 + 1
1030 M2 = 1000 - M8
1040 M7 = IG(M2) + 1
1050 IF M6 < M7 THEN GOTO 1120
1060 N9 = 1000 - M7
1070 N8 = 1000 - M6
1080 FOR I3 = N8 TO N9
1090 I = 1000 - I3
1100 X(I) = X(I - 1)
1110 NEXT I3
1120 J = IG(M2)
1130 X(J) = 0
1140 SUM = 0
1150 FOR I = 1 TO M6
1160 Z3 = MN - I + 1
1170 Z4 = MN - M2 + 2
1180 SUM = SUM + XX(Z3, Z4) * X(I)
1190 NEXT I
1200 XK = (AM(M2) - XM) / SUM
1210 FOR I = 1 TO M6
1220 X(I) = XX(I, M2) - XK * X(I)
1230 NEXT I
1240 SUM = 0
1250 FOR I = 1 TO M6
1260 IF ABS(SUM) < ABS(X(I)) THEN SUM = X(I)
1270 NEXT I
1280 FOR I = 1 TO M6
1290 X(I) = X(I) / SUM
1300 NEXT I
1310 M3 = M3 + 1
1320 IF M2 < > 1 THEN GOTO 1360
1330 FOR I = 1 TO M3
1340 VC(I, M) = X(I)
1350 NEXT I
1360 NEXT M8
1370 NEXT M
1380 RETURN
1390 Y1 = 100000!
1400 FOR I = 1 TO NN
1410 X(I) = 1
1420 NEXT I
1430 XM = -100000!
1440 FOR I = 1 TO NN
1450 SG = 0
1460 FOR J = 1 TO NN
1470 SG = SG + A(I, J) * X(J)
1480 NEXT J
1490 Z(I) = SG
1500 NEXT I
1510 XM = 0
1520 FOR I = 1 TO NN
1530 IF ABS(XM) < ABS(Z(I)) THEN XM = Z(I)
```

```
1540 NEXT I
1550 FOR I = 1 TO NN
1560 X(I) = Z(I) / XM
1570 NEXT I
1580 IF ABS((Y1 - XM) / XM) > D THEN GOTO 1600
1590 GOTO 1620
1600 Y1 = XM
1610 GOTO 1440
1620 X3 = 0
1630 FOR I = 1 TO NN
1640 IF ABS(X3) < ABS(X(I)) THEN X3 = X(I)
1650 NEXT I
1660 FOR I = 1 TO NN
1670 X(I) = X(I) / X3
1680 NEXT I
1690 RETURN
1700 N9 = N - 1
1710 FOR NX = 1 TO N
1720 DI = A(NX, 1)
1730 IF DI = 0 THEN PRINT "MATRIX IS SINGULAR"
1740 FOR NY = 1 TO N9
1750 Y9 = NY + 1
1760 A(NX, NY) = A(NX, Y9) / DI
1770 NEXT NY
1780 A(NX, N) = 1 / DI
1790 FOR NZ = 1 TO N
1800 IF NZ = NX THEN GOTO 1870
1810 O = A(NZ, 1)
1820 FOR NY = 1 TO N9
1830 Y9 = NY + 1
1840 A(NZ, NY) = A(NZ, Y9) - A(NX, NY) * O
1850 NEXT NY
1860 A(NZ, N) = -A(NX, N) * O
1870 NEXT NZ
1880 NEXT NX
1890 RETURN
1900 REM START OF MAIN PART OF PROGRAM
1910 REM PRINT:PRINT"TYPE IN THE NODE NUMBERS OF THE ZERO DISPLACEMENTS
IN ASCENDING ORDER"
IF NF = 0 THEN GOTO 2140
1930 FOR I = 1 TO NF: NS(I) = 0: NEXT I
1950 FOR I = 1 TO NF
1960 INPUT #1, SU
NS(I) = SU
1970 IF SU < 1 OR SU > NN THEN GOSUB 4130: STOP
2110 NEXT I
2140 REM PRINT:PRINT"TYPE IN THE NODAL COORDINATES"
2150 FOR I = 1 TO NN
2160 REM PRINT"TYPE IN X COORD. OF NODE ";I
2170 INPUT #1, X(I), Y(I)
2175 XG(I) = X(I)
2180 REM PRINT"TYPE IN Y COORD. OF NODE ";I
2195 YG(I) = Y(I)
```

```
2200 NEXT I
2240 REM PRINT:PRINT"TYPE IN THE NUMBER OF FREQUENCIES - THIS MUST BE
< =";N
2245 INPUT #1, M1
2250 IF M1 < 1 OR M1 > N THEN GOSUB 4130: STOP
2265 INPUT #1, CSPEED: REM  SPEED OF SOUND IN FLUID
2310 GOSUB 4160
2320 REM PRINT:PRINT"TYPE IN MEMBER DETAILS"
2330 FOR LE = 1 TO LS
2340 REM PRINT:PRINT"TYPE IN DETAILS OF MEMBER NUMBER ";LE
2350 REM PRINT"TYPE IN I NODE"
2360 INPUT #1, I, J, K
2380 IF I < 1 OR I > NN THEN GOSUB 4130: STOP
2400 REM PRINT"TYPE IN J NODE"
2420 IF J < 1 OR J > NN THEN GOSUB 4130: STOP
2425 REM PRINT"TYPE IN K NODE"
2435 IF K < 1 OR K > NN THEN GOSUB 4130: STOP
2450 IF I = J OR I = K OR J = K THEN GOSUB 4130: STOP
2460 I9(LE) = I: J9(LE) = J: K9(LE) = K
2575 NEXT LE
GOSUB 4500
2576 Z9 = 0: GOSUB 10000
2577 FOR LE = 1 TO LS
2578 I = I9(LE): J = J9(LE): K = K9(LE)
2580 PRINT : PRINT "ELEMENT NUMBER "; LE; " UNDER COMPUTATION": PRINT
2590 B(1) = Y(J) - Y(K)
2600 B(2) = Y(K) - Y(I)
2610 B(3) = Y(I) - Y(J)
2620 C(1) = X(K) - X(J)
2630 C(2) = X(I) - X(K)
2640 C(3) = X(J) - X(I)
2650 DL = X(I) * B(1) + X(J) * B(2) + X(K) * B(3)
2660 FOR II = 1 TO 3
2680 FOR JJ = 1 TO 3
2700 SM(II, JJ) = .5 * (B(II) * B(JJ) + C(II) * C(JJ)) / DL
2720 NEXT JJ, II
2780 FOR II = 1 TO 3
2790 FOR JJ = 1 TO 3
2800 SK(II, JJ) = 1
2810 NEXT JJ
2820 NEXT II
CN = DL / (24 * CSPEED ^ 2)
2830 SK(1, 1) = 2
2840 SK(2, 2) = 2
2850 SK(3, 3) = 2
2910 FOR II = 1 TO 3
2920 FOR JJ = 1 TO 3
2960 SK(II, JJ) = SK(II, JJ) * CN
2970 NEXT JJ
2980 NEXT II
2990 I1 = I - 1
3000 J1 = J - 1
3005 K1 = K - 1
```

```
3010 FOR II = 1 TO 1
3020 FOR JJ = 1 TO 1
3030 MM = I1 + II
3040 MN = I1 + JJ
3050 NM = J1 + II
3060 N9 = J1 + JJ
3070 A(MM, MN) = A(MM, MN) + SK(II, JJ)
3080 XX(MM, MN) = XX(MM, MN) + SM(II, JJ)
3090 A(NM, MN) = A(NM, MN) + SK(II + 1, JJ)
3100 XX(NM, MN) = XX(NM, MN) + SM(II + 1, JJ)
3110 A(MM, N9) = A(MM, N9) + SK(II, JJ + 1)
3120 XX(MM, N9) = XX(MM, N9) + SM(II, JJ + 1)
3130 A(NM, N9) = A(NM, N9) + SK(II + 1, JJ + 1)
3140 XX(NM, N9) = XX(NM, N9) + SM(II + 1, JJ + 1)
3147 NM = K1 + II: N9 = K1 + JJ
3149 A(MM, N9) = A(MM, N9) + SK(II, JJ + 2)
3150 XX(MM, N9) = XX(MM, N9) + SM(II, JJ + 2)
3151 A(NM, MN) = A(NM, MN) + SK(II + 2, JJ)
3152 XX(NM, MN) = XX(NM, MN) + SM(II + 2, JJ)
3160 MM = K1 + II: MN = K1 + JJ
3164 A(MM, MN) = A(MM, MN) + SK(II + 2, JJ + 2)
3166 XX(MM, MN) = XX(MM, MN) + SM(II + 2, JJ + 2)
3176 MM = J1 + II: MN = J1 + JJ
3178 NM = K1 + II: N9 = K1 + JJ
3179 A(MM, N9) = A(MM, N9) + SK(II + 1, JJ + 2)
3180 XX(MM, N9) = XX(MM, N9) + SM(II + 1, JJ + 2)
3182 A(NM, MN) = A(NM, MN) + SK(II + 2, JJ + 1)
3183 XX(NM, MN) = XX(NM, MN) + SM(II + 2, JJ + 1)
3185 NEXT JJ, II
3187 NEXT LE
IF NF = 0 THEN GOTO 3750
3190 FOR I = 1 TO NF
3200 N7 = NS(I)
3210 IF N7 = 0 THEN GOTO 3410
A(N7, N7) = A(N7, N7) * 1E+12
3410 NEXT I
3750 PRINT : PRINT "THE STIFFNESS MATRIX IS BEING INVERTED": PRINT
3760 GOSUB 1700
3770 FOR I = 1 TO N
3780 FOR J = 1 TO N
3790 VE(J) = 0
3800 FOR K = 1 TO N
3810 VE(J) = VE(J) + A(I, K) * XX(K, J)
3820 NEXT K
3830 NEXT J
3840 FOR J = 1 TO N
3850 A(I, J) = VE(J)
3860 NEXT J
3870 NEXT I
3880 D = .001
3890 GOTO 380
3900 CLS
3910 PRINT "COPYRIGHT OF DR.C.T.F.ROSS"
```

```
3920 PRINT
3930 PRINT "NOT TO BE REPRODUCED"
3940 PRINT
3950 PRINT "HAVE YOU READ THE APPROPRIATE MANUAL"
3980 PRINT
3990 PRINT "BY"
4000 PRINT
4010 PRINT "C.T.F.ROSS": PRINT
4020 PRINT "IF YES, TYPE Y; ELSE N"
4030 A$ = INKEY$: IF A$ = "" THEN GOTO 4030
4035 IF A$ = "y" THEN A$ = "Y"
4040 RETURN
4050 PRINT : PRINT
4060 PRINT "IF YOU ARE SATISFIED WITH THE INPUT OF YOUR LAST BATCH OF DATA,
TYPE Y; ELSE N": PRINT
4070 X$ = INKEY$: IF X$ = "" THEN GOTO 4070
4075 IF X$ = "y" THEN X$ = "Y"
4080 RETURN
4130 REM WARNING
4140 PRINT : PRINT "INCORRECT DATA": PRINT
4150 RETURN
4160 REM OUTPUT 1
4180 PRINT #2,
4190 PRINT #2, "TWO-DIMENSIONAL ACOUSTIC VIBRATIONS"
4200 PRINT #2, : PRINT #2, "NUMBER OF NODAL POINTS ="; NN
4210 PRINT #2, "NUMBER OF ELEMENTS"; LS
4220 PRINT #2, "NUMBER OF NODES WITH ZERO DISPLACEMENTS ="; NF
4230 PRINT #2, : PRINT #2, "THE NODAL POSITIONS OF THE ZERO DISPLACEMENTS
ARE:-"
4240 FOR I = 1 TO NF
SU = NS(I)
4270 PRINT #2, "NODE ="; SU
4320 NEXT I
4330 PRINT #2, : PRINT #2, "THE NODAL COORDINATES X0 & Y0 ARE:-"
4340 FOR I = 1 TO NN
4350 PRINT #2, "NODE ="; I
4360 PRINT #2, "X("; I; ")="; X(I); "    Y("; I; ")="; Y(I)
4370 NEXT I
4380 PRINT #2, : PRINT #2, "NUMBER OF FREQUENCIES ="; M1
4390 PRINT #2, "SPEED OF SOUND IN FLUID ="; CSPEED
4400 RETURN
4500 REM OUTPUT 2
FOR LE = 1 TO LS
I = I9(LE): J = J9(LE): K = K9(LE)
PRINT #2, "ELEMENT ="; LE; "I-J-K ="; I; "-"; J; "-"; K
NEXT LE
4580 RETURN
10000 IF Z9 = 99 THEN PRINT "EIGENMODE ="; IJ, "FREQUENCY ="; SQR(AM(IJ)) / (2 *
3.14159); "Hz"
10001 IF Z9 = 77 OR Z9 = 99 THEN GOTO 10300
10005 XZ = 1E+12
10010 YZ = 1E+12
10020 XM = -1E+12
```

```
10030 YM = -1E+12
10070 FOR EL = 1 TO LS
10080 I = I9(EL)
10090 J = J9(EL)
10100 K = K9(EL)
10130 IF X(I) < XZ THEN XZ = X(I)
10140 IF Y(I) < YZ THEN YZ = Y(I)
10150 IF X(J) < XZ THEN XZ = X(J)
10160 IF Y(J) < YZ THEN YZ = Y(J)
10165 IF X(K) < XZ THEN XZ = X(K)
10167 IF Y(K) < YZ THEN YZ = Y(K)
10170 IF X(I) > XM THEN XM = X(I)
10180 IF Y(I) > YM THEN YM = Y(I)
10190 IF X(J) > XM THEN XM = X(J)
10200 IF Y(J) > YM THEN YM = Y(J)
10205 IF X(K) > XM THEN XM = X(K)
10207 IF Y(K) > YM THEN YM = Y(K)
10210 NEXT EL
10250 XS = 350 / (XM - XZ)
10270 YS = 175 / (YM - YZ)
10280 IF XS > YS THEN XS = YS
10300 IF Z9 = 0 OR Z9 = 77 THEN SCREEN 9: PRINT "TO CONTINUE, PRESS Y"
10350 FOR EL = 1 TO LS
10360 I = I9(EL)
10370 J = J9(EL)
10375 K = K9(EL)
10377 FOR IV = 1 TO 3
10378 IF IV = 1 THEN II = I: JJ = J
10379 IF IV = 2 THEN II = J: JJ = K
10380 IF IV = 3 THEN II = K: JJ = I
10385 IF Z9 = 99 THEN GOTO 10440
10390 XI = (X(II) - XZ) * XS
10400 YI = (Y(II) - YZ) * XS
10410 XJ = (X(JJ) - XZ) * XS
10420 YJ = (Y(JJ) - YZ) * XS
10430 GOTO 10460
10440 XI = X(II) - XZ * XS: XJ = X(JJ) - XZ * XS
10450 YI = Y(II) - YZ * XS: YJ = Y(JJ) - YZ * XS
10460 LINE (XI + 40, 280 - YI)-(XJ + 40, 280 - YJ)
10465 NEXT IV
10470 NEXT EL
10650 IF Z9 = 77 THEN GOTO 10800
10660 A$ = INKEY$: IF A$ = "" THEN GOTO 10660
10670 IF A$ = "Y" OR A$ = "y" THEN GOTO 10720
10700 GOTO 10660
10720 SCREEN 0: WIDTH 80
10800 RETURN
11000 REM PLOT OF EIGENMODES
11010 FR(0) = 0
11020 FOR II = 1 TO N
11030 FR(II) = FR(II - 1)
11040 FR(II) = FR(II) + 1
11050 FOR JJ = 1 TO NF * 2
```

```
11070 IF FR(II) = NS(JJ) THEN GOTO 11040
11080 NEXT JJ
11090 NEXT II
11100 FOR IJ = 1 TO M1
11110 FOR II = 1 TO NJ
11120 X(II) = XG(II)
11130 Y(II) = YG(II)
11140 NEXT II
11150 Z9 = 77
11160 GOSUB 10000
11180 FOR II = 1 TO N
11190 UG(FR(II)) = VC(II, IJ)
11300 NEXT II
11310 SC = 15
11320 FOR II = 1 TO NJ
11330 X(II) = XG(II) * XS + SC * UG(II)
11340 Y(II) = YG(II) * XS + SC * UG(II)
11360 NEXT II
11380 Z9 = 99
11390 GOSUB 10000
11410 NEXT IJ
11430 RETURN
25000 REM END
```

INDEX

acoustic 1, 314, 315, 317, 318
ASCII 3
axisymmetric 2

beam 18, 19, 22, 24, 28, 32, 43, 45, 72, 89
bending moment 8, 21, 23, 24, 25, 27, 31, 41, 42, 46, 48, 114, 118, 126-128, 205, 262
bending stress 213
buckling 6, 7

cantilever 27, 155, 174, 192, 258, 263
conduction,
        heat 266, 289, 291, 302, 310
constant strain triangle 153, 217, 227
continuous reduction 92
convection 266, 294
cross-stiffened 111
cylindrical shells 219, 220, 223, 252

directional cosines 75, 76
doubly-curved shell 3, 217, 218

eigenmode 92, 110, 227, 228, 232-237, 243-249, 253, 254, 318-321
eigenvalue 95, 316
eigenvalue economiser 250
elastic foundation 139, 148
electrical field problems 268
electric potential 272
electromagnetic waveguides 1, 314
electrostatic 1, 269

field problems 264, 278, 293
fluid flow 284, 285, 305, 307
frame,
        pin-jointed 52
        rigid-jointed 31, 35, 39, 52, 72, 82, 86, 89, 94, 95, 98, 100

Gauss points 183, 238, 279
Gauss-Radau 200, 238
grid 111, 112, 127, 129, 134
grillage 111, 113-115, 118, 119, 124, 127-129, 131, 132
groundwater flow 268

harbour,
        oscillations 314

heat,
> flux 266
> transfer 1, 266, 274

Helmholtz's equation 314

hydrostatic load 5, 18, 20, 22, 26, 27, 34, 38, 39, 53

in-plane,
> stress 4
> plate 152, 227, 249

isoparametric 139, 152, 179, 192, 198, 278, 293, 310

joints,
> pin 43, 50, 58, 62, 69
> rigid 43, 50

lake,
> oscillations 314, 315

Laplace's equation 264, 278, 293

magnetic field 269

magnetostatics 272

medicine 1

membrane vibration 314

natural frequencies 89, 94, 98, 100, 131

out-of-plane 200, 217, 249

permeability 269

permittivity 268, 315

plane,
> strain 153, 160-162, 166
> stress 153, 155, 160-162, 166, 196

plate 4, 143, 144, 147, 148, 152, 153, 158, 160, 169, 174, 185, 190, 192, 196-200, 202, 203, 210, 223, 227, 237, 239, 240, 243

principal,
> axes 89, 94
> plane 76, 79

prop,
> elastic 28, 148

Rankine-Gordon 7

resonant frequencies 102, 134, 136, 227, 228, 237, 239, 242, 243, 249, 251, 252

right-hand screwrule 73, 112

screen-dumping 6

shearing force 41, 42, 52, 60, 61

shear wall 228

snow loads 5

shell 217-219, 249, 252
ship 124, 127, 128, 136-138
slab 139, 218
streamlines 313
stream functions 267, 284, 305, 307
stresses in solids 255
struts 7
supports,
        elastic 18, 26, 112
        rigid 18, 26, 52
tab key 3
tetrahedron 255, 260, 261
three-dimensional 62, 65, 66, 72
torsion,
        non-circular sections 265
truss 8, 9, 11, 13-15, 62, 65, 66, 69, 89

uniformly distributed loads 18, 34, 111

vibration 5, 89, 94, 100, 129, 227, 237, 249, 314-316

wind loads 5

Publisher's announcement

CONVERTING HEAT INTO WORK:
Engineering Thermodynamics of Gas and Steam Cycles
G. COLE, Professor of Engineering Design and Manufacture, University of Hull, Yorkshire

This timely and important book explains concepts behind the laws of thermodynamics for university undergraduates and graduates, and for industrialists who provide 98% of the world's electricity by converting traditional working fluids of gas (usually air) and steam. Its importance lies in the conservation of energy resources while expanding work available throughout the world. It is written for senior undergraduate and postgraduate degree courses in engineering science, mechanical engineering, engineering design, engineering manufacture, and applied physics. For professional engineers, physicists and applied mathematics, it will serve as a standby in the reference library.

Professor Cole's original approach differs from the traditional by demonstrating the unity between gas and steam systems for energy-work conversion as its centrally important theme. Modifications and improvements of the thermodynamic cycles are shown as immediate and direct consequences of the approach to the cycles, treated in detail from this single viewpoint. The book ends with a review of likely functions of air/steam systems within a planned global energy provision as we move towards the 21st century.

The text provides explanations of new and important modern concepts of engineering thermodynamics which are clear and easy to understand. This quality commends it not only as a didactic textbook for university study and research but also as one which will find its way into the industrial sector as a reference source for engineers dealing with thermodynamic problems, and those needing to keep up-to-date with new technological developments or wishing to explore new fields.

Features: New approach, set into general energy background. Attractive writing style for degree and post graduate study. Plentiful examples, problem-exercises & solutions/hints.

Chapter contents: 1.Sources for Energy; 2.Thermodynamics Background; 3.Gases as Working Fluids; 4.Steam as Working Fluid; 5.Gas-Steam Synthesis; 6.Applications Involving Combinations of Cycles; 7.Summary & Future Prospects.

Publisher's announcement

FINITE ELEMENT TECHNIQUES IN STRUCTURAL MECHANICS
CARL T. F. ROSS, Professor of Structural Dynamics,
Department of Mechanical and Manufacturing Engineering, University of Portsmouth

This advanced undergraduate and postgraduate text will serve for courses in mechanical, civil, structural and aeronautical engineering; and naval architecture. It will serve also for professional engineers in industrial or academic research. It is written in a step-by-step methodological approach so that readers can acquire knowledge, either through formal engineering courses or by self-study. For industrial engineers, the book will serve as a standby in the reference library.

Professor Ross writes in a style easy for comprehension and is well-known as the author of six successful engineering books. His 30 years of experience in teaching at polytechnics and universities is reflected and takes into consideration the difficulties experienced by most modern students with mathematics, which is deliberately non-rigorous at the level understood in introductory engineering courses.

He is also well-known for his work on the application of computers in finite element analysis and his important discoveries on buckling of cylinders and cones under external pressure, for which he was awarded a D.Sc. by the CNAA (Council of National Academic Awards, UK). The author's impressive industrial experience strengths the teaching quality of the text and conveys the reality of practical engineering problems, and points to their solution.

Contents:
1. Introductiont tp Matrix Algebra: covers course requirements, including solution of simultaneous equations and the inverse of matrices;
2. The Matrix Displacement Method: Statics of two- and three-dimensional pin- and figid-jointed frameworks. Continuous beams;
3. Finite Element Method: Introduction, via the method of minimum potential. Stiffness matrices developed for in-plane plates, and also for rods, beams, and torque bars. Distributed loads and thermal stresses;
4. Dynamics of Structures: Deriving mass matrices for in-plane plates, and also for rods, beams and torque bars. Applications to continuous beams, and two- and three-dimensional structures;
5. Elastic Buckling and the Non-linear Behaviour of Structures, together with non-linear vibrations;
6. Modal Method of Analysis, together with forced vibrations and impact of structures.
7. References.

Printed and bound by CPI Group (UK) Ltd, Croydon, CR0 4YY

03/10/2024

01040339-0015